QB632 .B76 2012
How to build a habitable planet :
33663005233164
HYS

DATE DUE

BRODART, CO. Cat. No. 23-221

HOW TO BUILD A HABITABLE PLANET

HOW TO BUILD A HABITABLE PLANET

• • ✷ • •

The Story of Earth from the Big Bang to Humankind

Revised and Expanded Edition

Charles H. Langmuir
and
Wally Broecker

PRINCETON UNIVERSITY PRESS

PRINCETON AND OXFORD

Copyright © 2012 by Princeton University Press
Published by Princeton University Press, 41 William Street, Princeton, New Jersey 08540

In the United Kingdom: Princeton University Press, 6 Oxford Street, Woodstock, Oxfordshire OX20 1TW

press.princeton.edu

All Rights Reserved

ISBN 978-0-691-14006-3
Library of Congress Control Number: 2012930678

British Library Cataloging-in-Publication Data is available

This book has been composed in Minion Pro with Gabriola

Printed on acid-free paper ∞

Printed in the United States of America

10 9 8 7 6 5 4 3 2 1

ACC LIBRARY SERVICES AUSTIN, TX

CONTENTS

PREFACE

This book is a revised and expanded version of the popular first edition of *How to Build a Habitable Planet* by W. S. Broecker, published by Eldigo Press in 1984. Many new discoveries have emerged in the last twenty-eight years. As of 1984, dark energy and dark matter were not yet discovered, the ocean ridges were barely mapped, hydrothermal vents on the sea floor were barely known, the Antarctic ice core had not been drilled, the "snowball Earth" hypothesis had not been fully formulated, global warming was not yet an urgent topic, and no extrasolar planet had been discovered. The first edition also did not have any discussion of life or Earth history, the rise of oxygen, and little discussion of volcanism and the role of the solid Earth in habitability. This edition includes the new discoveries and underrepresented topics of the original edition, while striving to maintain the conversational style of the original book, and attempting to be clear about what is known and what is unknown. We also emphasize a "systems" approach to the history and understanding of our planet, and emphasize the linkages of all parts of the Earth system, as well as the relationship of those parts to the solar system and universe. If there is one theme that we hope comes through in the book, it is of a connected universe in which human beings are an outgrowth and integral part.

The growth of knowledge and new topics have the unfortunate consequence of more than doubling the length of the book. We attempted to develop each topic from the ground up so that the material would be accessible to the interested reader who is not a scientist. Chapter 4 deals with basic chemistry and would be a rapid read for those with basic knowledge of that field. Other topics, such as short-lived radionuclides, isochron dating, phase diagrams, and the oxidation/reduction reactions that are so central to life and its planetary consequences, are more challenging.

The writing process has benefited from countless interactions with friends and colleagues, not all of whom can be thanked adequately here, and some of whom are no doubt forgotten over the nine-year gestation of this edition. James Kasting provided a formal review of the entire book. Colleagues at Harvard—Rick O'Connell, Ann Pearson, Andy Knoll, Francis Macdonald, David Johnston, and Peter Huybers—generously reviewed individual chapters in their discipline. Dan Schrag made the useful suggestion of breaking up the "Making It Comfortable" chapter into two chapters, one dealing with the early Earth and the other with glacial cycles. Useful comments or discussions were also provided by Felicia Wolfe-Simon, Candace Major, Dave Walker, Dennis Kent, John Hayes, Chris Nye, Bob Vander Hilst, David Sandwell, Thorston Becker, Raymond Pierrehumbert, Wasserburg student, Steve Richardson, Stephane Escrig, Jeff Standish, and Sarah Stewart. Professional writers Kirsten Kusek and Molly Langmuir carefully edited the first half of the book, improving the writing and pointing out sections that were unclear for the nongeologist.

A course of the same name as the book has been taught at Harvard for six years, and students in that course have provided valuable feedback on material that was not clear or overly difficult. Teaching fellows in that course also contributed greatly to refining the content, particularly Sarah Pruss, Michael Ranen, Susan Woods, Allison Gale, Carolina Rodriguez, and Francis Macdonald. Jean Lynch-Stieglitz taught a course at Georgia Tech using the draft of the book submitted to the publisher for review, and dozens of her students provided feedback on individual chapters, pointing out those that needed substantial revision. Jean also gave valuable insight on "what worked and what didn't" in her use of the book.

Invaluable secretarial help was provided by Christine Benoit, Rady Rogers, and Olga Kolas. Raquel Alonso assisted with editing, searching for figures, drafting figures, and making sure everything stayed organized. Her help was indispensable.

All of these interactions and comments as well as many others have made the book much better than it would been otherwise. The remaining errors and shortcomings are the sole responsibility of the authors.

Supplementary materials for teaching and self-learning, including color versions of many of the figures and potential course syllabi, are available on www.habitableplanet.org.

HOW TO BUILD A HABITABLE PLANET

Fig. 1-0: Earth from space. (Courtesy of NASA; image created by Reto Stöckli, Nazmi El Saleous, and Marit Jentoft-Nilsen, NASA GSFC)

CHAPTER 1

Introduction

Earth and Life as Natural Systems

At the moments when we are able to separate ourselves from our daily concerns and ponder deeply, most of us have encountered fundamental questions of our existence as human beings. Where do we come from? What happened before humans appeared on Earth? Where do the stars come from? Do we have a place in planetary evolution? Are there others like us out there somewhere?

These questions are common to all of us, irrespective of national origin or political persuasion. They are the stuff of myths, creation stories, philosophy, and religion throughout human history. Today major aspects of these questions are susceptible to rigorous scientific inquiry. In this book we explore these questions, the scientific story of creation, the history of the universe that has permitted planetary takeover by an intelligent civilization.

The story begins with the inception of our universe by the Big Bang, through the formation of the elements in stars, to the formation of our solar system, the evolution of our world that became home to life and ultimately to human beings who can question and begin to understand the universal processes from which we are derived. Viewed on the largest scale, this story is the central story of our existence. It relates us to the beginning, to all of natural history, and to everything we can observe. While this book has a primary aim to present some of the current scientific knowledge on these topics, a secondary aim is to encourage a mode of thinking that is often latent for us—how we are derived from and related to a larger world.

The approach to this understanding of the world we inhabit and to which we are inextricably linked requires a range of scales that is difficult for us to encompass, from the atomic to the universal. The story also cannot be told by reduction to its smallest parts. Relationships among the parts and evolution through time also are necessary, a "systems" approach to scientific understanding. From a systems perspective, stars, planets, and life have a set of properties in common that appear to be characteristics of many of the "natural systems" of which the universe is made.

Introduction

The origin and evolution of our inhabited world is both a single subject and a topic of enormous diversity. It is "natural science" in the sense the term was used hundreds of years ago—the understanding of nature—but with a vastly greater panorama of scientific fields and data. The fields include the fundamental sciences of physics, chemistry, and biology, and also the integrative and historical sciences of astronomy and earth science. Most subjects broached in this book could each occupy an entire term of study, so the task is daunting for all of us.

Our aim is to explore Earth's history in detail as an example of a habitable planet, and from that history try to deduce the likelihood of similar histories occurring elsewhere. Within this story there are large numbers of exciting scientific developments and outstanding questions. And the story is the grandest that can be told, the scientific creation story of the universe whose evolution has led to us, human beings who are able to question and investigate the origin of our existence and the universal laws that led to us and surround us.

One of the challenges we face is that the range of scales we will need to encompass is almost unfathomable, from the minute atoms of which we and our planet are made, to the grander scales of solar system and universe of which we are a minuscule part. The smallest scales pertain to how atoms are formed and how molecules combine. The smallest scale of concern to us is the size of the hydrogen nucleus—a starting point for all atoms—which is 0.000000000000001 meters (m) in size. Dealing with

Table 1-1
Exponential units and names

Decimal	10^n	Prefix	Symbol	Name
1,000,000,000	$1 \cdot 10^9$	giga	G	billion
1,000,000	$1 \cdot 10^6$	mega	M	million
1,000	$1 \cdot 10^3$	kilo	K	thousand
1	$1 \cdot 10^0$			one
0.001	$1 \cdot 10^{-3}$	mili	c	thousandth
0.000001	$1 \cdot 10^{-6}$	micro	μ	millionth
0.000000001	$1 \cdot 10^{-9}$	nano	n	billionth

very small (or very large) numbers is obviously cumbersome, so we will use exponential notation and abbreviations (Table 1-1). The hydrogen nucleus has a diameter of 10^{-15} m. At vastly larger scales, stellar distances are measured in light years—the distance light travels in a year. Since the speed of light is 3×10^8 m/sec multiplied by the ~3×10^7 seconds in a year, a light year is 9×10^{15} meters. The nearest star is 3 light years away, our galaxy the Milky Way is 100,000 light years across, and the universe is estimated to be billions of light years in diameter, or ~10^{26} m. Hence, our task encompasses 10^{26}m /10^{-15}m, or 41 orders of magnitude in terms of distance!

Similar large magnitudes exist for time. As we will discover in Chapter 2, the age of the universe is roughly 14 billion years (14 Ga), or 4.2×10^{17} seconds. The time of the atomic reactions that are involved in the creation of matter can be nanoseconds (10^{-9} seconds). Our range of time encompasses 26 orders of magnitude.

The challenge of working with these huge ranges of time and space is that our experience as human beings is so limited. The same size of figure on a page (Fig. 1-1) can be used to portray vastly different scales. We might tell the story of the evolution of the universe taking place over billions of years with the flavor of a story about our summer vacation—without recognizing the difference in scale between the story of everything and that of our small life. The journey through the story of the universe will make much more sense if we remain mindful at each moment of the scale of the phenomena we are investigating.

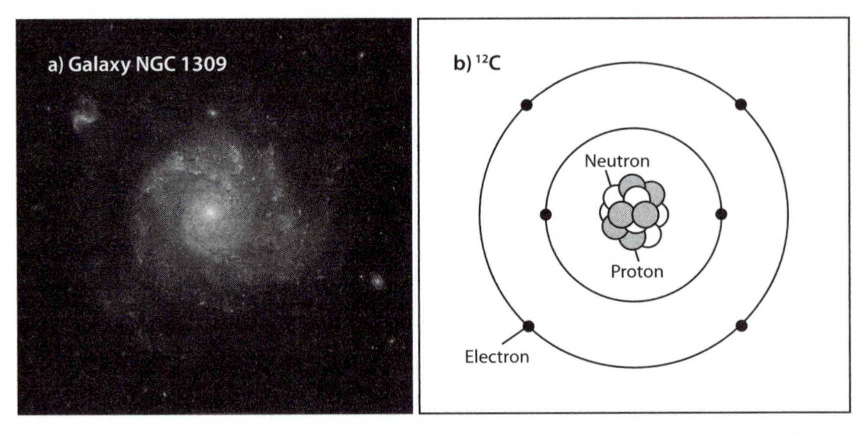

Fig. 1-1: Two subjects considered in this book, galaxies and atoms. In this figure they take up about the same space on the page but differ in actual size by more than 25 orders of magnitude. (*a*) Spiral galaxy NGC 1309 (http://hubblesite.org; NASA, ESA, the Hubble Heritage Team [STScI/AURA], and A. Riess [STScI]); (*b*) Schematic portrayal of the carbon atom. Carbon has the symbol C, atomic number 6, and mass 12. The size of the nucleus is greatly expanded in order to make it visible.

THE POWER AND LIMITATIONS OF SCIENTIFIC REDUCTIONISM

Our approach in this book is to try to relate the smallest of parts to the largest of systems. This may appear to contrast with the traditional scientific approach of so-called reductionism. Much scientific understanding has come about by discovering governing mathematical equations, or laws, that account for diverse phenomena. In this approach, understanding comes from our ability to "reduce" the whole to the fundamental laws of physics from which all phenomena arise. Then phenomena calculated at the most fundamental level can, at least in theory, explain and predict the whole.

The great scientific revolution of the seventeenth century exemplifies the power of the reductionist approach. Newton's mathematical expression of gravity was able to account for Kepler's laws describing the motion of the planets around the sun and the careful measurements of falling objects carried out by Galileo. The big idea stemming from Newton's successes was that fundamental physical laws, described by mathematical equations, can explain everything we see. Today we are probably only partially able to conceive of the wonder of these results when they first appeared—everything we observe, from the tiny scale of a rolling marble to the movements of celestial objects, governed by mathematical law discernible by the mind of man. As Alexander Pope wrote of Newton's discoveries:

> Nature and Nature's laws lay hid in night: God said, Let
> Newton be! and all was light.

Emerging from these startling successes came the idea of the "clockwork universe," where everything is explained by physical law rather than divine intervention. Once the laws are understood, everything can be accurately described and predicted through calculation. This is often understood as the fundamental scientific approach.

An aspect of this approach is the belief that understanding complex phenomena comes from breaking them down to their simplest parts. If we wish to accurately describe a crystal or a gas, the behavior of individual atoms provides the ultimate answer, and if we wish to understand the behavior of individual atoms, we must understand subatomic particles and quantum theory, and ultimately the strings of string theory. Understanding then comes from isolation of variables, improvement of the resolution and precision of observation, and discoveries of fundamental laws from which calculations are theoretically possible from first principles.

In this way, phenomena that appear to be miraculous become explained. Any human being prior to the scientific revolution, if they were to hear an amplified sound system or view images on a television screen, might believe they were in the presence of a miracle (or more likely the presence of the devil). But once the machine is taken apart and all the components are understood, the actions of physical laws appear. Understanding the operation of some of the electronic components would require reduction and observation to microscopic levels, ultimately down to the fundamental particles that make up the atoms involved. The same approach applies to the processes of life. The "miracle" of medications derives from understanding the processes of the body and the action of the drug on the molecular scale. The apparent miracle of evolution can be reduced to individual mutations of DNA molecules. Many of the topics of this book reflect the efficacy of this approach. Understanding how laws that operate at small scales manifest on much larger scales is one of the great triumphs of the scientific method.

Despite its obvious successes, however, reductionism falls short when we try to calculate or understand many natural phenomena. From the practical point of view, very few natural phenomena can actually be calculated from first principles. Let's take a simple example of calculating the atmospheric pressure at some point on the surface of Earth, say at the top of your head as you read this book. This is a straightforward, one-dimensional problem that involves simply summing the weight of the atmospheric column directly above you. We can measure pressure very precisely indeed—what would it take to calculate it as well as we can measure it?

To calculate it, we would need to know the density of the atmosphere at each point along the column. Thermodynamics helps with the general pressure-temperature-volume relations, but quantitative thermodynamic calculations apply best to closed systems, and the air over your head is in movement. The density of air also depends on the concentration of water vapor, which can vary both laterally and vertically. Winds are a response to pressure gradients, so pressure changes continually owing to movements and forces exterior to your personal atmospheric column. We could take an average temperature profile for this time of year, and assume a clear sky with constant relative humidity and no wind, but that gives an approximate pressure, nowhere near as precise as what we can measure. We could send a probe up through the atmosphere and measure the temperature and water vapor, but that is a bit like simply measuring the pressure, and by the time we got the data and processed it, the atmosphere could have undergone small changes from time of day and weather, leading to small errors.

This simple example illustrates the fundamental point that natural systems at any moment can never be specified completely. They are open systems without clear boundaries. Energy and material continually flow in and out; physical and chemical properties are not constant and homogeneous. For the gases of the atmosphere, the water of the ocean, the rocks of Earth's mantle, the liquid metal of Earth's outer core, or the plasma of the sun's interior, pressure and temperature change in space and time, mass and energy are constantly flowing in and out of the system, and we cannot make enough measurements to define accurately the state of the system on any but approximate scales, or as a long-term or broadly spaced average.

This state of affairs influences our ability to calculate and predict. Every calculation requires specification of initial conditions. Calculation of the weather tomorrow starts from our knowledge of the weather today. Initial conditions for real systems can never be measured simultaneously everywhere. Predictions become difficult for the immediate future, and progressively more uncertain for more distant times. This is particularly true for systems that contain "feedback," where movement in one direction causes a countervailing movement. This common characteristic of the real world often even leads to *chaos*.

CHAOS

Nowhere is the uncertainty of prediction more evident than chaotic systems. Chaos occurs in common equations where the outcome is so sensitive to the tiniest of changes in initial conditions or constants in the equations that long-term prediction is impossible. A simple illustration of this point comes from a time series generated by a "feedback" equation where the initial value of x is permitted to vary between 0 and 1. Subsequent values of x are then calculated with the following equation:

$$F(x_n) = Ax_n(1 - x_n) \tag{1-1}$$

A is a constant. To construct the time series, the equation is used repeatedly with the output from one step used as the input for the next step, i.e., $x_{n+1} = F(x_n)$. Whenever x is large, the Ax term gets larger, but the $(1 - x)$ term gets smaller, and vice versa. This is a negative feedback, where an increase in one term causes a decrease in the other. Negative feedbacks are very important in many natural processes. If $A = 3$ and we start with a value of $x = 0.5$, then $F(x) = 0.75$. For the next step, $x = 0.75$ and $F(x) = 0.5625$ and so on. This can easily be set up on a spreadsheet or a simple computer program to calculate what happens after large numbers of steps—a recommended exercise for the reader.

Equation (1-1) is an inverted parabola (see Fig. 1-2), and we can track the evolution of the time series as a path from one point on the parabola to the next. Figure 1-3a illustrates what happens to a time series for moderate values of A. When $A = 2$, the system proceeds rapidly to a

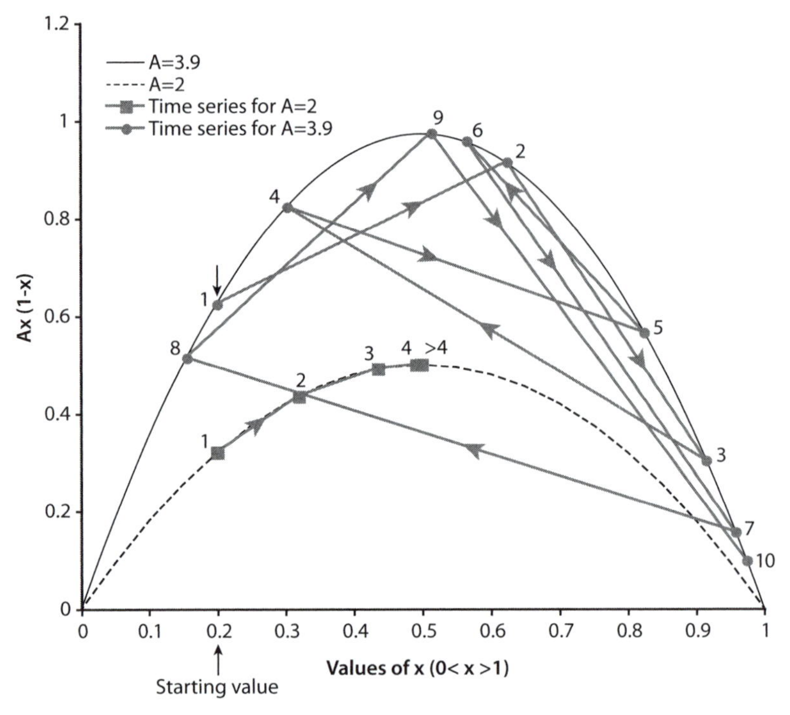

Fig. 1-2: Illustration of the behavior of time series controlled by the simple function generated by $Ax(1 - x)$ for different values of A. Both illustrated time series begin with the same starting value for x, $x_o = 0.2$. For A = 2, the time series (squares) marches along the curve to a constant value. For A = 3.9, the time series (dots) is chaotic and never reaches a fixed value.

steady-state value of 0.5, while for $A = 2.8$ the steady-state value is near 0.64. The steady state values for a particular value of A are independent of the initial value of x. The time series begins to exhibit more interesting behavior once A exceeds values of 3.0. For values of 3.2, for example (Fig. 1-3b) the time series proceeds to an oscillation between two states, also independent of the initial value of x. For $A = 3.9$, no such regularity appears in the result—no matter how many the number of steps.

To have a more comprehensive view of the behavior of this simple function, Figure 1-4 plots on the vertical axis the range of values of x obtained after a large number of steps, for different values of A plotted along the horizontal axis. For values of A less than 3, the time series arrives at a steady state value, independent of the starting value of x. Up to values of A less than (~3.45) the time series oscillates between two

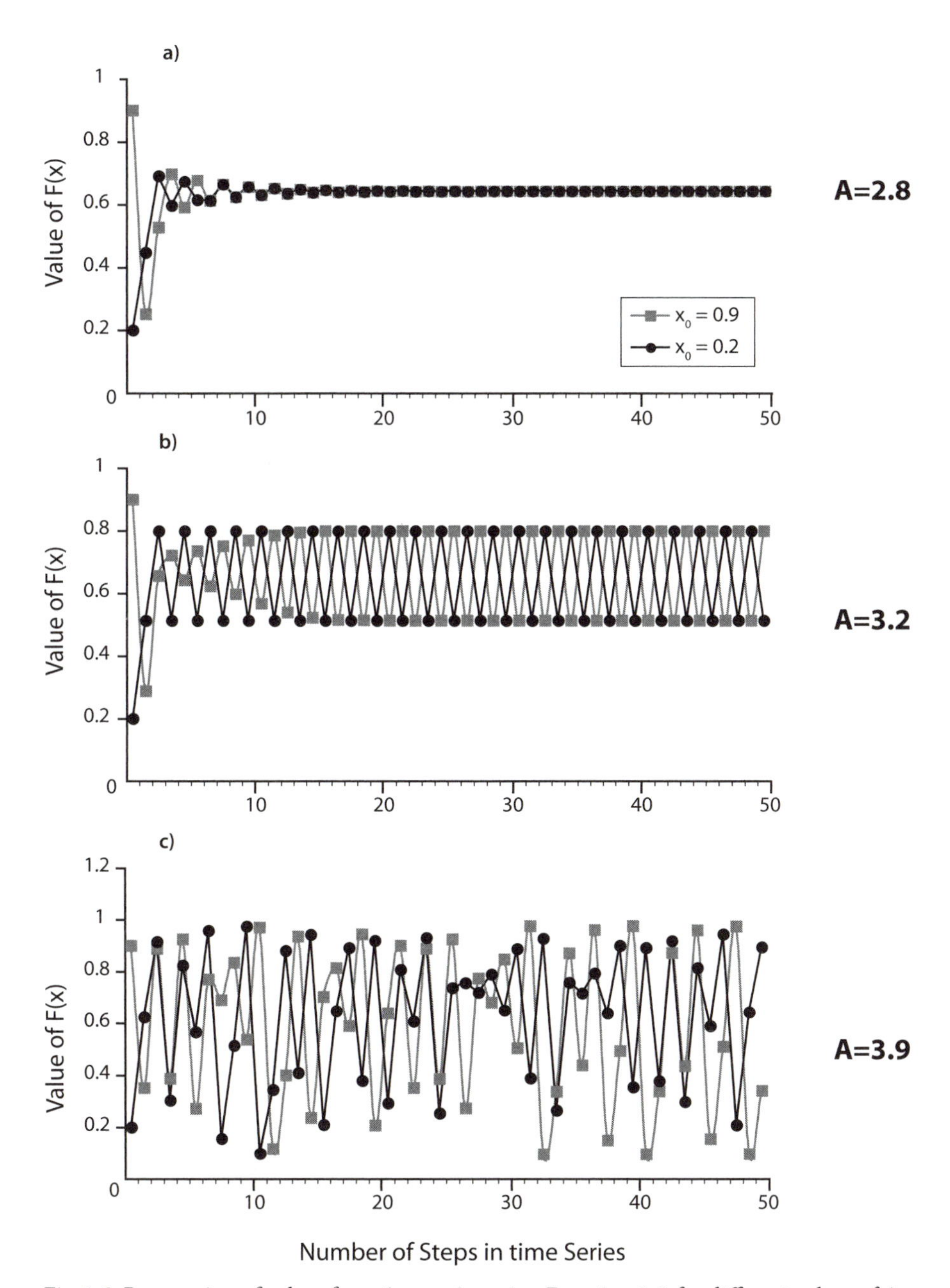

Fig. 1-3: Progression of values for a time series using Equation 1-1 for different values of A and two different starting values for x, $x_0 = 0.2$ and $x_0 = 0.9$; (a) for A=2.8, independent of starting value, the time series reaches a common value of $F(x) = 0.64$; (b) for A = 3.2, the time series reaches a steady state where it oscillates between two values; (c) for A=3.9 no steady state is reached, and the value after a fixed number of steps is very sensitive to even very small changes in the initial input value.

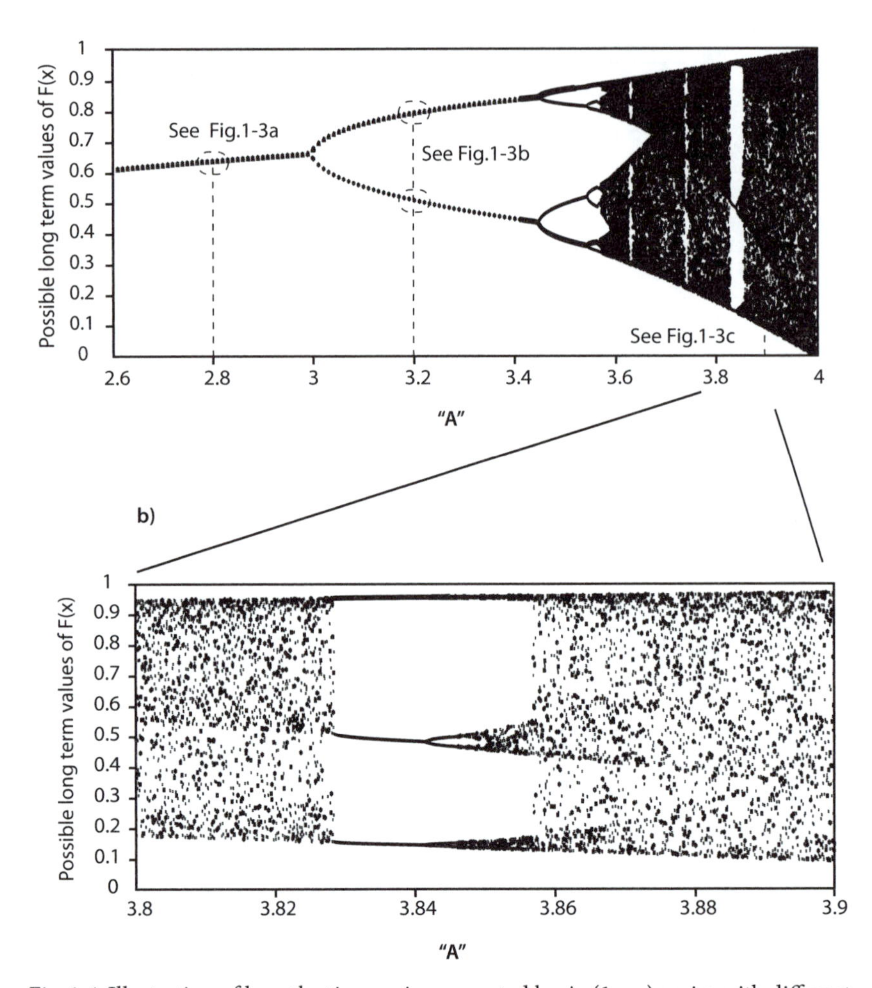

Fig. 1-4: Illustration of how the time series generated by Ax(1 – x) varies with different values of A. For A less than three, a constant steady state is obtained independent of initial conditions. Then there are two states, then four. At values of A greater than about 3.6, there is chaotic behavior. Within the chaos, there are limited regions where steady-state values again are obtained (blown up for clarity in the lower panel). The figures show the range of states reached after time series of >100 steps.

Table 1-2
Demonstration of chaos
A = 3.9

X_0 =	0.3	0.29	0.299	0.29999	0.299999	0.2999999
N = 1	0.819	0.803	0.817	0.819	0.819	0.819
...						
...						
98	0.974	0.918	0.845	0.411	0.203	0.954
99	0.098	0.292	0.511	0.944	0.632	0.170
100	0.346	0.807	0.974	0.205	0.907	0.551

Output values of x for A = 3.9 after iterations (H) of Equation 1-1 for values of X_0 near 0.3. After the first iteration, the values are all very similar to three decimal places (they differ at more significant figures). But after 98–100 steps the results are chaotic, and do not converge on the calculation for X_0 = .3 as X_0 gets closer and closer to 0.3 starting from 0.29. Similar chaos would result from infinitesimal changes in the value of A.

values. At values of A slightly above this number there are four stable oscillating states, and the number of stable states increases up to values of *A* of about 3.57, above which chaos begins. Then values vary over large ranges in almost random fashion between an upward and lower bound. Steady-state values reappear for $A = 3.83$, and then chaos reappears up to values of 4. For the chaotic states, the most minuscule change in the value of *A* leads to a completely different series of results, such that the state of the system at future time cannot be predicted.

In the chaotic regime, the result after a fixed number of steps also varies with minuscule changes in initial conditions—i.e., the initial value of *x*—and these changes are nonintuitive. Table 1-2 illustrates the result after 100 steps of the time series for $A = 3.9$. No matter how precisely the initial value is specified, while the first step is the same, the final value after a sufficient number of steps can vary across the entire range. "Extreme sensitivity to initial conditions" is the hallmark of chaotic systems and has been described by the so-called butterfly effect, where the beating of a butterfly's wings in China could lead eventually to a hurricane in the Atlantic.

The weather is a familiar example of a chaotic system. There are upper and lower bounds of temperature that characterize the seasons, and prediction is quite good for periods of hours. But last night's weather forecast in Boston predicted rain with one inch of snow accumulation overnight. Today the skies are gray, but no precipitation has fallen. What will the weather be two weeks from now? It will still be the same season, but accurate weather prediction is not possible. To the extent that natural systems are chaotic, their behavior cannot be calculated precisely, even if the governing equations are known exactly.

Another characteristic of natural systems is *self-similarity*, or their tendency to exhibit fractal behavior. A geometric object that is fractal looks exactly the same when it is examined over a very large range of scales. For fractal systems you cannot tell the size of an object without a scale bar. Great canyons can extend over thousands of kilometers, but the topography of a single streambed or a rivulet on a mud flat can have very similar ratios of width to depth and the same sinuosity (Fig. 1-5). Benoit Mandelbrot illustrated this concept beautifully in a paper entitled "How Long Is the Coast of Britain?" The coastline looks similarly irregular at small scales and large scales. And the length of the coastline depends entirely on the size of ruler used to measure it—the smaller the ruler, the longer the coastline. For example, if you can measure the distance around every pebble the coastline becomes very long indeed! There is no single answer to the question of length unless the ruler size is specified. Donald Turcotte has demonstrated the huge number of natural processes that exhibit fractal behavior.

Of course, these various challenges do not mean that scientific understanding is not possible. Great discoveries lie in the determination of fundamental laws and equations. Nor are the calculations useless. Even with Equation (1-1), increased precision of initial conditions means that more extended prediction is possible—the first few steps in the time series are increasingly well established. And even for long-time series there are upper and lower bounds that are not exceeded.

There is nonetheless a practical and philosophical dilemma for reductionism. In principle, everything can be reduced to the smallest phenomena from which everything else derives. In practice, even a straightforward phenomenon such as water flowing out of a garden hose cannot be predicted quantitatively from first principles. There is a gap between our practical experience of nature and the pure laws governing phenomena. The laws remain true and nature cannot defy them; but the gap can never, even theoretically, be filled.

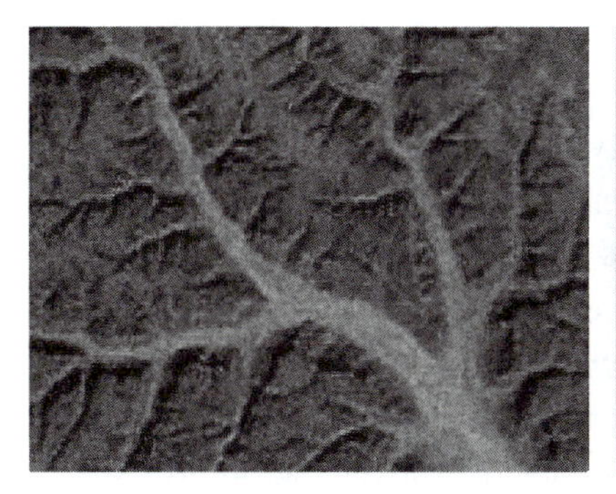

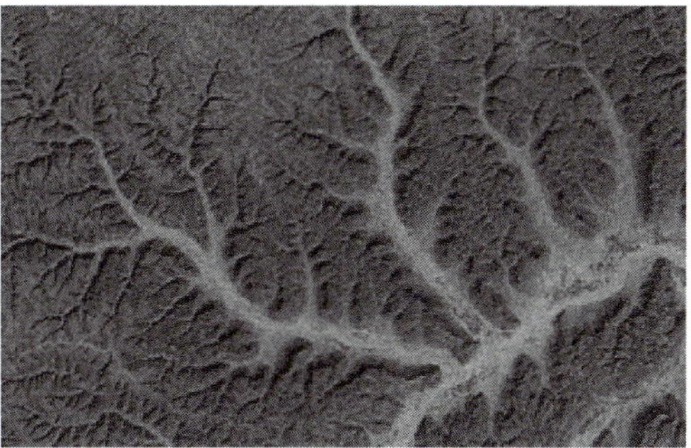

Fig. 1-5: Examples of dendritic drainage patterns. Zooming in on one small part of it gives similar pattern rules to the whole image. Simply from looking at the image, it is not possible to tell the scale because the pattern exists over a large range of scale. Can you find which image is a blowup from the other? (Copyright © 1995–2008 Calvin J. Hamilton and courtesy of NASA; http://www.solarviews.com/huge/earth/yemen.jpg)

"Systems"

On a more human level of what we perceive as understanding, reductionism also falls short. A cell is much more than a collection of reactions among chemicals. It has functions that cannot be inferred or understood from the atomic scale. It has a history, is descended from ancestors going back in time to the origin of life, and relates to its surroundings in ways that must be observed to be understood. It has a relationship both to the underlying molecular scale and the next larger scale (e.g., the organism) of which it is a part. Properties apparent from a larger scale often cannot be inferred from the individual parts of the smaller scale. An understanding of any natural phenomenon requires knowledge of its constituent parts and their relationship to each other, and to the larger world of which the phenomenon is a part.

Including relationship in the understanding of phenomena has come to be called *systems thinking*. This approach recognizes the necessity of accurately describing both the reduced components and the relationships among them. To put it simply, reductionism assumes that the whole can be reduced to the sum of the parts. Systems thinking asserts that the whole is greater than the sum of the parts, and that there are "emergent properties" that arise from the whole that could never be understood or predicted from a reductionist approach. The most obvious example is any living organism. For an individual cell, its DNA might be very precisely defined, but its living manifestation requires complex feedbacks and relationships on a variety of scales in an inevitably open system. *Systems biology*, for example, stems from the recognition that individual biological components such as DNA or RNA can only be fully understood through their relationship to the larger scales—organelles, cells, organs, and so on. This concept expands to larger scales. An animal can never be fully understood from even highly precise understanding of the properties of the individual cells. The relationships among the cells that make it up and its relationship to the larger ecosystem in which it lives must be included. An ecosystem can never be understood from isolated consideration of the individual plants and animals, and also exists in the context of various planetary conditions such as soils, elevation, and climate. The universe would not be able to be fully understood from a detailed understanding of individual galaxies.

Another important aspect of all these examples is that they are in movement. "Relationship" includes transport of matter and energy. Understanding of a cell, an organism, an ecosystem, or a planet is not fully revealed by a snapshot. Observation in action is necessary.

To illustrate these principles we can consider a simple object like a mechanical watch. If we spread the parts out on the table in front of us, there is much we can learn from investigating the gears, the nature of the spring, and so on. We can investigate the detailed chemical make-up of the pieces and even their atomic structure. However, seeing all these pieces separately, even understanding them down to the level of individual atoms, would not allow us to predict what functions they all served. Once the watch was put together a new understanding would emerge—we could see how all the parts relate to one another, the driving mechanism of the spring, the meshing of the gears, the place of the watch face. Only once the watch is put in motion, however, does the mechanism become clear—with a new appreciation for the details and complexity of the parts, and the appreciation of precise timekeeping. Still, however, the watch remains an isolated mechanical object, with no relationship to a larger world. When it is on someone's wrist, and used to monitor and guide the movements of the person's day, its function becomes still more evident, and we are forced to deal with the big idea of time. We might then also consider the history of watchmaking and time, and how the measure of time, represented by the watch, has related to the detailed development of human civilization.

This simple example illustrates some of the basic principles that apply to systems:

- The full significance of the watch could never be predicted (without advance knowledge) from the description of the parts, and particularly not from their description on an atomic scale.
- Understanding the relationships among the parts creates an entirely different phenomenon than the parts themselves.
- A fuller understanding is not possible without observing the watch in movement, and realizing that without movement it has no function.
- The function of the watch is still not evident until it is understood in its relationship to the larger system represented by the human being.
- There is evolution over time in this larger relationship that relates to development of a still larger system.

Looking from the largest scale down, the precise mechanisms of the watch can be understood and placed in an appropriate context. That is, vision downward in scale can show how the smallest scales relate to the largest. From the smallest scale, vision upward is very limited. The spring in the watch, for example, knows what it is connected to and that it moves back and forth—the full dimensions of the watch described above are not perceptible.

You will notice that there is no figure illustrating a system. Why not? Because systems involve relationship, cycles, feedbacks, and movement. They cannot be adequately represented by a static figure on a page.

Characteristics of "Natural Systems"

While systems thinking has very broad applicability, our concern relates to the natural systems of the world around us. Observation of these systems suggests a suite of shared characteristics.

NATURAL SYSTEMS ARE OUT OF EQUILIBRIUM

The drive toward *equilibrium*, a minimum energy state where there is no further tendency to change, is one of the fundamental principles of chemistry and physics. Falling objects come to rest at the lowest energy state; chemical reactions proceed to completion where no further reaction occurs. While this driving force is manifest everywhere, natural systems, even when they are at steady state, are usually far from equilibrium. At equilibrium, properties such as temperature and pressure are constant throughout the system, and the system is isolated from external influences. That is not the natural world! One of the benefits of laboratory experiences is that when we actually try to measure properties at equilibrium, it becomes evident that controlling the conditions for such perfection to appear is very difficult indeed. Natural systems are not isolated. Matter flows in and out; properties such as temperature and pressure change continuously. Natural phenomena do not exist in static equilibrium states, but are in movement at all scales.

Instead, a characteristic of most natural systems is *disequilibrium*. Often this condition reflects a balance of forces and fluxes that leads to a *steady-state disequilibrium* where the natural system remains within narrow bounds, perched at a disequilibrium state. All living organisms reflect this state, for example. Our body temperatures are maintained within a narrow range irrespective of the temperature of our surroundings, sustained by the food and air we take in and various feedbacks that maintain the temperature. Most of the molecules that make up life are out of equilibrium with oxygen and decay rapidly once they are no longer metabolically sustained. Earth's atmosphere is in a disequilibrium state, its temperature sustained by the continuous influx of solar energy and the warming effects of greenhouse gases coming from Earth's interior and modulated by life. The sun is not at equilibrium, but reflects a balance between gravitational forces leading to contraction, and expansion forces coming from heating of its interior by nuclear fusion. Energy flows continuously outward. The forces toward equilibrium are acting everywhere, but the state is one of disequilibrium.

Another characteristic of systems is that they can become increasingly complex and organized with time. This characteristic also appears contradictory from the point of view of equilibrium. Part of the force toward equilibrium is the inevitability of increased entropy, or randomness, or the decay of order. Two separate gases mix to homogeneity, temperature differences become homogenous, or potential energy is released to minimum energy states. How then can a relatively stable state far from equilibrium be maintained, with order preserved or increasing? Why doesn't the system simply move to its minimum energy, equilibrium state and rest there?

NATURAL SYSTEMS ARE MAINTAINED BY EXTERNAL ENERGY SOURCES

Equilibrium applies to isolated systems that move toward and then maintain a minimum energy state. To be maintained far from equilibrium, an external source of energy is necessary. The sun is maintained by the energy provided by nuclear fusion. Earth receives energy from the sun and also from decay of radioactive elements. Life on Earth is sustained by the sun. Absent the external energy, all of these would decay to a static equilibrium state. It is the external energy source that also permits evolution to increasing order within the system. In the largest context, disorder increases.

"STEADY-STATE DISEQUILIBRIUM" IS MAINTAINED BY FEEDBACKS AND CYCLES

Despite being out of equilibrium, natural states often occupy a narrow range of states that can persist despite significant changes in external conditions. How can such states be maintained?

There are various mechanisms that permit stability despite variations in energy input. For example, when we boil the water, the water is maintained at a steady-state temperature different from its surroundings by an external heat source. What makes the temperature difference with the room possible? The external energy source. And then what keeps the temperature of the water constant? Why doesn't its temperature rise to be the same as the flame that heats it?

When water boils, the liquid is converted into gas. This requires a great deal of energy—the energy of vaporization. While to heat one gram of water by one degree takes one calorie, to convert that gram of water to gas takes 539 calories. Once water begins to boil, any additional energy that is added does not raise the temperature but just converts the liquid water to gaseous steam. If we plot temperature vs. time as we add energy to the water, we get the plot shown in Figure 1-6. Note that there is a long stretch of time where the temperature is constant. If we reduce

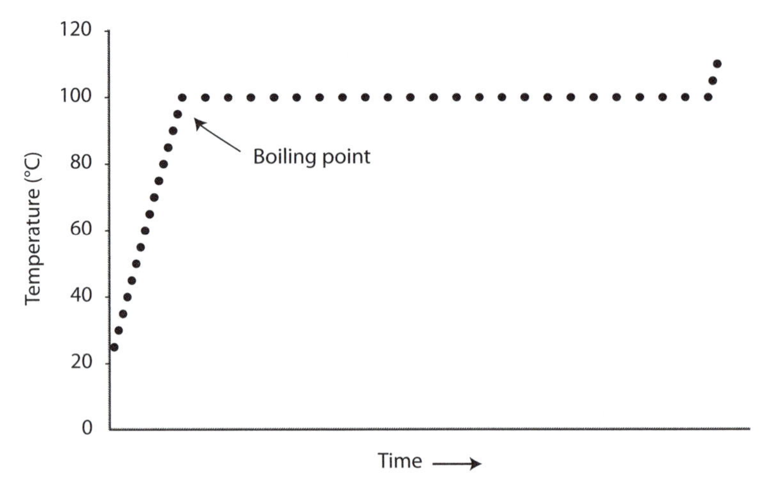

Fig. 1-6: The temperature of water subjected to external heat over a period of time. The water initially heats up rapidly, but once it reaches the boiling point, the temperature remains constant until all the water has turned to steam.

the heat input, the water boils more slowly; if we increase the heat, the water boils more quickly, absorbing all the additional energy being added. Hence the boiling water responds to changes in heat input in a way that maintains the constant temperature in the pot—a steady state despite changes in the energy source or external environment. Many more complex chemical phenomena also have such properties.

Feedback occurs when the system is maintained at steady state because the response "feeds back" to control the input. Another example from the kitchen is the oven's thermostat. Once it is set to a particular temperature, the heat source goes on when the temperature is below the set point, and turns off once the temperature is above it. The external energy source permits the oven to be far from equilibrium with the kitchen; the mechanism of the thermostat provides the feedback that maintains the temperature within a narrow range.

The thermostat is an example of negative feedback. The negative feedback maintains the system at steady state. Increased input triggers a response that counteracts or turns off the input. This principle of negative feedback is an essential component of natural systems.

There also can be positive feedbacks, where the response becomes amplified rather than damped. The oven has a positive feedback when

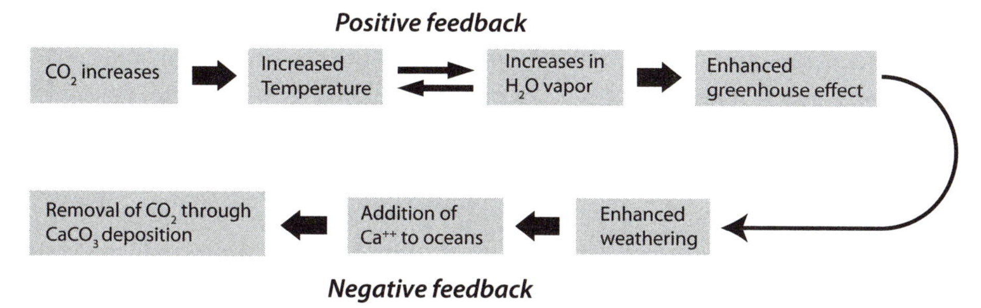

Fig. 1-7: Illustration of feedback influencing Earth's climate stability. Increases in CO_2 (e.g., from volcanic eruptions) lead to a small temperature increase, which causes increased water in the atmosphere. Water is also a greenhouse gas and causes temperature to increase further. This is a positive feedback—higher temperatures cause still higher temperatures. The increased temperature then enhances weathering, which adds Ca to the ocean causing removal of CO_2 through precipitation of $CaCO_3$ (limestone). This is a negative feedback—higher temperatures lower CO_2, which lowers temperatures. The negative feedback in this system occurs on long timescales.

a small spark of electrical current trips a switch that releases a large gas supply. An important natural example of positive feedback occurs in the climate system. Carbon dioxide is a greenhouse gas, but it is at low concentrations that absent emissions from humans would be about 300 parts per million (ppm) in the atmosphere. Increases in CO_2 increase the temperature of the atmosphere a little bit, causing more water vapor to be created. Water vapor is also a greenhouse gas, and there is lots of water vapor in the atmosphere. Because there is so much water, there is a much greater warming effect than would be caused by the CO_2 alone. Therefore, the water amplifies the temperature increase. The same process works in the opposite direction—lowering CO_2 lowers temperature slightly, which reduces water vapor, which lowers temperature much more. Positive feedback amplifies small changes.

Both positive and negative feedbacks are important in natural systems (Fig. 1-7). Positive feedbacks cause rapid response and sensitivity to small changes; negative feedbacks provide balance and stability. Both are important factors in maintaining steady-state disequilibrium, and natural systems often have complex interplays between positive and negative feedbacks, which can make them very difficult to model accurately. Increased water vapor, for example, while increasing the greenhouse effect also leads to more cloud cover, which is a negative feedback because clouds reflect sunlight back into space. Therefore, climate models will be very sensitive to the assumptions made about the relative importance of the positive and negative feedback associated with water vapor.

Chemical cycles are also a requirement for long-term stability far from equilibrium. Cycling in itself implies lack of equilibrium, because at equilibrium there is no movement and a static state, while in systems there is circulation in continual movement. For systems to have longevity, they must be able to persist over long periods of time. To persist over time and maintain a steady-state disequilibrium, materials cannot be exhausted, and hence natural systems must recycle.

Many parts of the Earth system illustrate this recycling. Rocks are formed, eroded, deposited as sediments, heated, melted, and erupted to form again (Fig. 1-8). On geological timescales, rocks are continually in movement. Water goes through the rock cycle, and it also has a cycle on a much shorter timescale (Fig. 1-9). If the ocean only evaporated, then it would rapidly diminish in size and increase in salinity, and the "waste product" of water vapor would have to be stored somewhere. There would be no steady state and no longevity. Instead, the water vapor turns into rain that flows back into the ocean, eroding the continents in the process. The ocean is maintained in volume and salinity, and the eroding power of the water contributes to the recycling of continental crust and the maintenance of continental volume and elevation in steady state. The various parts of the Earth system—rock, water, atmosphere—are all involved in interrelated cycles where matter is continually in motion and is used and reused in the various planetary processes. Without interlocked cycles and recycling, Earth could not function as a system.

This discussion leads to a list of shared characteristics of natural systems:

- Natural systems are invariably in movement.
- They are sustained by an external energy source and energy flow through the system.
- Matter cycles through the system, providing sustainability through recycling.
- The system is maintained over a narrow range of states, but these states are normally "steady-state disequilibrium."

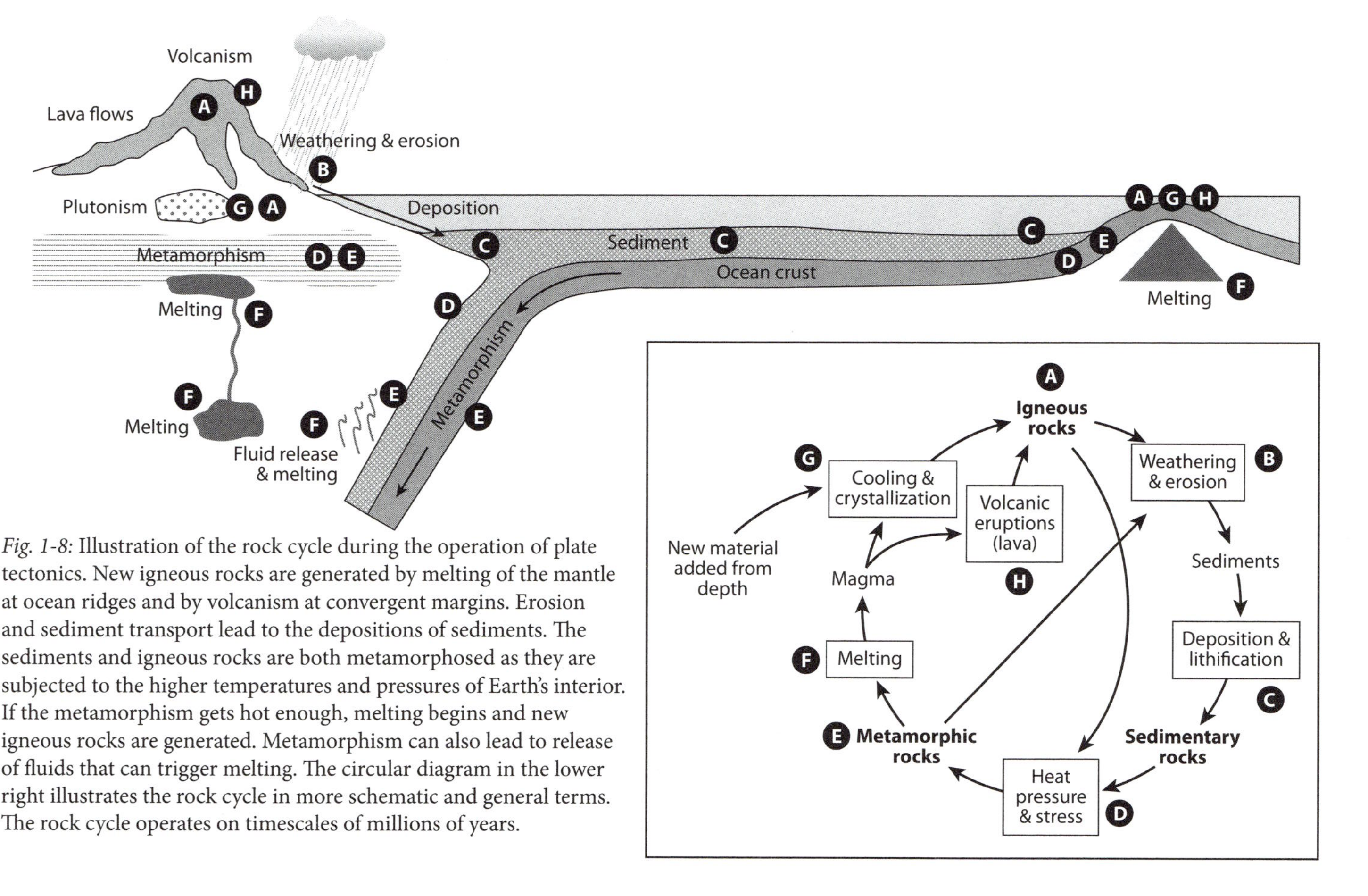

Fig. 1-8: Illustration of the rock cycle during the operation of plate tectonics. New igneous rocks are generated by melting of the mantle at ocean ridges and by volcanism at convergent margins. Erosion and sediment transport lead to the depositions of sediments. The sediments and igneous rocks are both metamorphosed as they are subjected to the higher temperatures and pressures of Earth's interior. If the metamorphism gets hot enough, melting begins and new igneous rocks are generated. Metamorphism can also lead to release of fluids that can trigger melting. The circular diagram in the lower right illustrates the rock cycle in more schematic and general terms. The rock cycle operates on timescales of millions of years.

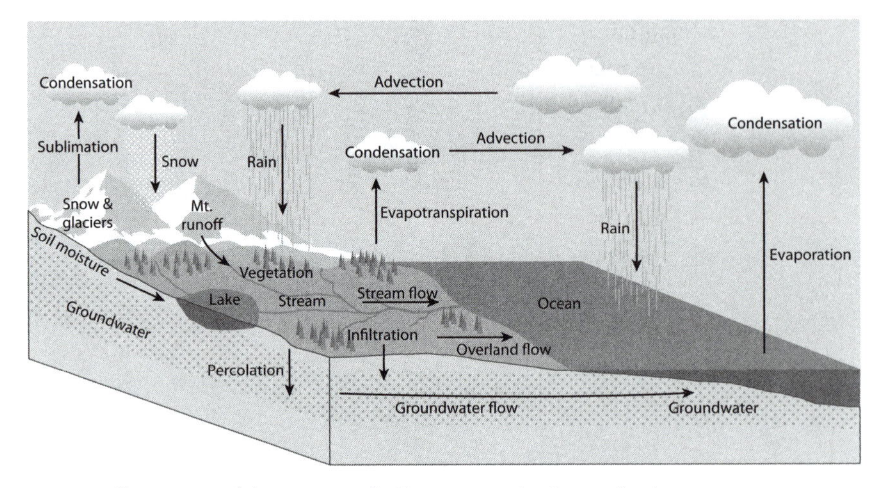

Fig. 1-9: Illustration of the water cycle. Evapotranspiration and rain operate on very short time scales (weather). The groundwater portion of the cycle has longer time-scales, >1000 yrs.

- Feedbacks operate to sustain steady-state conditions
- The system is nested within a larger scale and relates to or contains within it systems at a smaller scale.
- There is change with time—creation, long-term evolution, and eventual death.

This set of characteristics applies to a range of scales—cells, animals, ecosystems, planets, stars, and so on. Let us consider how Earth fits these characteristics:

- In the last fifty or so years we have come to recognize the movements in all Earth's layers, including the plates at the surface, the mantle, and the core as well as the atmosphere and ocean.
- Earth has two important external energy sources: the sun and the "battery power" of radioactivity stored in Earth since its formation. This energy flows continuously between the various components of the Earth system.
- All of Earth's reservoirs have chemical cycles where matter is added to, circulates through, and is removed from the various reservoirs. Earth recycles—elements are cycled through the parts of the system and reused.

- Earth has maintained a narrow range of states throughout most of its history. The temperature range of the atmosphere/ocean system has been above the freezing point and well below the boiling point of water for billions of years, maintaining conditions suitable for life. The tectonic plates have moved over the surface for a similar period of time, and for all of Earth's history that we have access to there have been continents as well as oceans, suggesting only modest changes in sea level.
- Each Earth reservoir is sustained by feedbacks.
- Earth is dependent on the larger-scale solar system, and contains within it smaller-scale natural systems (e.g., the ocean, ecosystems).
- Earth had a beginning at the dawn of the solar system, and evidence for evolution and change abounds. As our energy sources have changed, there has been evolution in the behavior of the solid earth, and striking evolution in the complexity of life and its relationship to ocean, atmosphere, and crust. Ultimately, as the energy sources from the sun and radioactivity decay, Earth will die.

James Lovelock has postulated that the steady-state disequilibrium that is characteristic of Earth's surface makes it a "living organism," for which he has suggested the name Gaia. This suggestion has raised some controversy, because there are obviously discrepancies between biological life and Gaia—Darwinian evolution and reproduction being the most obvious. What Earth and life do have in common is the shared characteristics of natural systems. These shared characteristics are not limited to the organic envelope of the Earth that Lovelock called Gaia. The Earth system includes the movements of the plates of the solid earth as well as the interior all the way down to the inner core. Therefore, if one asks of natural systems the general question, "On what principles does the system work?" then Earth and life work in similar ways. They both have the shared characteristics of natural systems. One might ask the question whether the universe taken as a whole operates with the same characteristics, out of equilibrium, powered by the Big Bang, with cycles of chemicals and energy in a long-term evolutionary process. From this perspective, then, there is a commonality extending from microcosmic to macrocosmic scales. Systems appear to be the way the universe works.

Summary

A diversity of scales and approaches is necessary to approach an understanding of the development of a habitable planet. The largest scale is that of the universe, billions of light years in size and more than 10 billion years of time. The smallest scale we will deal with is the construction of atoms from neutrons and protons, 41 orders of magnitude smaller where reactions happen in nanoseconds.

A reductionist approach to science implies that understanding and causality can be determined by reduction to the smallest scales. We need this approach to understand many of the materials and processes of stars, Earth, and life. Despite its explanatory power, reductionism is incomplete. For such calculations, initial conditions need to be specified and boundaries defined. For natural systems such as planets, initial conditions can never be specified completely and boundaries are open. Furthermore, many equations that describe natural processes exhibit chaotic behavior that prevents accurate calculation over long time periods.

Natural systems such as stars, Earth, and life have properties that are not revealed by an exclusively reductionist approach. They have states that are out of equilibrium and that are maintained at steady state by a balance of forces and continuous movement of matter and energy through the system. Systems thinking states that the properties of the whole cannot be inferred from the properties of the parts. Relationships among the parts and their evolution in time are essential. Most natural systems are also nested, containing within them systems of a smaller scale and being a part of systems at a larger scale. From a systems approach, Earth and life share many common characteristics, the characteristics of natural systems. Understanding of such systems requires knowledge of the constituent parts, the energy that drives the system, the cycles and feedbacks that relate the parts, the nested systems at larger and smaller scales, and the inevitable evolution over time that tells the story of our inhabited world. These are our tasks in the remaining chapters.

Supplementary Readings

Fritjof Capra. 1997. *The Web of Life.*New York: Anchor Books.

James Gleick. 1998. *Chaos.* New York: Penguin Books.

James Lovelock. 1995. *The Ages of Gaia.* New York: W. W. Norton & Co.

Benoit Mandelbrot. 1982. *The Fractal Geometry of Nature.* New York: W. H. Freeman & Co.

Fig. 2-0: Galaxy Cluster Abell 2218, about 2.1 billion light years from Earth, which is composed of thousands of individual galaxies. The large mass of the cluster has also been used by astronomers as a gravitational lens that magnifies and makes visible even more distant galaxies, which are distorted into the long thin arcs that are visible in the image. (NASA, ESA, Richard Ellis (Caltech), and Jean-Paul Kneib (Observatoire Midi-Pyrénées, France). Acknowledgment: NASA, A. Fruchter, and the ERO Team (STScI and ST-ECF))

CHAPTER 2

The Setting

The Big Bang and Galaxy Formation

Earth is a minor member of a system of planets in orbit around a star we call the sun. The sun is one of about 400 billion stars that make up our Milky Way. The light from this myriad of stars allows observers in neighboring galaxies to define our galaxy's spiral form. The galaxy is the basic unit into which universe matter is subdivided.

Like its billions of fellow galaxies, ours is speeding out on the wings of a great explosion that gave birth to the universe. That these major pieces of the universe are flying away from each other is revealed by a shift toward the red of the "bar code" of spectral lines of the elements in the light reaching us from distant galaxies. The close correlation between the magnitude of this shift and the distance of the galaxy from Earth tells us that about 13.7 billion years ago all the galaxies must have been in one place at the same time. The catastrophic beginning of the universe is still heralded by a dull glow of background light. This glow is the remnant of the great flash that occurred when the debris from the explosion cooled to the point where the electrons could be captured into orbits around the hydrogen and helium nuclei. The Big Bang was the impulse from which everything else in the universe has been derived. Contained in the galaxies lying within the range of our telescopes are about 400 billion billion stars. A sizable number of these stars are thought to have planetary systems.

Careful examination of the data from galaxies shows that a vast amount of matter is not accounted for in what we can see. This "dark matter" is not visible to us and is poorly understood, but it makes up almost six times as much total mass as the atomic matter that makes up stars, planets and life. Careful measurement of retreating galaxies shows that they are speeding up with time and that the universe will not ultimately contract into a "big crunch." To explain this phenomenon requires "dark energy" that has a repulsive force that counters the effect of gravity. Some 76% of the universe is believed by physicists to be made of dark energy, relegating the material that we know and understand to only 4% of what was created in the Big Bang. While the beginning event is well established, great mysteries concerning the contents and operation of the universe remain to be understood.

Introduction

Where the universe comes from and how it got here are the essential first questions as we delve to the starting point of Earth's history, deep in time even long before the formation of the Milky Way. Was there a beginning? When and how did it happen? In this chapter we will see that there was indeed a spectacular beginning to the universe—the origin of everything we can observe—and we can even determine when it happened. From there, everything else unfolds.

The Big Bang

The universe as we know it began about 13.7 billion years ago with an explosion that astronomers refer to as the Big Bang. All the matter in the universe still rides forth on the wings of this blast. Speculations as to the nature of this cosmic event constitute the forefront of a field called *cosmology*. What went on before this explosion is a matter that is not currently subject to scientific investigation, because every observable phenomenon in the universe (that we know of) dates from the Big Bang. No physical record of prior events remains.

To state that we know the age of our universe and its mode of origin is rather bold. Is this fanciful thinking or is there evidence? While it is remarkable that we have detailed knowledge of the beginning of the universe, the observations astronomers have made provide compelling

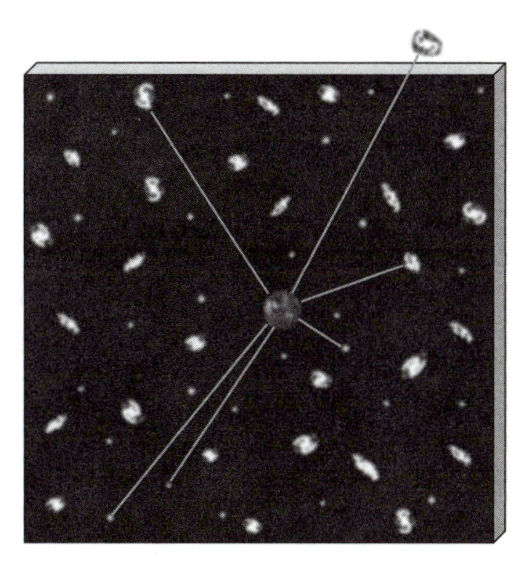

Fig. 2-1: Schematic illustration of one aspect of Olber's paradox. If the universe is infinite in space and time, then looking out into space from Earth, every line of sight ultimately hits a star or distant galaxy. If the line did not intersect light within the box, it could simply be extended further until there was an intersection. Therefore, the night sky should be filled with light, but it isn't.

support for the Big Bang theory of universe origin. On a scale of reliability that goes from 0 (idle speculation) to 10 (proven fact), this theory gets a 9.9!

Before presenting this evidence, let us consider a seeming paradox that confronted astronomers before the concept of an expanding universe was proposed. This paradox was articulated by Heinrich Olbers in 1826. To state it simply, no one was able to explain the fact that the nighttime sky is dark. The black background between the stars seemed to demand either that the universe has a finite extent or that the light from the most distant stars is being intercepted by dark matter in the voids of space. To understand this, one has only to envision a universe of infinite extent made up of luminous objects separated by empty voids. In such a universe, no matter where we looked we would see the light from some distant star (Fig. 2-1). The sky would be blindingly bright! The obvious alternative is that the universe is finite. In a finite universe, we could look between the stars into the black void beyond. Another possibility is, of course, that there are clouds of nonluminous matter floating in the voids between stars and that these clouds block the light from the more distant stars from our view.

The first alternative appeared unacceptable, because in a universe of finite extent there would be nothing to hold the stars apart. The mutual star-to-star gravitational attraction would lead to an unbalanced pull toward the "middle" of the universe. It would be as if we secured a series of balls on a great three-dimensional latticework and then connected each ball to each of the other balls with a stretched rubber band. While the balls near the center of the latticework would be pulled more or less equally from all directions, those near the edge of the lattice would be pulled toward the inside. If by magic we suddenly removed the latticework, leaving only the balls and stretched rubber bands, there would be a massive implosion as the balls streaked toward the lattice's center. Only if the lattice were infinite in extent would nothing happen. In this case the pull on every ball would be exactly balanced. The universe has no latticework to hold the stars apart, yet there they are. Hence, the finite universe explanation for the dark sky must be rejected as inadequate.

The second explanation—that the light from very distant stars is intercepted by dark clouds of dust and gas along its path to the Earth—is also unacceptable. In this situation the light from stars at intermediate distances would also be affected. We should see a glow of scattered light similar to that in the night sky over a great city or from headlights approaching through fog. No such glow is seen! So this explanation must also be rejected.

More than a hundred years passed before this cosmological puzzle was solved. In 1927 a Belgian astronomer, Georges Lemaître, proposed that the universe began with the explosion of a cosmic "egg." This clever concept neatly explained the long-standing paradox in that the force of the explosion prevents the gravitational pull from drawing the matter toward the center of the universe. It would be as if a bomb were to blow the balls on our lattice away, overpowering the pull of the rubber bands. In the absence of observational evidence, the Lemaître hypothesis would have received relatively little attention. Within two years of its publication, however, Edwin Hubble reported observations that turned the attention of the scientific world toward the concept of an expanding universe. Hubble reported a shift toward the red in the spectra of light reaching us from the stars in very distant galaxies. The simplest explanation for such a shift was that these distant galaxies were speeding away from ours at incredible speeds.

THE RED SHIFT: MEASURING VELOCITY

The light coming from the sun consists of a spectrum of frequencies. As these light rays pass into and out of raindrops, they are bent. Each frequency is bent at a slightly different angle separating the bundle of mixed light into a rainbow of individual color components. Each of these frequencies leaves a different imprint on our retina. We see them as colors.

Isaac Newton in the seventeenth century did a number of experiments with light, forming rainbows by passing sunlight through a glass prism. Light rays passing through such a prism are bent according to frequency. As shown in Figure 2-2, the red light (that with the lowest frequency detectable by our eyes) is bent the least, and violet light (that with the highest frequency detectable by our eyes) is bent the most. What we see as white light is actually a combination of all the colors in the visible spectrum.

Astronomers have long used prisms (and more recently diffraction gratings) in their telescopes as a means of examining the color composition of the light from distant galaxies. Rather than producing a continuous spectrum, the light from stars is broken up by dark bands that mar the otherwise smooth transition from red to orange, to yellow, to green, to blue, and to violet. The dark bands are produced by the absorption of

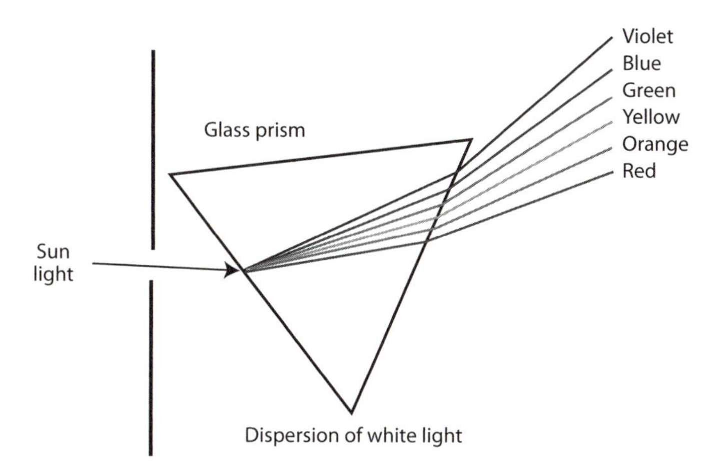

Fig. 2-2: Light rays passing through a prism are bent according to frequency. The red light (that with the lowest frequency detectable by our eyes) is bent the least, and violet light (that with the highest frequency detectable by our eyes) is bent the most.

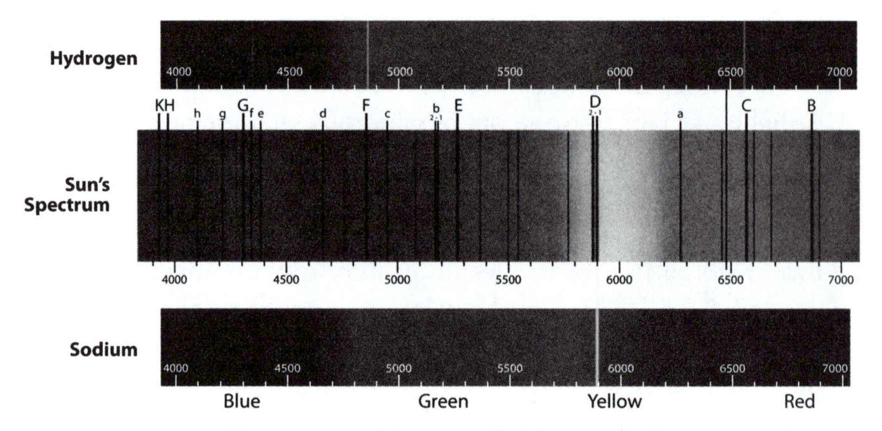

Fig. 2-3: A portion of the spectrum of the sun, called the *Fraunhofer spectrum* after its discoverer, compared to portions of emission spectra of hydrogen and sodium. The *C* and *F* lines of the Fraunhofer spectrum are created by hydrogen in the sun's atmosphere; the dark *D* lines in the yellow are the most prominent lines of the sodium spectrum. Other lines are produced by absorption by other elements. See also color plate 3.

certain frequencies of light by the element-containing halo of gas surrounding the star that is producing the light. A packet of light can interact with an atom only if it has just the right energy to lift one of the atom's electrons from one of its permitted energy levels to another. While transparent to some frequencies of light, the excitation of elements in the gas absorbs other frequencies so that specific wavelengths of light do not get through. Early spectra identified only the most prominent lines (Fig. 2-3). When examined in detail, thousands of lines became apparent. Most do not completely blacken the rainbow; they produce a weakening of the intensity of the light at that frequency. This weakening is the result of partial absorption of the departing light by the star's "atmosphere," depending on the abundance of the element.

Astronomers originally took an interest in these bands because they offered a means of making chemical analyses of the star's halo of gas. Unlike Earth, whose atmosphere has a composition that bears no relationship to that of its crust or interior, the composition of a star's atmosphere is close to that of its bulk. Each partially darkened line in the spectrum represents a single element. Using laboratory arcs as a means of calibration, astronomers were able to estimate the relative abundances of the elements making up the atmospheres of neighboring stars. Since

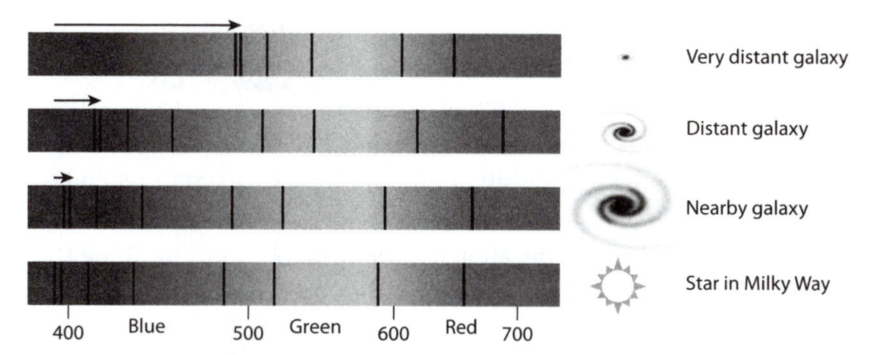

Fig. 2-4: Schematic illustration of how the spectral lines of distant stars shift toward the red (the right-hand ends of the spectra). The bottom spectra represents a nearby star in our galaxy and the spectral lines correspond with the wavelengths we could observe for the elements on Earth. For nearby galaxies, which appear large in a telescope, the spectra are shifted slightly toward the red. For distant galaxies, which appear very small in the telescope, the red shift is much greater. Arrows indicate the extent of the red shift for each spectrum.

all stars contain at least some of all the elements, the characteristic lines become a fixed "bar code," with spacings and relative intensities controlled by the fundamental characteristics of atoms (see Fig. 2-3).

As bigger and better telescopes become available, astronomers were able to extend their chemical analyses to more distant objects. It was here that the great discovery came. When astronomers looked at very distant objects, they found that the characteristic "bar code" shifted with respect to the rainbow background. For example, patterns of lines that were in the blue part of a spectrum taken from the sun would instead be found in the green part of the spectrum for the light from a distant galaxy; a line that appeared in the yellow part of the sun's spectra would appear in the orange part of the distant galaxy's spectrum, and so forth. The "bar code"—the spacing and relative intensity of lines—remained the same. But it looked as if someone had lifted the whole set of dark lines off the background rainbow, moved it toward the red end, and then replaced it. More startling was the finding that the more distant the object, the greater the shift toward red (see Fig. 2-4).

We can understand how this occurs if we first grasp what we will refer to as the "train-whistle concept" (physicists call it the *Doppler shift*). Those who have indulged in train watching may remember that most engineers of express trains blow their whistles as they roar through local stations. Anyone standing on the platform experiences a strange sensation as the train passes. The pitch of the whistle suddenly drops! It drops for exactly the same reason that the lines in the spectra for distant galaxies shift. Since the whistle situation is a bit easier to comprehend, we will consider it first.

Sound travels through the air at a velocity of 1,236 km/hr. If the train passes through the station at 123 km/hr, then the frequency of the sound impulses on the listener's ears would be 10 percent higher as the train approaches and 10 percent lower after it has passed by. This phenomenon is easily understood if we substitute for the train's whistle a beeper that gives off one beep each second. Were an observer to count the beeps from a train stopped down the tracks, he would get 60 each minute. Were he to count the beeps from a train speeding toward him at 123 km/hr, he would hear 66 each minute. Were he to count the beeps from a train speeding away from him at 123 km/hr, he would hear only 54 beeps per minute. The ear counts the frequency of sound waves hitting our eardrum. When the source of the sound is receding, each beat is further away and has to travel further to reach the eardrum, so the ear detects a lower frequency and sends to the brain a lower pitch.

If a source of light is receding, the "pitch" of its light is also lowered. However, as light travels at a staggering 1,080 million km/hr, the frequency of light reaching us from a speeding train is not significantly changed, because to have an effect the speed of retreat has to be a significant fraction of the speed of transmission. So if we observe a shift toward red in the spectrum of the light reaching us from a distant galaxy corresponding to a 10% reduction in frequency, the galaxy must be speeding away from us at the amazing speed of 108 million miles an hour!

MEASURING DISTANCE

As stated above, the more rapidly a galaxy is retreating, the larger will be the shift of its light toward the red. Examples of spectra observed from a series of galaxies are shown in Figure 2-5. The great discovery that followed the discovery of the red shift was that those galaxies that exhibit the greatest red shift are also the farthest away. This discovery required a series of developments that gradually led to a robust distance scale.

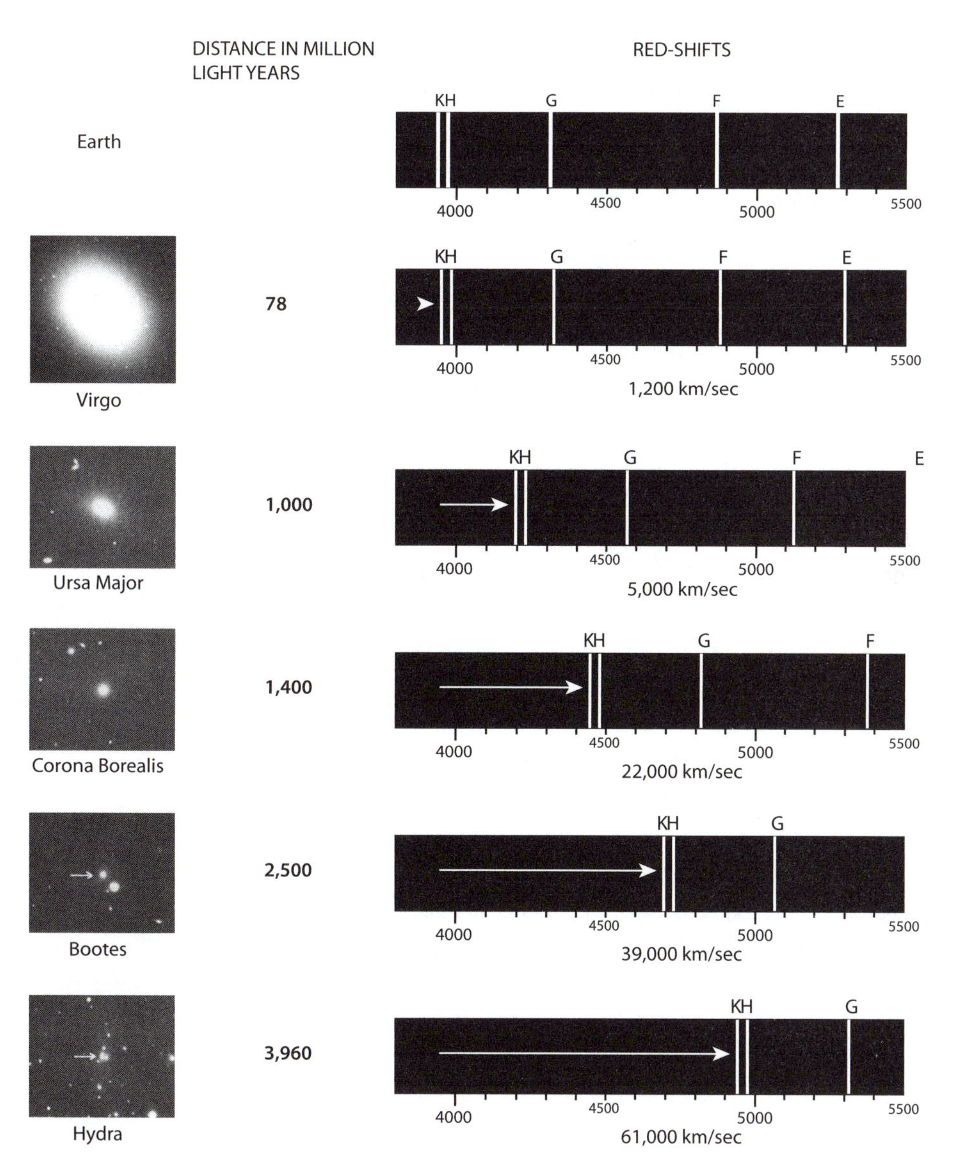

Fig. 2-5: Galaxies and their light spectra: On the left are shown photos of five galaxies taken with the Hale Observatory telescope. As these objects are probably similar in size, Virgo must be located much closer to Earth than Hydra. Also shown, on the right, are schematic light spectra from the galaxies, compared to spectral lines as measured on Earth. The horizontal white arrows show the displacement of an easily identified pair of dark lines from its position in a light spectrum for the Sum (or for a laboratory arc). The recession velocities corresponding to these arrow lengths are given. As can be seen, the more distant the object, the greater is recession velocity. (Images courtesy of California Institute of Technology)

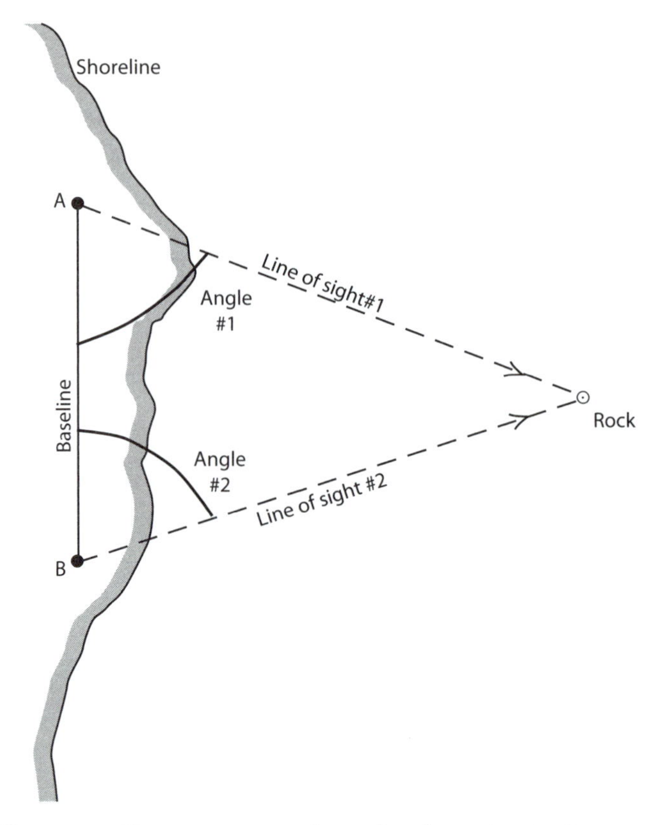

Fig. 2-6: Illustration of how geometry can be used to determine the distance to far away objects. The surveyor observes the distant rock out in the ocean from the ends of a baseline on the shoreline of known length and notes the angles between baseline and the lines of sight. He can then compute the distance through trigonometry.

Distances are far more difficult to measure than velocities—sufficiently difficult that it is beyond our task here to try to grasp exactly how it is done. A few paragraphs will suffice to show the general principles.

As do all surveying schemes, measurements out into space start with a baseline (Fig. 2-6). If a surveyor wants to measure the distance of an object that he cannot easily reach (like a rock out in a lake), he sets up a baseline on shore and measures its length. He then observes the rock from both ends of the baseline and notes the angle between the line of sight and the baseline. Simple trigonometry allows him to calculate the distance to the rock.

As can be seen from Figure 2-7, the ranges of distance confronting the astronomer are staggering! The astronomer boldly starts by using

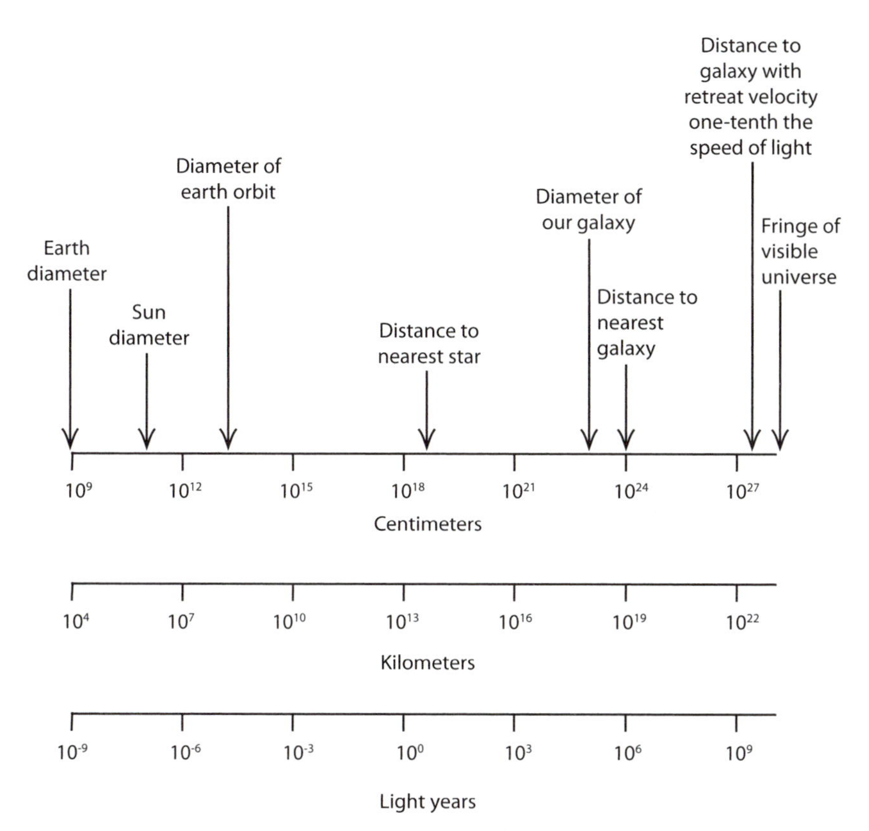

Fig. 2-7: Distance scales. Astronomers must cope with distances ranging over 19 orders of magnitude.

Earth's orbit about the sun as his baseline. By making observations of objects in the sky from the extremes of the orbit, the astronomer can use the triangulation method to measure the distance of "rocks" out in space. Even with this seemingly gigantic baseline, this proves to be a very tough task. The baseline is 3×10^8 km long. Even the nearest star is 4×10^{13} km away. Thus, it is akin to measuring the distance of a rock 10 km off the coast using a baseline only 1 cm long!

Through the use of a very accurate technique called *parallax*, the distance to a few thousand of our nearest neighbor stars can be determined using Earth's orbit as a baseline. This method, however, is limited to a very, very small portion of our own galaxy.

The baseline was greatly extended by showing that our own sun is rushing through our galaxy at a rather large rate of speed—6×10^8 km/yr.

In this way an ever-growing baseline far longer than Earth's orbit affords has been established. It would be as if a surveyor were driving along a shore road in a truck, periodically taking sightings on a distant island. From the truck's velocity and the elapsed time, he could determine the length of his growing baseline. In a related but more complicated way (called *statistical parallax*), astronomers have been able to measure the distances of stars out to about 3×10^{15} km. Even so, all these stars reside in our own galaxy.

Finding the distance to galaxies beyond our own posed a problem so formidable that the trigonometric approach had to be abandoned. Nature, however, provided an alternative approach, which astronomers hit upon and exploited. Some of the stars in our galaxy show regular pulsations in their luminosity. Hence, they are more like lighthouses than headlights. These stars show a range of blinking rates. The important characteristic is that stars that blink at the same rate have the same luminosity. It's as if the Coast Guard decided to have all its lighthouses use "light bulb" strengths related to their turning time. For example, all lighthouses with 100,000-watt bulbs would turn once each minute; those with 200,000-watt bulbs would turn twice a minute, and so forth. The variation in intensity of these stars is large—almost a factor of 10, so they are quite easy to spot even in galaxies other than our own (Fig. 2-8).

The astronomers jumped on this relationship and reasoned that blinking stars visible in nearby galaxies probably followed the same rules: from the blinking rate the luminosity of the star could be estimated. By comparing that luminosity at the source with the intensity of light seen from Earth, the distance of the star, and hence of its hosting galaxy, could be determined. This "headlight" method is a quantitative version of how we intuitively judge the distance of an approaching car on a dark highway. As automobile headlights have similar luminosities, we judge the distance of an oncoming vehicle by the brightness of its headlights. The distances to the nearby galaxies could be estimated by the differences between the intensities of light received from these distant blinkers and the intensity of light received from one of its cousins in our own galaxy, whose distance had been determined by trigonometry. Knowing the distances of these nearby galaxies, the astronomer could then from trigonometry determine their diameters. A "map" showing the Milky Way and its neighboring galaxies and clouds of gas and dust is shown in Figure 2-9.

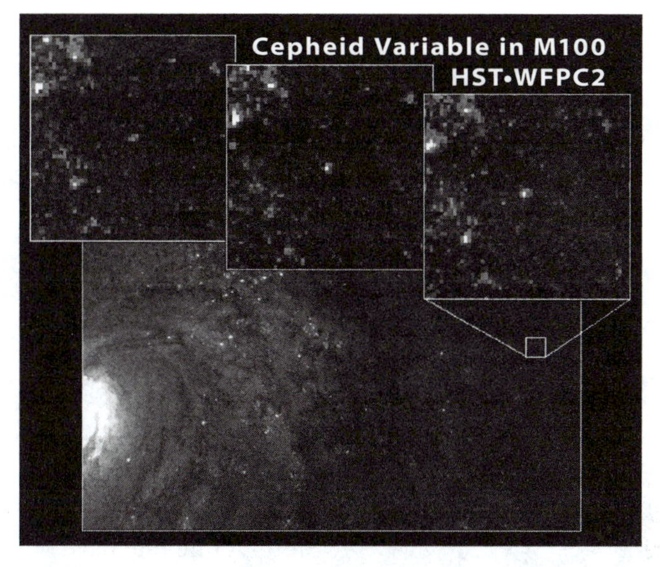

Fig. 2-8: Illustration of the change in magnitude (intensity of light) of a pulsating star in a nearby galaxy. The squares at the top are blowups at three different times of the small region in the galaxy identified by the small box in the lower image where one of these stars, centered in the boxes, can be identified. (Courtesy of NASA; http://apod.nasa.gov/apod/ap960110.html. Credit: NASA, HST, W. Freedman (CIW), R. Kennicutt (U. Arizona), J. Mould (NU))

Unfortunately, the galaxies that show significant red shifts are so far away that even our biggest telescopes cannot resolve individual stars, though much more distant galaxies have been able to be measured by this method using the Hubble telescope, which avoids the interference of Earth's atmosphere by residing in space. For the most distant galaxies, however, the whole galaxy appears only a bit larger than a nearby star. Thus, no individual pulsing stars can be identified and the lighthouse method is not applicable.

The last step out into space is then taken by using the size of the galaxy itself. More often than not, galaxies are found in clusters. Astronomers have carefully studied the sizes of the galaxies in nearby clusters. As with the sizes of people (and cars), they follow some simple rules. The assumption is made that the galaxies in very distant clusters have a similar spectrum of sizes and brightness as the ones in "nearby" clusters. For example, the distance of a car can be estimated not only from the brightness of its headlights but also how far apart they seem to us. While

Fig. 2-9: Illustrative map showing the Milky Way and its nearest neighbor galaxies and clouds of gas and dust. The Andromeda Galaxy, with one trillion stars, is 2.5 million light years from the Milky Way. (Courtesy of NASA/CXC/M.Weiss; http://chandra.harvard.edu/resources/illustrations/milkyWay.html)

there is some variation, we will not be far off with this method. Like the automobile driver, the astronomer infers the distance of these clusters by the sizes of their individual galaxies.

Recently, astronomers have improved on this approach by observing a certain class of supernova explosions in distant galaxies. Such events occur in any given galaxy roughly once each century. Thus, each decade about one in every ten galaxies is lit by such a strong flash of light that it can be observed. These flashes are thought to be excellent headlights.

THE VELOCITY-DISTANCE RELATIONSHIP: DATING THE BEGINNING

Having determined both velocity and distance for galaxies throughout the universe, astronomers can then make a graph with the distance of the galactic cluster on one axis and the velocity at which it is receding from us on the other axis. As shown in Figure 2-10, when the observations for various galactic clusters are plotted on such a graph, the points form a linear array. A factor of 10 increase in distance is matched by very nearly a factor of 10 increase in recession velocity. What is the significance of this striking relationship?

The significance of the distance-vs.-red shift relationship is that it is what would be expected for galaxies that were once all together at the same time and place. Consider, for example, a birthday party where

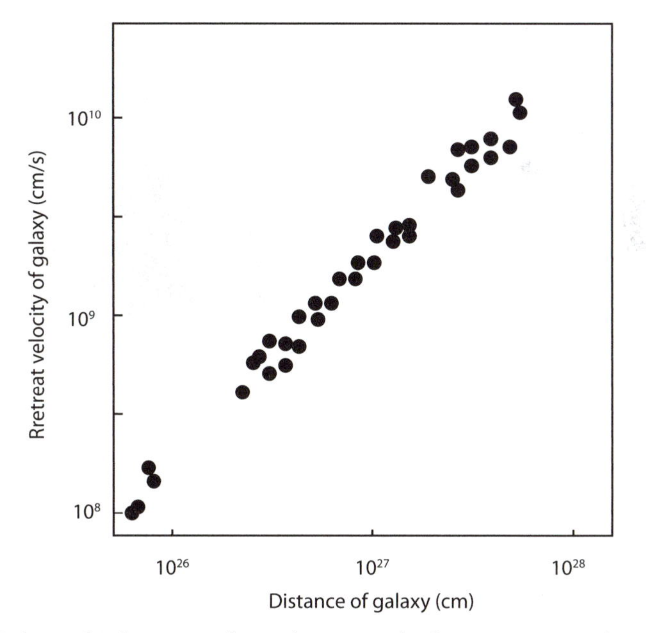

Fig. 2-10: Relationship between galactic distance and galactic recession velocity. Each point represents a distant galaxy (or cluster of different galaxies). Since the distances range over a factor of 100, a logarithmic rather than linear scale is used.

everyone leaves at exactly the same time. Some people walk home at 4 km/hr, others cycle at 10 km/hr, others drive at 50 km/hr, and one takes a helicopter at 500 km/hr. Imagine that they all travel for an hour in a straight line but heading off in different directions. After one hour, the walkers are 4 km away, the cyclists 10 km away and so on. Plotting the speed of their retreat vs. their distance from the party house would produce a straight line, and the slope of the line gives the time they left the party. It is also true that plotting the distance from any one of the groups would lead to the same result, so everyone who was at the party would produce a graph with the same slope, because they were all at the party and they all left together. The farther two groups are from one another, the faster they must be moving apart. The same thing happens in three-dimensional space. If we turn time around and move the various galaxies backward at the rates they are observed to be retreating, all come together at the same time! The actual date can be obtained from the distance (from our galaxy) and the rate of recession (from ours) of any distant galaxy.

So this simple diagram both shows us that all the galaxies were together at one point in time and gives us that time, which is the age of formation of the universe. Since one axis is cm and the other is cm/sec, the slope is time (or 1/time). The ratio of distance to recession velocity yields an age for the universe. The result is that the matter in the universe is flying outward on the wings of an explosion that occurred about 13.7 billion years ago.

In Figure 2-11 the evolution of the distance-velocity relationship is depicted. Were we to have lived only 5 billion years after the Big Bang, the line depicting the velocity-distance trend would have been about three times steeper than the one we obtain today. This is because the retreat velocity for any given galaxy remains nearly the same, while the galaxy's distance from us increases.

A natural question that arises from this reasoning is, Where is the center of the universe? A train analogy in Figure 2-12 shows why the velocity/distance relationship tells us nothing about this. Observers on night train A speeding along one track see a light mounted on top of train *B* speeding along another track. They also hear train B's whistle. They know this train left the central station at the same time their train did. From the strength of its light they determine the distance of train B.

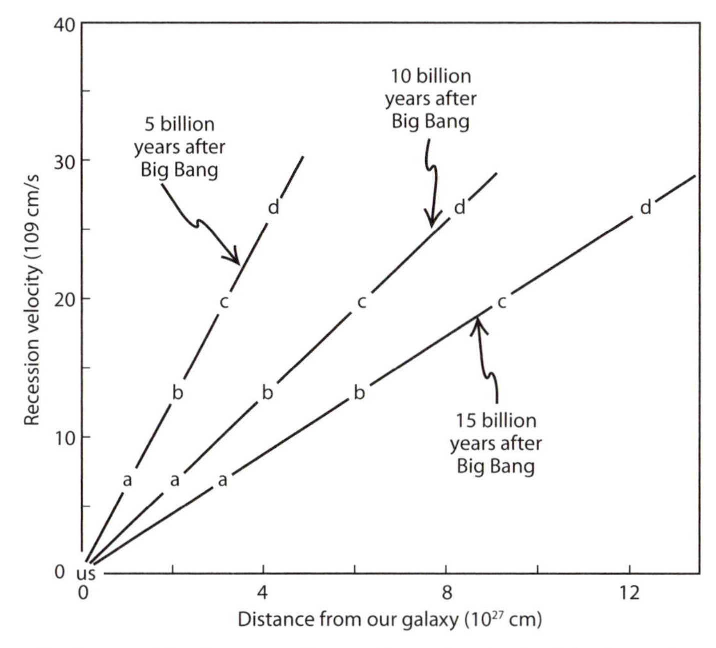

Fig. 2-11: Evolution of the distance-velocity relationship. Each of the four galaxies a, b, c, and d moves away from us at a different velocity. These velocities have remained nearly constant through time. However, as the universe becomes older, the distance separating these galaxies from us increases. Some 15 billion years after the explosion, they are three times as far away as they were 5 billion years after it.

From the pitch of its whistle they know that it is moving away from them and the exact speed of this recession. Knowing only this much, the observers could not determine where the central station is located. Neither can astronomers locate the center of the universe.

Added Support for the Big Bang Hypothesis

Additional support for the Big Bang was provided by the discovery that the universe has a nonvisible background glow. To understand this glow it is necessary to realize that all objects above absolute zero emit radiation that is diagnostic of their temperature (see Fig. 2-13 for a description of the various temperature scales). This radiation, called *blackbody radiation*, can be used to estimate the temperature of distant objects. The wavelength of emitted radiation decreases as the temperature increases.

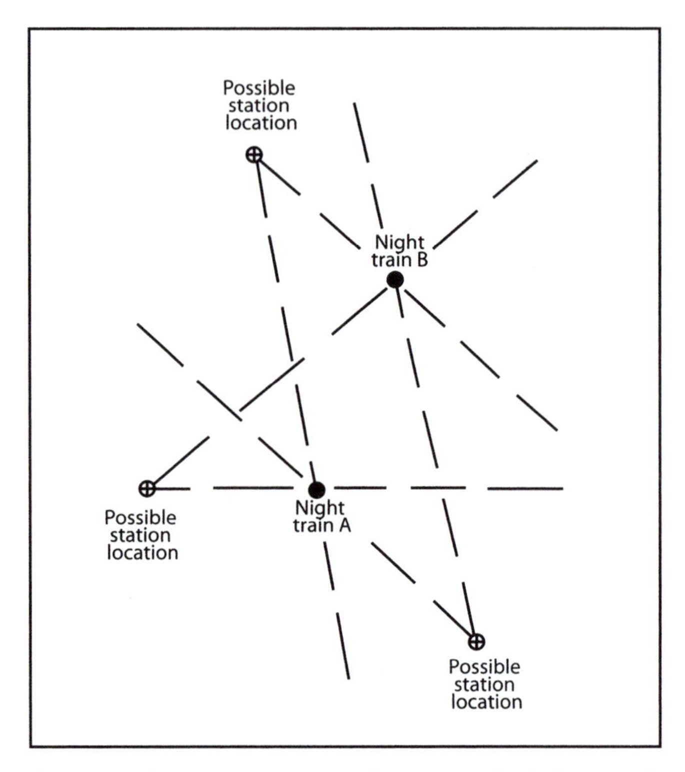

Fig. 2-12: The train analogy. Passengers on night train A sight the beacon of night train B, which left the central station at the same time they did. From its brightness they determine the distance between the two trains. They also hear the whistle. From its pitch they determine the speed at which the two trains are moving apart. However, unless they have some other information (e.g., the direction in which their train is going and the speed at which it moves along the track), there is no way for them to determine where the central station is located. Of the infinite number of possibilities, three are depicted here.

At very low temperatures, the radiation is not visible. But as the temperature gets above a few hundred degrees C, the wavelengths start to enter the visible range, and the object glows a dull red. At higher and higher temperatures the object turns orange, then white hot, and so on. It is not a single wavelength of radiation that is emitted, but a characteristic pattern that is completely diagnostic of the temperature of the object. This is apparent on the dark coils of an electric stove, for example. As the coils get hotter, the radiation emitted changes wavelength. At first the coils remain dark because the emission is in the infrared, where our eyes are not sensitive. Then they glow a dull red as some of the radiation

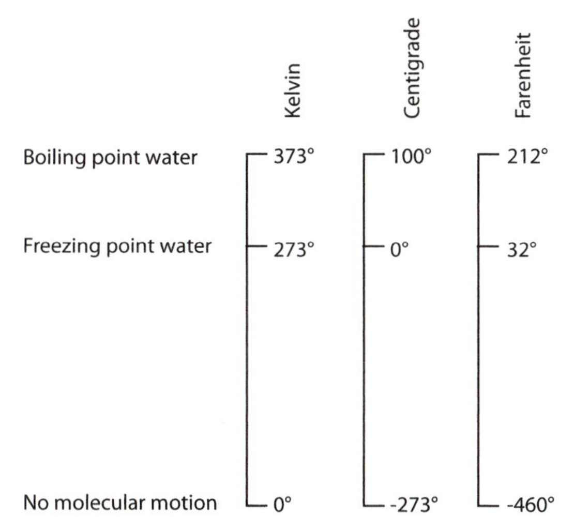

Fig. 2-13: Description of the three temperature scales. Absolute zero on the Kelvin scale is the temperature of no molecular motion. In everyday life in the United States used the Fahrenheit scale is used. The centigrade scale is used by most other countries in the world.

reaches the visible range. If they get too hot they are almost white, as many colors of the visible spectrum are emitted. Measuring the detailed pattern of radiation from a distance can tell us the temperature of the object. For example, Earth gives off radiation characteristic of its surface temperature of about 288°K. This radiation is centered in the infrared range. The sun gives off radiation characteristic of its surface temperature of 5700°K. This radiation is centered in the visible range.

That's the background. The surprising data was obtained by physicists Robert Wilson and Arno Penzias of the Bell Laboratory in New Jersey, who for other reasons were experimenting with an detector that was very sensitive to very long wavelength radiation—electromagnetic waves in the 0.1- to 100-cm range (i.e., microwaves) When they happened to turn their instrument toward the heavens, they found that although no visible light can be observed in the dark voids between stars and galaxies, there is a nonvisible glow. Looking at the detailed pattern of the radiation, they were able to show that this universal glow was the same as that emitted by an object whose temperature is 2.73° above absolute zero . After the Wilson and Penzias discovery, subsequent work, including very precise measurements from satellites, has demonstrated with

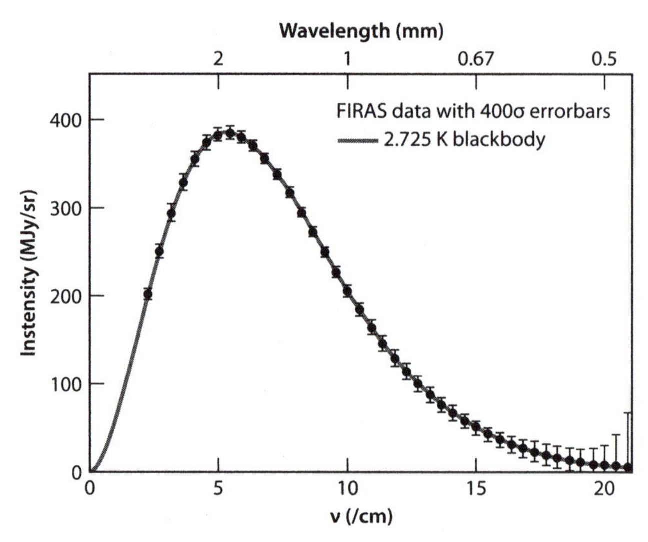

Fig. 2-14: The cosmic microwave background of the Universe determined by the far infrared absolute spectrophotomer on the COBE (Cosmic Background Explorer) satellite. The universe glows with radiation in the microwave range. By measuring the intensity of this glow at many wavelengths, it can be seen that the spectrum corresponds exactly with blackbody radiation from a material with a temperature of 2.725 + 0.002°K. The temperature corresponds extraordinarily well with the Big Bang theory. (Courtesy of NASA;http://lambda.gsfc.nasa.gov/product/cobe/firas_overview.cfm)

great precision that the relative intensities of the various wavelengths of radiation in this range are consistent with this very cold glow (fig. 2-14).

What is the source of the universal blackbody radiation? Shortly after the Big Bang, a great flash of light appeared when the protons and electrons in the expanding universe cloud cooled to the point where they could combine into neutral atoms. At that time the universe was only about 100,000 years old and the gas had a temperature of about 4,000°K. The reason why this light, which was given off from a gas at 4,000°K, appears now to have been given off by an object about 1,500 times cooler (that is, one with a temperature of 2.76°K) has to do with the expansion of the universe since that time. While the computation of the magnitude of this "cooling" is too complex to be described here, to the physicist it is exactly as expected. Hence, the discovery of the afterglow of the Big Bang is taken by physicists as a strong confirmation of the Big Bang hypothesis.

As we shall learn in the next chapter, universe matter right after the Big Bang consisted almost entirely only two elements, hydrogen (H) and helium (He). By careful modeling of the Big Bang, physicists have been able to calculate what the proportions of hydrogen and helium should have been from the atomic reactions that took place. The calculated proportion of 10:1 corresponds with the observed H/He ratio in the universe.

The combination of these different and independent lines of evidence—the velocity/distance relationship of galaxies, the background radiation of the universe, and the chemical composition of the universe all combine to support the Big Bang hypothesis of the universe's origin.

An Expanding Universe and Dark Energy

The universe has been expanding from the beginning, counterbalanced by an inevitable gravitational force. This led to the idea that gravity might be strong enough to gradually slow the expansion to zero, and then a grand contraction would occur, leading to a "big crunch" and even a possible oscillation of the universe. Could this be shown from observations? The launch of the Hubble Space Telescope in 1998 provided the necessary data, as well as the surprising and completely unexpected result—that the expansion has been accelerating. Theorists have striven to come up with possible explanations for this result, and the various ideas come under the name of *dark energy*. Dark energy is not a minor phenomenon; to explain the observations, it must make up some 70% of the universe! Furthermore, it has the opposite effect to what we perceive as matter and energy and exerts an expanding force on the universe that is able to overcome gravitational attraction.

A further problem is that all the visible matter in the universe is insufficient to account for the mass that is required to explain various cosmological observations. The remaining, invisible mass is referred to as *dark matter*. Dark matter is also not trivial, but makes up about six times as much mass as "normal" matter. Dark matter is not in stars, planets, or back holes. Physicists know what it isn't, but are not sure what it is!

All our discussion in the remainder of this book will be referring to what we call "normal" matter and energy. This world that we can see and discuss turns out to be only ~4% of the composition of the universe

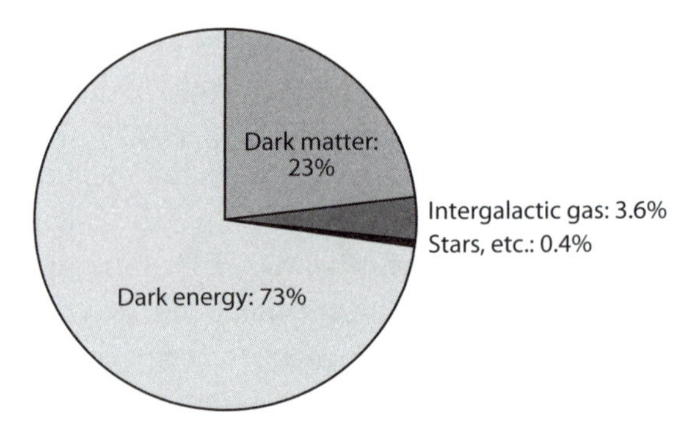

Fig. 2-15: Pie chart illustrating the composition of the universe. All the matter that we can directly observe and will be discussing in this book makes up only 4% of the universe.

(Fig. 2-15). As we continue with our discussion of what is known, and the questions about what is known, it is also useful to reflect on the fact that the unknown remains far greater than the known.

Aftermath of the Big Bang

About 100,000 years after the Big Bang, when the expanding matter had cooled to the point where the heretofore free electrons could become entrapped in orbits around the positively charged nuclei, helium and hydrogen gas formed. This gas was lighted only by the afterglow of the Big Bang. At this point the universe was a dull place indeed. No galaxies, no stars, no planets, and no life were to be found. There were only molecules of gas in a rapidly expanding cloud.

Then, for reasons as yet not entirely understood, the cloud began to break up into a myriad of clusters. Once formed, these clusters remained as stable units bound by their mutual gravitation. Each of these clusters in turn evolved into one or more galaxies. Within these galaxies the gas further subdivided to form many billions of brightly burning stars. The universe was no longer dark!

While these early stars are by now either dead or lost among their younger counterparts, we can be quite sure that they had no Earthlike planets. The reason is that Earthlike planets cannot be formed from hydrogen or helium. Elements not present in the young universe are required. Thus, the next step in our journey to habitability will be to see where and how the remaining ninety elements formed.

Summary

Human beings have always been interested in looking for knowledge and inspiration in the heavens. Natural curiosity and questions such as, "What is the spectrum of the sun and how does it compare to stars?" and "How far away are the stars?" led to unexpected discoveries. Distant galaxies have their element "bar code" of spectral lines shifted to the red, requiring that they be moving away from us at great rates of speed. Most surprisingly, the speed of retreat correlates with the distance, suggesting a common origin at the same time and place 13.7 billion years ago. This inference from direct observation obtained unexpected support from another observation based on curiosity, the answer to the question, "Does the universe emit any background radiation?" The blackbody radiation then turned out to be striking confirmation of the Big Bang. Subsequent understanding of nuclear physics led to another confirming prediction in the H/He ratio. All of these combine to make the Big Bang one of our fundamental pieces of knowledge of where we come from, and when it happened.

In the last decade, further observations from the Hubble Space Telescope show us that all the matter that we can observe makes up but a small fraction of the universe. Much remains to be discovered in our exploration of the universe.

Supplementary Readings

Frank Durham and Robert D. Purrington. 1983. *Frame of the Universe*. New York: Columbia University Press.

William J. Kaufman III. 1979. *Galaxies and Quasars*. New York: W. H. Freeman & Co.

Joseph Silk. 2001. *The Big Bang*, 3rd ed. New York: W. H. Freeman & Co.

Steven Weinberg. 1977. *The First Three Minutes*. New York: Bantam Books.

Richard Panek. 2011. *The 4 Percent Universe: Dark Matter, Dark Energy, and the Race to Discover the Rest of Reality*. Boston: Houghton Mifflin Harcourt.

Fig. 3-0: The Crab Nebula, a supernova remnant in the constellation of Taurus, 6500 light years from Earth. The nebula is the expanding cloud that began with the supernova recorded by Chinese and Arab astronomers in 1054. The expansion rate is about 1500 km/sec, and the nebula is currently about 11 light years across. The size of our solar system out to the planet Neptune (~0.001 light years) would only be the size of a tiny spot on the image. The nebula was the first to be associated with a historical supernova explosion. (Courtesy of NASA, ESA, and Allison Loll/Jeff Hester (Arizona State University). Acknowledgment: Davide De Martin (ESA/Hubble))

CHAPTER 3

The Raw Material

Synthesis of Elements in Stars

During the explosive birth of our universe only two elements were formed in abundance: hydrogen and helium. Were this the end of the story, no planets and life could have appeared in the history of the universe. Our planet and the sun contain all the elements, so the other ninety elements in the periodic table must have been produced over the history of the universe. Stars are the universe's element production factories. Stellar interiors are so hot that atomic nuclei can interact and fuse together, emitting massive energy and making heavier elements in the process. Nuclear fusion, however, can occur only up to nuclei with 56 atomic particles, which is the element iron (Fe). Stars that reach this point then explode, creating heavier elements in the process and casting forth into neighboring regions of the galaxy a mixture of all ninety missing elements. The frequency of these awesome explosions in galaxies like our own Milky Way is about one per thirty years.

Evidence that supports this origin is imprinted in the relative abundances of the elements that make up our solar system. For example, the high relative abundance of iron is consistent with the fact that it is the ultimate product of the nuclear fires at the centers of stars. Element production in stars is also demonstrated by the existence of the spectral lines imposed by elements with very short radioactive half-lives. The radioactive decay of 78-day half-life cobalt (Co) with 56 nuclear particles (^{56}Co) dominates the light given off after a supernova explosion, demonstrating element production of heavy elements during such an event. The element technetium (Tc) is also found in stellar spectra. Since all the isotopes of Tc are radioactive with short half-lives, Tc can be present only in matter freshly produced by a nuclear furnace.

Because supernovae happen relatively frequently, the history of individual explosions can be monitored. Chinese astronomers observed a supernova in 1054. The debris cloud from this explosion continues to expand and is now known as the Crab Nebula (see frontispiece). Through the course of our galaxy's history, the formation and demise of about 100 million red giants has converted about 2% of the galaxy's hydrogen and helium into heavier elements. Contained in this 2% are the ingredients needed to build planets and form life. The processes of element creation are common to stars in all galaxies, and the raw material for planets and life is omnipresent in the universe.

Introduction

By cosmic standards, our Earth and its fellow terrestrial planets are chemical mavericks. They consist primarily of four elements: iron (Fe), magnesium (Mg), silicon (Si), and oxygen (O). By contrast, we look out on stars that are made up almost entirely of two different elements, hydrogen and helium. For the universe as a whole, all elements other than hydrogen and helium are small potatoes; taken together they account for only about 2% of all of the 4% of matter that is not dark matter or dark energy.

Despite their rarity, the elements other than H and He are prerequisites for habitability. A habitable planet must have a solid or liquid exterior and an abundance of the element carbon (C). Objects made primarily of hydrogen and helium gas offer no solid base. Hence, high on our agenda must be an understanding of how elements heavier than hydrogen and helium were formed and how these elements were separated from the bulk gas and forged into rocky planets. In this chapter we tackle the first of these problems.

The Chemical Composition of the Sun

All stars form from the gravitational collapse of clouds of gas. Since the lion's share of the matter in the collapsing clouds ends up in the star itself, the star's chemical composition must be representative of the parent cloud. If we could somehow determine the chemical composition of the sun, we could constrain the composition of the galactic matter from which the sun formed.

As noted in Chapter 2, our information about the composition of stars comes from the dark lines in its spectrum that result from absorption by chemical elements in the atmosphere of the sun through which the light passes. The extent to which the light corresponding to each line is muted in the rainbow is a measure of the abundance of that particular element in the sun's atmosphere. Fortunately for stars like our sun, except for hydrogen and helium, the atmosphere is thought to have a composition nearly identical to that of the star's interior.

The strength of the lines can then be converted into the relative abundance of elements in the sun's atmosphere. By "relative abundance" we mean the ratio of the number of atoms of a given element to the number of atoms of a reference element. By convention, astronomers use silicon as the reference element. The relative abundance of an element is stated as the number of atoms of that element for each 1 million atoms of silicon. These abundances are plotted versus element number on the graph in Figure 3-1. This graph has a logarithmic power of 10 scale. For example, helium atoms with a relative abundance lying between 10^9 and 10^{10} on this scale are 10 billion times more abundant than bismuth (Bi) atoms with a relative abundance lying between 10^{-1} and 10^0.

Other than the dominance of the abundances of hydrogen and helium over those of the other ninety elements, a prominent feature of the graph is the general decline in abundance with increasing element number. Superimposed on this decline are several obvious features. One is that the abundance of the element iron is 1,000 times higher than would be expected were the decline smooth. A second is that the elements lithium (Li), beryllium (Be), and boron (B) have abundances many orders of magnitude lower than would be expected were the decline smooth. A third is that the abundance curve has a saw-toothed appearance because elements with an odd number of protons are generally less abundant than their even-numbered neighbors. These characteristics of the abundance curve provide hot clues regarding the mode of origin of the elements heavier than hydrogen and helium.

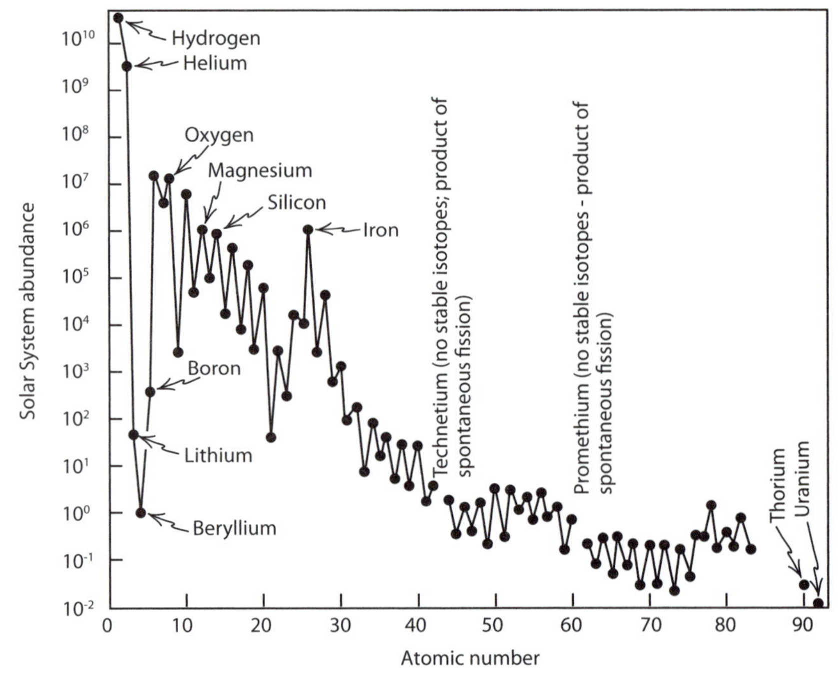

Fig. 3-1: Relative abundances of the elements in our sun: as the abundances range over 13 orders of magnitude, they must be displayed on a power of 10 (logarithmic) scale. The abundance of each element is expressed as the number of atoms per million (i.e., 10^6) atoms of the element silicon. The gaps in the sequence of technetium and promethium represent elements that have only radioactive isotopes and are, therefore, absent in a relatively low temperature star such as the sun.

Hydrogen, Helium, Galaxies, Stars

Physicists conjecture that at the instant of the Big Bang, all matter must have been contained in a very compact blob. The pressures and temperatures in this primordial blob were so high that stable combinations of neutrons and protons could not exist. Within seconds after the explosion, however, such combinations could and did form. At one time it was hypothesized that the blend of elements we see in our sun might have been generated entirely during the first hour of universe history. But subsequent work has shown that the only elements produced in significant amounts during this very early phase of universe evolution were hydrogen and helium. The others were produced billions of years later, inside giant stars.

The hydrogen and helium gas produced during the Big Bang eventually agglomerated into megaclouds. These megaclouds organized into the spiral and elliptical shapes we see in distant galaxies. Some of the gas in these newly formed galaxies in turn broke into far smaller subclouds that collapsed under their mutual gravitation into stars. So through their telescopes astronomers see a host of galaxies, each defined by billions of stars. Through careful observation, astronomers have been able to show that the process of star formation continues. They see new stars forming and old ones dying. By observing stars of all sizes and in all stages of their evolution, astronomers have been able to map out the history of these objects. Interwoven with this evolution is the conversion of hydrogen and helium to heavier elements. It is here, rather than in the Big Bang, that we must look for the production of the iron, magnesium, silicon, and oxygen of which Earth is largely comprised.

Again one might ask, How do scientists know that the elements heavier than helium were born in the centers of stars? As we shall see, a rather impressive case can be made. No jury could deny it. Like the Big Bang theory, the *stellar synthesis* theory of element origin gets a ranking of 9.9 out of a possible 10.

DESCRIPTIVE ATOMIC PHYSICS

In order to comprehend the case to be put forth in support of the stellar synthesis hypothesis, we need to consider some of the simple facts about the architecture of atomic nuclei.

Each atom has a compact nucleus made of neutrally charged neutrons and positively charged protons (e.g., see Fig. 1-1). This nucleus carries nearly all of the atom's mass and is incredibly small, only about 10^{-15} m in diameter. A fluff of negatively charged electrons fly in complicated orbits around the central nucleus and give the atom its size but add almost nothing to its mass. The diameter of this electron cloud is about 10^{-10} m (i.e., the atom is 100,000 times bigger than its nucleus). The electrons are held in captive orbits by the electrical attraction of the positively charged protons in the nucleus.

The power of the electrostatic force that keeps the protons and electrons together can be appreciated by comparing it to the power of gravity. A paper clip rests on a countertop because of the gravitational pull of the entire Earth. If a very small magnet is placed just above the paper clip, the paper clip jumps up to the magnet, because the electrostatic attraction of the little magnet is stronger than Earth's gravity. By measuring such forces accurately, physicists find that the electrostatic force is 10^{36} times stronger than gravity! The only reason that large objects like Earth do not exert a powerful electromagnetic force is because the negative and positive charges of atoms exactly cancel each other out, making gravity the potent force for very massive objects. For tiny objects like an atom, gravity has no power, and the electrostatic force is essential.

Just as opposite charges attract, charges that are the same repel—two similar poles of a magnet repel each other if we try to put them together. This repulsive force doubles for every factor of two decreases in distance. This feature of the atom prevents two positively charged nuclei from approaching one another during normal chemical interactions. The electrons manage because they occupy such a large volume and can avoid one another. Normal chemistry depends on attraction as well as repulsion.

So far so good, but a little reflection leads to quite a paradox. Given the power of the electrostatic force and its large increase with decreasing distance, how can the multiple positively charged protons in the nucleus stay together in such a small volume? The repulsive force must be enormous! Given the huge repulsive forces that exist in the nucleus, some much stronger force must exist to hold the protons together. This "strong force" is 138 times more powerful even than electromagnetism, but it operates only over very short distances. It is analogous to glue, a force that can operate only when two objects "touch" one another. Because of this property of the strong force, physicists gave the particle that carries the force the name *gluon*. So imagine, for example, two powerful magnets that repel one another, with superglue on their surfaces. As they get closer and closer they are repelled ever more strongly, but if we get the surfaces to touch, the superglue holds them together despite the electrostatic repulsion. The relative power of these forces in the universe is shown in Table 3-1. Gravity that is so important to us is so puny in comparison to other forces!

Table 3-1

The Four Fundamental Forces			
Name of force	Relative strength	Distances over which it operates	Where it is important
Strong force	1	10^{-15} m	In the atomic nucleus
Electromagnetic force	1/137	Infinite	Everywhere
Weak force	10^{-5}	10^{-17} m	Nuclear particles
Gravity	10^{-39}	Infinite	Well beyond the atomic scale, large masses needed

At low temperatures, the repulsive power of the nucleus keeps atoms separate, and interactions between atoms occur by the sharing of electrons to form chemical compounds. During such chemical reactions, only the character of electron orbits changes; the nucleus remains intact. These reactions occur at temperatures of tens to thousands of degrees centigrade. To make nuclear reactions occur, the nuclei need to get so close together that they "touch," allowing the strong force to exert its power. This can happen only if the nuclei have very high velocities. Velocity increases with temperature, so very high temperatures are necessary. To ignite the nuclear fires, temperatures of 50 million degrees or more must be achieved—no simple task for planetary beings. Only by accelerating charged particles in mighty cyclotrons or by setting off nuclear explosions can physicists create these high temperatures. This is why the alchemists dedicated to making gold from less valuable elements failed. They had no means by which to start a nuclear fire!

The places in the universe with natural furnaces with the temperatures required for nuclear fires are at the centers of stars. Every star must have such a fire at its core; otherwise, the star could not shine. Stars are the alchemists of the universe, where one element can be converted to another.

To understand which nuclides might be manufactured in stars, we must be aware that only certain combinations of neutrons and protons form stable units. The power of the repulsive force helps us to understand the important role of neutrons—they keep the protons separated from one another, reducing their electrostatic repulsion. Neutrons and protons also have a special relationship to one another, which is that they can be converted one to the other. Neutrons that are isolated are unstable, and they decay to a proton plus an electron (a hydrogen atom) with about half the neutrons decaying approximately every ten minutes. And given the right inducement, protons can convert to neutrons by capturing an electron—so the proton-neutron configuration of the nucleus is convertible in terms of the proportions of each. Neutrons are useful because they separate protons, but neutrons left to themselves decay. Protons are held together by gluons, but prefer to be separated. If the nucleus has too many neutrons, they decay to protons; if too many protons, they decay to neutrons. This balance causes stable nuclei to have roughly coequal amounts of both particles.

This balancing act in the nucleus leads to a *band of stability*, where there is no further preference for conversion of one nuclide to another. Figure 3-2 shows that out of all the possible neutron/proton combinations, relatively few are in this stable category. The rest are *radioactive*, and given enough time, will spontaneously transform into one of the stable combinations. The pathways for these transformations are shown in Figure 3-2.

There is one further aspect to stability of the nucleus. If the nucleus gets too large, there are so many protons that electrostatic repulsion becomes large enough for the nucleus to eject protons and neutrons. ^{209}Bi is the stable nuclide with the most neutrons and protons. All nuclei with more than 209 particles are radioactive. At first nuclear parcels, like a helium nucleus containing two protons and two neutrons, are ejected. For still heavier nuclei, the entire nucleus falls apart, in a process called *nuclear fission*.

Those nuclides which, left to themselves, will forever remain unchanged end up making planets and life. They form a *band of stability* that traverses the chart of the nuclides running from ^{1}H at one end to ^{209}Bi at the other. The course of this band represents the most favorable ratio of neutrons to protons. This ratio is near unity for the low proton number elements. Larger nuclei become more neutron rich, with the ratio of neutrons to protons reaching 1.5 for bismuth (Bi).

All nuclei outside the band of stability are radioactive and decay to the band. Nuclei that are too neutron rich convert neutrons to protons +electrons, called *beta decay*. Nuclei that are too proton rich capture

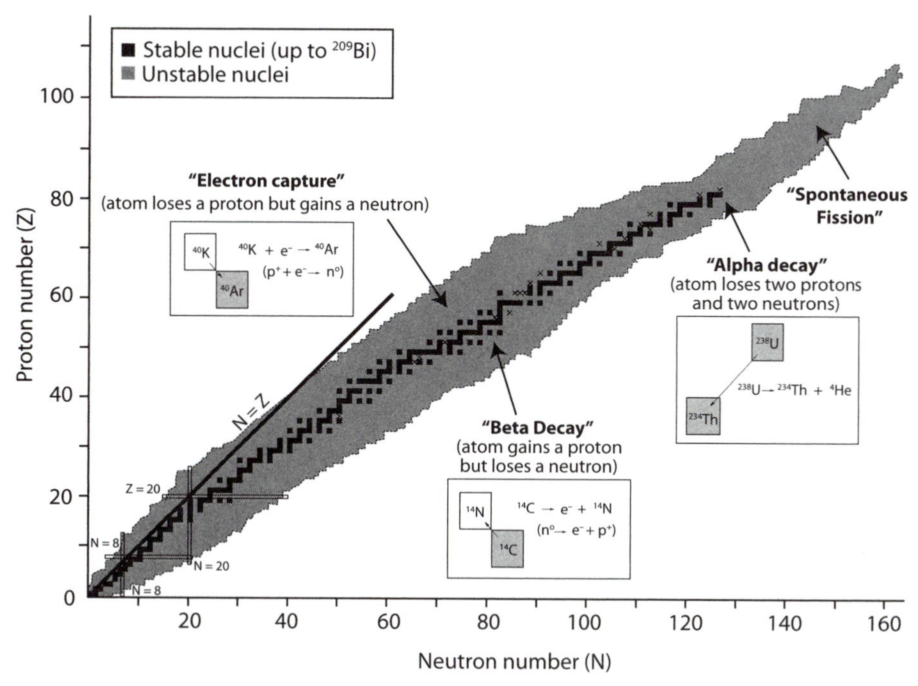

Fig. 3-2: The chart of the nuclides. Stable nuclides (the *band of stability*) are indicated by the solid black boxes. Radioactive nuclides that decay back to the band of stability with variable half-lives are indicated by the gray field. Very heavy isotopes decay by fission, where they spontaneously break up into pieces. Heavy masses decay by alpha decay, ejecting a helium atom of two protons and two neutrons. Neutron-rich isotopes decay by beta decay, converting a neutron to a proton without changing the number of nuclides in the nucleus. Proton-rich isotopes decay by electron capture, converting a proton plus electron to a neutron. The $N = Z$ line shows that at low masses the number of neutrons and protons is quite equal. Higher masses become neutron rich.

electrons to convert protons to neutrons in a process called *electron capture*. And nuclei that are too big eject a helium atom, called an *alpha particle* (because it was discovered first). Note that the first two processes do not change the number of nuclear particles in the nucleus, while *alpha decay* decreases the number of nuclear particles by four, two neutrons and two protons (Fig. 3-2).

All stable nuclides are found on Earth, in meteorites, and on other planets, so all must somehow have been produced from hydrogen and helium at the centers of stars. As we shall see, this buildup from small to large occurs in many steps. To produce a carbon atom requires only two

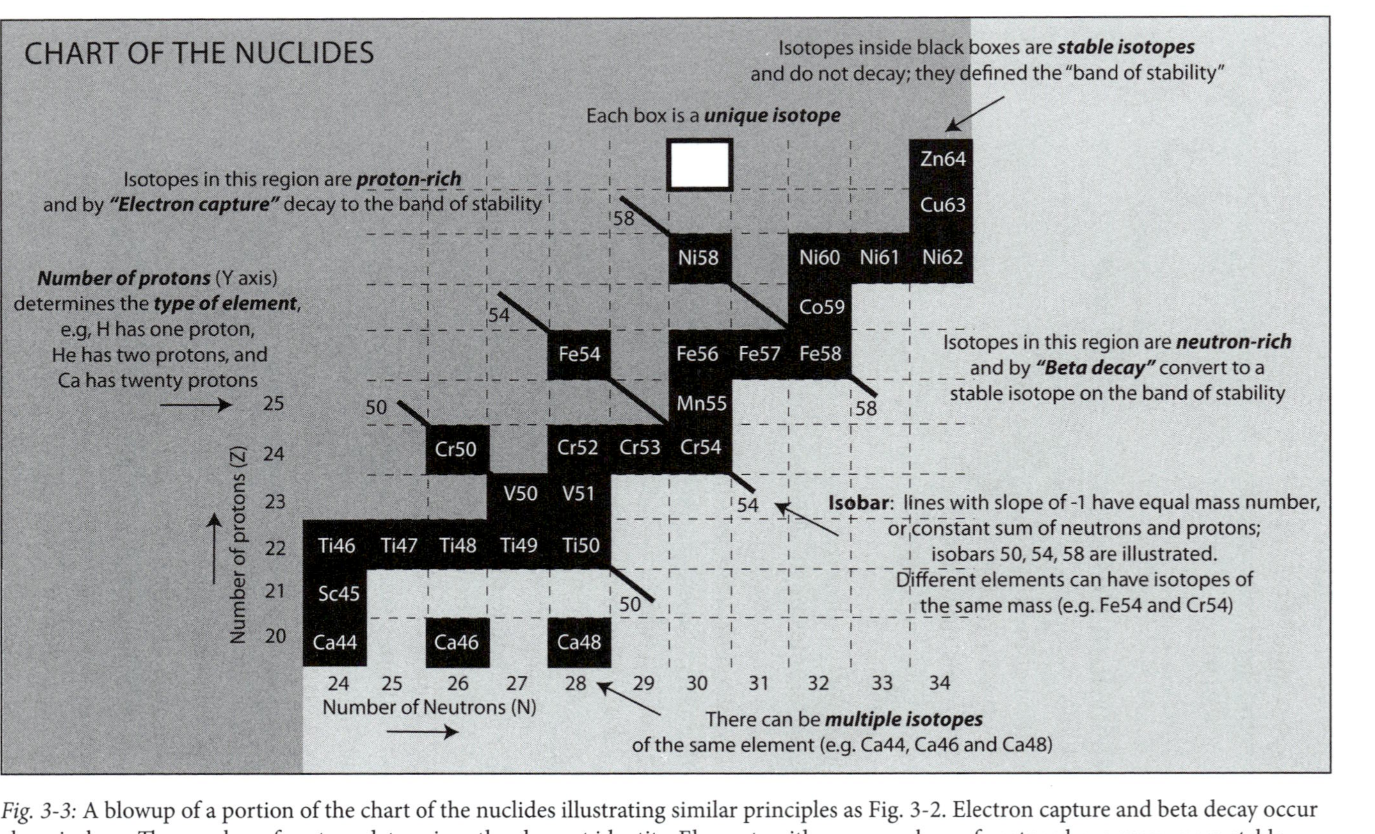

Fig. 3-3: A blowup of a portion of the chart of the nuclides illustrating similar principles as Fig. 3-2. Electron capture and beta decay occur along isobars. The number of protons determines the element identity. Elements with even numbers of protons have many more stable isotopes than those with odd numbers of protons, which usually have only one isotope. Note the very small number of stable isotopes that are "odd-odd."

steps; to produce an iron atom requires a few more steps; to produce a bismuth atom requires many more steps. It is because of this stepwise buildup that the "light" elements are produced in greater abundance than the "heavy" elements.

Element Production during the Big Bang

Let's examine what these steps are. In the fireball of the Big Bang, matter was largely in the form of neutrons. Once released from their dense confinement, neutrons underwent spontaneous radioactive decay to protons plus electrons, with half of the neutrons present decaying every 10.2 minutes—this time is referred to as the *half-life* of the decay. (For example, after three half-lives, one-eighth of the original atoms remain.) During those minutes of stability, many of the neutrons collided with a proton to make ^{2}H, an isotope of hydrogen consisting of one proton and one neutron called *deuterium*. Other collisions could lead to masses 3 and helium atoms of mass 4. At this point there is a remarkable aspect of nuclear stability, which is that no stable nucleus of mass 5 or mass 8 exists (Fig. 3-4). A helium nucleus colliding with the abundant protons or neutrons would produce no reaction. Or two helium atoms colliding would also produce nothing. Instead, only rare reactions could jump over mass 5 to produce masses 6, 7, or 8. For example, a proton and a neutron would have to have collided at the same time with a ^{4}He nucleus to produce ^{6}Li. As is the case on a pool table, in the expanding gas cloud "three-ball" collisions were far less frequent than "two-ball" collisions, so infrequent, in fact, that the number of nuclei formed that were heavier than ^{4}He was insignificant. Thus, at the end of Day One universe matter consisted almost entirely of the elements hydrogen and helium, with only very small amounts of the next three elements—lithium (Li), beryllium (Be), and boron (B). Further synthesis awaited the formation of galaxies and the formation of stars within these galaxies.

Physicists have made models of the collisions that would have occurred during the first day of universe history. They found that there would have been one helium atom for every ten hydrogen atoms.[1] This

[1] While there are 100 ^{4}He atoms for every 1,000 ^{1}H atoms, because the helium atoms are four times as massive, they account for $4 \times 100/1{,}400$, or 29% of the universe mass.

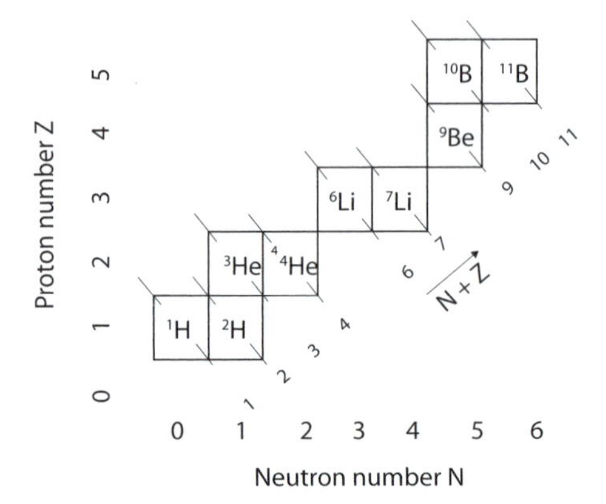

Fig. 3-4: The lower left-hand corner of the chart of the nuclides. Stable nuclides with a particle number in the 1 to 11 range. Note that no stable nuclide exists with a neutron plus proton number totaling 5 or 8. It is these two gaps in the chain that prevented element formation during the Big Bang from continuing beyond helium to any significant extent.

is about the fraction of helium seen in young stars from throughout the universe, which was the third line of evidence for the Big Bang hypothesis discussed in Chapter 2.

Element Formation in Stars

Stars are hot inside for the same reason that brake shoes on a stopping car are hot. When a moving vehicle is brought to a stop, the energy associated with its motion is converted to heat in its brake linings. During the collapse of a cloud of gas, gravitational energy is likewise converted to heat. The amount of heat produced is so vast, and the insulation provided by the enshrouding envelope of gas so effective, that the core of the protostar becomes hot enough to ignite a nuclear fire.

For the nuclei in a star to react, they must touch. To touch, they must fly at one another at such high velocities that they overcome the electrical repulsion exerted by one proton on another. It is much like trying to throw a Ping-Pong ball into a fan. A very high velocity is required to prevent the ball from being blown back in your face.

The hotter atoms are, the faster they move. Temperature is a scale for molecular motion. Touching a hot stove causes the molecules in the

Table 3-2
Conversion of mass to energy

Element	Element mass (g/mole)	Number of atoms	Total mass (g)	Mass loss	Calories of energy
H	1.008	4	4.032 g	0.029 gm	$0.26 \cdot 10^{-10}$ Joules
He	4.0026	1	4.002 g		
Si	28.0860	2	56.172 g	0.33 gm	$2.97 \cdot 10^{-10}$ Joules
Fe	55.8450	1	55.845 g		

skin of your finger to move so fast that the chemical bonds holding them in place are rent; we call this molecular damage a *burn*. For two protons to collide requires velocities equivalent to a temperature of about 60,000,000°C. Through a somewhat complicated series of collisions, four protons can combine to produce a helium nucleus (and two electrons). The helium nucleus contains two of the original protons and two neutrons. These neutrons come into being through the mergers of protons with electrons (for each proton in a star there must be one electron).

As first recognized by Einstein, for nuclear fusion to occur there must be a release of energy, and this energy release results in a reduction in mass. This lost mass reappears as heat. Indeed, the weight of a helium atom is just a little less than that of four hydrogen atoms (see Table 3-2), and this mass is converted to heat when the atom is manufactured in a star. As the proponents of fusion power are quick to point out, the amount of heat obtained in this way is phenomenal. So phenomenal, in fact, that once a protostar's nuclear fire is ignited, its collapse is stemmed by the pressure created by the escape of the heat generated. The star stabilizes in size and burns smoothly for a very long time. For example, our sun has burned for 4.6 billion years and won't run short of hydrogen fuel for several more billion years.

Most of the stars we see are emitting light created by the heat from a hydrogen-burning nuclear furnace. Thus, one might say that stars are continuing the job begun during the first day of universe history; they are slowly converting the remaining hydrogen in the universe to helium.

Our sun is small enough that hydrogen burning can take place for billions of years. Since helium nuclei have two protons, the force of electrical repulsion between them is four times the force of repulsion between two hydrogen nuclei. At the temperatures of hydrogen fusion, the velocities of the nuclei are insufficient to overcome this electrostatic repulsion. For this reason, fusion of helium atoms does not take place in small stars. Within the center of a large star, the gravitational attraction is larger, and to counter it the hydrogen fuel converts to helium relatively rapidly. So-called red giants run through their hydrogen supply in something like a million years. When the core of a red giant becomes depleted in hydrogen, the nuclear fire dims and the star loses its ability to resist the inward pull of gravity. It once again begins to collapse. The energy released by the renewed collapse causes the core temperature to rise and the pressure to increase. The higher temperatures reach the ignition temperature required for helium fusion. Then helium nuclei begin to combine to form carbon nuclei (three ^{4}He nuclei merge to form one ^{12}C nucleus). The mass of the carbon atom is less than that of the three helium atoms from which it was formed. This lost mass appears as heat. The heat from the rekindled nuclear fire stems the star's collapse, and its size once again stabilizes.

In large stars this cycle of fuel depletion, renewed collapse, core temperature rise, and ignition of a less flammable nuclear fuel is repeated several times (Fig. 3-5). A carbon nucleus can fuse with a helium nucleus to form oxygen, or two carbons can merge to form magnesium nuclei, and so forth. Each merger leads to a small loss of mass and to the corresponding production of heat. This entire process can continue as long as fusion to produce heavier nuclei produces a loss of mass and production of heat. The extra heat is needed to prevent the star from collapsing and to keep it in a stable state where expansion from heat production balances contraction from gravitational attraction.

The maximum mass that can be created by this process is isotope 56 of Fe (^{56}Fe). Above this mass, merger of nuclei does not lead to loss of mass, and instead heat must be added to nuclei if they are to merge. That is, since mass and energy are related by $E = MC^2$, the mass of nuclei heavier than iron proves to be slightly larger than the mass of the nuclei that are merged to form them. These reactions are heat sinks rather than heat sources, and therefore cannot stem the gravitational collapse of the star. For this reason, the nuclear furnaces of stars can produce only elements ranging from helium through iron. It should be noted that in-

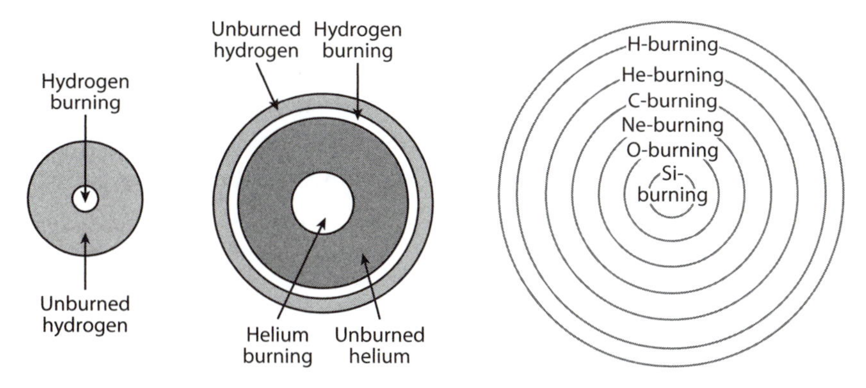

Name of process	Fuel	Products	Temperature
Hydrogen-burning	H	He	60×10^6 °K
Helium-burning	He	C, O	200×10^6 °K
Carbon-burning	C	O, Ne, Na, Mg	800×10^6 °K
Neon-burning	Ne	O, Mg	1500×10^6 °K
Oxygen-burning	O	Mg to S	2000×10^6 °K
Silicon-burning	Mg to S	Elements near Fe	3000×10^6 °K

Fig. 3-5: Three stars with progressively hotter nuclear fires. Like our sun, the star at the left burns hydrogen to form helium in its core; this core is surrounded by unburned fuel. The middle star is burning helium to form carbon and oxygen in its core. This core is surrounded by a layer of unburned helium. Outside of this is a layer in which hydrogen burns to produce helium. Finally, there is an outer layer of unburned hydrogen. The star on the right has a multilayered fire all the way up to Si-burning to create ^{56}Fe. The approximate temperatures required to ignite the successive fuels are also given.

cluded in this range are the elements carbon, nitrogen, oxygen, magnesium, and silicon.

We are then left with two problems. First, there are many elements more massive than ^{56}Fe. How can these be produced? And second, making elements in stellar interiors is not much use for planet building if the elements remain trapped inside. There must be some distribution mechanism that allows these elements to become broadly dispersed throughout the universe. We know this must be the case not only because of the compositions of the planets but also the composition of the sun itself. The materials from which the sun formed must have included all the elements, because the solar spectrum shows that all of these elements exist in the sun itself. It is not made up only of H and He.

Before discussing the solution to these two problems, let us briefly consider the fate of smaller stars like our sun. When the core of our sun runs out of hydrogen, several billion years from now, it will resume its collapse. However, our sun is just barely massive enough to generate the temperature necessary to start a helium fire. Then, after it has burned its core helium, it will collapse into a very dense object that will cool slowly until it gives off only a dull glow. A star to which this has happened is referred to as a *white dwarf*.

Element Synthesis by Neutron Capture

The solution to the two problems of heavy element creation and element distribution is solved thanks to the numbers of very massive stars that exist. These massive stars are ten to twenty-five times the mass of the sun and have such a large gravitational attraction that very high temperatures are required to prevent their collapse. They rapidly progress to the multilayered structure shown in Figure 3-5. Once Fe forms in their core, however, no further heat production is possible through fusion, and there is nothing to prevent their further collapse. The ensuing collapse is catastrophic, bringing the iron nuclei so close together that their nuclear shells begin to interpenetrate. The resistance for further compression generates a shock wave that pushes its exterior outwards. The result is like throwing gasoline onto a hot fire. An incredible explosion occurs, tearing the star asunder. Much of the interior material is blown free of the star's gravity into the galactic surroundings (see Figure 3-0). Astronomers call these explosions *type II supernovae*. A second kind of supernova (called *type I*) evolves when a white dwarf accretes material from its companion stars. When its mass reaches a certain limit, ^{12}C and ^{16}O fuse to form ^{56}Fe, leading to a gigantic nuclear explosion.

Nuclear reactions occurring during these explosions create elements heavier than iron. To understand these reactions we must consider the one nuclear reaction that can occur at "room temperature." It is called *neutron capture*. Because the neutron has no charge, it is not repelled by any nucleus it happens to encounter; it can freely enter any nucleus, regardless of how slowly it is moving. This ability of the neutron to react with nuclides under "room temperature" conditions lies at the heart of the principle of nuclear power generation.

During the explosion that marks the death of a massive star, a host of nuclear reactions occur that release free neutrons. In the close-packed conditions inside the exploding star, the neutrons encounter a nucleus long before they get around to undergoing spontaneous decay to a proton plus an electron. Many of these encounters will be with Fe nuclei. The Fe nucleus absorbs the neutron and becomes heavier. In the supernova explosion, these neutron hits will be like bullets from a machine gun. No sooner is an Fe atom hit by one neutron than it is hit by another. The Fe nucleus gets heavier and heavier until finally it cannot absorb any more neutrons. This very brief pause in growth ends when one of the extra neutrons that have been plastered on undergoes beta decay by emitting an electron. Each neutron that decays becomes a proton and increases the atomic number by one, while maintaining the same total number of nuclear particles in the nucleus. The decay of one neutron converts the Fe nucleus to cobalt (Co). This is the first step along the chain of production of the heavy elements. The Co nucleus can in its turn absorb neutrons one after another, until it too becomes saturated. It then emits an electron and in so doing becomes a nickel (Ni) nucleus. These would be the first steps along the road from iron to uranium.

This sequence is repeated over and over again, driving matter along the neutron saturation route (Figs. 3-6 and 3-7). Because of the rapidity of the impacts, radioactivity proves to be no barrier to this buildup. The buildup zooms past bismuth and even past uranium (U) and thorium (Th), stopping only when the nuclei get so big that the neutron impacts cause them to fission. The fragments produced by a fission event are caught up in the bombardment and begin once again to move along the saturation route. The process of such rapid addition of neutrons that there is no time for decay has led to the name *r-process* (*r* for "rapid") for this mode of heavy element creation that occurs during supernova explosions.

Since we are dealing with an explosion, this bombardment is as brief as it is intense. The flux of free neutrons stops suddenly, and no more neutrons are available to be added to nuclei. All the nuclei that have been generated, however, are very neutron rich and far from the band of stability. These neutron-rich isotopes convert one neutron after another to proton plus electron until they have achieved a stable neutron-to-proton ratio (Fig. 3-7). Those nuclides heavier than bismuth emit alpha particles (He nuclei) as well as electrons, moving toward stability as isotopes of the element lead (Pb). While for most nuclides this adjustment process is quickly completed, for a few nuclides with long radio decay

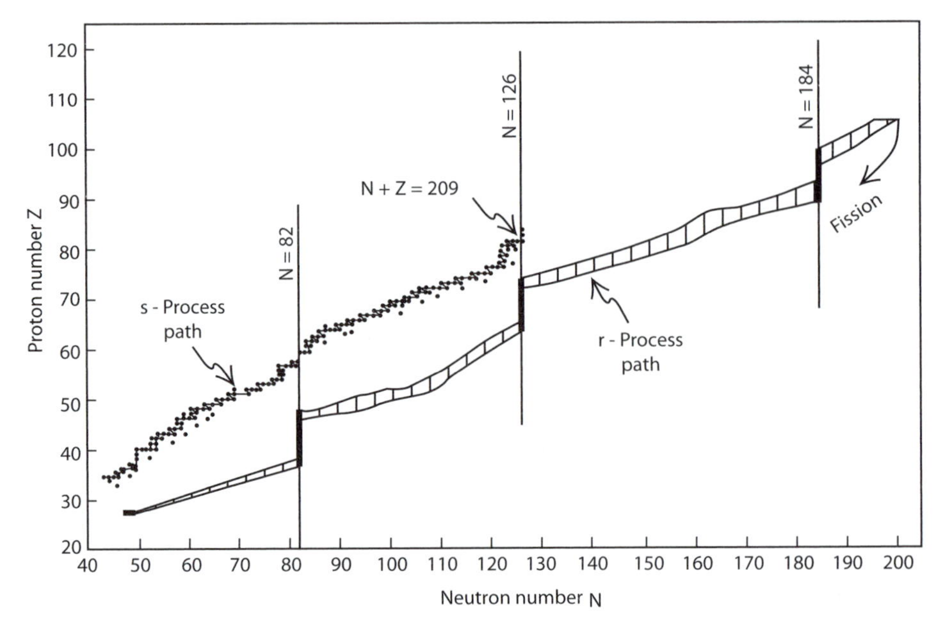

Fig. 3-6: The elements heavier than iron were built by neutron irradiation. Two quite different processes contributed to this production. One, the s-process, proceeds in a controlled way. Neutron hits are spaced out in such a way that the nuclides have time to achieve stability through beta decay. Thus, the buildup path follows the belt of stability shown in Fig. 3-2. For the same reason it terminates at ^{209}Bi, the heaviest stable nuclide. The r-process (rapid process) occurs during the supernova explosion. No sooner has a nuclide absorbed one neutron than it is hit by another. No time exists between hits for radio decay. Instead, radio decay occurs only when the nuclide becomes so neutron rich that it cannot absorb any more, as indicated by the vertically hachured band in the figure.

half-lives, the adjustment process still goes on today. As we shall see, the radioactivity of these remaining long-lived isotopes plays a very important role in the evolution of planetary interiors and will provide us with timescales for planetary processes.

It turns out that the rapid events during supernova explosions are not the only time that free neutrons that can be added to nuclei are produced. As part of the steady nuclear burn that characterizes most of a star's history, side reactions occur that release neutrons. These neutrons also can build lighter elements into heavier elements, but they do it very slowly, one neutron after another, during the relatively long process of stellar evolution. Because the neutron addition is slow, it is called the

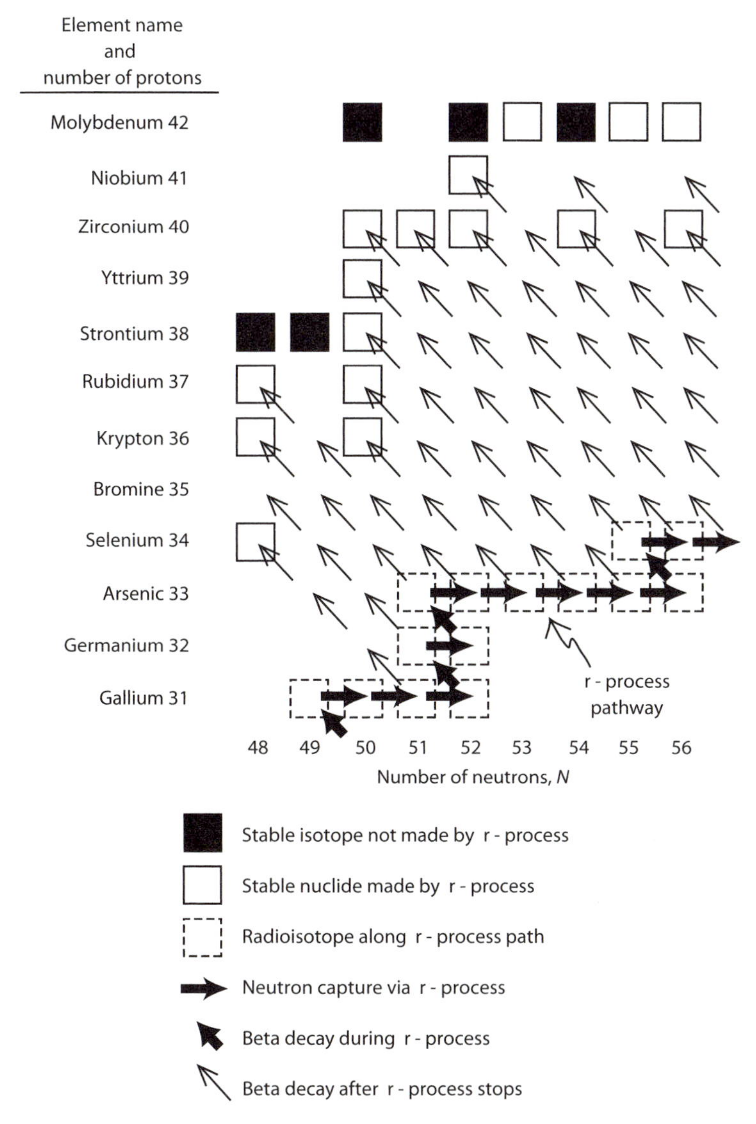

Fig. 3-7: A segment of the r-process pathway. Rapidfire neutron bombardment adds neutrons until a nuclide cannot hold any more. Only then does the nuclide undergo beta decay to become the next heavier element. This process—neutron capture to saturation followed by beta decay—is repeated over and over again, producing successively heavier elements. The r-process buildup occurs during the explosion that destroys the red giant. Hence it ends abruptly. The neutron flux stops and the highly radioactive isotopes on the r-process pathway emit beta particles one after another until stability is achieved. Note that in the case of those isobars for which two stable nuclides exist, only the neutron-rich nuclide of the pair is produced by the r-process.

s-process (*s* for “slow”). For the r-process, the frequency of neutron hits is extraordinarily high—so high, in fact, that even those nuclides with extremely short half-lives do not have a chance to undergo radio decay before being hit again with a neutron. In contrast, the neutron bombardment associated with the steady nuclear fires of stellar cores is far more leisurely. Adequate time exists between hits for all but the radioisotopes with the longest half-lives to undergo radio decay (Fig. 3-8). The s-process produces most of the stable nuclides not produced by the r-process.

The s- and r-processes combined create much of the complexity of the band of stability. As can be seen in Figures 3-2 and 3-7, there are generally two stable nuclides on each even-numbered isobar (as opposed to only one on each odd-numbered isobar). Of these two, the r-process produces only the stable nuclide with the most neutrons (Fig. 3-7). Some isotopes can be produced by both r- and s-process. Those which are neutron rich and separated from other isotopes are r-process only. Those isotopes which are shielded by a stable nucleus that is richer in neutrons along the same isobar are s-process only.

Careful inspection of the chart of the nuclides shows that there are a few nuclides that are produced by neither the r- nor the s-process. For example, in Figure 3-8, ^{58}Ni, with 28 protons and 30 neutrons, is not intersected by either the r-process or s-process path. These isotopes are found in much lower abundance than their r- and s-process neighbors. They can be produced by the *p-process* of proton addition, or by disintegration of heavier nuclei made by the r- and s-processes.

To sum up, we have seen that a diversity of processes have combined to form all the elements. The governing diagram is the chart of the nuclides, with the band of stability revealing what nuclei can survive without decay. The Big Bang makes the raw material of H and He, with small amounts of Li, Be, and B. Fusion of nuclei in stellar interiors makes more He, and the elements from C to Fe. The larger the star, the shorter the lifetime and the heavier the elements that are formed, up to Fe. Within these stellar interiors, the s-process can create some heavier elements. For the most massive stars, collapse and explosion occur, leading to the r-process, formation of all the heavy elements from Fe to U, and distribution of the elements into space where they become available for stars of subsequent generations and the planets that surround them.

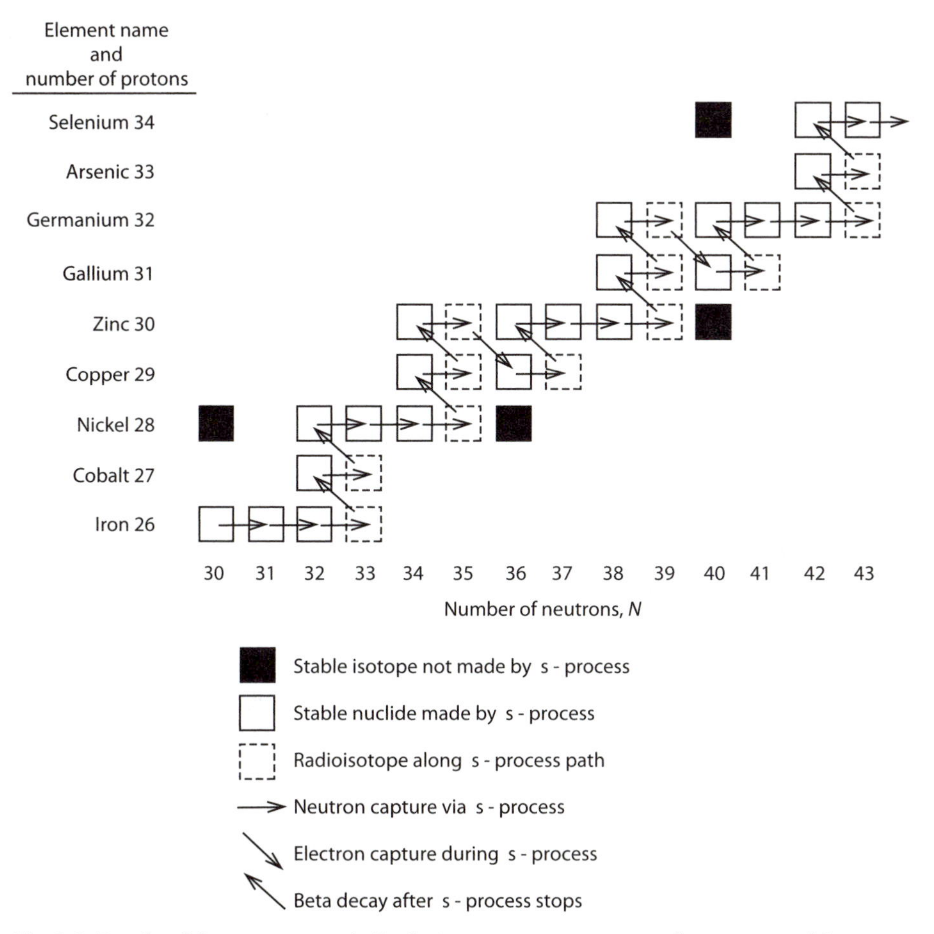

Fig. 3-8: Details of the s-process path. Each time neutron capture produces an unstable, radioactive isotope, decay occurs changing either a neutron into a proton or a proton into a neutron. Not all of the stable isotopes found in solar system matter can be produced in this way. Those stable isotopes lying below the path are produced by the r-process. Those stable isotopes lying above the path are produced by proton bombardment.

Evidence Supporting the Stellar Hypothesis

Shall we accept the progressively intensifying nuclear fires and catastrophic explosions proposed by astrophysicists to explain the synthesis of the ninety elements heavier than helium? Can this scenario be defended with hard evidence, or might it be merely a cosmic fairy tale?

Clearly, no one has ever sent a probe into the core of a star, so we have no direct evidence regarding these phenomena. There are six indirect lines of evidence, however, that are compelling. First, the only conceivable source of energy that could keep stars burning and so hot is nuclear. The pressures and temperatures of the cores of very large stars are adequate to permit not only hydrogen to burn but also helium and even heavier elements. Second, explosions of large stars have been observed (Fig. 3-9).

A third line of evidence comes from the element technetium (Tc), which is not present in Earth because this element has no stable isotopes. Nor are its dark lines present in the spectra of the light reaching us from our sun or from distant stars. The reason is that all these objects are old enough that any technetium produced in stellar interiors has long decayed away. However, the dark lines characteristic of the element do appear in the spectra associated with supernovae explosions. Technetium has two isotopes with moderately long half-lives: ^{97}Tc (2.6×10^6 years) and ^{98}Tc (4.2×10^6 years). These isotopes persist for millions of years after their production. They would have completely disappeared, however, during the 4.5×10^9 years that have elapsed since our solar system formed. The presence of the dark lines of technetium in the atmospheres of a class of objects referred to as *AGB stars* provides powerful support for the hypothesis that elements are being formed in stars.

A fourth line of evidence comes from gamma rays emitted by ^{56}Co formed from ^{56}Fe during the r-process onslaught during a supernova. These gamma rays light up the nebula created by the explosion. We know this because the nebular glow decreases exponentially following the 78-day half-life of ^{56}Co!

A fifth line of evidence comes from the relative abundances of the elements. Using experiments carried out in particle accelerators, astrophysicists have accumulated substantial data on the stability of nuclei and the forces that hold nuclear particles together. Elaborate calculations have been carried out to determine what the proportions of elements and isotopes should be if elements were produced in massive stars. These calculations reproduce very nicely the important features of the element-abundance curve.

Finally, nuclear physicists are able to produce many of the same reactions postulated for stellar interiors in particle accelerators, and of course in the hydrogen bombs, where the conversion of hydrogen to

Fig. 3-9: Evidence for supernova explosions. Photographs taken before (*top left*) and after (*bottom left*) a supernova explosion. Panel on the right shows a closeup of a supernova in 1985 and 2007, revealing the rapid expansion of the cloud. (Photos on left courtesy of Hale Observatories. Right panel courtesy of NASA; http://science.nasa.gov/science-news/science-at-nasa/2008/14may_galactichunt)

helium on even a small scale has very powerful effects. The detailed reactions proposed to occur during stellar nucleosynthesis are largely susceptible to experimental confirmation.

All of these lines of evidence make nucleosynthesis in stars one of the established facts of nature, meriting a 10 on our theory scale.

The characteristics of the abundance curve are also central to the future habitability of planets, because planetary processes must be based on the elements that are abundant. For these reasons it is useful for us to examine the abundance curve in more detail.

In Figure 3-10 the nuclide abundances are plotted as a function of the number of nuclear particles. While there is great richness and complexity in the details of the element abundances, here we point out a few that are very important and easily grasped. One is the peak associated with iron (at particle number 56 in Fig. 3-10). If the stars that explode have

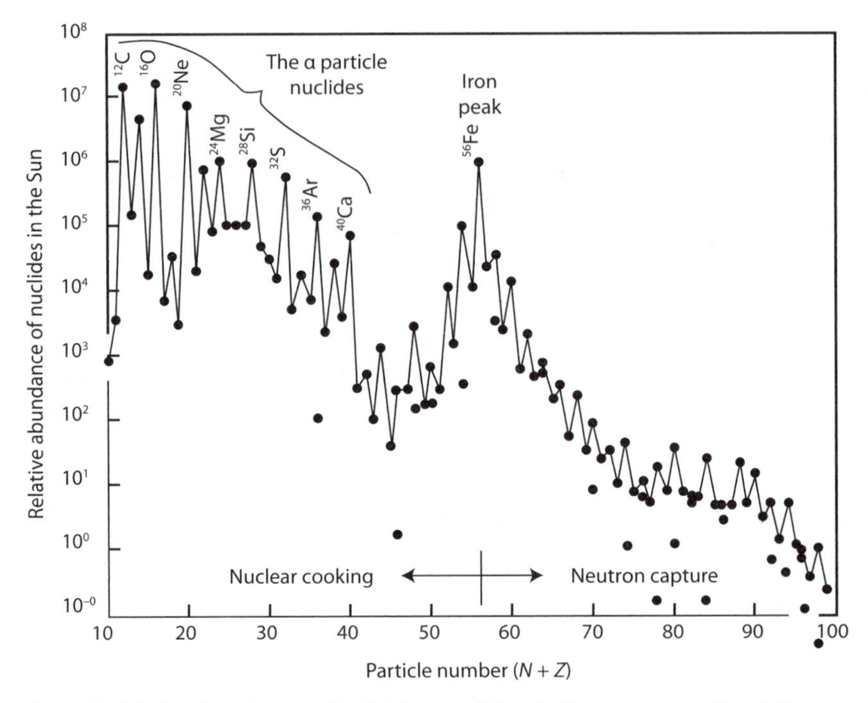

Fig. 3-10: Relative abundances of individual nuclides: In the mass range 10 to 100, nuclides with particle numbers divisible by 4 (i.e., 12, 16, 20, 24, 28, 32, . . .) have abundances far above those of their neighbors. They are referred to as the *alpha particle* nuclides. In the particle number range 50–100 the abundances of nuclides with an even particle number stand about a factor of 3 above those for their odd-numbered neighbors. Where more than one point is shown at a given mass number, two different nuclides with the same neutron plus proton number exist.

cores of iron, then it is not surprising that the major nuclide of iron (^{56}Fe) is more abundant than its neighbors in universe material. Since ^{56}Fe is the end of the nuclear fusion assembly line, its abundance will stack up as more and more stellar material is processed. One might ask, in fact, why is the iron peak not even more prominent? If all the material at the star's interior were converted to iron, then elements such as carbon, oxygen, magnesium, and silicon should be absent in the debris from supernovae. While this is the case for the star's core, it is not for the layers of gas surrounding the core. When the cores collapse to form supernova, the outer layers are still in earlier states of nuclear fusion, forming lighter elements.

Another feature of the abundances is the prominence in the mass range 10 to 40 of those nuclides with mass numbers divisible by 4. These nuclei are aggregates of the very stable ^{4}He nucleus, and are the primary products of nuclear fires. They are called *alpha-particle* nuclides, since they are made up of multiples of helium atoms.

Supporting evidence for nucleosynthesis by the processes we have described can be seen in the upper portion of Figure 3-11. Two humps appear in the element-abundance curve that punctuates the smooth decline with increasing proton number, one centered at about 55 protons and the other at about 80 protons. The same humps appear when plotted against the total number of nuclides in the nucleus (lower portion of Fig. 3-11) near masses 138 and 208. These peaks results from what physicists refer to as the magic neutron numbers 82 and 126. For example, the isotopes ^{138}Ba (56 protons, 82 neutrons) and ^{208}Pb (82 protons, 126 neutrons) are unusually abundant. It turns out that nuclear configurations involving either 82 or 126 neutrons are particularly stable. One expression of this stability is their lowered propensity to gobble passing neutrons. Thus, once formed during the s-process, nuclides with 82 or 126 neutrons are less likely to capture more neutrons and move further along the buildup chain. Because of this they were produced in higher abundance than their neighbors. In the r-process, radioactive nuclides with 82 and 126 neutrons also produce a bottleneck that causes them to be produced in greater abundance. Because the r-process nuclides are far from the band of stability, once the intense blast of neutrons is turned off, the extra neutrons are one by one converted to protons via beta decay. For example, one of these nuclides is the radioactive ^{124}Mo (42p, 82n). As it decays back to the band of stability, 8 of its neutrons are converted to protons generating the stable nuclide ^{124}Sn (50p, 74n). For this reason, the abundance peak for the r-process nuclides does not fall in the same place as that for the s-process nuclides. Rather, it is shifted toward lower particle number by 8 to 12 mass units. The existence of these twin peaks in the abundance curve provides strong support for the existence of both the r and the s neutron buildup process.

Finally, one last characteristic of the abundance curve merits comment. Both the element abundance and mass abundance plots in Figure 3-11 show a distinctive saw-toothed pattern. Odd elements and odd-particle-numbered nuclides have lower abundances than their even-

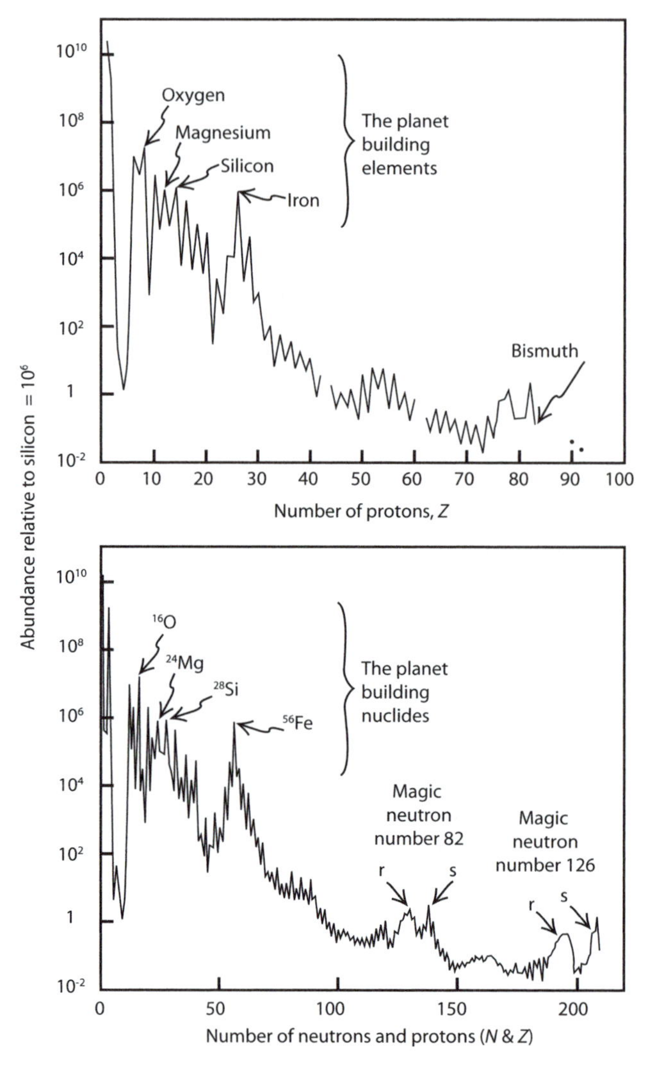

Fig. 3-11: Top panel: Abundances of elements produced during nucleosynthesis in stars. The trough for Li, Be, and B exists because these elements were produced in small amounts during the Big Bang and then partially consumed in stellar interiors. The "sawtooth" is the result of nature's preference for even-numbered elements. The highest peaks are the alpha particle nuclides that will become the raw materials for planets and life. Slight peaks at high mass number reflect the preference for nuclei with 82 or 126 neutrons. *Bottom panel:* The relative abundance of the isobars (i.e., some isobars have more than one element). Only two isobars of nuclear number less than 208 are not represented in nature, those of mass 5 and mass 8. The double peaks for magic neutron numbers are evidence for s-process and r-process operating during element formation.

numbered neighbors. This pattern reflects a preference in nuclear architecture for even numbers. Nuclides with both an even number of neutrons and an even number of protons are strongly favored. Except for ^{2}H (1p, 1n), ^{6}Li (3p, 3n), ^{10}B (5p, 5n), and ^{14}N (7p, 7n), no stable nuclide with both an odd number of neutrons and an odd number of protons exists in nature. Other odd-odd nuclides formed in stars subsequently undergo radio decay to form the preferred even-even nuclides (by converting a neutron to a proton).

Summary

After the Big Bang made hydrogen and helium, nucleosynthesis in stars produced the remaining elements. Particularities of nuclear stability had far-reaching consequences for the universe. The lack of stability of masses 5 and 8 prevented production of heavy elements during the Big Bang, creating the possibility for later development of stars and stellar evolution. The enhanced stability of alpha-particle nuclides during nuclear fusion led to the formation of certain elements in great abundance. These elements then became the raw material for planets and life later in the history of the Universe. The fact that ^{56}Fe is the most stable nucleus and that nuclear fusion cannot occur beyond that point is what leads to instability of massive stars, the formation of heavier elements and the distribution of all the elements through the galaxies. All of these consequences that are central to the operation of the universe and its ultimate habitability are the result of the detailed laws of relative stabilities of atomic nuclei.

The size of stars is central to their role in the overall evolution of galaxies. Large stars have massive gravitational forces leading to intense nuclear fires and very bright, short lifetimes. These stars produce all the elements and distribute them through supernova explosions. Such stars provide the elements required for planets and life, but would not themselves form habitable solar systems because of their short lifetimes and expolosive deaths. Smaller stars like the sun, with their lesser gravitational contraction, are rendered stable with lower-temperature nuclear fires produced by fusion of hydrogen to form helium, and can have lifetimes of billions of years. This leads to long-lived solar systems where planets have a long and stable environment sufficient for complex planetary evolution. Both types of stars are necessary for a habitable universe.

Since the sun contains all the elements, it cannot have been among the first stars to form in the universe. Early stars contain only the hydrogen and helium produced by the Big Bang. Instead, the sun must be a latecomer. Its pallet of elements has been enhanced by the explosive deaths of large numbers of red giants that formed and died in our galaxy long before the sun advent. The multiple processes over galactic history have led to the abundances of stable elements as well as long-lived radioactive isotopes present on Earth (Fig. 3-12).

From what we have learned here, heavy element formation must have gone on in all the galaxies making up our universe, and spectra from these distant regions show evidence for the same elements that make up our sun. The ingredients for rocky planets and life surely are available

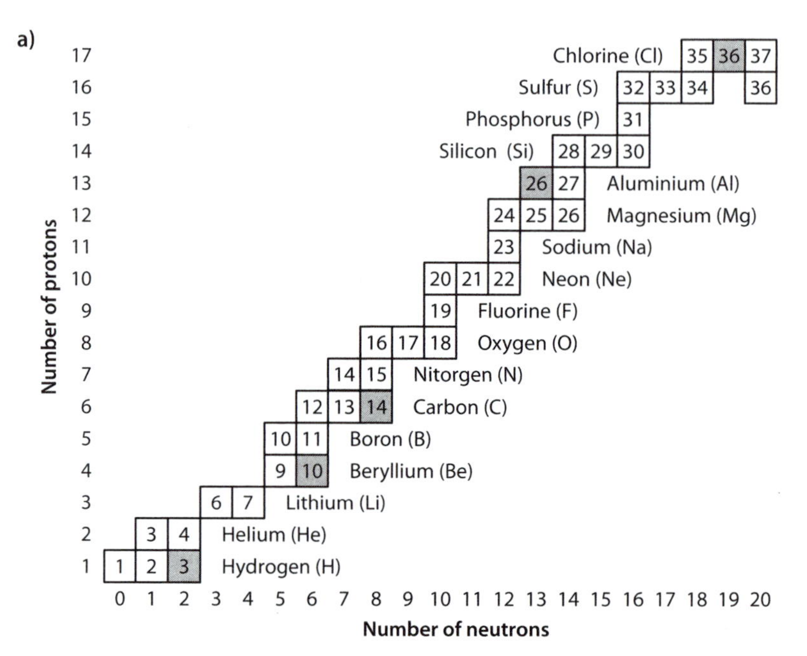

Fig. 3-12: Chart of the nuclides. Shown in this series of diagrams are all the nuclides present in nature. The shaded squares represent radioactive isotopes. Some of these are long-lived remnants of element production in stars. Others are being produced in very small quantities by cosmic rays bombarding our atmosphere. To avoid confusion, the decay chains of long-lived thorium and uranium isotopes are shown separately (see panel d).

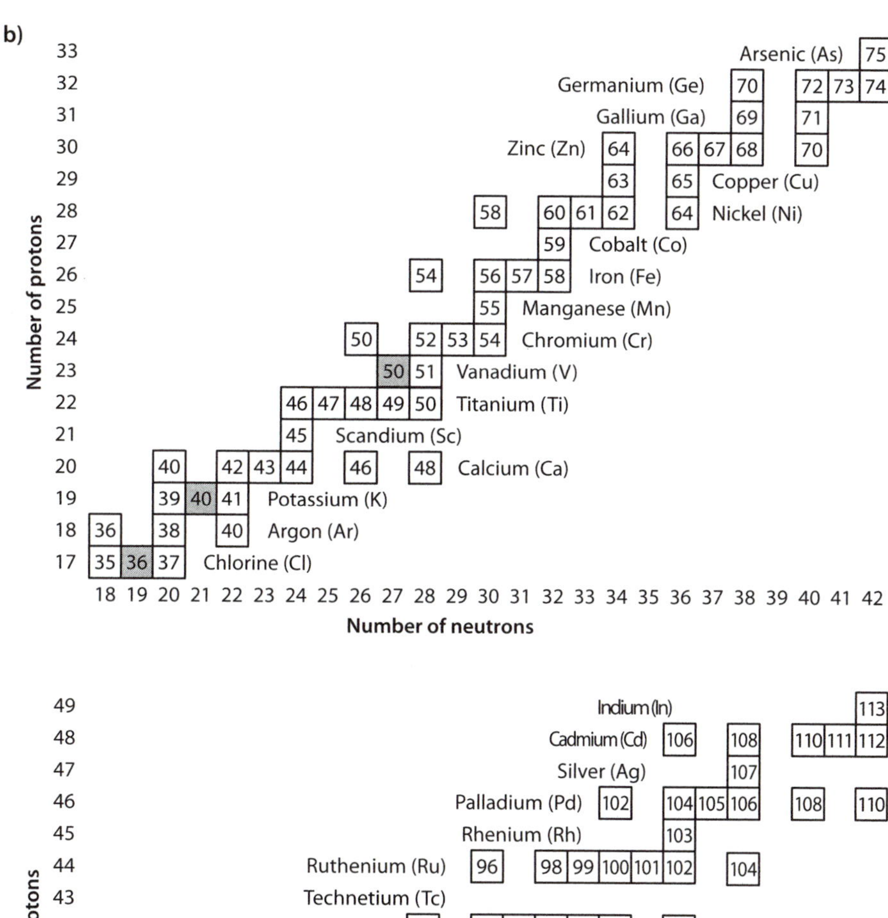

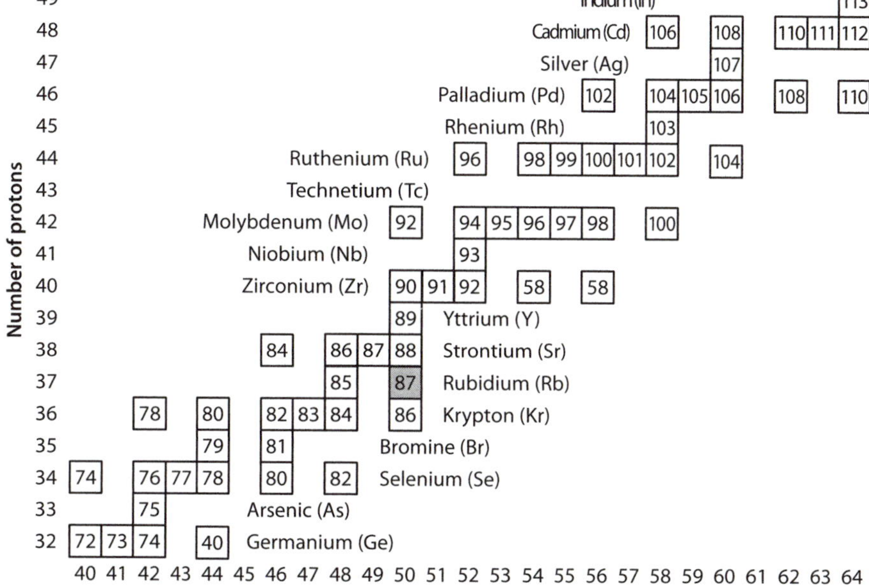

Fig. 3-12: Continued

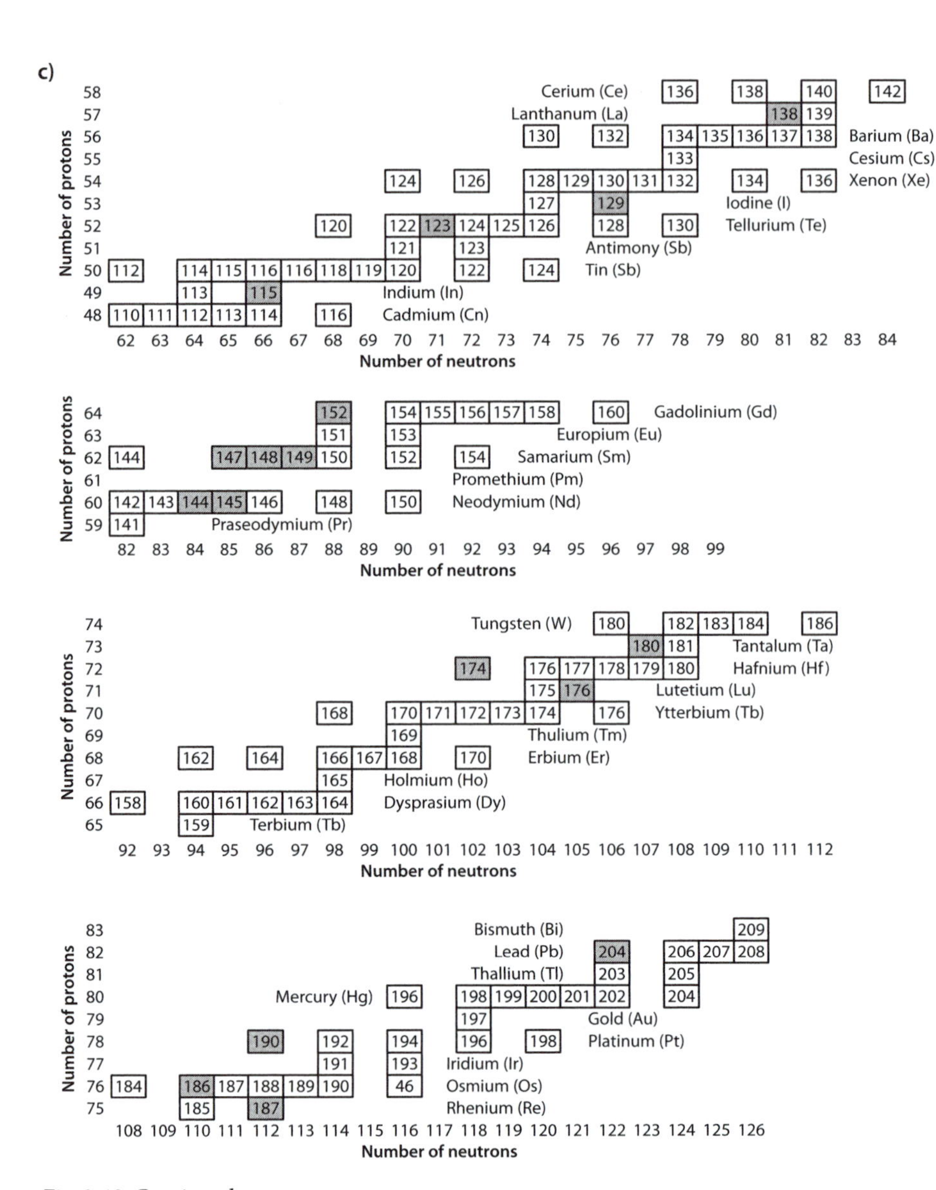

Fig. 3-12: Continued

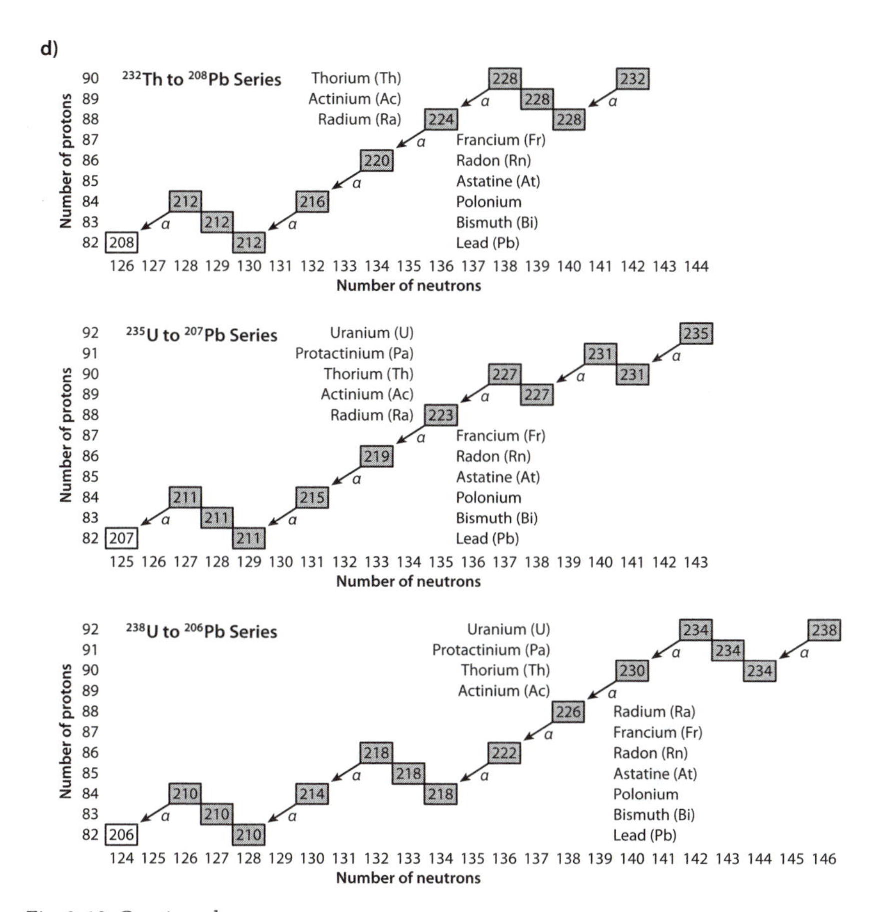

Fig. 3-12: Continued

everywhere in the universe. Hence, the generation of Earthlike planets has not been impeded by a lack of raw material!

Supplementary Readings

C. A. Barnes, D. D. Clayton, and D. N. Schramm, eds. 1982. *Essays in Nuclear Astrophysics*. Cambridge: Cambridge University Press.

R. J. Tayler. 1972. *The Origin of the Chemical Elements*. London: Wykeham Publications, Ltd.

D. D. Clayton. 1983. *Principles of Stellar Evolution and Nucleosynthesis*. Chicago: University of Chicago Press.

Fig. 4-0: The atomic structure and physical form of the halite mineral. The transparent box in the ball diagram shows the cubic "unit cell," which is apparent in the cubic form of the visible mineral in the photograph. The symmetry of the mineral reflects the structure on the atomic scale. Minerals are the materials of which solid planets are made.

CHAPTER 4

Preliminary Fabrication

Formation of Organic and Inorganic Molecules

In stellar interiors all the important reactions involved the nuclei themselves, and one atom being transformed to another was the routine activity. But outside of stars, energies fall by orders of magnitude and a different set of laws operates. Atoms become the fundamental, unchangeable building block of matter. The number of protons in the tiny nucleus of the atom controls the number of electrons required for charge balance, and this electron cloud extends 100,000 times farther than the size of the nucleus. Interactions among atoms involve interactions among the electron clouds. The laws of electron cloud interaction control the formation of molecules in interstellar space, the formation of planets, and all the processes that subsequently take place on them. With the exception of the rare radioactive nuclides that retain the vestiges of their stellar origin, everything that happens on Earth deals with reactions among the electron clouds. For stars and atomic chemistry, the fundamental unit was the nucleus that made an isotope with a single mass, and the chart that summarized our knowledge was the Chart of the Nuclides. For planets, the fundamental control lies in the configuration of the electron clouds—elements rather than isotopes become the fundamental chemical substance. The Periodic Table, organized around electron shell structure and with all isotopes of the same element combined, presents in concise form the fundamental organization of electron clouds. Electron cloud interactions cause atoms to combine to form molecules, and almost all the chemical reactions we deal with involve interactions among molecules.

The first molecular building blocks are constructed in the vast clouds of interstellar space to form the inorganic molecules known as minerals, and also the simplest organic molecules. The minerals will become the building blocks for solid planets, and the organic molecules will contribute to larger, gaseous planets and also serve as the first building blocks for life.

Introduction

Our discussions in Chapter 3 dealt with stellar processes at temperatures of millions of degrees. At these temperatures, the positively charged nuclei move so quickly that they collide and react, following the laws of nuclear physics. At this stellar level, events that are foreign to our human experience become normal. Atoms are created and destroyed; no molecules exist; there are no such things as rocks or minerals; life as we conceive of it is impossible.

Outside of this stellar realm, temperatures drop from millions of degrees to less than thousands of degrees. At these temperatures, positively charged nuclei are much less energetic and become surrounded by negatively charged electron shells. The nuclear chemistry of stellar interiors is no longer applicable, and we arrive in the realm of "normal chemistry" that we see all around us on Earth.

The basic understanding of planetary chemistry came of age in the eighteenth and nineteenth centuries, when chemists explored matter and tried to break down materials into their fundamental components. For centuries alchemists tried in vain to produce precious metals (gold and silver) from more common materials like lead and copper. They failed, but their attempts showed that other substances, such as water and air, could be separated into components with very different masses and properties. Gradually it emerged that there were fundamental building blocks of matter with specific masses and chemical affinities that could not be split into parts. The relative masses of these building blocks could be determined by seeing what mass of iron or hydrogen would react with a specific volume of oxygen, for example. In this way, the physical and chemical properties of each substance could be uniquely determined. The substances that could not be broken apart became known as *elements*, made up of individual, indivisible particles called *atoms*. "Atoms are the fundamental building block of matter that can neither be created nor destroyed" came to be a guiding principle of the new chemistry. The commonsense point that since atoms exist, they must have been created somewhere, was relegated to the class of philosophical questions that were beyond the realm of observation and scientific laws.

Some of the newly discovered elements had similar chemical behavior, and it was possible to divide them into groups based on their chemical affinity. Lithium (Li), sodium (Na), and potassium (K), for example, all could make similar salts when combined with fluorine (F), chlorine (Cl) or bromine (Br). Classes of elements with similar chemical affinities also showed quite regular increases in mass. For example, for the triad Li, Na, and K, the mass of Na (~23) is the mean of the weights of Li (~7) and K (~39). Similarly, strontium (Sr, mass 88) is the mean of calcium (Ca, 40) and barium (Ba, 136) and so on.

In 1869 the Russian chemist Dmitri Mendeleev (1843–1907) suggested the elements as a whole reflected an organized system: "The properties of the elements, such as their forms and the way they combine with other elements, are a periodic function of their atomic weight." He constructed a table, now known as the Periodic Table, where the sixty-three elements identified at that time were arranged with those of the same affinity in the same column in order of increasing mass, and in each row the masses increasing regularly to the right. The regularity of the table was interrupted by some gaps, and Mendeleev predicted that other elements would be discovered to fill the gaps. Over the next short interval of time, many of these elements were discovered, offering convincing proof of the regularity that Mendeleev had been able to see. The Periodic Table of the Elements (Fig. 4-1) elegantly and concisely describes the most fundamental principles of chemistry. Spectra from other stars show that exactly the same elements exist everywhere in the universe, and the elements studied here on Earth reflect the same laws and processes that happen over the vast universal reaches of time and space.

The original conception of atoms was that they were indivisible particles with no internal structure. This idea gave way after the discovery of radioactivity by Henri Becquerel and Marie and Pierre Curie in the 1890s. Some of the heaviest atoms were emitting energy and were dubbed "radio-active." This energy had an electric charge, so it could be focused in a beam. Ernest Rutherford directed such a beam at thin gold foil to see what would happen. A majority of the radiation passed straight

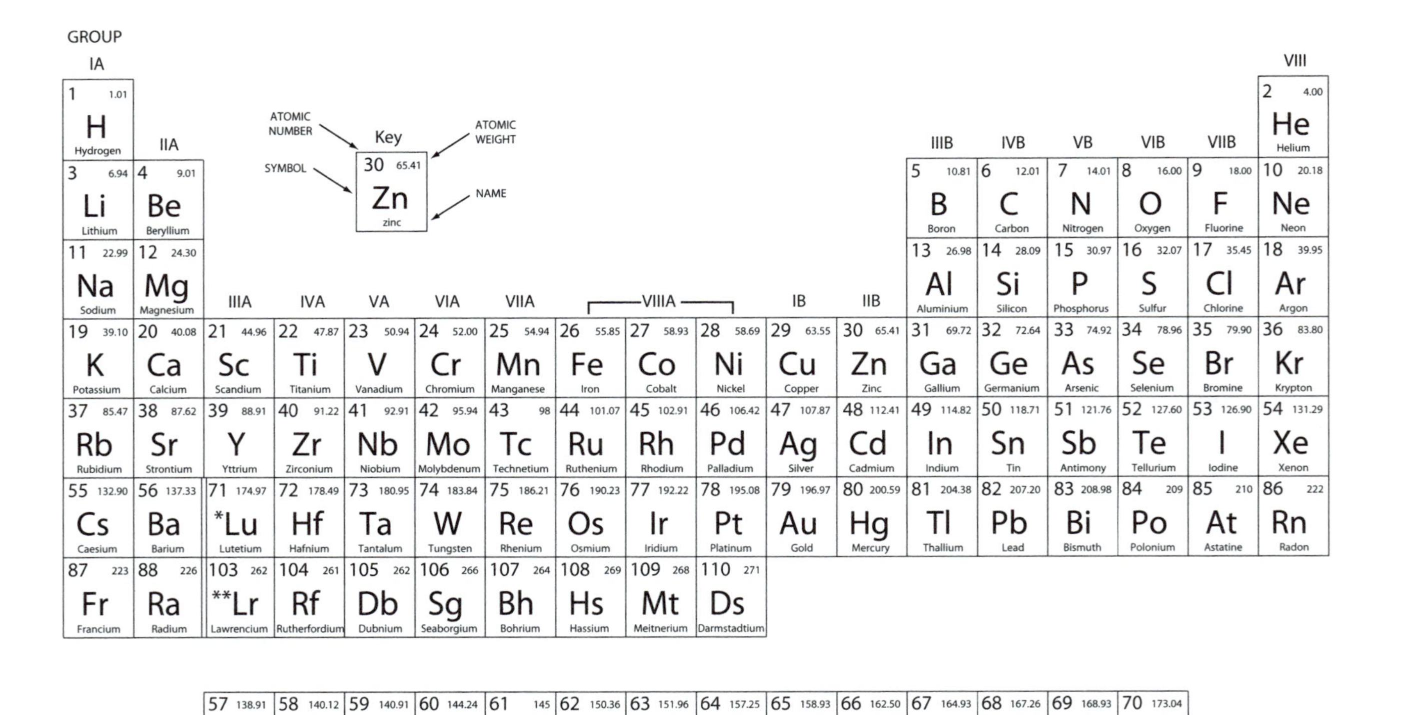

Fig. 4-1: The modern rendition of Mendeleev's Periodic Table of the Elements. Each row in the table corresponds with a particular electron shell and orbital, so the table is a kind of symbol of the electron shell structure of the atom.

through the foil, and some of it bounced back! This could be most simply explained if the beam was a collection of particles (called *alpha particles*) of positive charge that encountered something else with positive charge that would repel them. This experiment showed that atoms were not indivisible objects but had distinguishable parts. Since almost all of the alpha particles passed through the gold foil, areas of positive charge in the gold foil were extremely rare. Therefore, atoms were discovered to be mostly empty space, with a very small positively charged nucleus surrounded and balanced by an equivalent number of negatively charged electrons. Essentially, the entire volume of the atom is taken up by the surrounding electron cloud. The size of the nucleus relative to the electron cloud is about 1:100,000. If the nucleus took up the distance between the sun and Earth, the electron cloud would extend beyond the distance to the nearest star, Alpha Centauri. Or if the nucleus were the size of an apartment building, the electron cloud would be the size of the whole Earth!

Research in the early twentieth century revealed the more complex modern concept of the atom, including the discovery of neutrons as a component of the nucleus, and the concentric "electron shells" that hold gradually increasing numbers of electrons—2, 8, 18, 32, 50. Each of the outer shells was also shown to consist of multiple types of orbitals, leading to complex shell configurations that took some time to sort out in detail.

The number of electrons and the configuration of the electron cloud control all interactions between atoms. The number of electrons that surround a neutral atom are controlled by the number of protons in the nucleus; hence the number of protons determines the identity and chemical behavior of the element. As we learned in Chapter 3, while the number of protons within the nucleus of a given element is always the same, the number of neutrons may vary, giving rise to different *isotopes* of that element. Because the electron clouds of the different isotopes with the same number of protons are the same, the isotopes are chemically almost identical. (As we shall see in later chapters, however, the slight mass differences cause isotopes of the same element to have very slightly differing chemical behavior, which is a very useful tool to use in understanding Earth processes.) In the Periodic Table, all the isotopes of a single element are combined and averaged, which is the main reason that the atomic weights of many of the elements, particularly those with even atomic numbers, are not close to being exact integers.

The modern presentation of the Periodic Table reflects in its structure the details of the electron shell configuration for the various elements. The rows reflect the number of electron shells around the atom. The first row has two elements, after which the first electron shell is filled. The second row has eight elements, corresponding to the eight electrons of the second shell. The innermost orbitals of the third shell become filled after the addition of eight more electrons. The next two electrons are added to the fourth shell. Then, from Sc to Zn, ten electrons are successively added to the other orbitals of the third shell. Each shell can accommodate more and more electrons, and hence the outer shells become more complicated in their behavior than the inner ones.

The columns of the Periodic Table are organized around the configuration of the outermost electrons, which is where reactions with other elements take place. Elements in the first column have one electron in the outermost shell. Those in the second have two electrons. The large filled interior shows atoms where the additional electrons are filling one of the inner shells rather than the outermost shell. On the far right are the noble gases, where the outermost layer is completely filled. The Periodic Table thus not only provides useful data but is also a kind of symbol that summarizes knowledge about all the elements found in nature. It ranks a 10 on our theory scale of knowledge of the universe.

Molecules

Because electron shells are energetically more stable when they are complete, atoms share, donate, and receive electrons to combine in ways that increase their stability under the conditions in which they find themselves. For this reason, with the exception of the noble gases, most elements on Earth appear in molecules that satisfy the needs of the elements to have complete electron shells. Lithium, with one electron in its outermost shell, combines very well with fluorine, which lacks one electron in its outer shell, because donation and reception lead to a "win-win" condition where both atoms achieve a more stable electron shell structure (Fig. 4-2). In the water molecule two hydrogen atoms each pro-

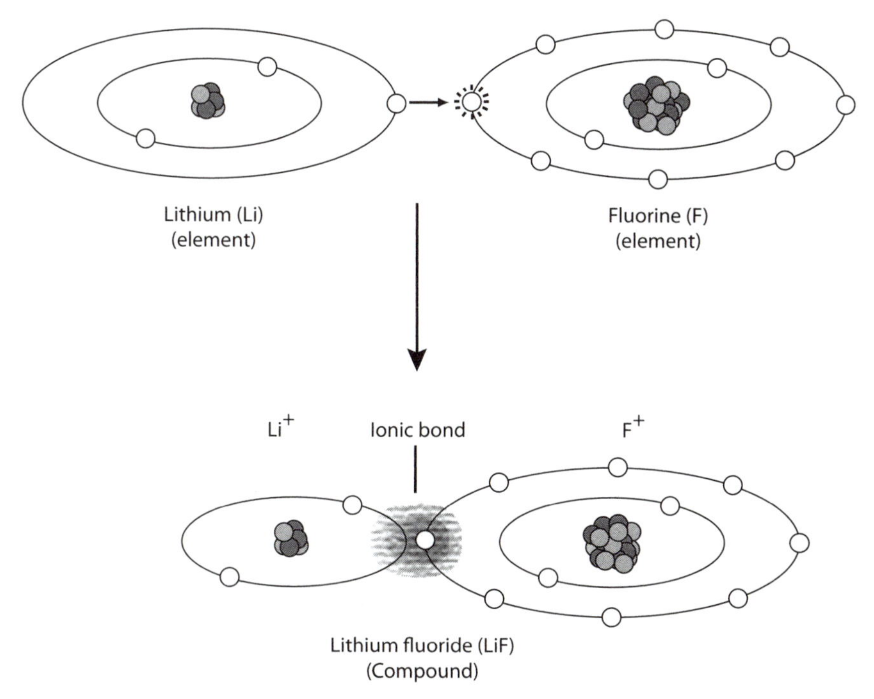

Fig. 4-2: Ionic bonds are created when excess electrons in the outer shell of one atom are donated to fill vacant spaces in the shells of another atom, as in LiF. White circles represent electrons, dark gray circles neutrons, and light gray circles protons. Nucleii are greatly enlarged in order to be visible.

vide an electron to the oxygen atom, providing it with a filled shell. Such bonding between elements leads to the formation of *molecules*, ranging from very simple ones such as NaCl and H_2O to huge organic molecules that can contain thousands of atoms.

The only atoms with completely filled electron shells are the noble gases, found on the far right of the Periodic Table. Their electron shell stability means they have no tendency to react with other elements, so each atom of a noble gas is content to exist in complete isolation. Argon (Ar) in air, while rather abundant, reacts with nothing. The oxygen in air is easy to identify because oxygen reacts vigorously with many elements, as we see when iron rusts, or wood burns, or, for that matter, with every breath we take. The lack of reactivity of all the noble gas atoms is what made them so difficult to discover in the first place. Lack of reactivity is a kind of invisibility.

Atoms are even willing to lose their charge balance in order to have a filled shell structure. This leads to the creation of *ions*. Sodium easily loses an electron to become a positive ion, called a *cation*, with a single charge. Oxygen readily accepts two electrons to become a negative ion, called an *anion*, with two negative charges, and so on. Charged ions play a very important role in chemical reactions.

A few hundred thousand different molecules are known, but the number of different molecules that is possible is essentially infinite, and new molecules are still commonly being discovered or created in the laboratory. The most common molecules, however, are relatively simple combinations of the reactive elements (i.e., not the noble gases) that are created most abundantly in the universe. These elements (see Chapter 3) are (1) the primordial element from the Big Bang—hydrogen; (2) the alpha-particle nuclides that are not noble gases—carbon, oxygen, magnesium, silicon, sulfur, and calcium; and (3) the most stable nucleus that is the final product of nuclear fusion—iron. Nitrogen is also relatively important, especially to living organisms. Of these elements, hydrogen is the most abundant. Apart from hydrogen, the remaining six elements make up more than 98% of the reactive matter in the universe. Therefore, molecules involving these elements will be dominant, and the molecules in naturally occurring substances that are most familiar to us—rock, water, air, and life—are primarily made up of these elements.

States of Matter

Solid, liquid, and gas are the three states of matter most familiar to us, and elements and molecules can occur in all three of these states. At sufficiently high temperatures, no molecules can exist, and there is a fourth state of matter, *plasma*. Plasmas can be described as ionized gases where the elements are stripped of their electrons and there is a chaotic mixture of ionized nuclei and negative electrons. Plasma is the most common state of matter in the universe, and we observe it in our terrestrial experience through the solar aurora, the northern lights, neon lamps, and flames.

Our intuition about states of matter comes from our experience on Earth's surface, where the pressure is uniformly low and variations in

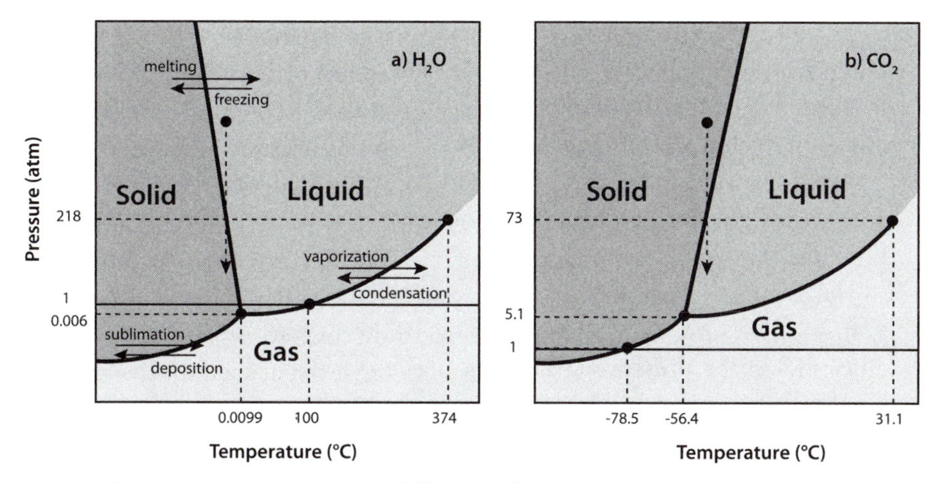

Fig. 4-3: Phase diagrams for (a) H_2O and (b) CO_2, showing the importance of both pressure and temperature in influencing the state of matter. Different shades of gray show the fields of solid, liquid, and gas. At constant temperature, pressure changes can convert materials from solid to liquid to gas, or vice versa. At very low pressures a direct solid-to-gas transition occurs, called *sublimation*. Vertical dashed lines with arrows illustrate phase changes produced at constant temperature by changing pressure.

temperature cause changes in the state of matter from solid to liquid to gas. Therefore, when we think of melting or boiling or the creation of plasma, we intuitively assume it reflects an increase in temperature. This bias comes from the fact that we live in a very constant pressure environment. Even small changes in pressure, such as those we experience when under water or on high mountaintops, can have very large effects on our metabolism. But the pressure changes we experience are trivial compared to the pressure range of the overall planetary environment. Since pressure is controlled by the weight of overlying material, pressures increase rapidly with depth. Imagine the pressures generated by the weight of rock a mile thick! For this reason a planet's pressure ranges are enormous—from essentially zero pressure in space to pressures of millions of atmospheres (megabars) in planetary interiors.

A diagram illustrating this reality is presented in Figure 4-3, showing states of matter for two common substances—water and carbon dioxide. At a temperature of 25°C and pressure of one bar, water of course is liquid and CO_2 is gas. At these pressures CO_2 is never a liquid—at very cold temperatures, it is solid, and upon heating, it sublimates from solid to gas. But at higher pressures, liquid CO_2 is stable—as is the case in many fire extinguishers, for example, where the CO_2 is strongly compressed. Water, on the other hand, goes from solid to liquid to gas at one atmosphere, but you can see from the diagram that at very low pressures it would also sublimate, and that the melting and boiling points of water, considered by us to be very fixed quantities, vary substantially as the pressure changes.

The potential importance of pressure can also be seen from the diagrams where CO_2 and H_2O can change their state at constant temperature (T) as the pressure (P) changes. These effects of T and P can be qualitatively understood by recognizing that melting or boiling requires the atoms to become progressively freer of their neighbors. In solid crystals the elements are strongly bound to one another and do not move readily. In liquids the atoms or molecules are more energetic and are more loosely bound, and liquids easily deform. In gases, connections between elements or molecules are even more tenuous as the particles move randomly and chaotically, bouncing off each other. Increasing temperature increases the energy of the molecule, ultimately leading to a gaseous state. Increasing pressure tends to push the molecules closer together, making a higher density state more favorable. For almost all substances (water is the notable exception), the crystal is denser than the liquid, and therefore increased pressure makes solids more stable. Therefore, either lowering pressure or raising temperature can often have similar effects on the state of matter.

Plasmas are also sensitive to both temperature and pressure, as shown in Figure 4-4. Low temperature plasmas are possible and common in space, where the pressure is exceedingly low. Thus, for all elements, the four states of matter are traversed in various ways as pressure and temperature change.

VOLATILITY

Volatility determines whether a molecule is a solid, liquid or gas under particular conditions of temperature and pressure. Highly volatile elements have very low melting and boiling points, such as all the noble gases and N_2. These substances are gaseous even at very low temperatures. Refractory elements have very high melting and boiling points.

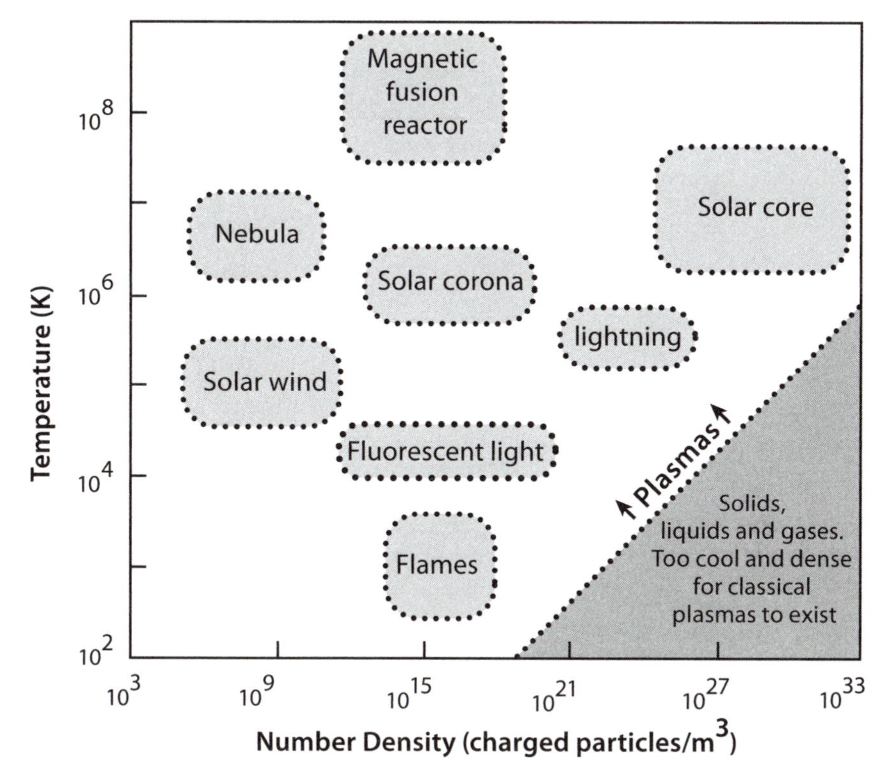

Fig. 4-4: Illustration of the pressure temperature fields for the four states of matter. At very low pressures (low number density) or high temperatures, a fourth state of matter, plasma, becomes important. While plasmas are somewhat unusual on Earth, in the universe they are a very common state of matter. Note the logarithmic scale and the extremely high temperatures of many plasmas.

Refractory materials such as alumina (Al_2O_3) and magnesia (MgO) are used in the walls of blast furnaces, because they melt at temperatures higher than 2,000°C. They remain in the solid state even when other somewhat less refractory materials, such as metallic iron, melt. The refractory materials permit the molten iron to exist in a solid container.

There is a large range of volatility between these extremes, and volatility can be ordered on a relative scale (see Table 4-1). Water is less volatile than carbon dioxide. Grease is less volatile than water, thereby permitting liquid grease to be stable at higher temperatures than boiling water, which gives us the difference between boiled potatoes and french fries.

Table 4-1
Physical constants of common molecules at one atmosphere pressure, ordered from most volatile to most refractory

Compound	Melting point of solid (°C)	Boiling point of liquid (°C)
CH_4	−182.47	−161.48
NH_3	−77.73	−33.33
CO_2*	−78.46	No liquid state
Hg	−38.83	356.62
H_2O	0	100
Fe	1538	2861
SiO_2	1713	2950
Mg_2SiO_4	1897	—
Al_2O_3	2054	2977

* *Sublimation point*

Iron and aluminum are still less volatile, permitting the two metals to exist in solid form even as water boils and grease melts.

DENSITY

Another important property of molecules is their *density*. The density of different elements varies substantially because the diameters of individual atoms vary only by a factor of 4, while the masses of the atoms vary by the number of total neutrons plus protons in the nucleus, from 1 in hydrogen atoms to 238 in uranium. As a good first approximation, the heavier an element, the greater the density of the substances the element makes. For example, solid lithium has a density of about 0.5 grams per cubic centimeter (gm/cm^3), iron about 6 gm/cm^3, gold 12 gm/cm^3, and uranium 18 gm/cm^3.

The same overall regularities apply to molecules made up of combinations of elements. Water (H_2O), which has a density of 1.0 gm/cc,

Table 4-2
Densities and number of nuclear particles per molecule of common terrestrial components.

Substance	Formula	Average nuclear particles per atom	Density (gm/cm^3)	Density/ particle
Water	H_2O	6.0	1.0	0.167
Gypsum	$CaSO_4 2(H_2O)$	14.3	2.32	0.161
Calcite	$CaCO_3$	20.0	2.71	0.135
Olivine	Mg_2SiO_4	20.0	3.27	0.165
Magnetite	Fe_3O_4	33.1	5.17	0.156
Iron	Fe	55.9	6.98	0.125
Gold	Au	197.0	17.1	0.087
Uranium	U	238.0	18.9	0.079

consists of two hydrogens with mass 1 and one oxygen with mass 16, so 18 nuclear particles in the molecule, or an average of 6 nuclear particles per atom. Forsterite (Mg_2SiO_4) has an average 20 particles per atom and a density of 2.8 gm/cc. Iron has 56 particles per atom and a density of 7.5 gm/cc. The proportionality between the number of nuclear particles per atom and the density is a little less than 1.0 because the size of the heavier atoms increases slightly with the increased numbers of electrons in the outermost shells. Other examples are given in Table 4-2. This regular behavior will be important for us as we try to determine the chemical compositions of distant planets for which we can determine density but as yet have no samples to measure.

The Two Great Classes of Molecules: Inorganic and Organic

In general, molecules occupy two grand groups with rather different characteristics—organic and inorganic. *Organic molecules* contain carbon combined with hydrogen, and often oxygen, nitrogen, phosphorous, and traces of other elements (there are a few exceptions that have C-N bonds and no C-H bonds). The molecules are called organic because it was originally thought they could be created only by living organisms. They are all rather volatile—even the high-temperature carbon compounds we call plastics tend not to be stable above temperatures of a few hundred degrees. Most organic chemistry takes place at temperatures close to room temperature. *Inorganic molecules* are all those that are not carbon bearing, as well as carbon compounds with no C-H bonds (e.g., CO_2, $CaCO_3$). Inorganic molecules that occur in nature in solid form are called *minerals*, and virtually all solid inorganic materials (e.g., rocks) have minerals as their fundamental constituents. Some knowledge of the structure and nomenclature of both organic molecules and minerals is necessary before we consider the construction of planetary bodies and their organic components.

MINERALS

A mineral is defined as a naturally occurring, inorganic solid with an ordered atomic structure, distinct physical properties, and a chemical composition that can be written as a molecular formula. Well-known examples of minerals are quartz (SiO_2), pyrite (FeS_2), magnetite (Fe_3O_4), diamond (C), and muscovite mica ($KAl_3Si_3O_{11}[OH]_2$). These and all other minerals can be clearly identified by their chemical formula. Each one also has distinctive physical properties. For example, mica has a prominent cleavage that allows it to break easily along parallel planes, and quartz has no cleavage. Instead, when a quartz crystal is split apart it has what is called a *conchoidal fracture*. All minerals have a specific hardness, or "scratchability." Diamond is the hardest of all minerals and will scratch any other; hence it makes the ideal jewel but easily scratches glass covers of copy machines! Graphite, another mineral with the same chemical formula as diamond but a different atomic structure, is one of the softest materials and can be scratched even by our fingernails. Other physical properties include the density, color, luster, streak (the color a mineral leaves when it is scratched on a hard surface), and whether or not the mineral is magnetic. The existence of a specific set of properties associated with each mineral often permits its identification in hand specimen without undertaking a chemical analysis.

The molecular *unit cell* of a crystal contains all the essential structural properties apparent in the macroscopic specimen and is what leads to beautiful, symmetric crystals. Minerals permit the microscopic molecule to have its essential characteristics viewed in macroscopic form. In a way, crystals make the invisible visible.

What does it take to make a mineral? To make a geometrically stable structure, atoms must fit together in terms of their size and charge. Atoms with many electrons are very large, whereas those with small electron clouds are very small. The atoms have to fit together so that their electron shells can interact with one another and produce a neutral molecule. For these reasons the sizes and electron shell structure of atoms determine what element combinations are possible and the geometrical form that various minerals take.

Because electrons are donated and received, it is the *ionic radius* of the element that controls the size and determines how atoms fit together in minerals. Size differences for ionic radii are larger than for neutrally charged atoms, because cations (positively charged ions) lose an electron and have their electron clouds pulled in by the positively charged nucleus, and anions (negatively charged ions) gain electrons, causing their electron clouds to expand. So Si^{4+} has a radius 0.32, while Cl^{-} has a radius of 1.72. These characteristics mean that most of the volume of minerals is made up of the anions. Figure 4-5 shows the sizes of various ions arranged in accord with the Periodic Table. Three trends are apparent. First, the anions on the right of the table are very large. Second, ions with the same charge (columns in the Periodic Table) show an increase in size with increasing atomic number down the columns (because their electron cloud gets larger). Third, ions with the same electron shell configuration, the rows in the Periodic Table, show a decrease in size with increasing positive charge to the right. As the charge increases, there are the same number of electrons in the outer shell, but an increasing number of protons in the nucleus. Hence the positively charged nucleus exerts more of a pull on the electron shell, making the atom smaller. For the same reason, as the *oxidation state* of individual atoms increases (i.e., they have more net positive charges), the atoms get smaller. The up-

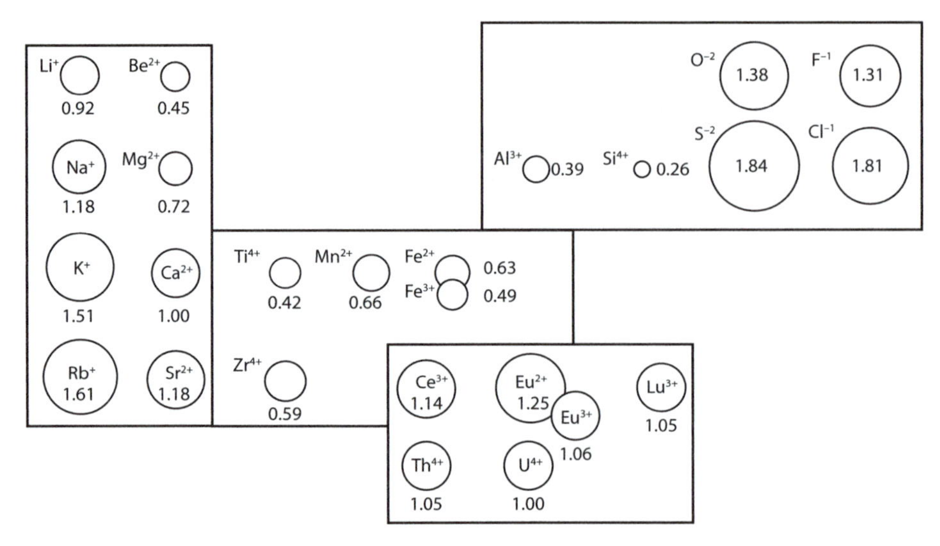

Fig. 4-5: Illustration of the different sizes of common elements in minerals. Note that the anions are large. Cations increase in size with increasing numbers of electron shells and decrease in size with increasing charge of the cation. Numbers in the circles represent the ionic radii of the ions. (Ionic radii from R. D. Shannon, *Acta Cryst* A32 (1976):751–67)

shot is that ions of K^+, Na^+ and Ca^{2+} are larger than the ions Mg^{2+} and Fe^{2+}, which are in turn larger than the ions Al^{3+} and Si^{4+}. The 2– anions are larger than then 1– anions, and so on.

Linus Pauling, the Nobel Prize–winning chemist, noted some simple regularities in the architecture of oxide minerals. Pauling's rules stem from the concept, now confirmed through x-ray imaging, that negatively charged oxygen ions are arranged in polyhedra around positively charged metal ions. The number of oxygen atoms in the polyhedra controls the size of interior spaces where the cation resides (Fig. 4-6). Using some simple trigonometry, Pauling calculated that a (small) Si^{4+} ion could hold apart four oxygen ions in a tetrahedron, but not six oxygen atoms (in an octahedron) because the interior space would be too large. Mg^{2+} and Fe^{2+} are larger than Si^{2+} and fit inside an octahedron. Because Si, Mg, Fe, and O are the most abundant elements that make minerals, most minerals are three-dimensional arrangements of tetrahedra and octahedra, with O^- making up the polyhedral container and the cations occupying the interstices.

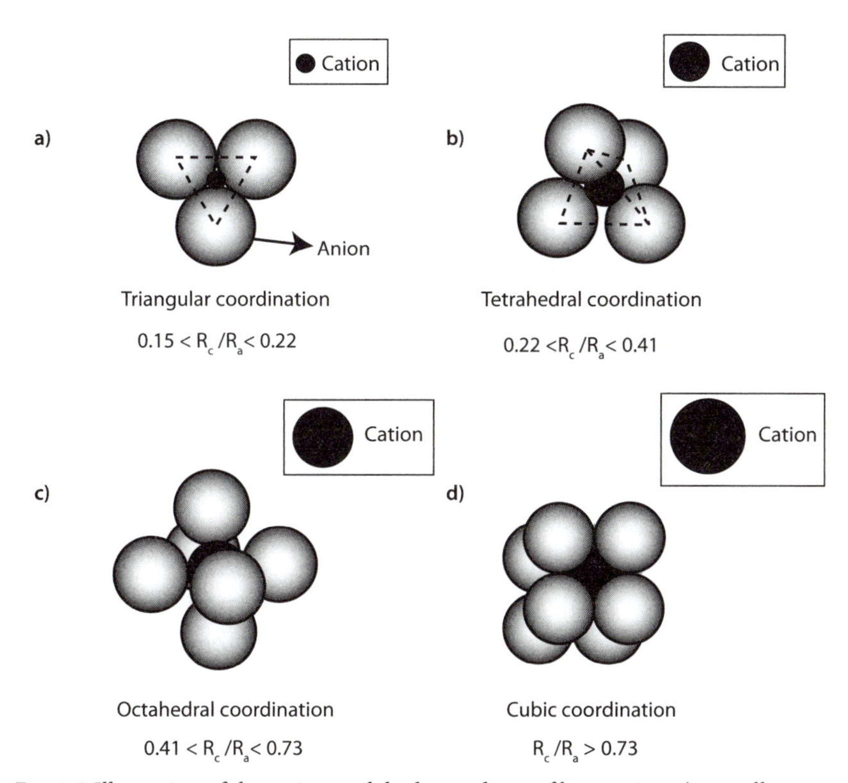

Fig. 4-6: Illustration of the various polyhedra made up of large anions (normally oxygen) that leave small interior spaces for cations to fill. The larger the coordination number, the larger the size of the interior cavity and the larger the cation that can be accommodated. R_c and R_a refer to the radii of the cation and anion, respectively.

In addition to size and charge constraints, in order to grow in a regular pattern the particular arrangement of the atoms must be infinitely repeatable. If two molecules combine and create a surface template where the next atoms satisfy their electron shell requirements in exactly the same way, then growth can take place indefinitely. If atoms were combined in a way that did not permit the addition of the same molecular structure, no ordered growth could occur. An extreme example of this is the noble gases, where each individual atom remains separate because no connections with other atoms are possible. Ordered growth is also energetically more stable than a disordered collection of molecules, giving ordered structures a great advantage in their development.

Infinite repeatability is the governing principle of the laws of symmetry, and all minerals are symmetrical. Symmetry permits a pattern to be infinitely repeated, as, for example, in tiling a floor or building a wall. What tile shapes can you buy that will ensure no gaps imposed by the tile shape? Why, for example, are there no pentagonal tiles, despite the beautiful form of pentagons? Such practical questions, as well as the appeal of symmetrical patterns, have made the study of symmetry a human passion for thousands of years, understood by the ancient Egyptians and extensively developed by Islamic culture. The mosaics of the Alhambra, the Islamic palace in Granada, southern Spain, built in the thirteenth century, contain almost all possible symmetries (Fig. 4-7).

The remarkable aspect of symmetry is that throughout the universe only symmetries of twofold, threefold, fourfold, and sixfold rotations are possible, as well as mirror images. The same principles for tiling a floor or building a wall apply equally to minerals, but in three dimensions. In all there are only thirty-two symmetry groups that are possible in all of nature, and every mineral belongs to one of them.

The common minerals are those made up of the most abundant elements that form solid substances, and these are the oxides of Si, Fe, Mg, Ca, and Al. The Si^{4+} cation needs to share four electrons to satisfy charge balance. Because of its small size, it must be tetrahedrally coordinated. Four O^- anions, the largest of the common elements, fit well around Si^{4+} in the shape of a tetrahedron, with each silicon sharing one electron with each oxygen, leaving the oxygens with a need for one more electron to fill their outer shell. This creates the possibility for stable building blocks of silica tetrahedra that can either bond with each other or with another metal atom. Silica tetrahedra are the fundamental building blocks of most minerals, equivalent to the central role that carbon with its four bonds plays in the world of organic chemistry. Carbon atoms form the backbone for almost all the molecules of life. Silica tetrahedra form the backbone for almost all the molecules of rocks. The fourfold coordination of each of these atoms, and their ability to create structural units that can bond both with themselves and with other atoms, are what permits symmetrical three-dimensional structures to be created.

The silica tetrahedron has a number of ways it can be organized, and these create the great classes of silicate minerals (see Fig. 4-8). If the tetrahedra remain isolated from one another with no Si-O-Si bonds, then

Fig. 4-7: Top: An example of tiles from the Alhambra in Spain, illustrating threefold symmetry with no mirror planes or twofold axes. *Bottom:* a classic quilting pattern exhibiting fourfold symmetry.

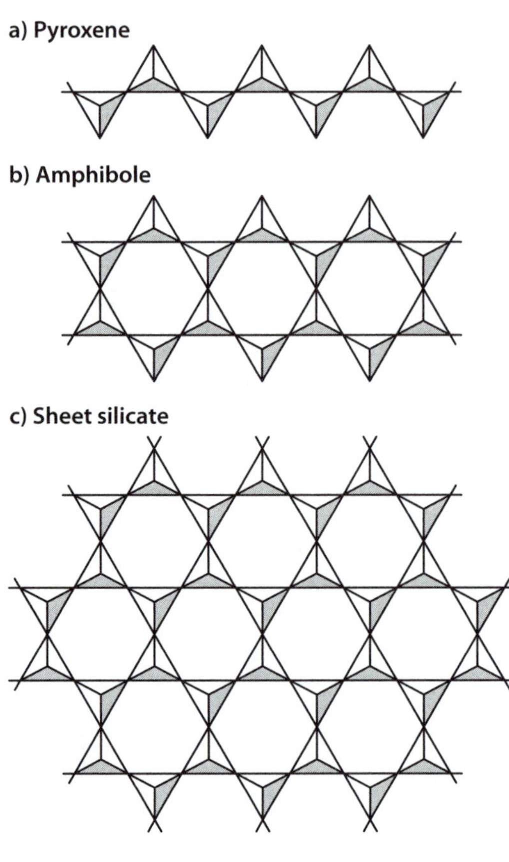

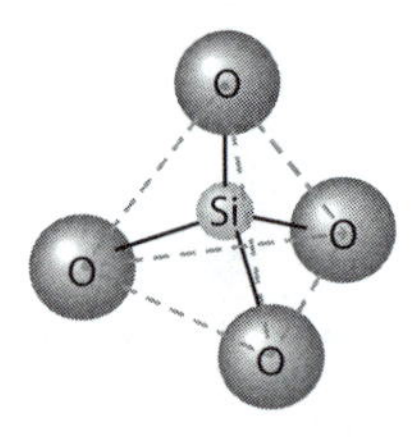

Fig. 4-8: The silica tetrahedron is the fundamental building block of the most common rock-forming minerals, the silicates. Tetrahedra can either be isolated, forming the olivine group, or combined in single chains (pyroxenes), double chains (amphiboles), sheets (micas), or three-dimensional frameworks (quartz and feldspars).

each oxygen atom bonds with another metal. This is the structure of the *olivine* group. Olivine, consisting of mixtures of forsterite (Mg_2SiO_4) and fayalite (Fe_2SiO_4), is the most abundant mineral in Earth's upper mantle. The next major group has as its structural backbone individual chains of silica tetrahedra. Each tetrahedron is joined to two others. Other metals then form the connecting bonds between the chains. Single-chain silicates are called *pyroxenes.* A large variety of different metals and minerals can fit into these flexible structures and form part of the pyroxene class. When two chains join together, the *amphibole* group is formed. The large hole in the center of the double chains permits the inclusion of larger cations such as K into the amphibole structure, whereas K is effectively excluded from all pyroxenes because of its large size.

The next group occurs when silica tetrahedra form continuous two-dimensional sheets. Because the sheets of silica tetrahedra are tightly bonded within the sheet, but the bonds connecting the sheets are weaker, they have a very characteristic sheety cleavage that we readily associate with the *sheet silicate* group of minerals, which includes the micas. Finally, the tetrahedra can form three-dimensional *framework* structures. Quartz is the most common example, but the *feldspar* class of minerals, which is the most abundant mineral group in Earth's crust, also has a framework structure.

An important aspect of most silicate minerals is their potential to make solid solutions. Just as there can be liquid solutions (like salty water or alcoholic drinks) that mix together to form a single substance, there can also be solid solutions if atoms have similar sizes and charges. The great geochemist Goldchmidt pointed out that if two atoms have their charge and radii within 15% of each other, extensive solution is possible. For example, Fe and Mg both have a charge of +2, and their ionic radii are 0.32 and 0.35, respectively. Because Mg^{2+} and Fe^{2+} are very close in size and have the same charge, they freely substitute for one another in minerals. As size difference increases, substitution becomes increasingly difficult, because the structure has to deform to accommodate the larger atom, and ultimately it is no longer possible to continue the symmetrical structure. For example, pyroxenes with only Mg and Fe coordinating with the chains of silica tetrahedra form more symmetrical minerals than pyroxenes that incorporate significant amounts of the larger metal calcium (Ca). The amount of Ca is limited to 50%—greater amounts cause such disruption that pyroxene symmetry is no longer possible.

There are also nonsilicate mineral groups, although these are of lesser abundance on Earth. They include the *sulfides* (the most common of which is pyrite, FeS_2 which makes "fool's gold"), the *oxides* (such as magnetite, Fe_3O_4), *halides* (such as salt NaCl) and the *carbonates* ($CaCO_3$), which are the dominant constituents of limestones. Carbonates are the rocks that link the geological and organic life cycles on Earth.

To sum up, the important minerals are determined by the overall abundances of elements available, the electron shell requirements for their combinations, the relative sizes of those elements, and the symmetry of the structures they can create. These features lead to the overwhelming importance of silicate minerals in creating the solid substrate of Earth.

ORGANIC MOLECULES

Organic molecules are those where carbon atoms are joined with hydrogen. Organic molecules also often have a carbon backbone and combine with other elements such as oxygen, nitrogen, phosphorous, and others. They were originally identified as those molecules which were made exclusively by living processes. But in the early nineteenth century experiments were done that showed that organic molecules could be made by normal physicochemical processes, demonstrating that the bridge between the inorganic and the organic could be made in the absence of life. Furthermore, some molecules such as methane (CH_4) are made in abundance in space and by inorganic reactions involving rocks, CO_2, and water. Experimentation has also shown the existence of thousands of organic molecules that do not exist in nature. Therefore, "organic" now refers to the broad class of molecules that contain carbon and hydrogen, many of which are created and used by living organisms, as well as many others not created by biological processes. This definition covers most organic molecules and suffices for our purposes, but it is not perfect, because certain carbon compounds that are definitely associated with living processes, such as urea, have C-N bonds but no C-H

```
                                    C
                                    |
                   H            H — C — H
                   |                |
       H       H — C — H        H — C — H
       |           |                |
   H — C — H   H — C — H        H — C — H
       |           |                |
       H           H                H

    Methane      Ethane          Propane
```

Fig. 4-9: The three simplest hydrocarbons. Methane is also known as *natural gas* and is the most abundant organic molecule on earth. All hydrocarbons can burn to form water and carbon dioxide. For example, $CH_4 + 2O_2 \rightarrow CO_2 + 2H_2O$.

bonds. That said, organic molecules are the stuff of life, and inorganic molecules are generally the stuff of rocks and minerals, and the distinction is important and necessary.

An important aspect of organic chemistry is the inorganic solvent, water. Many organic molecules contain water and undergo transformations associated with the incorporation or release of water. Most reactions among organic molecules take place in the presence of water—the average cell is about 80% water. Therefore, while water is an inorganic molecule and ice a mineral, the world of organic molecules, at least on Earth, is very much dependent on the hydrous environment. Even in interstellar space, where liquid water does not exist, ice may play an important role. And it also appears that much of the synthesis of organic molecules in space requires an inorganic mineral substrate as a catalyst. Therefore, in general organic molecules depend very much on the inorganic molecules that either host them or give rise to their creation.

The simplest organic molecule, and one of the most abundant in the universe, is methane (CH_4) (Fig. 4-9). Methane is the simplest form of *hydrocarbon*—organic molecules made up of a carbon backbone connected to hydrogen atoms. Many more complex hydrocarbons exist—oil and gas are largely made up of a complex assemblage of hydrocarbons. All hydrocarbons burn in the presence of oxygen, which, if the burning is complete, converts them to the inorganic molecules CO_2 and H_2O.

More complex organic molecules that have essential roles in biological processes can generally be divided into four groups—*carbohydrates, lipids, proteins, and nucleic acids.* Carbohydrates contain carbon combined with oxygen and hydrogen in the same proportions as water. Common examples are glucose, starch, and cellulose. Lipids have much less oxygen than the carbohydrates and a much higher energy content per gram. They are very efficient storage molecules for energy and include fats in animals and oils in plants. Proteins are by far the most diverse group of organic molecule. They are long chains of *amino acids.* The amino acids have at one end an amine group of a nitrogen atom bonded to two hydrogens and a carbon atom, and at the other end an acid (COOH), connected to what is referred to as an *R group.* The identity of the R group is what varies from one amino acid to another. A huge number of amino acids are theoretically possible and can be synthesized in the laboratory. Remarkably, only twenty of them are used to build the proteins of terrestrial life. But just as twenty-six letters allows a large number of words, having twenty different amino acids permits remarkable diversity in proteins. Some 100,000 different proteins are known to occur in living organisms, but the potential number of proteins that could be constructed is vastly greater. Some protein molecules are huge collections of amino acids—hemoglobin, for example, is a structure made up of 10,000 amino acids, and there are many even larger protein molecules.

The *nucleic acids* are long double chains that form naturally as double helices. The backbones of the nucleic acids are alternating groups of sugar and phosphate molecules. The links between the chains consist of pairs of bases—adenine, guanine, cytosine, and thymine or uracil. The combination of the base with the sugar and phosphates is called a *nucleotide.* The nucleotides then combine into very long nucleic acid chains. Of course, such chains are the fundamental carriers of information—they form the genes of all life—and are the means by which all cells can replicate.

The organic molecules on Earth generally are associated with the liquid state of matter. Even mammals are largely liquid, and it is the inorganic parts of our anatomy—bones—that give us some rigidity. Because of the liquid medium, organic processes generally take place as interactions among individual molecules, and the general principles discussed above for crystals and symmetry in minerals rarely find application in the organic realm.

Environments of Molecular Construction

As the universe cooled down after the Big Bang, gaseous hydrogen and helium were the only elements of significant abundance in the universe. The first generation of stars, therefore, was made up exclusively of these elements, and there were no solid particles or planetary systems in the universe, and no organic or inorganic molecules! Only after nucleosynthesis and distribution by supernovas were the elements available to create the molecules essential for planets and life. After formation these elements rapidly become surrounded by electron shells and combine into the simplest molecules. Hydrogen as the most abundant element is an important ingredient, as in CH_4. Because oxygen is made in abundance in these massive stars, there is sufficient oxygen to make oxides of all the important metals. The oxygen left over combines with hydrogen to make water. Therefore, the earliest and most abundant molecules apart from hydrogen and CH_4 are oxides, such as CO and H_2O, and also oxides of all the other metals in accord with their abundance created during nucleosynthesis.

The metallic oxides then combined with SiO_2 to make tiny grains of silicates. The existence of grains of minerals such as olivine in interstellar clouds is now verified by astronomical observations from telescopes in orbit around Earth. These grains are so small that they would hardly qualify as dust if we observed them on earth—they are more like silicate smoke. These silicate grains would generally be surrounded by tiny mantles of ice, since H_2O would be far below its melting point in the very cold environment of interstellar space. Later production of reduced species in the absence of oxygen would add molecules such as CN, CH, and HCN to the mix. Mixing in space would not necessarily lead to chemical equilibrium among all these species, because the temperature is just a bit higher than absolute zero and particle densities are lower than the most extreme vacuum that can be generated in a terrestrial laboratory. Chemical reactions that we consider normal, therefore, are rare.

The interstellar environment has two additional factors that make it very different from molecular construction on Earth. In addition to the starlight that we see in the visible part of the spectrum, stars also emit stellar winds and ultraviolet radiation. Stellar winds became known through the discovery of the *solar wind* in the second half of the twentieth century. Theoretical calculations by Eugene Parker in 1958 showed that the intensely hot solar corona should emit high velocity particles. The theory was confirmed by spacecraft that measured energetic particles at high velocities streaming through space from the sun. The mean composition of these particles is very similar to the mean composition of the sun. Even more energetic particles were discovered as well, showing that other stars also emit such particles, giving rise to a general cosmic radiation. Some of this radiation is far more energetic than the solar wind, coming from much more energetic objects, such as large stars and supernovas.

The atoms involved in our everyday life are not particularly energetic. Molecules of air, for example, have low energies and move as fast as bullets, though still quite a bit slower than a typical satellite. When such molecules collide, they bounce off each other and do not affect the electron shell structure of the colliding molecules. The cosmic radiation is far more energetic, with speeds up to one-tenth the speed of light. The energies are high enough that collisions can knock molecules apart or knock off electrons and create ionic species. Therefore these particles can have an important effect on what chemical species exist.

Stars also emit ultraviolet radiation. We learned in Chapter 2 about blackbody radiation, and how the radiation emitted by an object becomes increasingly short in wavelength (more energetic) as the temperature increases. Hot massive stars, therefore, emit much more energetic radiation. The total amount of radiation emitted also goes up exponentially with temperature, so stars of many solar masses emit huge amounts of ultraviolet radiation, and this radiation can have important effects on chemical species, causing them to be ionized or break apart. Massive stars, through their winds, UV radiation, and ultimate explosions as supernovae, dominate energy injection into the interstellar medium.

These sources of elements, energetic particles, and UV radiation are not isolated in space, because most stars are created in massive clouds that serve as "stellar nurseries" where thousands of new stars are born. Some of them are the massive stars emitting radiation so intense that they clear space around them. Supernovae are common in these environments, spewing newly made elements that will then be incorporated into small stars and planetary systems. The interstellar clouds (see frontispiece of Chapter 5) are the largest factories in the universe, not only for element creation but for the evolution of molecules.

Where the radiation is most intense in the cloud, complex molecules cannot form because they are continually broken apart. When the cloud becomes denser, however, the thicker dust protects the interior particles and gases from much of the radiation. In these environments, given the long time frames of interstellar space, complex reactions can take place and hundreds of different molecules are created. An important aspect of these reactions is that they differ substantially from reactions observed on Earth. On Earth, for example, virtually all organic reactions take place in the presence of water. But in space the pressures and temperatures are so low that water exists only as ice. The sluggish chemical reaction rates that one might expect there, however, are overcome by the energy supplied by the ultraviolet light from nearby stars. The rarity of interactions in the high vacuum is overcome by reactions that take place on the surfaces of the interstellar dust. Some molecules form when atoms stick to the surface of an olivine crystal and through diffusion encounter each other. Others require a combination of particles and radiation. For example, carbon dioxide can form when an ice containing carbon monoxide is irradiated with UV light in the presence of oxygen atoms.

The net result of these reactions has only been observable in the last few years, with the launching into space of telescopes that are able to explore wavelengths previously obscured by Earth's atmosphere. These studies have revealed more than a hundred different molecules, and the list grows every year. These molecules include not only the major silicates but a host of organic molecules, including "prebiotic" molecules that would likely be those required to start life on Earth—such as water, methanol, formaldehyde, and hydrocyanic acid. Thus the interstellar clouds are not only incubators for the formation of new stars, they also make the molecular raw material that ultimately combines to form planets.

Through the study of comets, which are the relics of icy planetesimals that accreted in the outer solar nebula, we can infer that the processes that are now observed in interstellar clouds in our galaxy were also pertinent to the formation of our own solar system.. The last decade has permitted astronomical study of the compositions of comets, and many of the same molecules discovered in protoplanetary disks in distant space are found in the comets as well. While this fast-moving area of science is by no means fully developed, it is clear that all the molecules necessary for the beginning of a habitable planet are abundant and widespread throughout the universe. Processes that we can infer for the solar system at the time of its creation are beginning to be observed happening now in other parts of the galaxy (see frontispiece to Chapter 5).

Summary

Elements do not remain isolated but come together to form molecules shortly after they are formed by nucleosynthesis. The laws that govern their combination relate to their size and the detailed structure of their electron shells. Large anions form the polyhedral framework of inorganic molecules that come together to form minerals, and the size of the polyhedra controls what cations can reside in the interior. The most important structural unit for minerals is made up of the two alpha-particle nuclides Si and O, forming the silica tetrahedron. The silica tetrahedra can join in a variety of ways to produce an astonishing number of silicate minerals that we will see are the main building blocks of planets. Organic molecules are those where carbon combines with itself and with hydrogen, nitrogen, and other elements to form the basic building blocks that ultimately will come together to form primitive life. The physical properties of volatility and density will play an important role in how these molecules distribute themselves during subsequent events in the universe. Volatile compounds remain as gas even to low temperatures, and cannot accrete as solid materials. They will become the primary constituents of planetary atmospheres. Refractory materials such as silicates remain solid even at high temperatures and will become the solid materials of planets. Density will then control at what levels within planets the various molecules reside.

Huge interstellar clouds are the likely nurseries where both element formation by nucleosynthesis and the early molecules and their reactions with one another take place in an environment that is very foreign to us as residents of a benign planetary surface. Supernovae distribute new elements and create massive energy fluxes through the nebular cloud that contribute to molecular construction. Smaller concentrations of matter come together to form smaller stars such as our sun that can give rise to long-lived planetary systems necessary for life.

Fig. 5-0: Panoramic photograph of the Orion Nebula. It is one of the largest pictures ever assembled from individual images taken with NASA's Hubble Space Telescope, showing star birth, element, and molecule formation in a stellar nursery. This figure is a composite image of the center of the Orion Nebula assembled from individual images taken by the Hubble telescope. The Orion Nebula is within the Milky Way galaxy, 1,500 light years away from Earth in the middle of the constellation Orion's sword. The size of the sky portrayed is only a portion of the entire nebula, but it still is about 5% of the area of a full moon. The composite image shows a star factory, where stars have formed from collapsing clouds of interstellar gas within the last million years. The four hottest and most massive stars, called the Trapezium, lie near the center of the image. In addition, some 700 other young stars have been observed by astronomers in the nebula at various stages of their formation. The mosaic also contains more than 150 embryonic solar systems, lending support to the general process of star and solar system formation similar to the processes believed to have created our own. Some of the curved features are enormous jets of gas generated by supersonic shock waves at 150,000 km/hr. The large variety of stars of different masses in the same region show that smaller stars like our sun may form in the intense conditions where short-lived stars have short but violent lives distributing elements and radionuclides into the nebular cloud. The high levels of radiation would also contribute to environments of molecular construction, and the necessary organic and inorganic precursors for planets and life can be formed in this nebular environment. See also color plate 4. (Photograph courtesy of NASA. Credit: NASA, C .R. O'Dell and S. K. Wong (Rice University). GIF and JPEG images, captions, and press release text are available at http://www.stsci.edu/pubinfo/PR/95/45.html)

CHAPTER 5

The Heavy Construction

The Formation of Planets and Moons from a Solar Nebula

Formation of the solar system took place as a coherent series of events stemming from a collapsing cloud of raw material present in a solar nebula. Most of the material in the nebula was drawn to the center to form the sun, giving it a composition very close to that of the original cloud. It contains 99% hydrogen and helium and only about 1% of the remaining ninety elements. A small amount of the matter in the cloud ended up in a nebular disk around the newly formed sun. A small percentage of this material condensed to solid form and then aggregated to form planets, moons, asteroids, and comets. The remaining gas was likely blown away by a violent solar wind. Fragments of materials from the early solar system occasionally reach Earth as meteorites. The composition of the primitive "chondritic" meteorites is very similar to the sun except for the very volatile elements. Other "achondritic" meteorites reveal much about early planetary differentiation.

The various objects in the solar system formed from a complex series of processes, understood well in general terms but with many details remaining to be clarified. The inner and outer planets differ greatly in size and density. The inner planets have densities greater than rocks, and based on evidence from meteorites, are comprised of a rocky exterior and an inner metallic core. The outer planets have very low densities, requiring a preponderance of ices and gases in their composition. The difference between inner and outer planets is largely the result of differential volatility. In the early solar system there would have been a *temperature gradient with distance from the sun. Closer to the sun, volatile molecules were gas and refractory molecules were solid. In those parts of the disk, only the least volatile molecules such as Fe, FeS, FeO, MgO, Al_2O_3, and SiO_2 were in solid minerals that could collect together to form dense planets. Farther from the sun, beyond the "snow line," ices also crystallized and almost all the elements except hydrogen and helium were in solid form, leading to large planets of low density. Their large mass also allowed them to accumulate vast gaseous atmospheres. Because of differential volatility, the comets and planets of the outer solar system have a chemical composition very different from the asteroids and planets of the inner solar system. Materials accreted in cooler parts of the solar nebula, where volatiles were more abundant, are likely to have impacted Earth early in its history, bringing with them the volatile components (and possibly organic molecules) that were essential for the inception of life.*

Solar system creation very likely took place in an interstellar cloud of gas and dust, which served as a "stellar incubator" (see frontispiece). Such clouds are common in the universe, and we can observe new star systems forming elsewhere in our galaxy. There is also evidence for the existence of many planets around other stars, indicating that creation of planetary systems is a normal and common event in our galaxy.

Introduction

Human beings have been looking at the sun and planets and pondering their significance for millennia. What was apparent to the earliest observers is still easily observable today (at least outside a city)—the sun, moon, and planets all move around the same narrow band of the sky, called the *ecliptic*, which is also the sun's equator. The apparent position of the sun relative to the stars changes progressively—in winter and summer Earth is on opposite sides of the sun, so to an Earth observer the sun appears against different stellar backgrounds that vary regularly throughout the year. These backgrounds have stellar groups associated

Table 5-1
Dynamic characteristics of the planets in the solar system

Planet	Orbital radius (Au)	Orbital period (yrs)	Inclination to orbit (°)	Orbital eccentricity	Rotation period (days)*
Mercury	0.39	0.24	7.00	0.206	58.64
Venus	0.72	0.62	3.40	0.007	−243.02
Earth	1.00	1.00	0.00	0.017	1.00
Mars	1.52	1.88	1.90	0.093	1.03
Asteroid Ceres	2.74	4.60	10.60	0.080	0.40
Jupiter	5.20	11.86	1.30	0.048	0.41
Saturn	9.54	29.46	2.50	0.054	0.43
Uranus	19.22	84.01	0.80	0.047	−0.72
Neptune	30.06	164.80	1.80	0.009	0.67

**Negative rotation period for planets with a retrograde rotation, opposite to other planets.*

with them, called *constellations*, which gave rise in ancient times to astrology. The planets pass through the same swath of the sky, also with varying stellar backgrounds. They all move in the same direction but at different rates depending on their particular position relative to the sun as they rotate around it. These fundamental observations, apparent to any regular observer of the night sky, demonstrate that the sun and planets are all coplanar. Therefore from earliest times the planets and sun have intuitively all seemed to be connected to one another. They are not rotating in random directions and orientations around the sun; they all rotate together, in the same plane and in the same direction. Furthermore, this direction is consistent with the spin of the sun. Hence the association of the word *planet* with "plane."

There are other striking regularities within the solar system. Using the surveying methods discussed in Chapter 2, astronomers have been able to determine the distance from the sun to each of the planets (see Table 5-1). Considering the asteroid belt between Mars and Jupiter as a failed planet, the spacing between the planets shows a remarkable regularity—the distance between orbits of successive planets increases by roughly a

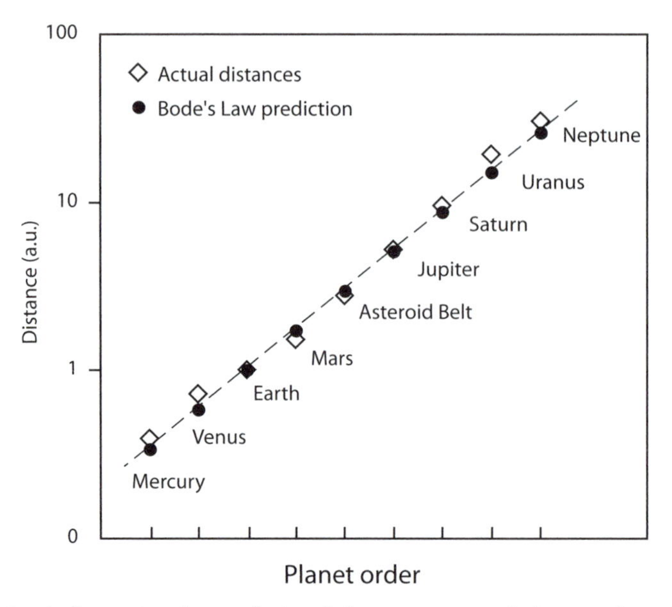

Fig. 5-1: Graph illustrating the regularity of planet spacing with distance from the sun. Moving outward from the sun, each planet is about 1.7 times farther from the sun than its closest neighbor: This regular relationship is known as Bode's Law.

factor of 1.7, an observation that was named after its discoverer—Bode's Law (Fig. 5-1).

There are also significant differences between the inner and outer planets. Mercury, Venus, Earth, and Mars are all small; Jupiter, Saturn, Uranus, and Neptune are huge. Therefore the planets occur in two regular and organized groups (Fig. 5-2).

This clear organization of the solar system gave rise in the eighteenth century to the nebular hypothesis of Kant and Laplace. They suggested that the planets and sun formed from a single, spinning flat cloud; the spin of the planets and the existence of the ecliptic conform to the original plane and spin of the cloud. If instead the orbits were helter skelter, and the big and small planets randomly distributed, an alternative model would be tenable, where for example the planets formed elsewhere and were subsequently captured. The Kant-Laplace model was strenuously attacked over the centuries, but their fundamental ideas are retained in more detailed and quantitative modern models of solar system formation.

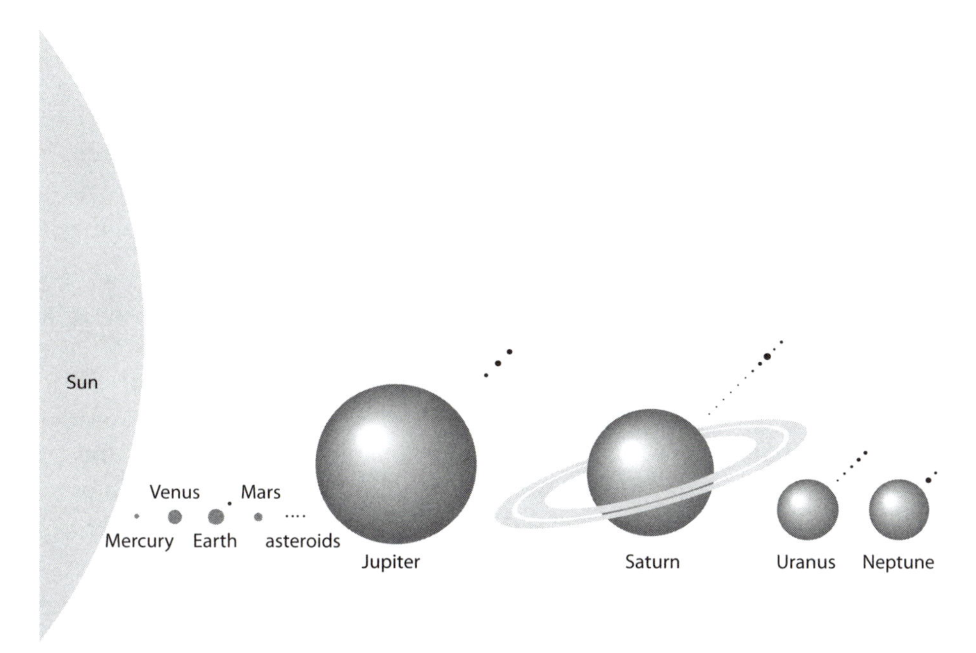

Fig. 5-2: Illustration of the approximate sizes of the planets relative to each other. The planets are ordered in their sequence from the sun. The sizes are proportional, but the distances in the figure do not correspond with their actual positions.

Planetary Vital Statistics

An understanding of the origin of the planets clearly must depend on knowledge of their physical and chemical characteristics. While we have a multitude of ways to determine these characteristics for Earth, it is more difficult for the other planets, because we only have samples or direct surface measurements from the moon and Mars. However, measurements from satellites that orbited around the various planets can be combined with basic astronomical observations to provide us with substantial information.

PLANETARY MASS

The mass of a planet can be determined from the gravitational influence it exerts on the orbits of its moons, on other planets, and on the space probes sent out from Earth (Table 5-2). The primary method used is a clever one, based on a law established in the early 1600s by Johannes Kepler, a German astronomer and mathematician. Kepler's Laws were later explained by Newton, making use of a constant of gravitation, *G*. Using Newton's laws, there is a straightforward relationship between the distance at which a moon orbits its host planet (*R*), the velocity of the moon around the planet (*V*), the mass of the host planet *Mp* and the universal constant of gravitation. The equation is:

$$M_p = RV^2/G \qquad (5\text{-}1)$$

Note from Equation (5-1) that the right hand term of the equation with the orbital velocity is independent of the mass of the orbiting object. All objects orbiting at the same distance from a planet will orbit at the same velocity. This is why astronauts need not be afraid to leave their spaceships or let go of their wrenches. Once the value of *G* was able to be measured accurately in 1811, the mass of planets with moons could be determined simply from by measuring the time it takes a moon to make one rotation of the planet, and the distance of the moon from the planet's center.

For those planets without moons (Mercury and Venus), the masses had to be calculated by a more elaborate scheme. The orbit of each planet is influenced by the gravitational attractions of neighboring planets. By careful observation of the perturbations caused by neighboring planets of known mass, the masses of Mercury and Venus have been able to be determined.

As can be seen from Table 5-2, the planetary masses change greatly as the distance from the sun increases. Mercury, closest to the sun, is the least massive. Venus, Earth, and Mars are larger (but note that Earth is ten times as massive as Mars), and then there is a huge jump to the giant planets Jupiter and Saturn. Uranus and Neptune, while somewhat less massive, are still far more massive than any of the four inner planets. So we see that the process responsible for planet building led to objects differing widely in size. The size variation does not vary linearly with distance from the sun. Instead there is clear demarcation between the inner planets, with masses of less than 6×10^{27} gm, and the outer planets, with masses greater than 88×10^{27} gm (see Table 5-2).

While these first order regularities about the solar system strongly suggest that the objects within it are co-genetic, they do not indicate with any specificity how the planets formed. Greater understanding awaited the discovery of the chemical compositions of the various planets.

Table 5-2
Physical characteristics of the planets in the solar system

Planet	Equatorial radius (10^8 cm)	Volume (10^{26} cm^3)	Mass (10^{27} gm)	Density (gm/cm^3)	Corrected density* (gm/cm^3)	Moons	Most abundant atmospheric gases
Mercury	2.44	0.61	0.33	5.43	5.40	—	Minimal
Venus	6.05	9.29	4.87	5.24	4.30	—	CO_2, N_2
Earth	6.38	10.83	5.97	5.52	4.20	1	N_2, O_2
Mars	3.38	1.63	0.64	3.93	3.70	2	CO_2, N_2
Jupiter	71.49	14,313	1,898.6	1.34	<1.3	63	H_2, He
Saturn	60.26	8,271	568.4	0.69	<0.69	61	H_2, He
Uranus	25.56	683	87.0	1.28	<1.28	27	H_2, He
Neptune	24.76	625	102.40	1.64	<1.64	13	H_2, He

* *Densities are corrected to one bar pressure. Only maximum densities are known for the outer planets.*

PLANETARY DENSITIES

The densities of the various planets (Table 5-2) can be determined by dividing the mass of the planet by its volume. We learned in Chapter 4 that the density of an object provides much information about the elements that make up the object—high atomic number elements are generally much denser than low atomic number elements. Hence the density of a planet has much to tell us about the planetary composition.

There is one more complication, however, before we can relate density to planetary composition—density also depends on pressure. As materials are squeezed, the atoms that make them up pack together more tightly and the density goes up. To know the densities of materials at high pressure requires careful experiments. For example, experiments have shown that Fe with a density of 7.5 gm/cm^3 at Earth's surface has a density of 9.8 gm/cm^3 at Earth's center. To compare planet densities in terms of the materials of which they are made, we need to know the density the planet would have if all the material were at the same pressure. The choice is arbitrary, so we choose one atmosphere, the pressure at Earth's surface, which we can call the *uncompressed density*. Uncompressed densities are always less than the actual density, because squeezing always increases density. This correction can be done well for the inner planets, but we cannot give accurate corrected densities for the outer planets because we do not know the densities of ices at the very high pressures that exist in the interiors of these bodies. We do know, however, that the uncompressed densities would be less than the observed density, as indicated in Table 5-2.

While the uncompressed densities give us an estimate of the average number of nuclear particles contained by the atoms of a given planet, there are a wide variety of combinations of elements through which this could be achieved. To determine which of these possible combinations is the correct one, we must use additional information. The situation is similar to that faced by someone about to open a birthday present. From the heft of the box many possibilities can be eliminated. If the box is quite heavy for its size, it could be a book or a mineral specimen. If it is quite light for its size, it could be a sweater or a very small present in a big box of tissue paper. While providing an indication, the heft leaves many options open!

PLANETARY COMPOSITION

Additional clues for planetary composition come from knowledge of nucleosynthesis in stars and the compositions of interstellar clouds. As we learned in Chapter 3, some elements, such as H, He, the alpha particle nuclides (C, O, Ne, Mg, Si), and Fe are created in far greater abundance than other elements. We can observe from the sun's spectrum that these elements are also the most abundant in the sun, and therefore in the cloud of gas and dust from which the sun and planets formed. The candidate molecules for planets are then constrained to be made largely of these elements, and these molecules are also those observed in interstellar clouds. The molecules fall into three different classes—ices, oxides, and metals, with very different densities and volatility (Table 5-3). What then are the planets made of? To match the density of the outer planets, with their uncompressed densities of less than 1.7 (see Table 5-2), ices must be predominant. The inner planets, on the other hand, with corrected densities higher than oxides, must consist of some mixture of oxides and metal.

By examining the melting points given in Table 5-3, it is also clear that the light molecules that make up much of the outer planets are very vola-

Table 5-3
Densities and melting temperatures of possible planet-forming solids

	Compound	Number of nuclear particles per atom	Density gm/cm^3	Melting point °C
ICES	CH_4	3.2	0.42	−182.5
	NH_3	4.2	0.7	−78
	H_2O	6	1.0	0
OXIDES	SiO_2	20	2.7	1710
	Mg_2SiO_4	20	3.2	1200
METAL	Fe	56	7.9	1540

tile, and in order to accrete as ices the temperatures must have been very cold. The materials that make up the inner planets, on the other hand, remain solid at very high temperatures. This leads to the straightforward idea that the solar nebula around the sun had a range of temperatures.

Close to the sun the temperatures were hot, and oxides and metal were the only solid materials that could collect together to form planets. The volatile compounds were all gases, and the small gravitational fields of the early *planetesimals* (small precursors to planets) would not be great enough to retain them. Far from the sun, temperatures were much colder. There, ices as well as silicates and metals were solid and could collect together to form planets. Ices were also present in far greater amounts, since they are made up of elements such as C, H, and O that are about ten times more abundant than the heavier elements that make up metals and oxides (see Fig. 3-10, showing the abundances of elements made during nucleosynthesis). The outer planets then got an equal share of oxide and metal, and a huge additional mass of icy volatile compounds. Their large mass also allowed them to retain a massive atmosphere. Both these effects lead to their great size and low density.

These considerations based on differential volatility account for the first-order differences between inner and outer planets. Planets form from solid materials. Near the sun, temperatures were so high that only the most refractory elements were solid. Far from the Sun, the solar nebula was cold and most elements were solid. It's a bit as if a machine were able to sweep up all the solid materials in the room you are sitting in at very different temperatures. At a thousand degrees, all the wood, paper, plastic, and living matter would burn and turn into gas, and the only solid materials that could be swept up would be rock and metal, which could be collected into a small pile, with a high mean density. At the normal temperatures we live in, the sweeping machine would collect not only the rock and metal, but also the wood, plastic, paper, mice, insects, and people, making the pile much larger and the density lower. At colder temperatures, the water in the air would freeze and ice would be added to the pile, making it even bigger and less dense. And at still colder temperatures the carbon dioxide, oxygen, and nitrogen would also become solid, making the pile larger still and even less dense. Volatility is an important control on what can be accreted.

Evidence from Meteorites

While differential volatility explains the gross differences between inner and outer planets, we would like to have more information about their compositional differences. One approach is to look at our own planet—rocks exposed on Earth. Measuring the densities of the rocks at Earth's surface we find they have densities between 2.7 and 3.0 gm/cm^3, far less than the uncompressed densities of all the inner planets (e.g., Earth's uncompressed density is 4.2 gm/cm^3). The inner planets' high densities mean there must be materials denser than rocks in the planetary interiors.

More constraints on planetary interiors come from the compositions of meteorites that fall from the sky. Most meteorites are pieces broken off from asteroids during violent collisions with one another. Since the asteroids were broken apart, the meteorites include both surface and interior fragments. In addition, a few rare meteorites have been shown to be hunks of rock blasted off the surface of the moon, and one or two are believed to be rocks blasted off the surface of Mars by the impacts of large objects from the asteroid belt. Meteorites provide invaluable information about the compositions of planetary interiors and also the exterior portions of Mars.

Our collections of meteorites come not only from objects whose fiery passages through the atmosphere have been seen by humans but also from the Antarctic ice sheet. The white ice collects the black meteorites on the surface, and they are carried by the ice for long periods of time.

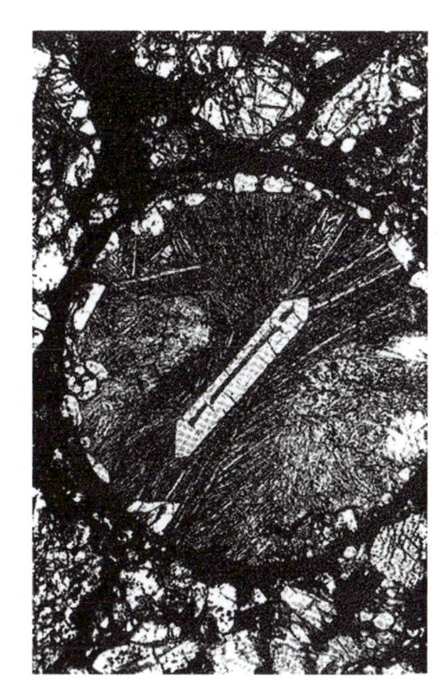

Fig. 5-3: Chondrule from a chondritic meteorite. Tieschits H-3 chondrule courtesy of Martin Prinz, American Museum of Natural History.

In certain places wind scour and evaporation remove more material than falls as snow. Meteorites contained in the ice are left on the surface as a lag, just as pebbles are left behind in desert areas whose sands are being swept away by strong winds. Over the last decade thousands of meteorites have been recovered from Antarctica by expeditions of American and Japanese scientists.

Two crucial characteristics of meteorites help to advance our understanding of the early solar system. First, meteorites are the oldest known objects whose age we are able to measure directly, and their age is consistent with that of Earth. Second, about eighty percent of meteorites appear to be fragments of solids from the solar nebula that somehow avoided being accreted into planets. They then provide us with primordial planetary material.

Evidence for this assertion comes not only from their age, but also from a very characteristic texture that is never found in rocks on Earth. The characteristic texture is the presence of millimeter-sized spheres called *chondrules* (Fig. 5-3). These chondrules are unique to meteorites.

Much attention has been given to these little spheres, and the conclusion is nearly universal: they were once molten rock droplets that condensed in space. Some think these drops were formed like raindrops as a hot nebular gas cooled; some think they were formed when small fragments of nebular dust were heated beyond their melting point. No one, however, has conceived a reasonable way to manufacture these spheres inside a planet—they appear to have formed in the very low pressures of space. Since the meteorites containing chondrules are of similar age to Earth and the moon, and the same composition as the sun, the logical conclusion is that these materials survived unchanged since the earliest history of the solar system. Chondrule-bearing meteorites, called *chondrites*, provide us with materials that formed in the solar nebula.

As the chondrules were once liquid, they formed at the high temperatures where rock and metal can melt—>1,100°C. The matrix between the chondrules, however, appears to have formed at rather different temperatures in the various classes of chondrites. For this reason the volatile abundances of different classes of chondrites can vary. One particularly important class is the *carbonaceous chondrites* because of their high carbon content. In these meteorites, chondrules coexist with minerals that are only stable below 100°C, so they contain materials that formed in both high- and low-temperature environments and then collected together.

Further evidence for the primordial attributes of carbonaceous chondrites comes from the fact that they have compositions very similar to the nonvolatile element composition of the sun. The chemical composition of carbonaceous chondrites is compared to that of the sun in Figure 5-4. Included in this figure are only elements that do not have high volatility (i.e,. not H, He, the noble gases, etc.). Not only does this agreement lend confidence to the spectral method of chemical analysis of stars, it also fortifies the conclusion that carbonaceous chondrites in particular provide a chemically unbiased sample of the solar system's less volatile elements. Thus these rock fragments likely carry the chemical information about the actual compositions that accreted to make the planets.

What are the characteristics of this composition? The most striking feature is the dominance of three metals: silicon, iron, and magnesium. As summarized in Table 5-4, these elements constitute 91% of the materials present in ordinary chondrites. Four elements—aluminum, calcium, nickel and sodium—form a group that runs a poor second, and six other elements a group that runs an even poorer third. Since all of these form minerals in combination with oxygen, oxygen is also among

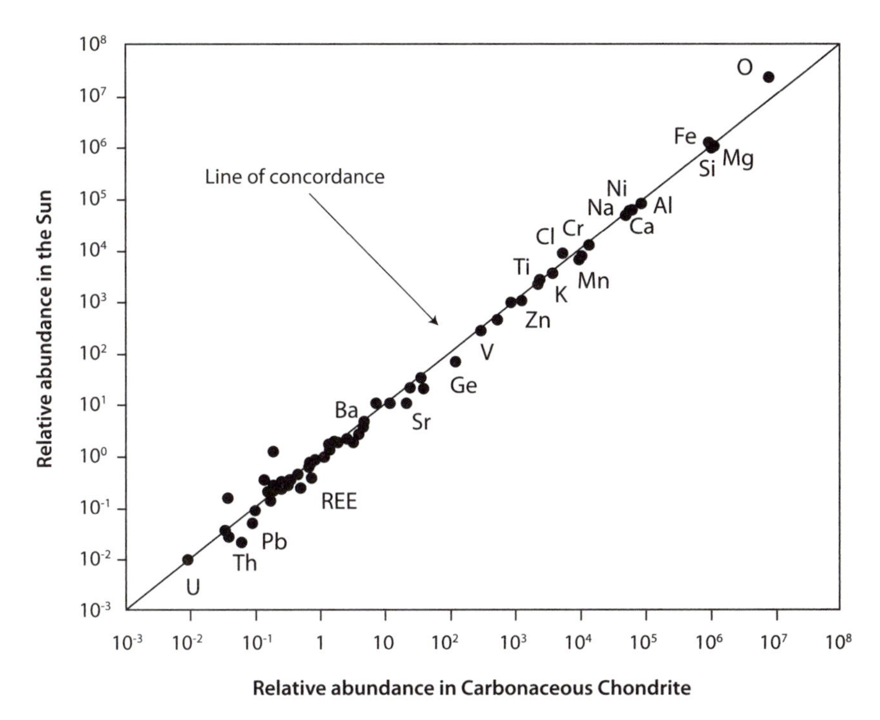

Fig. 5-4: Comparison of the relative abundances of elements of low and moderate volatility in the sun's atmosphere with those in carbonaceous chondrites, relative to 10^6 atoms of Si. For these elements, carbonaceous chondrites provide a chemically unbiased sample of bulk solar system matter except for volatiles such as H, C, N, O, and noble gases. (Data from Anders and Grevesse, *Geochim. et Cosmochim. Acta* 53 (1989):197–214, and Anders and Ebihara, *Geochim. et Cosmochim. Acta* 46 (Nov. 1982):2363–80)

the most abundant elements. These elements must then make up most of the inner planet's composition.

The chondrites also contain key clues about the origin of the high densities of the inner planets. Microscopic observation of chondrules shows that they consist of the silicate minerals olivine and pyroxene, sulfides, and iron metal. The olivine and pyroxene contain iron in the form of iron oxide, the sulfides have iron combined with sulfur. Therefore, it is apparent that in the solar nebula iron can occur in three different forms, as metal, as oxide (which combines with silicon and magnesium oxides to form silicate minerals), and as sulfide. Metallic iron has a density of 9, while silicates have a density of about 3. Varying proportions of iron in its different forms could give rise to large density differences indeed.

Table 5-4
Metallic element abundances in chondritic meteorites

Elements	% of metal atoms
Magnesium (Mg)	32
Silicon (Si)	33
Iron (Fe)	26
Aluminum (Al)	2.2
Calcium (Ca)	2.2
Nickel (Ni)	1.6
Sodium (Na)	1.3
Chromium (Cr)	0.40
Potassium (K)	0.25
Manganese (Mn)	0.20
Phosphorus (P)	0.19
Titanium (Ti)	0.12
Cobalt (Co)	0.10

The importance of both rock and metal is confirmed by study of other meteorites that do not contain chondrules, called *achondrites*, or "differentiated meteorites." Some of these meteorites, the basaltic achondrites, are lavas that clearly have undergone melting and processing in planetary objects in the early solar system. These rocks are rather similar to volcanic rocks erupted on Earth. Another class of nonchondritic meteorites is made of iron metal. When cut and polished, these objects show beautiful hexagonal patterns consisting of alternating bands of Fe-Ni alloys (Fig. 5-5) unlike anything that can be found or manufactured on Earth. Metallurgists recognize this pattern as one that forms when an alloy of iron and nickel is cooled very slowly—but they are unable in the laboratory to cool anywhere near as slowly as would be required to give rise to the patterns seen in the iron meteorites. Such slow cooling is what would be expected deep in the interior of a planetesimal where a core of metal was surrounded by a thick insulating blanket of rock.

The separate classes of silicate achondrites and iron meteorites if combined back together would be similar in composition to the chon-

Fig. 5-5: Polished section of an iron meteorite showing the characteristic banding of Fe-Ni alloys, called a *Widmanstätten pattern*. Such patterns are observed only in meteorites and result from a very slow cooling rate in the cores of ancient planetesimals. (Photo courtesy of Harvard Museum of Natural History)

drites. In fact, if we were to separate the metal out of chondrites with a magnet, we would create the compositions of two of the major classes of differentiated meteorites—those made almost exclusively of iron and those made up only of silicates.

These observations collected together suggest a rather straightforward scenario—that metallic and silicate achondrites were formed through the melting of chondrites. Because liquid iron and liquid silicate cannot mix together—like oil and water—the denser metal separates downward and the lighter silicate floats on top, creating a differentiated planetary object with a metallic core and a silicate mantle. Breakup of these objects would then lead to the creation of the achondritic meteorites.

The actual proportions of silicate and metal depend on the proportions of iron in its three stable forms: as metal, as iron oxide, and as iron sulfide. The amount of iron in each of these forms depends on the amounts of silicon, magnesium, oxygen, and sulfur available. The oxygen first combines with the Si and Mg. The available sulfur, which is relatively minor in abundance, combines with iron to make iron sulfide. If there is any oxygen left over it combines with Fe to make FeO. The remainder of the iron is in the metallic state. Therefore, a gradation in oxygen content leads to different proportions of iron oxide to sulfide plus metal in chondrites. In this way solar system processes could lead to planets with slightly different densities, depending on the fraction of Fe tied up in silicate or metal.

These clues then lead us to be able to interpret the relative densities of the inner planets. The high densities of the inner planets relative to the silicate rocks we find at the surface could be due to deep-seated metal that separated to the planet's interior to form a planetary core. Mercury, with the highest density of the inner planets, would have proportionately the largest core. Mars, with the lowest density, the smallest core. This scenario then gives a first-order account of many important observations—the compositions of chondritic and achondritic meteorites and how they tie in to the compositions of the four inner planets.

Scenario for Solar System Creation

These various lines of evidence, combined with increasingly detailed observations of nascent planetary systems in interstellar clouds and improved modeling thanks to increased computer power, provide an overall scenario for the early history of the solar system (Fig. 5-6).

When matter contracts toward the center of a protoplanetary disk, it heats up because of the conversion of gravitational energy to heat. This leads to a radial distribution of temperature, where material near the future star is hotter than material that is farther away. Recent estimates put the temperature at this early stage of solar system formation in excess of 1,000°K at the distance of Earth from the sun, and as low as 200°–100°K at the distances of Jupiter and Saturn. At the high temperatures appropriate for Earth, no volatiles would be in the solid state, and the dust would be made up of silicate and metal. Farther from the sun, the temperature was low enough that volatile elements would precipitate as ices. The two are separated by a "snow line" controlled by the temperature distribution around the Sun.

The solid particles within the disk then begin to stick together, forming small solid objects (a very small number of which were preserved to become meteorites and comets much later in solar system history). In a few tens of thousands of years, irregular solid objects varying in size

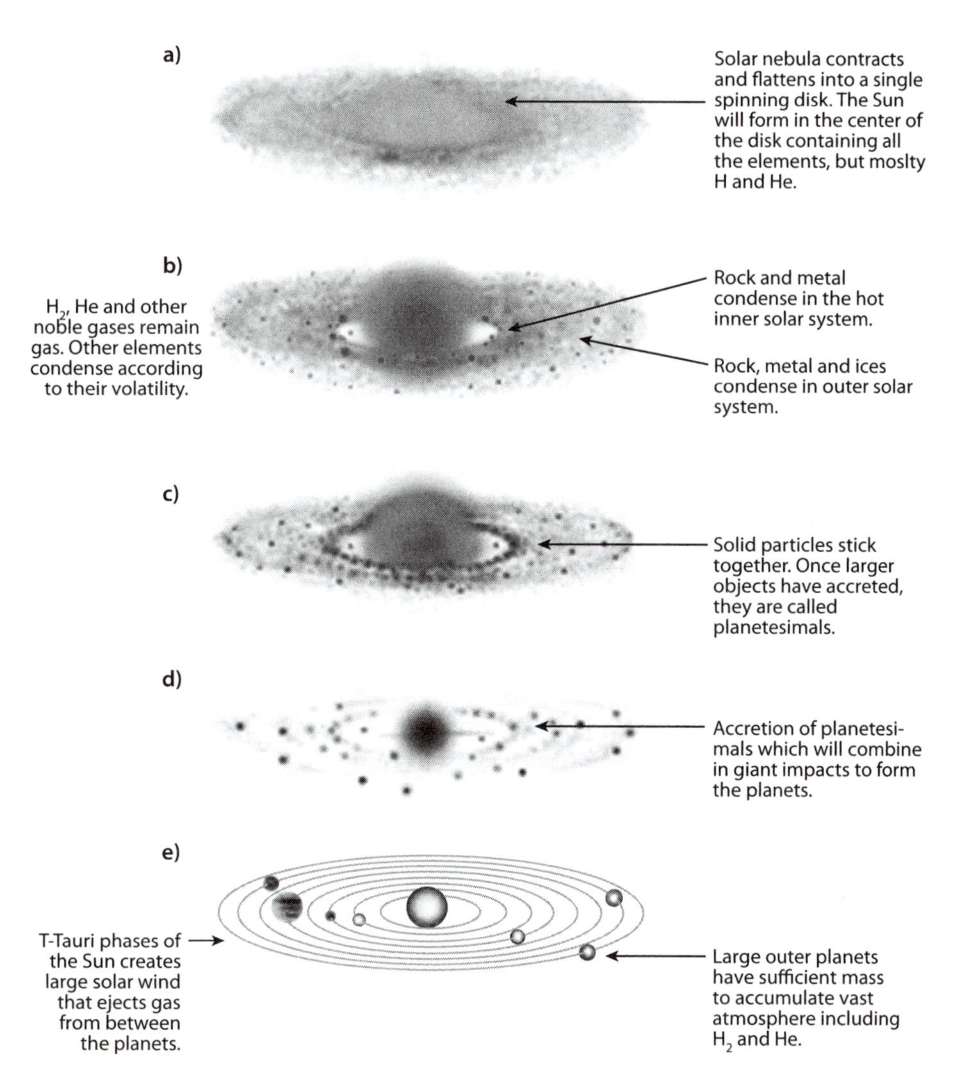

Fig. 5-6: Schematic illustration of the steps in solar system formation from an initial nebula of dust and gas, to a collection of solids in the nebula, to the formation of planets and the ejection of excess gases from between the planets. Modified from Lifengastronomy web (lifeng.lamost.org)

from 1 to 10 km likely appeared. Such small accumulations, called *planetesimals*, all rotated in the same direction around the Sun and hence collided gently with one another to make larger and irregular objects. The largest ones would then have had sufficient gravitation to attract the smaller, and further growth no longer depended only on accidental impacts. This led to more rapid growth through increasingly energetic impacts to form the *protoplanets*—those objects that would eventually combine to become the planets. The later stages of solar system formation would have been marked by giant impacts among the protoplanets, which as we shall see in Chapter 8 had far-reaching consequences for Earth.

The outer reaches of the early solar system had a much higher quantity of solid material, because the lower temperatures made many of the volatile compounds solids. The protoplanets of the outer solar system were thus substantially larger than present-day Earth. These objects then had such a large gravitational field that they could attract and retain an immense gaseous atmosphere. Does this mean that Jupiter lacks iron and silicate? Not at all—the large circumference and gravitational attraction made Jupiter actually accumulate *more* silicate and metal then Earth; estimates are that Jupiter has a mass of silicate and metal 30 times the mass of Earth. This giant also accumulated the vast amounts of ices as well as hydrogen and helium leading to its low density and a total mass some 300 times that of Earth. Similar events happened for the other outer planets.

At this early time the solar system remained a very crowded space. There may have been many dozens of protoplanets in the inner solar system, many more planetesimals, and even vaster numbers of comets reflecting perturbed orbits of icy planetesimals from the outer reaches of the solar system. Small impacts of the planetesimals and huge impacts of the protoplanets eventually gave rise to single "winners," the surviving planets. Of course, small numbers of impacts with comets and asteroids would continue, and indeed continue today.

There is a record on two of the inner planets of these final huge impacting events. As we will see in more detail in Chapter 8, a late impact of a Mars-sized object (about one-tenth Earth mass) with the proto-Earth likely led to the creation of the moon. Such an impact would explain why the moon has no core and has a composition that looks very much like Earth's mantle, and why the moon is so large relative to the size of Earth. Mercury also seems to reflect a late large impact. If the Earth impact were grazing but the Mercury impact direct, then in one case the moon might have formed, and in the other much of the silicate mantle might have been lost to space, explaining the large core and high density of Mercury.

Two difficulties were encountered in early versions of the solar system scenario:

> (1) One of the most puzzling properties of the solar system is the distribution of angular momentum. Just as a skater spins rapidly by starting to spin slowly and then collapsing her limbs around her body, the collapse of the solar nebula to the center should have transferred most of the rotation of the nebula to the center and made the sun spin very fast. The planets, staying far from the center, should orbit slowly. Yet while the sun has more than 99.9% of the solar system's mass, it has only 2.0% of the total angular momentum. How could the sun be spinning so slowly?
>
> (2) The second and apparently unrelated difficulty was that the amount of solid dust that could accumulate to form planets made up just 0.2–2.0% of the mass of gas. Hydrogen and helium were by far the most abundant elements, and yet would never be in the solid state. Space between the planets is now empty—what happened to the most abundant elements that must have been the dominant components of the solar nebula?

A possible solution to both problems has been proposed based on observations of the early life cycle of stars. During a small star's earliest history, prior to the onset of fusion of hydrogen to make helium, violent winds emanate from the star that would push away the gas and particles that surround it. This effect, named *T-Tauri* after the first star where it was found, would have ejected a significant amount of mass from the early sun. Just as the skater stops her spin by throwing out her arms, the outward movement of material from the sun causes the spin of the sun to slow down. The T-Tauri wind also might blow away in very short periods of time the remaining gas and dust that have not accreted into protoplanets. The solar system would then be left with early planets in gasless space, orbiting a slow-spinning sun. This idea will be able to be tested further by more detailed analysis of stellar nurseries such as the

Orion nebula (see frontispiece), where solar systems in various stages of formation can be observed.

We have called this section a "scenario" because the understanding of formation of planetary systems is complex and occurred some 4.5 billion years ago. The meteorite record is incomplete; we have few samples from other planets; there is little data from comets; and modeling of such a complex process requires major simplifications and assumptions. New wrinkles in solar system formation come from increasingly complex models of accretion that show large numbers of protoplanets and suggest that early in the history of a solar system planets are not formed in fixed and stable orbits, but tend to migrate toward or away from their star. In particular the outer planets of our solar system are proposed to have migrated outwards from the sun in early solar system history. In other models, large planets can migrate inward toward their star, wreaking gravitational havoc in some early solar systems. It is heartening that new data from space-based telescopes have now identified other planets and planetary systems forming around other stars. While the first discovered planets were all massive, like Jupiter, the most recent results have found smaller planets as well (see Chapter 21). These new discoveries are for the first time giving us other examples of solar systems where solar system formation can be observed at various stages of development in other parts of the galaxy. Rather than having to rely only on inferred events billions of years ago, new constraints will come from direct observation of other solar systems. The interplay between models and observations will lead to substantial advances in coming years.

Understanding the Chemical Compositions of the Terrestrial Planets

We are now in a position to try to understand in more detail the compositions of the inner planets and how they came about. Let us run through the list of elements in order of increasing atomic number in Table 5-5 and see why inner planetary compositions are dominated by relatively few elements.

The first element on the list is hydrogen. Most hydrogen is in the form of hydrogen gas, while some is present as gases of carbon (CH_4, CHN), of nitrogen (NH_3), or oxygen (H_2O). Earth and its fellow terrestrial planets accreted where none of these gases were solid. Only a small fraction of the least volatile of them arrived with later impacts from materials formed in the outer solar system. Therefore hydrogen should be scarce.

Helium exists only as a gas and furthermore is so light that it can escape from the top of Earth's atmosphere even today. The very small amounts of helium we find today are derived mostly from the radioactive decay of uranium and thorium.

The next three elements—lithium, beryllium, and boron—were produced in very small abundances by nucleosynthesis in stars. Their overall abundance in the universe is too small to permit them to be major constituents of planets.

Carbon and nitrogen, in the presence of the large amounts of hydrogen gas in the planetary nebula, would have been in the form of CH_4, NH_3, and CO—gaseous compounds that did not accrete.

The element oxygen is even more strongly attracted to the various metals than it is to hydrogen. In the nebular cloud there were five times as many oxygen atoms as all metal atoms taken together. Most of the oxygen combined with hydrogen and carbon, but there was enough oxygen left that most metals combined with oxygen to make oxides. Because most metals combine with oxygen, oxygen is present in great abundance and became a major planetary constituent.

After oxygen on the list in Table 5-5 come fluorine and neon. Fluorine is volatile and has a strong tendency to combine with hydrogen in the form of hydrofluoric acid (HF), a molecule that is also volatile under the conditions of the inner solar system. Neon is a noble gas like helium, remaining always in the gaseous state and therefore would not accrete.

So of the first ten elements, six formed gases and were largely lost. Three others had such small universal abundances as to be unimportant. Only oxygen was sufficiently abundant and prone to form solid phases that it became a major contributor to the terrestrial planets.

The next five elements on the list are all metals that prefer chemical unions with oxygen. Four of them (magnesium, aluminum, silicon, and phosphorous) were in solid form, while sodium is moderately volatile and thus less abundant. Both silicon and magnesium are alpha-particle nuclides produced in greater abundance in stars, and hence they are far more abundant than sodium, aluminum, and phosphorous. SiO_2 and MgO are major planetary constituents.

Table 5.5
Relative abundances of the first 28 elements and their fates during the formation of the terrestrial planets

Element number	Element name	Solid	Gas	Relative abundance in Sun*	Fate**	Relative abundance in Chondrites
1	Hydrogen		H_2	40,000,000,000	(1)	—
2	Helium		He	3,000,000,000	(1)	Trace
3	Lithium	Li_2O		60	(3)	50
4	Beryllium	BeO		1	(3)	1
5	Boron	B_2O_2		43	(2)	6
6	Carbon		CH_4	15,000,000	(1)	2,000
7	Nitrogen		NH_3	4,900,000	(1)	50,000
8	Oxygen		H_2O***	18,000,000	(2)	3,700,000
9	Fluorine		HF	2,800	(1)	700
10	Neon		Ne	7,600,000	(1)	Trace
11	Sodium	Na_2O		67,000	(2)	46,000
12	Magnesium	MgO		1,200,000	(3)	940,000
13	Aluminum	Al_2O_3		100,000	(3)	60,000
14	Silicon	SiO_2		1,000,000	(3)	1,000,000
15	Phosphorus	P_2O_5		15,000	(3)	13,000
16	Sulfur	FeS	H_2S	580,000	(2)	110,000
17	Chlorine		HCl	8,900	(1)	700
18	Argon		Ar	150,000	(1)	Trace
19	Potassium	K_2O		4,400	(2)	3,500
20	Calcium	CaO		73,000	(3)	49,000
21	Scandium	Sc_2O_3		41	(3)	30
22	Titanium	TiO_2		3,200	(3)	2,600
23	Vanadium	VO_2		310	(3)	200
24	Chromium	CrO_2		15,000	(3)	13,000
25	Manganese	MnO		11,000	(3)	9,300
26	Iron	FeO, FeS, Fe		1,000,000	(3)	690,000
27	Cobalt	CoO		2,700	(3)	2,200
28	Nickel	NiO		58,000	(3)	49,000

**Relative to 1,000,000 silicon atoms.*

***(1) Highly volatile; mainly lost; (2) moderately volatile; partly captured; (3) very low volatility; largely captured.*

**** Plus metal oxides.*

Next on the list is sulfur (S). Its situation is akin to oxygen. It can form the gas H_2S and also combine with iron to form a solid FeS. The evidence from meteorites is that a significant proportion of S would have been captured in combination with iron.

The next two elements on the list, chlorine and argon, were largely lost as gases. Chlorine was in the form of the volatile hydrochloric acid, HCl, while argon is a noble gas. Then come two more metallic elements, potassium and calcium. Calcium in the oxide form has a very low volatility. Potassium like sodium is moderately volatile and hence was less efficiently captured. (Despite its low abundance, potassium has an important role in Earth studies, however, because one of its isotopes, ^{40}K, is radioactive).

So we see that in the second group of ten elements, five of them (Mg, Al, Si, S, Ca) were largely captured. Three were partially captured and two were lost. Of the captured five, Mg and Si are ten to twenty times more abundant in the solar nebula than Al, Ca, and S and hence make up far greater proportions of planets.

Between calcium and iron there is a big sag in the abundance curve of elements created by nucleosynthesis (see Fig. 3-10). Although most of the elements in this interval are metals of low volatility that combine with oxygen; none has a sufficiently high cosmic abundance to be particularly important. In contrast, the abundance of iron, the ultimate product of nuclear fires, stands well above that of its neighboring elements and is close to Mg and Si in cosmic abundance. It occurs in three chemical forms, none of which are particularly volatile. As its cosmic abundance is similar to Mg and Si, it is also one of the major constituents of the inner planets.

Beyond iron, the abundance of the elements drops rapidly with increasing proton number. Only nickel is sufficiently abundant to be important.

Thus we see that overall chemical abundances are controlled by a combination of nuclear physics, which determines the cosmic abundances of the elements, and inorganic chemistry, which determines the molecular combinations and volatility of chemical compounds. Rocky planets like Earth consist primarily (>90%) of the "big four" elements—O, Mg, Si, and Fe. A secondary group that makes up most of the rest consists of Ca, Al, Ni, and S.

The relative abundances of most of the remaining elements in the table are strongly influenced by their volatility. Figure 5-7 shows how

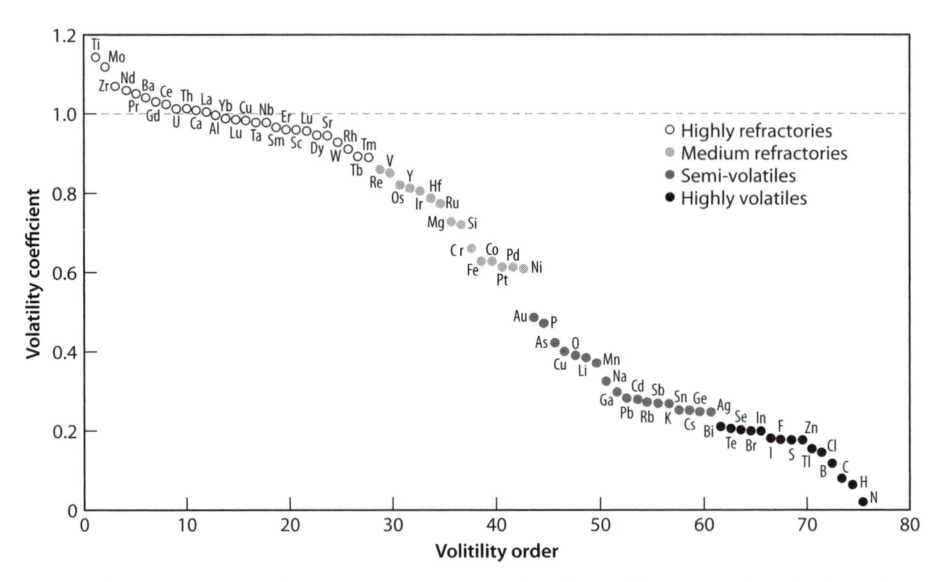

Fig. 5-7: Depletion of volatile elements in ordinary chondrites. The greater the volatility of the element, the more it was depleted during solar system processes that led to the creation of the ordinary chondrites. The terrestrial planets underwent even more severe volatile element depletion. (C. J. Allègre personal communication)

the ordinary chondrites compare to the carbonaceous chondrites that appear to have formed further out in the solar nebula, where they retained greater proportions of the available volatiles. Elements that are highly refractory, such as Mg, Ca, Al, and Ti have abundances that are very similar in both types of chondrites (i.e., ratios very near 1). Elements of progressively greater volatility are increasingly depleted in the ordinary chondrites.

From what we know of the compositions of the inner planets, they all have similar relative proportions of the most refractory elements, and the volatile elements are all depleted relative to carbonaceous chondrites. In detail, however, the specific amount of volatile depletion is quite variable. A proxy for this variability is the ratio of a slightly volatile element, K, to a highly refractory element, U. The elements K and U are particularly useful for this purpose, because both of them tend to travel together during geochemical processes, and both have long-lived radioisotopes that emit gamma rays. These powerful electromagnetic radiations can be detected by instruments dropped toward the planetary sur-

Table 5-6
K/U ratios for solar system materials

Component	K/U ratio
Venus	7,000
Earth	12,000
Moon	2,500
Mars	18,000
Ordinary chondrites	63,000
CI Carbonaceous chondrites	70,000

face. Hence, unmanned spacecraft landed on Venus have been able to send back estimates of the K/U ratio of the surface rocks, which can be compared to direct measurements of chondrites, Earth, and the moon, as well as to meteorites that are believed to have come from Mars. These ratios are shown in Table 5-6, from which it is apparent that the extent of volatile depletion can be quite variable.

The table is put in sequence of increasing distance from the sun. Formed farthest from the sun and with the highest ratio are the carbonaceous chondrites. Ordinary chondrites are only about 10% depleted in K relative to U. On the basis of current data, which is sparse for Mars and very uncertain for Venus, the three inner planets then become progressively more depleted in K passing from Mars to Earth to Venus. This result is consistent with an increasing nebular temperature toward the sun. Volatile depletion explains why Earth's Na content, for example, is about 10% of its Ca content, despite the similar abundance in chondrites evident in Table 5-3. The moon stands out as being very volatile depleted, an important fact that bears on its origin, as we shall see in Chapter 8.

The greatest uncertainties in inner planetary compositions relate to the origin of the most highly volatile elements. Noble gases are always in the gaseous state in the solar nebula, and yet they are present on Earth in small but significant amounts. Furthermore, the ratios of Earth's noble gas isotopes are clearly not the same as those of the solar wind, so direct trapping of gas from the nebula is not a possibility. While CO_2 is of minor abundance in the atmosphere, the total amount on Earth, now mostly residing in carbonate rocks, is substantial, as is the amount of water, as discussed at length in Chapter 9. The appearance of these volatiles on Earth is essential for the subsequent development of life, and hence is of more than minor interest. One possibility that seemed promising was that cometary impacts and the dust from comet tails may have contributed to the total volatile budget. This seemed evident given calculations that there would be huge numbers of comets in Earth-crossing orbits in the early solar system.

A test of this idea became possible with recent missions to comets that would enable us to learn more details of their composition. One important measurement is the ratio of the two stable isotopes of hydrogen, ^{2}H (deuterium) and ^{1}H. The ratio $^{2}H/^{1}H$ is referred to as the *D/H ratio*. The mass difference between these isotopes is so large that they can become separated during chemical processes, and the D/H ratio in various solar system materials varies by many tens of percent or more. If comets were the source of Earth's volatiles, comets should have the same D/H ratio as Earth. But the initial measurements showed that the measured ratios in comets were not consistent with Earth's ratio. Could earlier comets have come from a different part of the solar system with appropriate isotope ratios? New measurements in 2011 showed that at least one comet has a D/H ratio consistent with Earth's value. New observations will throw increased light on the origins of Earth's vital volatiles.

There is one other perplexing aspect of element abundances that has been difficult to explain, which is the variable ratios of refractory elements such as Mg to Si to Fe in the terrestrial planets, and why these ratios can differ slightly but significantly from those observed in chondritic meteorites. Some of the variation is now thought to reflect the importance of late impacts of large protoplanets. This might explain the exceptionally high Fe in Mercury, for example, if part of the silicate mantle of the protoplanet were lost to space. The Si/Mg ratio of Earth, however, also differs from the chondritic value. Since both elements reside entirely in silicate phases at low pressures, this cannot be explained by impacts. One possibility that is being actively considered is that at the very high pressures of Earth's deep interior, some Si can be dissolved in the Fe-core, but this conjecture remains unproven. Hence, while the broad points of planetary accretion are quite well understood, much remains to be discovered.

Summary

The solar system formed from a collapsing cloud of gas in an environment likely to have been very similar to the interstellar clouds where newly forming stars and planetary systems can now be observed elsewhere in our galaxy. Formation of solar systems therefore appears to be a regular and lawful process, not a random occurrence. The large differences between inner and outer planets can be explained well by the different thermal environments that existed in the early solar system. The hot regions of the nebular cloud close to the sun would precipitate solids made up only of metal and silicate, while the colder regions beyond Mars would also precipitate the major ices containing nitrogen, carbon, and hydrogen. Accretion of this dust into planetesimals and then protoplanets took place rapidly, likely followed by removal of the remaining gases by the hurricane of solar winds during the T-Tauri phase of the early sun. The protoplanets and remaining planetesimals then collided to form the planets observed today. The overall compositions of the inner planets can be well understood from the combined knowledge of nucleosynthesis, differential volatility, and increasingly detailed and quantitative models of solar system formation. Significant puzzles remain, particularly with respect to the abundances of the most volatile elements that are essential for the development of a stable climate and life.

Supplementary Readings

N. McBridge and I. Gilmour, eds. 2004. *An Introduction to the Solar System.* Cambridge: Cambridge University Press, 2004.

W. K. Hartmann, ed. 2005. *Moons and Planets*, 5th ed. Pacific Grove, CA: Thomson Brooks/Cole.

Fig. 6-0: Folds and "angular unconformities" in the geological record require great expanses of geological time. The older strata were first deposited horizontally, then buried and folded, then uplifted and eroded to create a land surface, after which the younger strata were deposited. Such evidence was used by James Hutton to enunciate his principle of "no vestige of a beginning, no prospect of an end." This image is of the Carboniferous-Triassic unconformity on the southwestern coast of Portugal. (Reprinted by permission of Filipe Rosas)

CHAPTER 6

The Schedule

Quantifying the Timescale with Radionuclides

Understanding any process requires knowledge of the time involved. Because our timescale as human beings is limited to centuries, early timescales for Earth and universe were thousands of years, time that seemed unimaginably long to human experience and imagination. Even Newton and Descartes, giants of modern physics, believed Earth formed in roughly 4000 BC. Early geologists, however, looking at the evidence apparent in the rocks around them, questioned these ideas. They noted that processes operating in the present could give rise to observed rocks and features of the landscape, but since these processes occurred very slowly, long times must be involved—perhaps many billions of years. Nineteenth-century physicists disputed these claims by calculating that Earth's current heat flow was not consistent with such a long timescale. The discovery of radioactivity provided an additional heat source for Earth's interior and revolutionized studies of Earth's history by permitting quantitative dating. Radiogenic "parent" isotopes decay to stable "daughter" isotopes with a characteristic half-life—the time it takes for one-half of the parent atoms to decay. Measuring isotope ratios, then, constrains time. Long-lived radionuclides, such as uranium, thorium (Th), potassium, and rubidium (Rb), permitted the dating of ancient events and showed that the meteorites and Earth all formed as a group some 4.55 billion years ago. Some radiogenic parents are created in the atmosphere by cosmic rays today. The most famous of these is ^{14}C*, which with a 5,730-year half-life is used to date very recent events.*

Radioactive isotopes constrain time in other ways as well:

(1) Theoretical calculations of the production rate of elements in stars, coupled with measurement of present-day abundances of the uranium and thorium isotopes, permit calculation of the age at which stars began to distribute elements in the galaxy. The timescale of roughly 10 billion years prior to solar system formation is consistent with the age of the universe inferred from the Big Bang.

(2) Extinct radionuclides, those with short half-lives so that the parent isotope no longer exists, constrain events in the early history of the solar system. ^{26}Al*, a radioactive isotope of Al created in supernovas with a half-life of less than one million years, produced isotope variations in its daughter Mg isotope that persist in chondritic meteorites today. The presence of* ^{26}Al *in the solar nebula when meteorites formed indicates that supernovas exploded nearby at the initiation of the solar system. This suggests solar system formation within an interstellar cloud where many stars were forming and exploding. Extinct radionuclides would also have been an important heat source in the early solar system, and may have facilitated rapid planetary heating and differentiation.*

Radioactive dating provides the essential timescale to understand Earth's development as a habitable planet. It allows us to determine when Earth began, when life began, the rates at which life evolved and Earth's tectonic plates move, and the duration of Homo sapiens on the planet. All the subsequent chapters in this book will make use of this timescale, deduced from sophisticated measurement of the decay products of minute atomic nuclei largely created in exploding stars. The quantification of geological time reveals that there is a long path from planetary accretion to advanced life.

Introduction

Knowing what happened in the evolution toward a habitable planet requires a timeline. What is the appropriate timescale? Is it thousands, millions, or billions of years? When exactly did events happen? How long did they take? What is the sequence of events occurring at different places? All such questions, fundamental to the unraveling of Earth's history and understanding our planet, depend on time.

All of us experience time personally, and we have a sense for how long things take and what "long" and "short" times are. Our personal experience is subjective—for most people time seemed to pass more slowly when they were children than as adults, and even as adults periods of intense activity, rich impressions, or great discomfort can make short periods of time feel longer. Objective time is also evident to us, as measured by the movements of the sun, moon, planets, and stars. A sense of time, of history, of ancestry, is intrinsic to us as human beings. Legends of our ancestors' ancestors and various creation myths about the beginning of time have been an integral part of most cultures.

The modern recognition of the vast billion-year expanses of time depends on quantitative dating with radioactive isotopes, fully developed only in the second half of the twentieth century. The first precise age of Earth was not made until 1956. While this age has been abundantly confirmed, meriting a 10 on our theory scale, casual internet search of "age of the earth" reveals the intense activity by some groups, particularly in the United States, to convince others that geological time is limited to a few thousand years. The thousand-years timescale comes from a careful reading of the Bible's "begat" paragraphs in Genesis (e.g., Adam begat Cain and Abel and so on). The most precise estimate was declared by Bishop James Ussher in 1640: the world began at 9 a.m., on a Monday, October 23, 4004 BC. Most scientists of the seventeenth century believed in this timescale. Descartes, for example, a great early physicist and one of the founders of modern science, supported the few-thousand-year timescale, and Newton did as well.

Then the geologists got involved. By the eighteenth century it was recognized that many rocks now on land formed under water. The new study of fossils also showed that many of fossils had no living representatives. Geologists observed processes creating rocks today similar to those observed in older strata, and the observed processes were very slow. Sediments were formed by slow deposition of erosion products carried by rivers. Vast thicknesses of volcanic lava were made up of thousands of flows, each one of which had the same appearance as recent flows

Fig. 6-1: A further, famous example of "angular unconformities" in the geological record, from the Grand Canyon of the western United States. Older strata with no fossils are overlain by younger, flat-lying, fossiliferous strata. See also color plate 1. (Geology pictures by Marli Bryant Miller, University of Oregon)

known in human history. Relative times could be determined, since lavas or sediments at the bottom of a formation that is right side up are the oldest. Folded rocks underlying flat-lying strata must be still older (Figs. 6-0 and 6-1). This led James Hutton, in his classic work of 1788, to suggest a vast expanse of geological time, with "no vestige of a beginning, no prospect of an end." Fifty years later, Charles Lyell enunciated the *principle of uniformitarianism*—that processes observable today, acting over long periods of time, created all the geological manifestations we see on Earth. Since the rates of these processes could be measured, it seemed likely that Earth was billions of years old. Quantitative timing of events, however, seemed an impossible task.

Responding to this challenge, physicists then took up the issue of Earth's age. From the high level of Earth's heat flow they calculated that a maximum timescale was tens of millions of years. How could the qualitative estimate of geologists be reconciled with the quantitative calculations of the physicists? Controversy reigned (see sidebar, next page).

Measuring Time with Radioactive Decay

The discovery of radioactivity provided a new internal heat source for Earth, and the understanding that solid rocks can move by convection (discussed at length in Chapter 11) provided a mechanism to actively transport heat close to the surface. Both of these contribute to the high heat flow the physicists were trying to explain, and vindicated the deductive reasoning of the geologists of the nineteenth century.

Radioactive isotopes still exist in Earth's interior only because some of the radioactive isotopes produced during nucleosynthesis take a very long time to decay. These remnants of supernovae remaining in Earth are Earth's long-lived battery power or internal heater and also provide us with Earth's timescale. Since these isotopes are present only in trace amounts in rocks, their importance is rather remarkable.

It took a long period of technological development, reasoning, and discovery to read the memory of time that was contained in the radioactive isotopes and their daughter products. While it was known at the turn of the century that uranium decayed to lead, isotopes were not discovered for another twenty years, and without knowing that elements consist of various isotopes, reliable dating was not possible. Equipment to accurately measure the abundances of the isotopes (mass spectrometers) did not undergo major improvement until World War II. After 1950, these new machines opened up the world of isotope geochemistry, and radioactive dating became a growth industry. The radioactive elements and their daughter products contain the memory of what happened when—the task of the isotope geochemist is to learn how to read that memory.

Dating with radioactivity is possible because radioactive elements decay regularly. Any one atom has a fixed probability of decay. Because there are such huge numbers of atoms, this means that a fixed proportion of decays occurs in each time period, and the number of atoms decays exponentially. This behavior is conveniently described by the

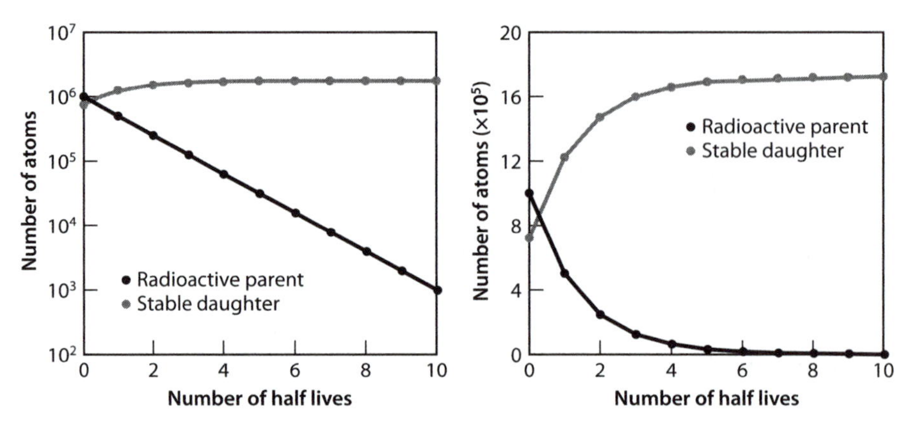

Fig. 6-2: Illustration of how parent and daughter isotopes change during radioactive decay. The choice of initial numbers of atoms is arbitrary. The left diagram shows the log linear relationship from exponential decay, 99.9% of which occurs after ten half-lives. Right-hand diagram shows changes on a linear scale.

half-life of a radioisotope. The half-life is the time it takes for one half of any population of the parent isotope to decay to its daughter product. After one half-life, half the atoms have decayed, after two half-lives three quarters have decayed, and so on (Fig. 6-2). This is conveniently illustrated on a semilogarithmic diagram where a constant half-life leads to straight lines for the radioactive parent isotope (Fig. 6-2a).

THE NINETEENTH-CENTURY DEBATE OVER THE AGE OF THE EARTH

The geological evidence for an ancient Earth ran into strong headwinds from nineteenth-century physicists, and particularly Lord Kelvin, the most famous and esteemed English scientist of his day. Physicists had developed precise equations to model the flow of heat, and Kelvin was able to make a sophisticated and accurate calculation of the cooling of Earth. If a body the size of Earth were at the highest temperature one could imagine, completely molten, how long would it take to cool enough to account for the present-day thermal gradient at Earth's surface? His calculation came out to a maximum of 20 to 40 million years. He did another calculation for the sun and came up with 50 million years, in rather good agreement with his estimate for Earth. The certain and definite conclusion, based on rigorous theory, data from Earth's heat flow, and quantitative calculation was that the geologists with their "qualitative deductions" were simply wrong.

Even in the face of the rigorous arguments of the physicists, the geologists would not back down, and this led to a lively and lengthy debate. Two developments led to the refutation of Kelvin's conclusions. The discovery of radioactivity and nuclear fusion showed that there was another source of heat for the sun and for Earth's interior. And the discovery that at high temperatures rocks could flow (convect) permitted much higher temperatures closer to the surface, accounting for Earth's relatively high heat flow. These permitted an old age and also a much more constant flow of heat through time, permitting long-term stability of the temperatures of surface rocks.

If you were transported back to Lord Kelvin's laboratory and told him that atoms were not immutable objects but that iron could be transformed to gold, that the heat source of the sun was something unknown and unimaginable to him, and that stars created the elements, what would have been his reaction? These statements would seem impossible and nonscientific based on knowledge at that time. It was not that Lord Kelvin was making faulty calculations or conclusions. There were simply other forces at play that were beyond his knowledge, and made the impossible possible. Will similar stories be told about the great scientists of today a hundred years from now?

The debate between Kelvin and the geologists did have the positive result that as the calculations were questioned and refined, geologists made more careful observations and tried to quantify their age assessments. The debate also led to significant improvements in the quantitative calculations of the cooling of solid objects, which had important practical applications in the late 1800s. Ultimately, physics and geology converged as the quantitative understanding of convection and radioactive dating led to a consistent understanding of Earth's longevity.

Notice in Figure 6-2 that after ten half-lives only one atom per thousand remains of the parent isotope, and very few atoms are produced by further decay. This limits the utility of radioactive isotopes to about ten

Table 6-1
Radionuclides of stellar origin found in meteorites

Radionuclide	Half-life (yr)	Stable daughter product
^{40}K	$1.25 \cdot 10^{9}$	^{40}Ca and ^{40}Ar
^{87}Rb	$48.8 \cdot 10^{9}$	^{87}Sr
^{138}La	$1.04 \cdot 10^{11}$	^{138}Ce and ^{138}Ba
^{147}Sm	$1.06 \cdot 10^{11}$	^{143}Nd
^{176}Lu	$3.5 \cdot 10^{10}$	^{176}Hf
^{187}Re	$4.6 \cdot 10^{10}$	^{187}Os
^{232}Th	$1.401 \cdot 10^{10}$	^{208}Pb
^{235}U	$0.7038 \cdot 10^{9}$	^{207}Pb
^{238}U	$4.4683 \cdot 10^{9}$	^{206}Pb

half-lives. If we want to study processes over billions of years, such as the age of Earth, we need isotopes with half-lives of at least hundreds of millions of years. Most radioactive isotopes produced in stars have very short half-lives and decay rapidly back to the band of stability. Fortunately, there are also a few long-lived radioactive isotopes, such as ^{238}U, ^{235}U, ^{87}Rb, ^{40}K, and ^{147}Sm, that have long half lives (see Table 6-1), and substantial numbers of parent isotopes persist throughout the processes of formation of molecules, accretion of planets, and later planetary processes.

The long-lived radioisotopes are useful for long timescales, but they are problematic for short timescales (e.g., thousands of years) because not enough decay takes place to be measurable. For shorter timescales, elements with short half-lives are necessary. The choice of isotope tool depends on matching the half-life with the process of interest. It would seem we would be out of luck with short timescales because all the short-lived radioisotopes produced in stellar interiors would have decayed away long ago. As luck would have it, some elements that do not have long half-lives are produced continually by cosmic radiation in Earth's atmosphere. These radioactive isotopes are called *cosmogenic radionuclides*, they are all around us and in us, and they are of great use to study relatively young events on Earth.

It is also apparent upon reflection that simply measuring the number of parent and daughter atoms does not provide age information. For example, if we measure a million atoms of the parent isotope, has no time passed because the rock started with a million atoms? Or have ten half-lives gone by because we started with a billion atoms? And if we measure the number of daughter atoms, how many were there to start with and how many have been produced by radioactive decay? For a precise date we need to know two things—the number of atoms at the beginning and the number now.

The most famous cosmogenic radionuclide is ^{14}C, with a half-life of 5,730 years that decays to the stable isotope ^{14}N. ^{14}C generated by cosmic radiation today mixes with the much more abundant ^{12}C and behaves chemically in the same way because it has the same electron shell structure. Plants and animals incorporate ^{14}C and ^{12}C into their tissues in the proportions they are present in the atmosphere, as long as they are alive and metabolizing carbon. Once the plant or animal is dead, the ^{14}C decays away to ^{14}N, which escapes as gas to the atmosphere, and the ratio of $^{14}C/^{12}C$ in the dead matter decreases progressively with time. To date using ^{14}C we need to estimate the $^{14}C/^{12}C$ when the plant or animal died long ago. Since ^{14}C is produced continually in the atmosphere, the assumption was made that the ratio observed in Earth's atmosphere has stayed constant through time, which would provide a universal starting point. In that case, the ^{14}C /^{12}C ratio of the sample divided by that of the atmosphere gives the age (Fig. 6-3). ^{14}C dating is the best tool we have to date organic materials, such as bones and charcoal. The rule of ten half-lives limits the ^{14}C method to materials younger than about 60,000 years.

How solid is the assumption that the cosmogenic formation rate of ^{14}C was constant? Edouard Bard and colleagues tested this proposition by measuring another short-lived isotope system that was not subject to such uncertainties. They found that the ^{14}C ages could be off by as much as 10% for the oldest samples. Therefore, a distinction is now made between the "^{14}C age" and the real age. Note that the discrepancies are small, consistent with small changes in cosmogenic nuclide production.

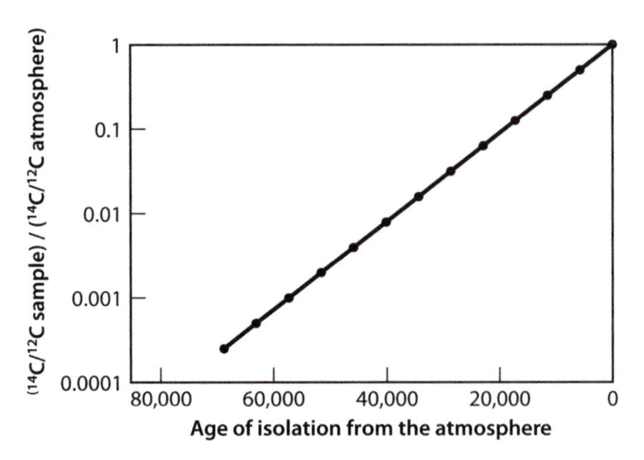

Fig. 6-3: Illustration of the principle of ^{14}C dating. The line shows the $^{14}C/^{12}C$ ratio in the sample today, assuming that when it was formed the sample had the $^{14}C/^{12}C$ ratio of today's atmosphere. After the organism stopped growing or breathing, it no longer incorporated carbon. If that happened long ago, very little remains and the sample today has a very low $^{14}C/^{12}C$ ratio. If death occurred today, the sample has the atmospheric ratio. Note the logarithmic scale of the vertical axis.

For almost all isotope systems other than $^{14}C/^{12}C$, we do not have a starting reference point like $^{14}C/^{12}C$ in the atmosphere. We do, however, have the advantage that it is possible to measure both parent and daughter isotope ratios (The ^{14}N produced by ^{14}C decay is lost as a gas). Using both parent and daughter isotopes, isotope geochemists developed another clever technique to determine accurate time information—the *isochron* method.

THE ISOCHRON TECHNIQUE OF RADIOACTIVE DATING

Can we then just measure the amount of the parent and daughter and obtain an age? Unfortunately, it is not that simple. Consider, for example, the parent/daughter isotopic system ^{87}Rb-^{87}Sr. In common terrestrial rocks there are roughly twenty times more ^{87}Sr atoms than ^{87}Rb atoms. Could all of these ^{87}Sr have been created by radioactive decay of ^{87}Rb? For that to be true, some 95% of the original ^{87}Rb would have had to decay away, since every decay of an atom of ^{87}Rb gives rise to one atom of ^{87}Sr. Decay of 95% of the atoms takes more than four half-lives. Since the half-life of ^{87}Rb is 49 billion years, that would require some 200 billion years, far older than the estimate of the age of the universe in

Chapter 2! There must have been a substantial amount of ^{87}Sr already existing prior to the start of the decay—a fact that would in any case be predicted from nucleosynthesis in stars. To date accurately it is necessary somehow to determine the isotope ratios at the starting point.

The clever method developed by isotope geochemists can be illustrated with a simple numerical example. Consider two minerals, phlogopite and plagioclase, that both start with 700 atoms of ^{87}Sr and 1,000 atoms of another isotope of Sr, ^{86}Sr, that is not produced by radioactive decay on Earth. The $^{87}Sr/^{86}Sr$ ratio of the two minerals will be exactly the same (e.g., 0.700) when they form. Because of its crystal structure, phlogopite is able to take in much more Rb than can plagioclase. In this example we will assign 1,000 atoms of ^{87}Rb to phlogopite and 100 atoms of ^{87}Rb to plagioclase. At any subsequent point in time, a fixed percentage of the ^{87}Rb in both minerals will have decayed away. When 5% of the ^{87}Rb has decayed, for example, 50 atoms of ^{87}Rb will have decayed to produce 50 atoms of ^{87}Sr in the phlogopite, while only 5 atoms of ^{87}Sr will have been produced in the plagioclase. Measurement of the minerals at this time would then yield $^{87}Sr/^{86}Sr$ of 0.750 in phlogopite and 0.705 in the plagioclase. When 10% of the ^{87}Rb has decayed, much later in time, then phlogopite has $^{87}Sr/^{86}Sr$ of 0.800 and plagioclase 0.710. Therefore, the difference in Sr isotope ratio of the two minerals increases progressively with time. Then we can deduce the starting conditions by backtracking to determine the time when the $^{87}Sr/^{86}Sr$ ratios of both minerals would have been exactly the same. This is the age of formation, illustrated in Table 6-2 and Figure 6-4.

This behavior can be represented mathematically by considering the radioactive decay equation:

$$N(t) = N_0 e^{-\lambda t} \tag{6-1}$$

where e is the constant 2.718 and λ is the decay constant for the radioactive element. $N(t)$ refers to the number of atoms at any time t, and N_0 is the number of atoms at the beginning. If we define the amount of ^{87}Sr and ^{87}Rb at the start as $^{87}Sr_0$ and $^{87}Rb_0$ and the amount today as $^{87}Sr(t)$ and $^{87}Rb(t)$ then the amount of ^{87}Sr added is related to the amount of ^{87}Rb that has decayed, and hence:

$$^{87}Sr(t) = {}^{87}Sr_0 + {}^{87}Rb_0 - {}^{87}Rb(t) \tag{6-2}$$

Table 6-2
Evolution of $^{87}Sr/^{86}Sr$ of three minerals from a uniform reservoir

Time	Time of formation	5% decay	10% decay
Phlogopite			
^{87}Rb	1000	950	900
^{87}Sr	700	750	800
^{86}Sr	1000	1000	1000
$^{87}Sr/^{86}Sr$	0.7	0.75	0.8
$^{87}Rb/^{86}Sr$	1	0.95	0.9
Feldspar			
^{87}Rb	100	95	90
^{87}Sr	700	705	710
^{86}Sr	1000	1000	1000
$^{87}Sr/^{86}Sr$	0.7	0.705	0.71
$^{87}Rb/^{86}Sr$	0.1	0.095	0.09
Pyroxene			
^{87}Rb	50	47.5	45
^{87}Sr	70	72.5	75
^{86}Sr	100	100	100
$^{87}Sr/^{86}Sr$	0.7	0.725	0.75
$^{87}Rb/^{86}Sr$	0.5	0.475	0.45

From the radioactive decay equation:

$$^{87}Rb(t) = {}^{87}Rb_0 \, e^{-\lambda t} \tag{6-3}$$

therefore

$$^{87}Rb_0 = {}^{87}Rb(t) \, e^{\lambda t} \tag{6-4}$$

Substituting from one equation to the other, we obtain:

$$^{87}Sr\,(t) = {}^{87}Sr_0 + {}^{87}Rb\,(t)\,(e^{\lambda t} - 1) \tag{6-5}$$

If we divide all terms by ^{86}Sr we obtain:

$$^{87}Sr/^{86}Sr\ (t) = {}^{87}Sr/^{86}Sr + {}^{87}Rb/^{86}Sr\ (t)\ (e^{\lambda t} - 1) \qquad (6\text{-}6)$$

$$y \qquad = \qquad b \quad + \qquad x \qquad\qquad m$$

This equation is the equation of a straight line ($y = b + mx$) with a slope determined by the age and an intercept of the initial $^{87}Sr/^{86}Sr$ ratio. The line is called an *isochron* because samples that plot on it all formed at the same time. In each mineral, the $^{87}Sr/^{86}Sr\ (t)$ and $^{87}Rb/^{86}Sr\ (t)$ are values in rocks today and can be measured using mass spectrometry, defining a point on the isochron diagram. Slope and intercept are unknowns in this equation, so there are two unknowns. That means at least two measurements are required (i.e., two points define a line), so at least two different minerals must be measured to determine the isochron. Figure 6-4 illustrates the initial condition and evolution with time of an isochron.

Of course, there are things that can go wrong with this approach. For example, some of the Rb or Sr might be lost or added by a more recent event. To verify the validity of a date, it is important to measure as many minerals as possible to make sure they are all colinear and lie on the same isochron. The date from any one decay system can then be checked with isochrons from other parent/daughter systems. Since all elements behave differently, agreement among several methods leads to a high degree of confidence in the age.

What does the "age" really mean? The isochron technique dates the time when various materials formed from a uniform reservoir with no subsequent rehomogenization, loss, or gain of the parent and daughter isotopes. If the rock containing the minerals is remelted, for example, then all the Sr atoms mix together and re-homogenize the $^{87}Sr/^{86}Sr$ ratio, which would serve as a new initial value when new minerals crystallized. For this reason the rocks on Earth, which are continually being reprocessed, do not give an isochron age for Earth. An isochron for a granite that formed 100 million years ago gives an age of 100 Ma. To date the beginning of the solar system and Earth's formation, we need materials that formed at that time and have been isolated ever since. Such materials are the chondritic meteorites, which escaped all planetary processing and have remained isolated in space until their recent arrival on Earth.

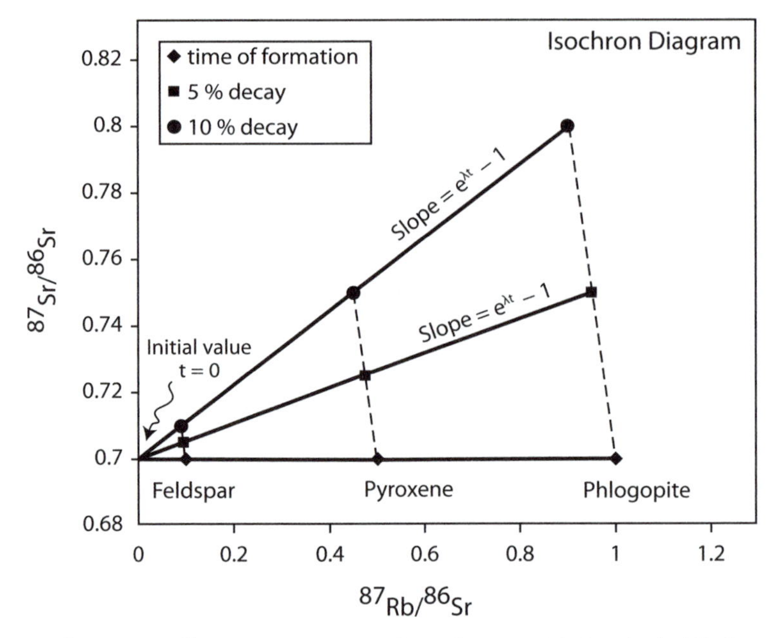

Fig. 6-4: Illustration of how isotopic composition changes with time in the Rb-Sr system. At the time of formation, three minerals form with uniform $^{87}Sr/^{86}Sr$ but variable ratios of ^{87}Rb to ^{86}Sr. As time passes, the ^{87}Rb decays to ^{87}Sr, increasing the $^{87}Sr/^{86}Sr$ ratio. The higher the $^{87}Rb/^{86}Sr$ ratio, the greater the change. At any time the three minerals plot on a straight line, called an *isochron*, whose intercept is the initial $^{87}Sr/^{86}Sr$ ratio. As the ^{87}Rb decays, the slope of the line increases progressively with time. Therefore, slope gives time and intercept gives initial value. The intercept has the same value as a hypothetical mineral that formed with no Rb atoms.

AGE OF THE CHONDRITES AND EARTH

Because of their unique importance for determining the age of the solar system, chondrites and other meteorites have been the subject of many dating investigations. Dates are summarized for nineteen different meteorites in Figures 6-5 and 6-6. As can be seen from the figures, the dates for these meteorites are in agreement and give an age of 4.56 billion years. Furthermore, many isotope systems have now been applied to the meteorites. These are summarized in Table 6-3. The agreement of the many separate meteorites for a single decay system, and the agreement among all the independent decay systems, makes the age of chondrite

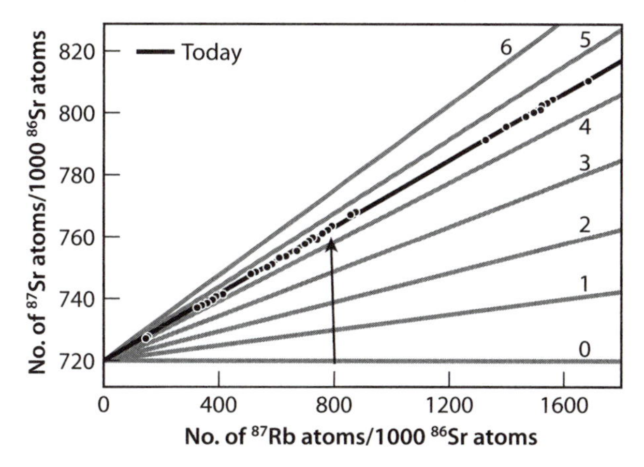

Fig. 6-5: The light lines show the evolution with time (billions of years) of ^{87}Sr and ^{87}Rb in minerals from meteorites. At the time the solar system formed, all the minerals had compositions falling along the line marked zero, i.e., they had a range of ^{87}Rb/^{86}Sr ratios but all had a ^{87}Sr/^{86}Sr ratio near 0.7. With time each grain increased in ^{87}Sr content and decreased in ^{87}Rb content. All chondritic meteorites plot along a single isochron of 4.56 billion years.

formation very well constrained. The agreement demonstrates that all chondritic meteorites formed at very nearly the same time from a well-mixed reservoir—the solar nebula—and fortifies our confidence in the ancient age of these objects.

Recall that the importance of the chondrites is that they contain chondrules that formed from a nebular cloud of dust and gas and hence reflect primitive objects in the solar system that escaped subsequent processes of planetary differentiation and have been preserved in the almost perfect vacuum of space. Therefore, the age of these objects reflects the time of nebular condensation.

To know if this age is also the age of Earth's formation requires another step in reasoning. Because Earth underwent extensive differentiation at its formation, and is continuing to undergo melting, magmatism, erosion, etc., the isotopic systems are continually being reset. Since no rock on Earth has remained isolated like the chondrites since the early solar system, how can we determine whether Earth is made up of objects that formed at the same time as the meteorites?

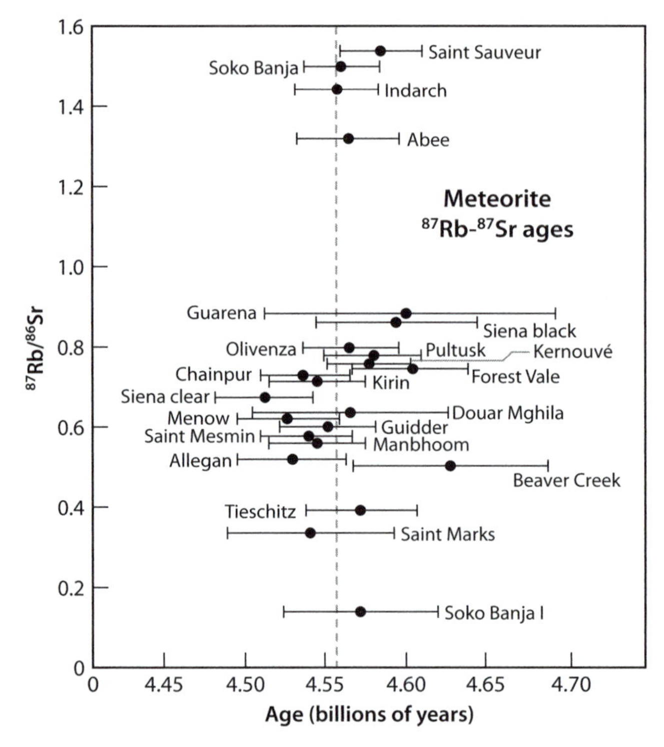

Fig. 6-6: Summary of Rb-Sr ages obtained on nineteen different meteorites. As can be seen, all the results lie between 4.52 and 4.63 billion years. The mean of all measurements is 4.56 billion years (shown by a vertical black line). As the uncertainty in each measurement (shown by horizontal bars) in all but three cases spans this mean, there does not appear to be significant differences in age among these objects.

Table 6-3

Meteorite ages based on different isotope systems

Isotope system	Age (Ga)	Uncertainty
Rb-Sr	4.56	0.05
Sm-Nd	4.55	0.33
Pb-Pb	4.56	0.02
Lu-Hf	4.46	0.08
Th-Pb	4.54	0.04
U-Pb	4.54	0.04

If we consider Earth to be one very big meteorite, then the mean composition of Earth should plot on the same isochrons as the meteorites. Various lines of evidence show that this is indeed the case.

The evidence came from a clever realization that applied to the Pb isotopes. The isochron diagrams such as Figure 6-5 require knowledge of both element and isotopic concentrations. But for the U-Pb system, there are two possible isochrons, one for ^{235}U-^{207}Pb and one for ^{238}U-^{206}Pb. If these isochron equations are divided one by the other, the concentrations of uranium cancel out, and slopes on a diagram of $^{206}Pb/^{204}Pb$ vs. $^{207}Pb/^{204}Pb$ (^{204}Pb is the nonradiogenic Pb isotope equivalent to ^{86}Sr for the Rb-Sr system) can be used to determine an age. Rama Murthy and Claire Patterson determined these ratios for a large number of chondrites and also for oceanic sediments from Earth. The oceanic sediments were ideal because they reflect well-mixed erosion from all Earth's continents, as well as some Pb contribution from the ocean crust, and are a pretty good Earth average. The initial value for the solar system could also be determined from the iron meteorites, which have lots of Pb and no U. What they found was that Earth's sediments plotted exactly on the isochron coming from the meteorites, supporting a common origin and the same age (Fig. 6-7). Earth is one big collection of meteorites.

Age of the Elements

The relative abundances of the long-lived radioisotopes in meteorites also have something to tell us about when the heavy elements in the universe were produced. While this argument is somewhat complicated, the reasoning illustrates important principles that will be useful in our later discussion of Earth processes.

In general, when materials are created at a constant rate and decay at an exponential rate, they reach a steady state value—i.e., a value that persists through time relatively unchanged even though the system is dynamic and material and energy continue to flow through the system. Such a characteristic is very widespread throughout natural systems.

Let's take a simple example to see how such a process develops. Imagine that you receive a paycheck that nets you $100 per week, and you

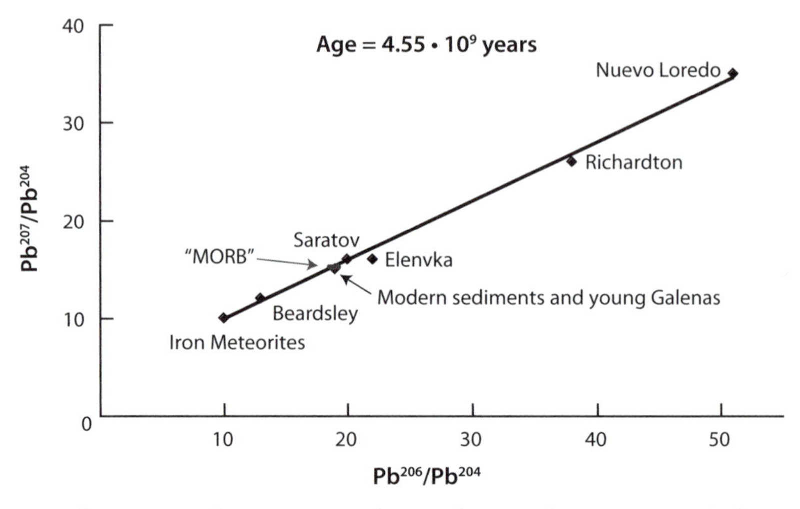

Fig. 6-7: Pb isotope results comparing sediments from Earth to meteorites. Each name refers to a different meteorite. *Young galenas* are lead sulfide minerals formed from fluids that may provide a good average for the continental crust. MORB are mid-ocean ridge basalts that represent the composition of the upper mantle. The fact that all the data plot along the same line shows that Earth and chondrites formed from the same reservoir at the same time.

decide to rigorously adopt the discipline that you will always spend exactly half of what you have in your bank account.

At the end of the first week, you spent \$50, then a new \$100 brings you to \$150, so the second week you spend \$75. The next \$100 brings you to \$175 so you spend \$87.50 and so on. Ultimately, when you reach \$200 in your bank account, you will be at steady state, spending and receiving \$100 per week. While funds are steadily flowing through your bank account, and you have a life working and spending, a person examining the balance at the end of the week would see a constant amount. During the approach to steady state, the person could also determine the number of weeks since you started working by looking at the amount of money in your bank account (Fig. 6-8).

The same kind of principle can be applied to the abundances of radioactive elements. They are produced at some rate in stellar interiors, and half of the total amount existing decays in a fixed period of time—linear production and exponential decay, which would lead eventually to

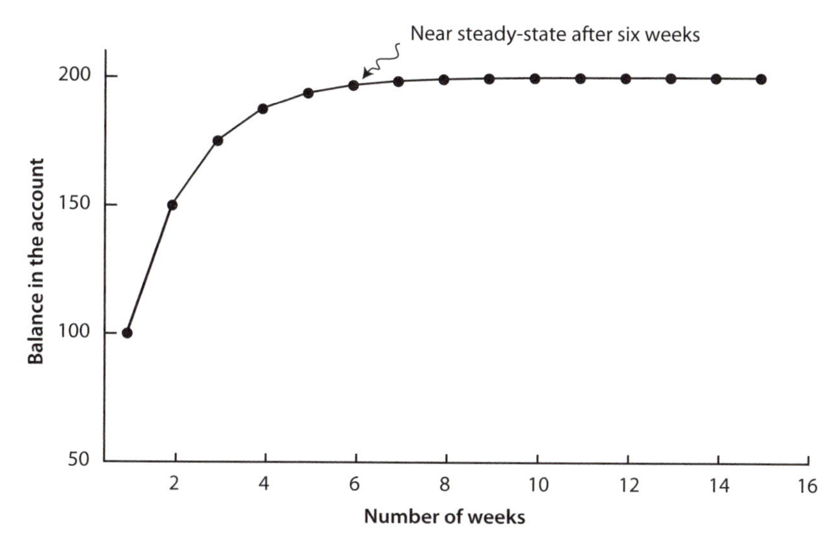

Fig. 6-8: Change in bank account balance following the rules given in the text. After six or so weeks, the account has reached steady state. Before that time, the balance in the bank account can be studied to give a good estimate of the amount of time that has passed since the bank account began.

steady state values in the universe. But the length of time it takes to arrive at steady state depends on the half-life (e.g., in the bank account example, if weeks are changed to decades, it would take six decades to reach steady state rather than six weeks). So if we examine elements with sufficiently long half-lives that the abundances have not yet arrived at steady state, we could date the time when elements first started to be created.

But there remains one additional complication. We do not know the actual production rate for individual radioisotopes. We are able to estimate, however, the production ratios of production rates of different isotopes, because the ratios can be calculated from nuclear physics. Can we then obtain the same sort of time information?

Another example will, perhaps, illustrate this more complex case. Consider a special exhibit at the Louvre in Paris, such as the Leonardo da Vinci drawings. A constant flow of people arrives all day, a third of whom are artists and two-thirds of whom are tourists. This influx is steady throughout the day. We will then (arbitrarily) assign the rule that

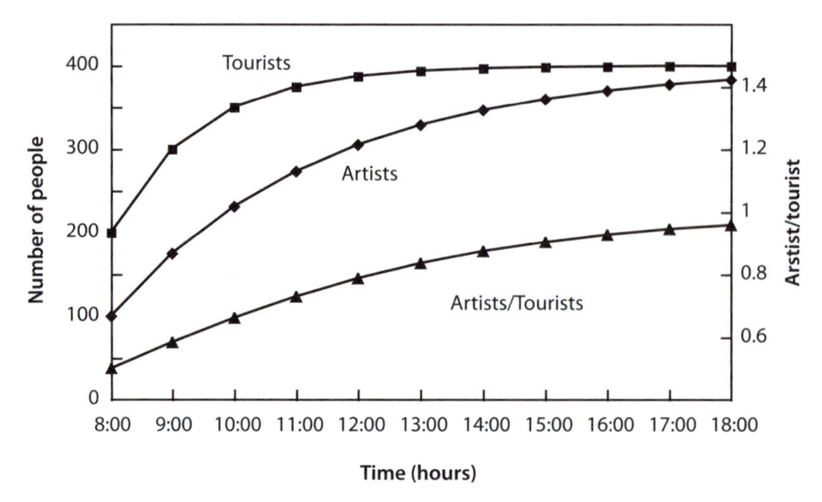

Fig. 6-9: How proportions of artists and tourists change with time of day for the example given in the text. Because the half-life of tourists is shorter than that of artists, they approach steady state more rapidly. Because steady state is never reached, the ratio of artists to tourists can provide the time since the exhibit has been open.

at the end of each hour one quarter of the artists who are present at the exhibition leave, but one-half of the tourists leave. These rules impose a steady, linear rate of new people and two different rates of exponential decrease, depending on whether the people are tourists (who have a short half-life in the museum) or artists (who have a longer half-life). This sets up the museum for an approach to steady state. Figure 6-9 illustrates how the number of people in the exhibition changes during the day. The number of tourists approaches steady state faster than the number of artists, owing to the shorter half-life of the tourists, and the ratio of artists to tourists increases progressively but never reaches its steady state value of 1 because the exhibit is not open long enough for this to occur. If you knew these rules and walked into the museum at any point, you could tell exactly how long the exhibit had been open by looking at the ratio of artists to tourists. For example, if the ratio of artists to tourists was 0.8, the exhibit had been open for four hours.

An analogous situation occurs for long-lived radioisotopes ^{235}U, ^{238}U, and ^{232}Th. All these isotopes are produced continually in constant ratios to one another by supernova explosions. Once formed, they decay away according to their half-lives. We appear on the scene and can measure

Table 6-4
Different values of ^{235}U, ^{238}U, and ^{232}Th in the early solar system

Values	^{235}U	^{238}U	^{232}Th
Half-life of each isotope (Ga)	0.704	4.47	14
Production relative to ^{232}Th in stars	0.79	0.525	1.00
Steady-state proportions	0.041	0.167	1.00
Proportions in early solar system	0.122	0.424	1.00

their ratios. By that measurement we can determine when they first started to appear—i.e., when element production began. We must use the proportions at the time of formation of the solar system rather than the values today because Earth has been isolated from new contributions by supernovas since the solar system formed (fortunately!). While element synthesis continues in the galaxy, there is no mechanism by which the elements produced over the last 4.55 billion years could have become incorporated into the sun or its planets. Therefore once the solar system formed, we became isolated from the great processes of element creation, supernovas and mixing that we observe taking place elsewhere in our galaxy in the great interstellar clouds.

To calculate the values at the start of the solar system, we use the radioactive decay equation (6-1) given earlier in the chapter. We know the values of U and Th today, $N(t)$, and we know the age of the solar system, t, so we can calculate the amounts present at the formation of the solar system, N_0. The proportions in the early solar system as well as production ratios, half-lives and steady-state proportions are given in Table 6-4, with the proportions normalized to an arbitrary value for ^{232}Th of 1.0. It is evident that the proportions in the early solar system are in between the production rates in stars and the steady-state values. This then permits a time to be estimated for the beginning of element production.

Figure 6-10 demonstrates how the isotope abundances will change with time during the evolution of the galaxy—note the similarity to the artists and tourists in Figure 6-9. In Figure 6-11 the solid curves show how the ratios change with time since element formation began. The dashed lines show the values for these ratios at the time of formation of the solar system. Where the curves intersect is the time elapsed between

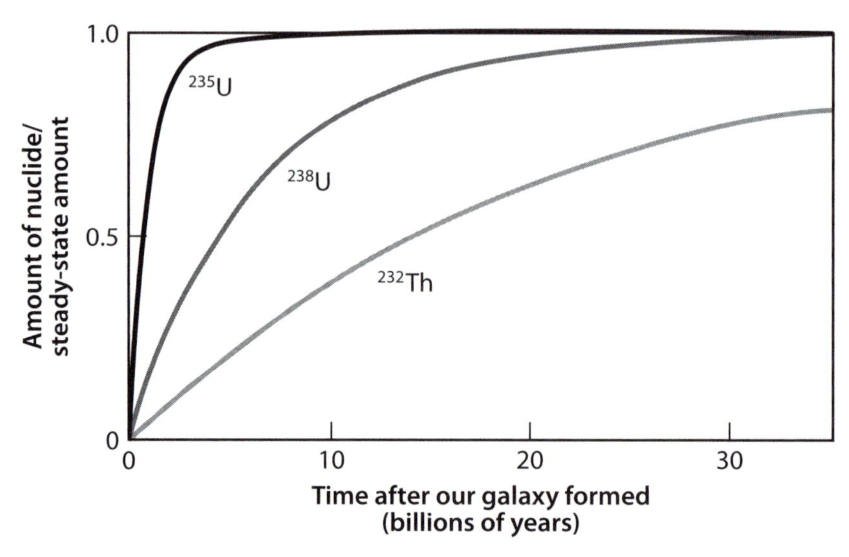

Fig. 6-10: The evolution of the amounts of the three isotopes in our galaxy. The assumption is made that supernova events have occurred regularly over the entire history of the galaxy. The steady-state amounts occur when the isotope decays away at the same rate that is produced. Note the similarity in form to Figure 6-9.

the formation of our galaxy and the formation of our solar system. For the ^{238}U/^{235}U ratio, the crossover is at about 12 billion years. For the ^{232}Th/^{235}U ratio, the crossover is at about 9 billion years. While both results have rather large uncertainties, they both suggest that element production began about 10 billion years before our solar system came into being. Since the age of the solar system is 4.55 billion years, that gives a total age of about 13.5–16 billion years since the creation of the first heavy elements.

The various isotope data then give an overall chronology that ties in well with the ages inferred from the red-shift distance relationship in Chapter 2, as summarized in Figure 6-12. The universe began about 13.7 billion years ago. Our galaxy formed sometime in the first billion years of universe history. Over our galaxy's entire history, the largest stars have steadily synthesized and distributed elements that have been mixed together in interstellar clouds. After about 9 billion years of galactic history, our solar system formed and became isolated from the stellar processes 4.56 billion years ago. The radioisotopes incorporated into the

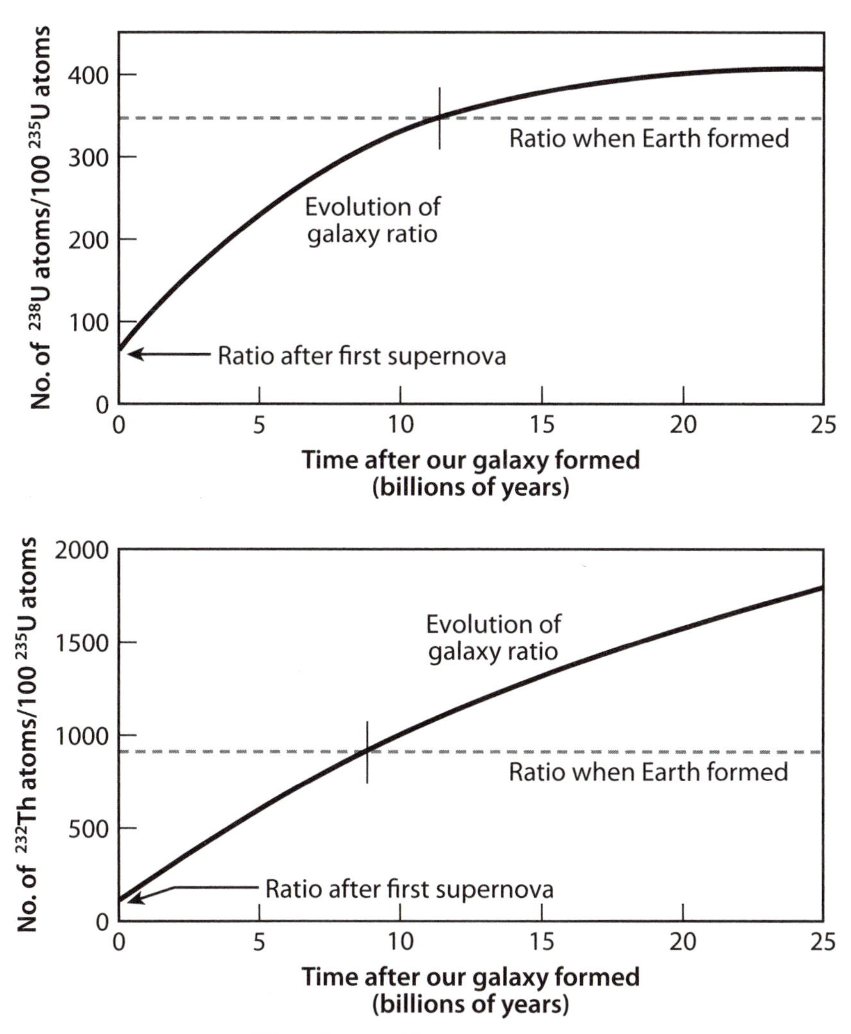

Fig. 6-11: The evolution of ^{238}U, ^{235}U, and ^{232}Th if heavy element production occurred at a constant rate. Early in the galaxy's history the ratios were equal to the production ratio in stars. With time the ratio changed favoring the longer-lived isotopes. The horizontal dashed lines correspond to the ratios at the time the solar system formed. The intersection of the dashed line with the solid evolution curves gives the time elapsed between the formation of our galaxy and the formation of the solar system. About 10 billion years elapsed between galaxy and solar system formation.

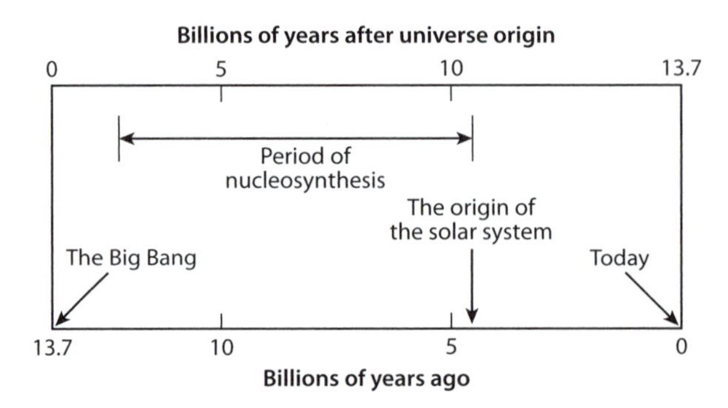

Fig. 6-12: Summary of the chronology of universe events recorded in the isotopes measured on Earth. The period of nucleosynthesis refers to the time interval over which the elements heavier than H and He that are found in our solar system were produced. For the galaxy as a whole, the period of nucleosynthesis extends right up to the present. The matter in the solar system was isolated from the galaxy 4.56 billion years ago.

planet then continued to decay to give us the values that we can measure in the laboratory today.

Unlocking the Secrets of Ancient Short-lived Processes with Extinct Radionuclides

Elements with short half-lives can give information for time spans only of about ten times their half-lives. This limits their utility as dating tools for long-lived processes, but there is still important information that can be obtained. The key is that the presence of the radioactive parent is preserved in variations in the isotope ratios of the daughter element even after all the radioactive parent has decayed away. For example, consider the Rb-Sr system that is now familiar to us. If we waited hundreds of billions of years until all the ^{87}Rb had decayed away, there would still be large variations in the remaining ^{87}Sr/^{86}Sr of the minerals if they had not been subsequently homogenized. The ^{87}Rb would be "extinct"—the daughter element isotope variations would remain, showing us that ^{87}Rb once was present. In terms of the isochron diagram (Fig. 6-4), after a very long time the isochron points straight up along the y-axis. No further Sr isotope variations would grow in, because all the ^{87}Rb would have decayed away.

A large number of short-lived radionuclides once existed and are now extinct. Because they decayed rapidly they created large isotopic variations in relatively short periods of time. Study of the daughter elements that record the presence of their highly radioactive parents has led to a surprising wealth of information concerning the early solar system and the processes that took place there.

^{26}AL AND THE PRESENCE OF SUPERNOVAS IN THE VICINITY OF THE SOLAR NEBULA

^{26}Al is one of the most important of the extinct radionuclides, in part because Al is a major constituent of stony meteorites and terrestrial planets. Unlike Rb, Sm, or U, which have concentrations of only a few parts per million or less in common rocks, Al has a concentration of 3 to 20% and is an essential constituent of many of the most common minerals. A small proportion of this Al would be ^{26}Al if it were present.

^{26}Al decays to ^{26}Mg with a half-life of 0.73 million years (fig. 6-13). Therefore, it is a very sensitive indicator of processes happening within 10 million years after it was created. After that, only its daughter product ^{26}Mg would remain to give evidence of its prior existence. ^{26}Al is created by the r-process in supernova explosions. Evidence for ^{26}Al would lead to the conclusion that a supernova explosion took place within 5–10 Ma of the time when the rocks containing evidence for ^{26}Al were formed. The evidence would be in the isotope variations of the daughter element Mg.

Gerry Wasserburg and colleagues at the California Institute of Technology set out in 1974 to test whether any ^{26}Al had existed in the nebular materials from which the meteorites formed. If the Mg isotopes showed that ^{26}Al had been present, then three things would be clear about early solar system development:

(1) A supernova event occurred nearby just before the solar system formed.
(2) The meteorites must have formed very rapidly after the supernova.
(3) Extinct radionuclides would have been present in the early solar system, providing a powerful heat source for early planetary bodies.

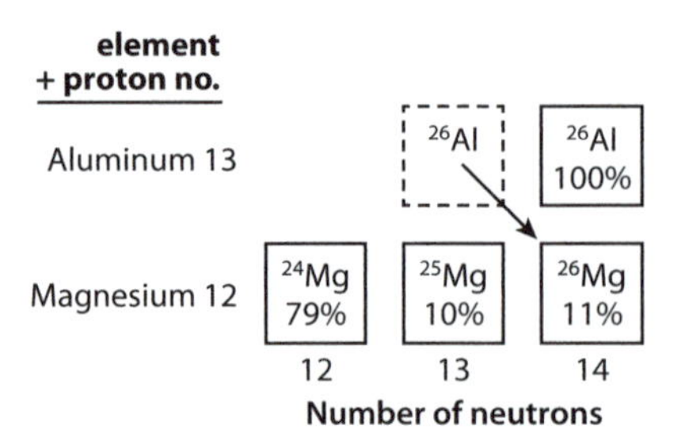

Fig. 6-13: The isotopes of the elements Al and Mg and their relative proportions on Earth today. Aluminum has only one stable isotope and magnesium three. Early in solar system history, however, a second aluminum isotope was present—radioactive ^{26}Al with a half-life of 0.73 million years. This isotope is now extinct, having long since decayed to ^{26}Mg.

The critical materials to measure were those that were very high in Al and very low in Mg. Because ^{26}Al is chemically identical to the stable ^{27}Al, ^{26}Al would have been incorporated along with the stable Al into aluminous minerals. (Aluminum foil from ^{26}Al would look exactly the same as ordinary foil, but would be very deadly!). The lower the quantities of Mg, the more ^{26}Mg would be created relative to the stable ^{24}Mg. So, just as dating with the isochron diagram works best for large variations in parent/daughter ratios, finding evidence for the extinct Al would work best with a very large range of Al/Mg ratios. Those minerals that had the highest ^{26}Al/^{24}Mg (e.g., Ca-feldspar ($CaAl_2Si_2O_8$) would after complete decay have the highest ^{26}Mg/^{24}Mg. Minerals with low Al/Mg would have the lowest ^{26}Mg/^{24}Mg. If there were no differences between these minerals, no ^{26}Al was present when they formed.

The candidate material for this investigation was the largest carbonaceous chondrite ever found, named Allende, which fell in Mexico in 1969. By making very precise measurements of the ^{26}Mg/^{24}Mg ratio in the trace amounts of magnesium present in the feldspars and comparing them with similar measurements in high Mg mineral grains, Wasserburg and co-workers were able to show that the ^{26}Mg/^{24}Mg ratio was higher in the magnesium from the feldspar. The proof came from measuring several different minerals with very different Al/Mg ratios (Fig. 6-14), and a correlation was found between the Al/Mg ratios and the ^{26}Mg/^{24}Mg ratios. This clever diagram makes use of the fact that the

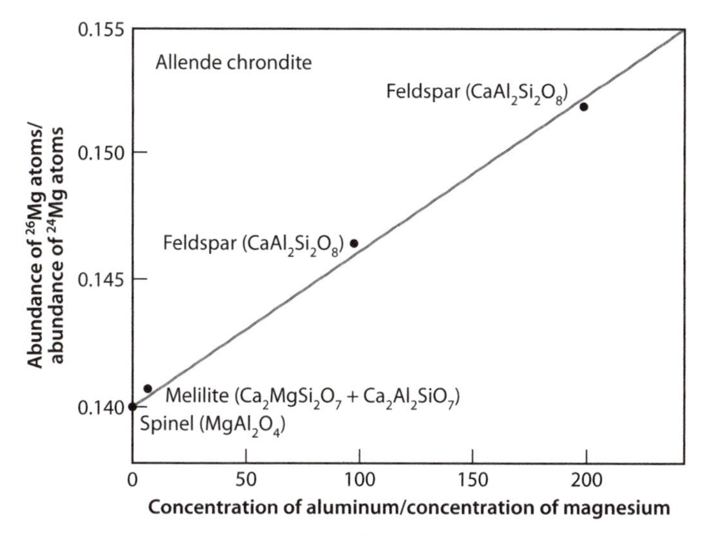

Fig, 6-14: Relationship between the ratio of ^{26}Mg to ^{24}Mg and the ratio of aluminum to magnesium in mineral grains from a chondritic meteorite. Feldspar grains with aluminum as a primary constituent and magnesium as a trace constituent have higher $^{26}Mg/^{24}Mg$ than minerals for which the Al/Mg ratio is lower. The $^{26}Mg/^{24}Mg$ ratio of average Earth material is 0.1394. This is not surprising, since the Al/Mg ratio in the Earth is about 0.1. (Based on Lee, Papanastassiou, and Wasserburg, *Geophys. Res. Lett.* 3 (1976):109–12)

present Al/Mg ratio of the minerals would correlate perfectly with the original ^{26}Al /Mg ratio. Therefore, the higher the present-day Al/Mg, the more ^{26}Mg relative to ^{24}Mg should have been produced.

The original amount of ^{26}Al that would be needed to explain these results seemed too high, and there was substantial controversy over the interpretation of these results. Technical developments in instrumentation, however, have made it possible to measure isotope ratios that would be generated by many different extinct radionuclides in individual grains in meteorites, and these results conclusively show the importance of supernova products in the grains of the early solar system. In fact, the isotopic diversity of the tiny grains retained in the most primitive meteorites shows a contribution from many different types of stars and stellar explosions to the solar nebula material. All this evidence confirms the hypothesis of the solar system being generated in a huge cloud of interstellar gas and dust where star formation and element production were common occurrences. One or more supernovas occurred in close proximity in space and time to the region where the sun and its planets were originally formed.

Summary

Radioactive elements are ticking clocks that reside in the molecules that make up planetary materials. As these elements decay, they create daughter isotopes that change the isotopic makeup of their respective elements. Measurement of these isotope ratios permits the study of the timescales of a vast number of Earth processes. Cosmogenic radionuclides such as ^{14}C can be used to date young organic materials. The long-lived radionuclides provide the tools needed to date ancient events that occurred in the early solar system and throughout Earth's history. Because there are a number of independent systems, important measurements such as the age of the meteorites can be double and triple checked by independent means. All the data are consistent with an age for the early solar system of 4.56 billion years. The coherence of Earth with these measurements shows that Earth formed from the same types of materials that are observed in meteorites still impacting Earth's surface from space. The long-lived radionuclides also permit an exploration of some of the fundamentals of a steady-state condition, and the timing of the beginning of element creation. This timing corresponds well with the estimates from the red shift/distance relationship from Chapter 2, and permits an overall chronology for Earth starting with the Big Bang and continuing with formation of elements in the galaxy through solar system formation to the present day. Short-lived radionuclides provide another set of tools that reveal processes that took place in the early solar system. Evidence for ^{26}Al and other extinct radionuclides shows that the environment of solar system creation was a very active one, with contributions to the nebular dust from diverse stars and nearby supernovas. Earth appears to have formed in a stellar incubator similar to those observed by astronomers elsewhere in our galaxy, and our solar system appears to simply be one example of a common universal process. The extinct radionuclides confirm the very short timescale of early solar system history, with only a few million years between supernova explosions and formation of the sun, meteorites, and planets. The short-lived radionuclides decayed with such intensity that they would have provided a powerful and short-lived heat source that would have aided rapid heating and differentiation of early planetary objects.

Supplementary Reading

Gunter Faure and Theresa Mensing. 2005. *Isotopes, Principles and Applications*, 3rd ed. New York: John Wiley & Sons.

Claude Allègre and Christopher Sutcliffe. 2008. *Isotope Geology*. Cambridge: Cambridge University Press.

Fig. 7-0: Photograph of a pallasite meteorite. Dark areas are olivine crystals. Lighter areas are metal. (Courtesy of Harvard Museum of Natural History)

CHAPTER 7

Interior Modifications

Segregation into Core, Mantle, Crust, Ocean, and Atmosphere

*After the planets and moons formed from planetesimals, they underwent profound internal changes that created their primary internal structure. The grand scheme of planetary differentiation can be summarized as a progressive stratification of the planet, where dense materials sink to the interior and light materials rise to the surface. Earth, for example, has become layered with an Fe-metal core, a silicate mantle, a solid crust that differs between ocean and continent, an ocean, and an atmosphere. Core separation is a consequence of the immiscibility of metallic and silicate liquids and the much higher density of metal, leading to a metallic core underlying silicate mantle. At the very high temperatures of Earth's interior, the solid silicate mantle convects, bringing hot, deep material toward the surface. This ascent causes the crust to form by melting of Earth's mantle at shallow depths, where melting points are lower. The melt is lighter than the mantle that surrounds it and rises buoyantly to the surface. Melts of the mantle form the rocks rich in Mg and Fe (*mafic rocks*) of the ocean crust. Further igneous processing leads to the creation of the rocks rich in feldspar and quartz (*felsic rocks*) of the continents. Both oceanic and continental crusts are lighter than the underlying mantle and float on top of it. The lower density and greater thickness of the continental crust cause the continents to float at higher levels than the ocean crust. The outermost layers of liquid ocean and gaseous atmosphere likely formed largely by degassing of the mantle, but may also have been influenced by the continuing influx of volatile-rich objects* *from space. Evidence from short-lived radionuclides shows that core and atmosphere formation happened in the earliest few tens of millions of years of Earth's history. The crust that we see today formed much later. The ocean floors are geologically young (<160 Ma) because they are continually being created and destroyed. The continents preserve a longer record, but there are only tiny remnants as old as 4,000 Ma, leaving Earth's earliest history with no direct record, apart from the important clues from meteorites. Achondritic meteorites that are older than 4,000 Ma suggest that the overall process of immiscibility, melting, and degassing to produce distinct compositional layers stratified by density is a common planetary process.*

The net effect of interior stratification is to distribute the elements according to their chemical tendencies. Siderophile (metal-loving) elements end up in the core. Most lithophile (rock-loving) elements end up in the mantle. A small group of lithophile elements concentrate into magmas (magmaphile elements) and are concentrated at the surface. Magmaphile elements include the volatiles (H_2O, CO_2, N_2) and P, Na, K, Cl—an assemblage that is focused to the surface and will then provide the environments and molecules for the establishment of a stable climate and the origin and evolution of life.

Introduction

Segregation of planets from the solar nebula explains features such as the bulk density, the low relative abundances of many volatile elements, and the preponderance of the big four planet-forming elements in terrestrial planets. Planets today, however, are not homogenous mixtures of matter. They have structure and have been differentiated into layers with distinct compositions. This is readily apparent from the meteorites, many of which provide us with samples from the interiors of disrupted parent bodies circling around the solar system. Some of these nonchondritic meteorites are Fe-rich metals, some are mixtures of metal and rocks, and others are volcanic rocks that reflect partial melting of planetary interiors. Processes of melting and separation of metal and silicate within planetary bodies clearly operated in early solar system history and therefore must have affected Earth and the moon as well. The meteorites provide clues as to what may have happened within Earth.

Since Earth has not been disrupted into little fragments, we have no direct access to its interior. The deepest drill holes only penetrate 10 km or so, a trivial distance compared to Earth's 6371 km radius. Therefore, while we can determine the compositions of the liquid ocean and gaseous atmosphere by direct measurement, and we have rocks from the exposed surfaces of the crust, we must rely on other lines of evidence to determine Earth's structure and its total composition.

Earth Structure

As we learned in Chapter 5, the first line of evidence comes from the estimate of Earth's density. To determine density (mass/volume), we need to know the volume and mass of Earth. Volume is readily measured. To determine mass requires the application of Newton's Laws. One method based on the moon's orbit was discussed in Chapter 5. An alternative is possible from surface measurements. A mass, m, near Earth's surface falls toward the surface with an acceleration commonly referred to as g. The gravitational force driving this acceleration is then :

$$F = mg \qquad (7\text{-}1)$$

This force is also related to Newton's third law that gives the gravitational force between two masses:

$$F = G\, m\, M_e/R^2 \qquad (7\text{-}2)$$

R is the radius of the earth, M_e the mass of the earth, and G the universal constant of gravitation. Since the two forces are equal,

$$M_e = g\, R^2/G \qquad (7\text{-}3)$$

R and g are easily measured, but G requires a much more difficult measurement of the gravitational attraction between two objects of known mass. Through painstaking experiment, Lord Cavendish determined a value for G in 1798. He then calculated the density of the earth to be 5.45 grams per cubic centimeter, close to the value of 5.25 gm/cm^3 determined by more accurate modern methods. Common rocks at Earth's surface have a density of about 2.7 gm/cm^3 (water is 1.0 gm/cm^3), so there must be very dense materials in Earth's interior to make the average so high. What is this dense, interior material, and where is it located?

This question can be approached from the ellipsoidal "shape" of the Earth. Since Earth spins around an axis, the equator is spinning at high speeds of 1,668 km/hour, while the North and South Poles are stationary. The centrifugal force caused by the great velocities at the equator causes the equatorial regions to bulge relative to the poles, making Earth slightly ellipsoidal in shape. The size of the equatorial bulge depends on how the mass is distributed within Earth. If the mass is concentrated toward the interior, the bulge is less. One can understand this by imagining (or actually trying) to swing a weight around in a circle over one's head. If the weight is on the end of a long string making one revolution per second, there is substantial pull on the arms. If at the same revolutions per second the weight is close to the body, there is much less pull (and of course the weight is moving at a much slower velocity). This property has the name *moment of inertia*. Earth has a moment of inertia that is about 20% less than if it were a sphere of uniform density. If the mass were uniformly distributed, Earth would have a larger equatorial bulge. Therefore, dense material must be concentrated toward Earth's center.

The mean density and moment of inertia can be combined to infer Earth's general density distribution. The result is that Earth must have a core with a density of about 11 gm/cm^3 that makes up roughly half Earth's radius. Few elements that we know of have such high densities at Earth's surface. Iron, with a density at one atmosphere of 5.6 gm/cm^3, is too light. Only very heavy metals such as gold and silver, with densities of 10–12 gm/cm^3 seemed to fulfill the requirements. Could it be that Earth has a core of solid gold?! This conundrum was resolved once it was realized that under the high pressures of Earth's interior solids are compressible, and their density at depth in Earth is greater than their surface density, just as we saw for the densities of the outer planets in Chapter 5. The effect of pressure on density leads to a gradual increase of density with depth for all of Earth's layers.

Much more clarity about internal earth structure came about in the early twentieth century from the field of seismology, the study of earth-

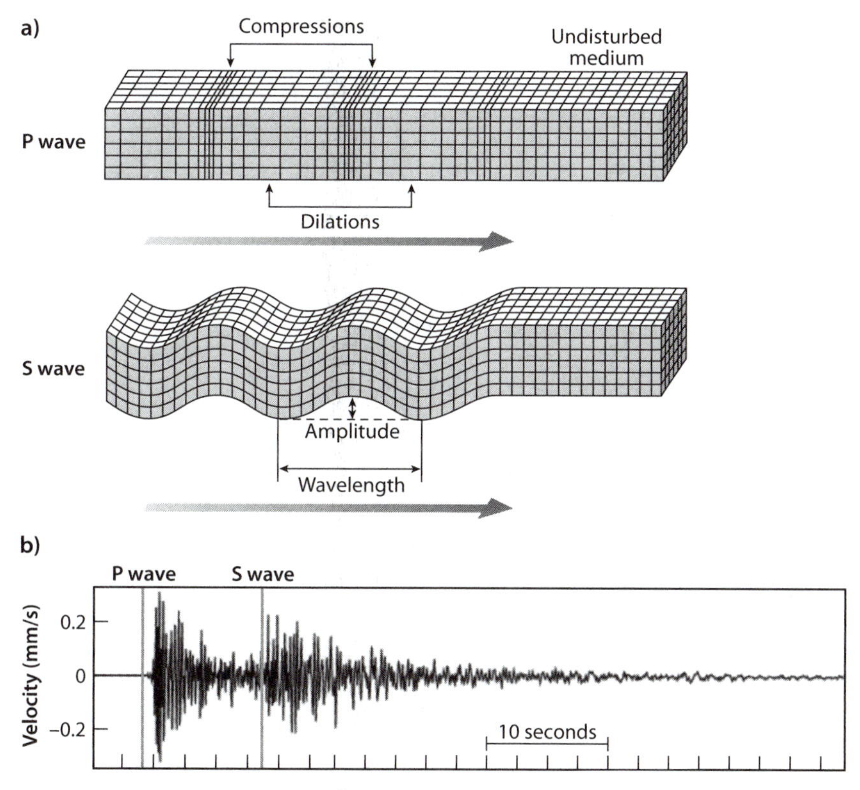

Fig. 7-1: (a) Illustration of the difference between compressional waves, where matter displaces in the direction of motion, and shear waves, where matter moves perpendicular to the direction of motion; (b) seismogram record showing the arrival of P and S waves produced by an earthquake. P waves have higher velocity and arrive first. Horizontal axis is time. (Courtesy of U.S. Geological Survey)

quakes. The shock of an earthquake causes Earth to ring like a bell and creates waves with so much energy that they traverse around and through the entire globe. It was discovered that these waves could be recorded with very precise pendulums (now called *seismometers*), which produced seismograms that recorded the detailed pattern of the waves (Fig. 7-1).

The timing of the arrival of the waves recorded at different places around the globe gives information about the velocity of the waves through the planet. This velocity is dependent on the physical properties of the material through which the wave travels, including the density. With the seismic velocity data, the density structure of Earth was able to be determined (Fig. 7-2). This structure shows large regions where the

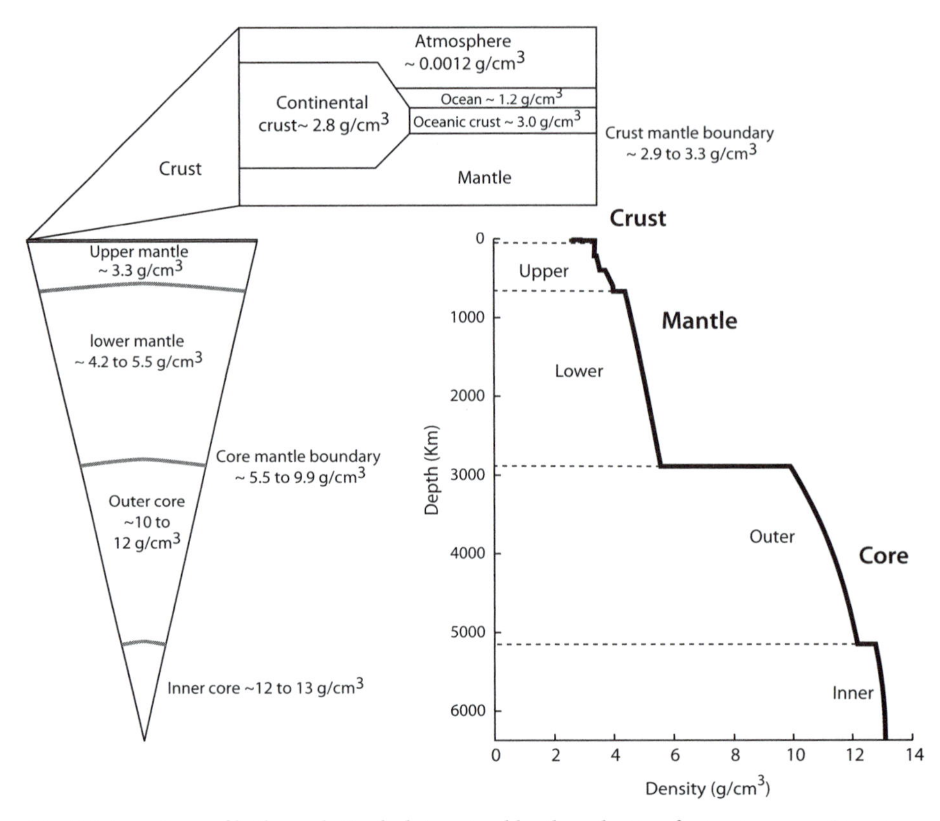

Fig. 7-2: Density profile through Earth determined by the velocity of seismic waves. Density increases progressively in each layer largely due to compression. Abrupt changes in density occur where the material composition changes abruptly. Some of the small changes in the upper mantle are due to changes in mineralogy for the same peridotite composition.

density increases gradually with depth and some depths where abrupt jumps in density show major changes in chemical composition. The biggest jump, from ~6 to 10 gm/cm^3 was used to define the core/mantle boundary.

As the complex patterns of recorded waves became better understood, seismologists realized from the seismograms that earthquakes create three major types of waves: compressional waves, where the material is moving forward and back in the direction the wave is moving; shear waves, where the material is moving perpendicular to the direction of motion; and surface waves, which pass around Earth's surface rather than through its interior (see Fig. 7-1). Surprisingly, shear waves abruptly disappear from what is seen by seismometers that are located a

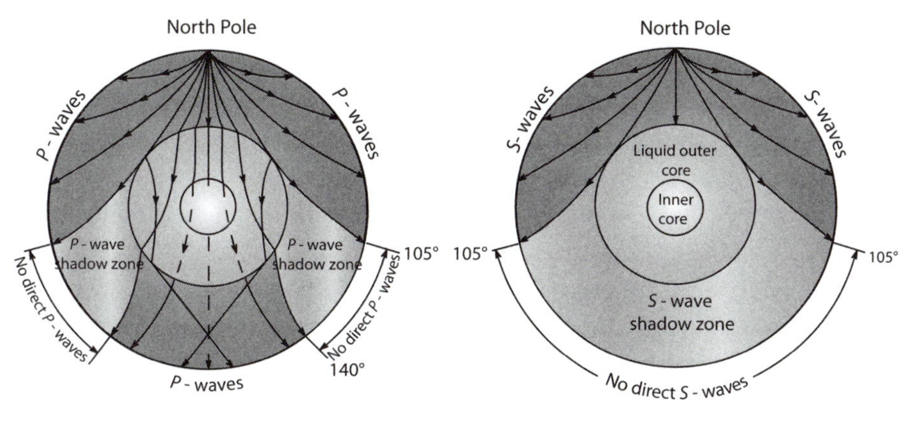

a) Compressional waves, *P* b) Shear waves, *S*

Fig. 7-3: Paths of earthquakes through Earth for an earthquake located at the North Pole. The waves bend depending on the changes in the density of the material through which they pass. The solid lines show the paths of the waves. The regions where no direct waves arrive are generally referred to as *shadow zones*.

little more than halfway around Earth from the location of the earthquake! The region where no shear waves appear became called the *shadow zone* (Fig. 7-3).

The shadow zone can be understood from the recognition that shear waves are unable to propagate through a liquid. Compressional and shear waves both propagate through solids, at slightly different velocities. In liquids, however, shear waves die out because fluids do not sustain shearing—you can bend a stick or a metal rod, but not a fluid because it does not have the strength to sustain the shearing force. Whales and dolphins can communicate at very long distances by sound waves, which are compressional, but the back and forth movement of their huge tails dissipates rapidly. The shadow zone showed shear waves disappear in a portion of Earth's interior, and that portion must then be liquid. The very systematic spatial distribution of the shadow zone permitted precise mapping of the interior liquid layer, which begins exactly at the abrupt change in density that defines the core/mantle boundary (Fig. 7-3).

The definition of layers of distinct density combined with information about whether the layer was solid or liquid provided the basic description of Earth's interior stratification. At the surface is the crust, with a thickness of about 35 km beneath continents, and about 6 km in the ocean basins. The base of the crust is defined by a change in seismic velocity called the *Mohorovicic discontinuity* (the "Moho"), where the density abruptly increases from about 2.7–3.3 gm/cm^3. Beneath the crust is the solid mantle, which extends all the way to a depth of 2,900 km, where the *Gutenberg discontinuity* in velocity defines the core/mantle boundary. Below that depth lies the 2,100 km of the outer, liquid core. The base of the liquid core is defined by the *Lehman discontinuity*, where the density jumps again to the $>$1,000 km solid inner core.

The next step was to determine the chemical composition of these layers. This step required knowledge of the density and seismic velocities of materials under the appropriate temperatures and pressures (Table 7-1). A long period of careful experimentation provided the densities and seismic velocities of a wide variety of minerals that could be used to calibrate the seismic results. Earth's core consists of Fe and Ni with a small percentage of some lighter elements (whose identity is still in dispute) that could produce slightly lower seismic velocities than pure Fe-Ni. Earth's crust is amenable to direct inspection, and its composition corresponds to the observed seismic velocities. The continental crust, with a density of about 2.7 gm/cm^3 is made up primarily of the minerals quartz and feldspar—the principal constituents of granite, with a small amount of Fe-Mg–bearing minerals such as pyroxenes and

Table 7-1
Common rocks of the crust and mantle

Rock	Location	Density at low pressure (gm/cm)	Principal minerals	~ Chemical composition				
				SiO_2	Al_2O_3	MgO	FeO	CaO
Granite/rhyolite	Continent	2.70	Feldspar, quartz	~70	~16	~1	~3	~6
Diorite/andesite	Continent/ island arc	2.85	Feldspar, quartz, pyroxene	~55	~18	~2	~5	~8
Gabbro/basalt	Ocean crust/ flood basalts	3.00	Feldspar, pyroxene, olivine	~49	~15	~8	~10	11
Peridotite	Mantle	3.30	Olivine, pyroxene	~44	~4	~39	~8	~3

amphiboles. The oceanic crust has no quartz, about 50% feldspar, and a much higher proportion of mafic minerals. Its mean density is about 3.0 gm/cm^3.

Earth's mantle proved the most difficult material to identify with certainty. Because the seismic data were insufficient to fully constrain the mantle composition, observations of rare mantle rocks exposed at the surface, experimentation and geochemical reasoning provided necessary constraints. The upper mantle is now well constrained to be predominantly made up of the rock type *peridotite*, consisting of the mafic minerals olivine and pyroxenes, with a density of about 3.33 gm/cm^3. The evidence is as follows:

(1) The understanding of nucleosynthesis, meteorite compositions, and the density constraints given above show that Earth must consist dominantly of the big four inner planet-building nuclides—Fe, Mg, O, and Si. Although much Fe is in the core, substantial Fe is left for the rest of the earth, so the mantle must consist of some combination of MgO, SiO_2 and FeO. To be consistent with meteoritic ratios for these elements, the mantle would have to consist, at low pressure, of olivine and pyroxene.
(2) The places where mantle has been thrust up to the surface along faults reveal the rock peridotite, consisting of about 55% olivine, 35% pyroxene, and 5–10% of phases that contain CaO and Al_2O_3.
(3) Some rare rocks called *kimberlites* erupt explosively from Earth's interior and contain rock fragments captured at various depths below the crust. Some fragments contain diamonds (kimberlites are the source of all natural diamonds). Because diamonds only form at pressures far below the crust, the kimberlites must come from and sample the mantle. The rock fragments from these depths, called *ultramafic nodules* are dominantly peridotite.
(4) At ocean ridges, where the crust is very thin, the volcanic rocks must form by partial melting of the mantle. The compositions of these rocks require that peridotite be the material that is melted.

All this information points to a mantle composition of peridotite. Experiments on olivines and pyroxenes have shown that these minerals change structure as pressures increase at greater depth, and this explains why the density curves for the upper mantle in Figure 7-2 are not perfectly smooth. When there is a conversion to a more dense mineral structure, the seismic velocity rises.

For the core, an obvious question that emerges is how there could be a layer of molten metal (the outer core) surrounded by solids above and below it, producing a kind of internal liquid metal ocean. At the core/mantle boundary, one possibility is that the temperature increases substantially. Alternatively, at the pressures that exist at a depth of 2,900 km, metallic iron may melt at a lower temperature than magnesium silicate, permitting the metal and silicate at the same temperature to be liquid and solid respectively. The primary cause of the outer core being liquid is the lower melting temperature of Fe metal compared to lower mantle rock, but as it turns out the temperature also jumps at the core/mantle boundary (Fig. 7-4).

The deeper solid/liquid boundary between the inner and outer core has another explanation. The melting point of all rocks and metals increases substantially with pressure, because melting involves expansion and breaking of chemical bonds, and higher pressures make this more difficult to accomplish. The pressure is sufficiently high in the inner core that despite its higher temperature the Fe metal once again becomes solid (Fig. 7-4). The liquid/solid boundary between inner and outer core results from the effects of pressure on melting temperature.

The diverse evidence and lines of reasoning then leads to a clear definition of Earth's inner structure (Fig. 7-5). This structure is supported by all geophysical data and is as close to an established fact as possible in the absence of actually being able to directly penetrate Earth's deep interior. It ranks a 9 on our theory scale. While the major structural elements have this certainty, the details of deep Earth structure in the core and lowermost mantle, such as their exact composition and mineralogy, remain to be fully elucidated.

Chemical Composition of Earth's Layers

The various elements of the periodic table are not evenly distributed among the four major Earth layers of core, mantle crust, atmosphere/ocean. To understand where the various elements reside, we need to consider their affinities for different types of materials and states of matter.

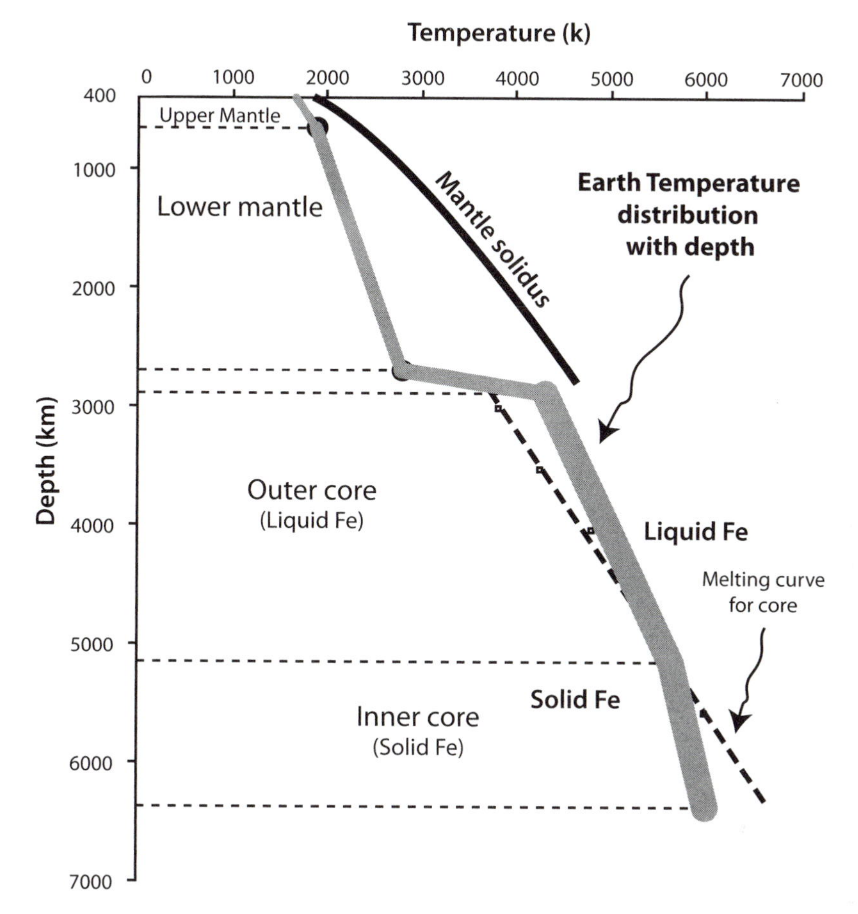

Fig. 7-4: Temperature profile through Earth. The figure also illustrates that the state of matter of Earth's interior depends on the different melting points of rock and metal and how they vary with pressure. The inner core can be solid and the outer core liquid even though the temperature of the inner core is higher than the temperature of the outer core, because of the increase in melting temperature with depth. The outer core is liquid while the mantle above it is solid because the temperature of melting of Fe is lower than that of silicate at great depths in Earth, and there is a large jump in temperature at the core/mantle boundary. (Data from Lay et al. *Nat. Geosci* 1 (2008):25–32; Madon, Mantle, in *Encyclopedia of Earth System Science,* vol. 3 (San Diego: Academic Press, 1992), 85–99; Alfé et al., *Mineralogical Magazine* 67 (2003):113–23; Duffy, *Philosophical Transactions of the Royal Society of London A* 366 (2008):4273–93; and Fiquet et al., *Science* 329 (2010):1516–18)

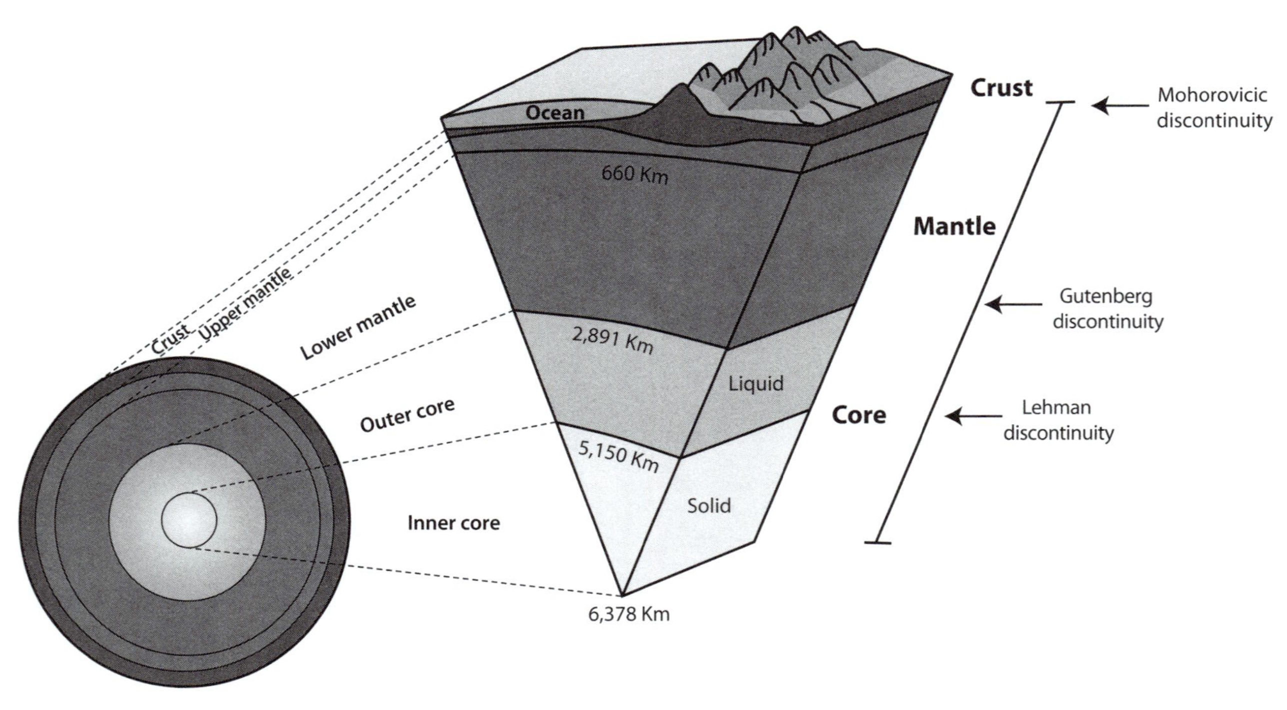

Fig. 7-5: Illustration of Earth's major layers and how they are distributed with depth.

CHEMICAL AFFINITIES OF THE ELEMENTS

It is convenient to divide the elements into four major groups (Fig. 7-6). The *atmophile* elements are those which are very volatile and tend to occur as gas or liquid molecules under the conditions on Earth. (*Phile* is a suffix that means "has an affinity for." A francophile is someone who loves France.) Atmophiles include the noble gases (such as helium, neon, and argon), water, carbon dioxide, and nitrogen. These elements have very low density and are overwhelmingly concentrated in the ocean and atmosphere.

The *lithophile* elements are those which prefer to be in silicate rocks. These include silicon, magnesium, oxygen, calcium, aluminum, titanium, etc. These elements are overwhelmingly in Earth's mantle and crust.

The *siderophile* elements are those which prefer the metallic state. These are the metals we are all familiar with—nickel, gold, silver, copper, iron, platinum and so on.

The *chalcophile* elements are sulfur loving and occur in sulfur-bearing minerals. They include lead, copper, zinc, platinum, and arsenic. There is substantial overlap between chalcophile and siderophile elements. Iron is unique because it lies in all three groups. Whether it is metal or silicate depends on the oxygen content. It also forms the most common sulfide, pyrite, otherwise known as "fool's gold."

In addition, there is a subset of lithophile elements that need special consideration—those that are concentrated into silicate liquid. Because molten rock is called *magma*, these elements can be called *magmaphile*. Magmaphile elements are so large that they do not fit readily into silicate minerals and are strongly concentrated into the liquid phase when a rock melts, since the liquid has more flexible "sites" that are able to accommodate the larger elements. Magmaphile elements and molecules tend to become concentrated in Earth's crust, which, we will see shortly, is created by partial melting of Earth's interior. Many of them then end up in the ocean or atmosphere. The magmaphile elements are generally those lithophile elements toward the bottom of the periodic table, where the high atomic number generally creates large ionic sizes (e.g., Rb, Cs, Ba, Sr, La, Pb, Th, and U), as well as the atmophile elements and molecules such as CO_2 and H_2O. Some siderophile and chalcophile elements (e.g., W, Sb) are also magmaphile.

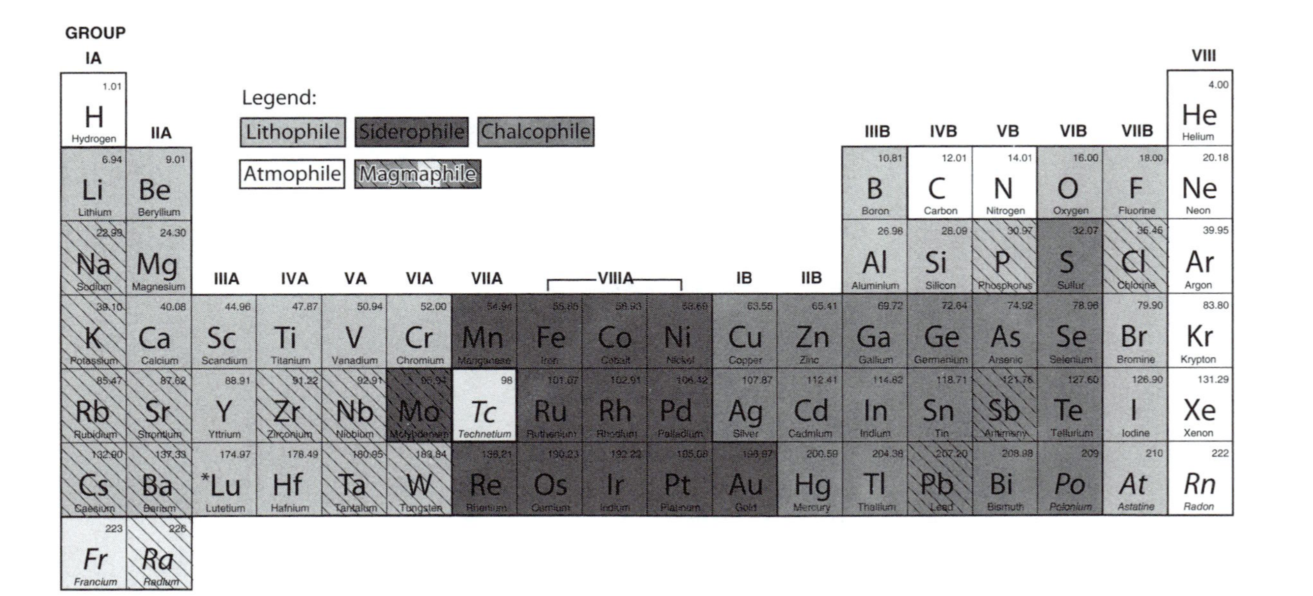

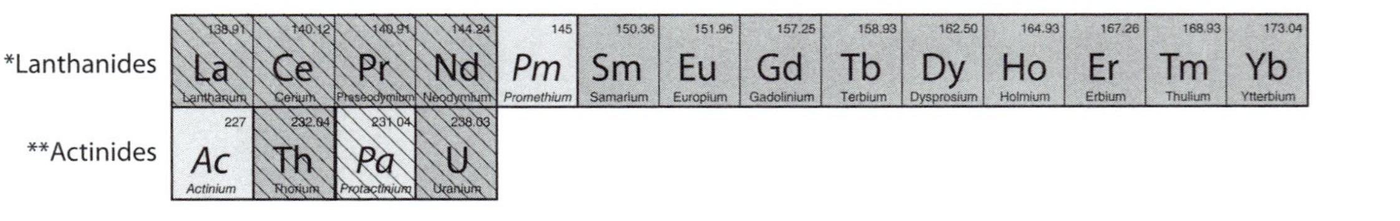

Fig. 7-6: Periodic table with element affinities based on Goldschmidt's classification. Elements are grouped according to their preferred host phases into lithophile (silicate loving), siderophile (iron loving), chalcophile (sulfur loving), and atmophile (gas loving). Magmaphile elements, which preferentially enter the silicate liquid when both solid and liquid are present, are indicated by diagonal lines. Elements in italics are short-lived radionuclides.

These affinities predict where most of Earth's elements reside. For example, although nickel is as abundant as calcium and aluminum in meteorites, most of Earth's nickel is in the core because of its siderophile tendencies. Gold, silver, platinum, and tungsten (W) were also depleted from the mantle by core formation, making these already rare elements even more precious. As we shall see in a later chapter, because Earth's iridium resides largely in the core, it has been possible to show that a large iridium-rich asteroid hit Earth 66 million years ago. Lithophile elements reside largely in the mantle, with the exception of the subgroup of magmaphile elements such as K and the volatiles that are largely in the crust, ocean, and atmosphere.

These considerations allow us to estimate a bulk composition of Earth and where the various elements reside (Fig. 7-7). As an example, consider the abundance of an element such as Fe. The mass of Earth's core is 1.87×10^{27} gm. That of the mantle is 4.02×10^{27} gm. Together the masses of Earth's thin outer crust, ocean, and atmosphere come to only 0.029×10^{27} gm. The iron content of the crust can be measured directly, but there is so little crust that it is a negligible contribution to the total. The iron content of mantle material can be estimated about 8% by weight. The iron content of the core is estimated from density and seismic velocity calibrated with experimental data to be 85% by weight. From these various results the iron content of the planet can be estimated:

Mass of iron in core	1.6×10^{27} gm
Mass of iron in mantle	0.26×10^{27} gm
Mass of iron in crust	0.002×10^{27} gm
Total mass of iron	1.86×10^{27} gm
Total mass of Earth	6.0×10^{27} gm

This yields an iron content for the bulk earth of 31.9 %. The abundances of most other elements can also be estimated by a similar approach that makes use of the geochemical behavior of the different elements (Table 7-2).

Also, we can now reconsider the hypothesis that Earth formed from meteoritic materials. If that hypothesis were true, then we would expect the silicate earth for which we can make direct measurements (Earth excluding the core) to have chondritic proportions of refractory lithophile

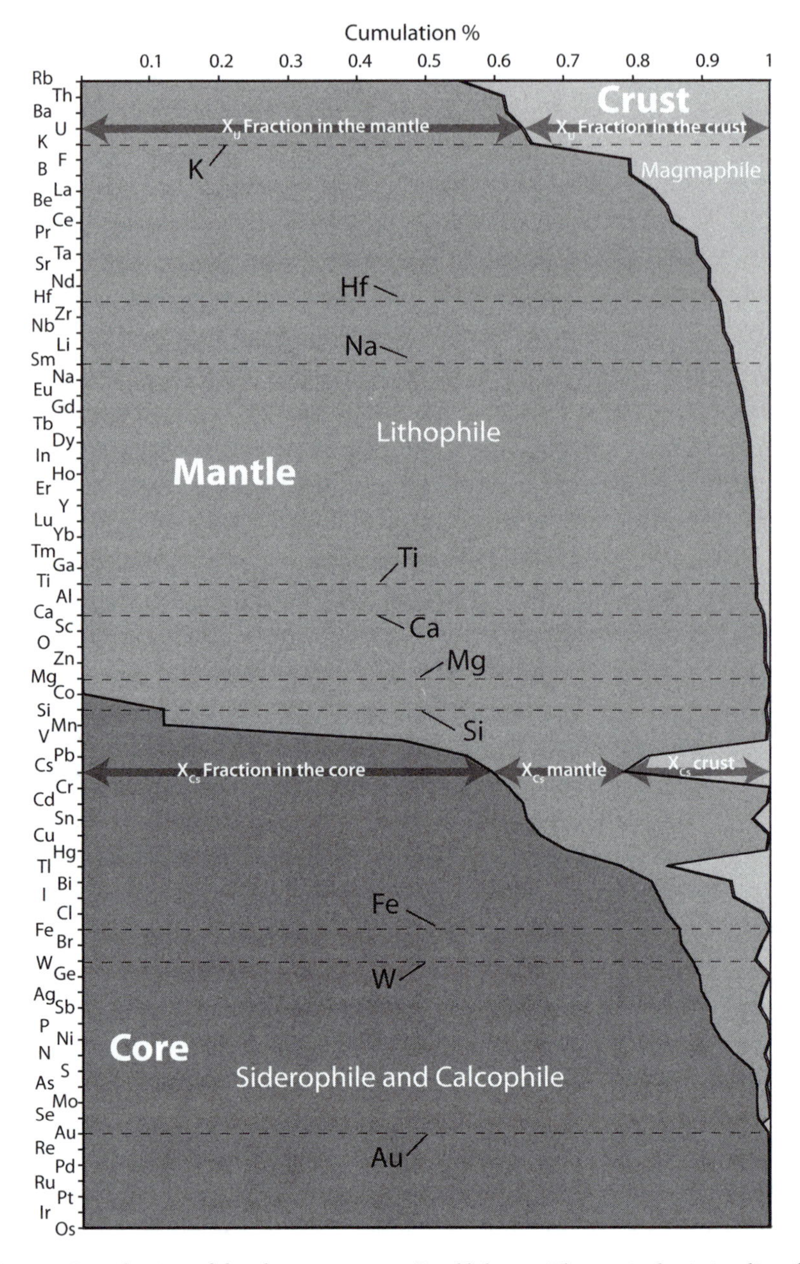

Fig. 7-7: Distribution of the elements among Earth's layers. The vertical axis is a list of 63 elements. The horizontal axis shows the proportion of the element in each layer normalized to 1, considering the relative mass of each layer: the mass of Earth: 5.997; mass of the crust: 0.0223; mass of the mantle: 4.043; mass of the core: 1.932. Chalcophile and siderophile elements have large proportions in the core; lithophile elements are in Earth's silicate layers, with the magmaphile elements concentrated in the continental crust. (Data from W. F. McDonough, *Chemical Geology* 120 (1995):223–253)

Table 7-2
Composition of the bulk Earth*

H	260	Zn	40	Pr	0.17
Li	1.1	Ga	3	Nd	0.84
Be	0.05	Ge	7	Sm	0.27
B	0.2	As	1.7	Eu	0.1
C	730	Se	2.7	Gd	0.37
N	25	Br	0.3	Tb	0.067
O (%)	29.7	Rb	0.4	Dy	0.46
F	10	Sr	13	Ho	0.1
Na (%)	0.18	Y	2.9	Er	0.3
Mg (%)	15.4	Zr	7.1	Tm	0.046
Al (%)	1.59	Nb	0.44	Yb	0.3
Si (%)	16.1	Mo	1.7	Lu	0.046
P	1210	Ru	1.3	Hf	0.19
S	6350	Rh	0.24	Ta	0.025
Cl	76	Pd	1	W	0.17
K	160	Ag	0.05	Re	0.075
Ca (%)	1.71	Cd	0.08	Os	0.9
Sc	10.9	In	0.007	Ir	0.9
Ti	810	Sn	0.25	Pt	1.9
V	105	Sb	0.05	Au	0.16
Cr	4700	Te	0.3	Hg	0.02
Mn	1700	I	0.05	Tl	0.012
Fe (%)	31.9	Cs	0.035	Pb	0.23
Co	880	Ba	4.5	Bi	0.01
Ni	18220	La	0.44	Th	0.055
Cu	60	Ce	1.13	U	0.015

* Data from W. F. McDonough, *Chemical Geology 120 (1995): 223–253*. Concentrations in ppm unless otherwise noted.

elements and strong depletions in the siderophile and chalcophile elements that would have been segregated into the core. In general this corresponds with the observations. However, precise estimates of bulk silicate earth compositions are models that make use of the assumption that Earth is chondritic. Within the uncertainties in the data, Earth could deviate slightly from chondritic proportions. At a level of high precision, much remains to be understood about Earth's exact composition.

The combination of moment of inertia, seismology, geological observations, experimentation, geochemistry, and cosmochemistry thus provides substantial knowledge of the compositions of Earth's layers. This knowledge in its broad aspects is universally accepted because it accounts for a large variety of observations, from seismic velocities to siderophile element abundances. The composition of crust and upper mantle within a small range of error bounds rank an 8–9 on our certainty scale because there are so many lines of evidence, including direct measurement and experimentation. We also know with high certainty that the lower mantle is silicate and the core is Fe-Ni metal because seismic data and bulk earth properties constrain their composition. The precise compositions of lower mantle and core rank only a 7–8 on the scale because there are no direct chemical measurements, and the very high pressures limit the experimental constraints.

Origin of Earth's Layers

Armed with knowledge of the existence, physical properties, and overall composition of Earth's layers, we now face the question of how these layers came about. As we learned in Chapter 5, some meteorites appear to be the remnants of small planetary objects that were broken apart by subsequent impacts. These various meteorites show that planetary bodies other than Earth ended up with discrete layers of metal, silicate, and volcanic material. Venus and Mars, like Earth, also have greater mean densities than their silicate crusts and must have large amounts of metal in their interiors, have evidence of crust formation by volcanism, and an atmosphere. Taking this body of evidence, it appears that the process of planetary differentiation to core, mantle, crust and volatile-rich surface is a general phenomenon—how does it happen?

SEPARATION OF CORE FROM MANTLE

A critical question is whether the layers were created as the planet formed, with a metal core coming together first, followed by the other layers, or whether metal and silicate were initially all mixed together and later segregated into the layers we see today. The first model can be called *heterogeneous accretion* because the materials that were initially added to the Earth would have been very different from one another and varied with time. The metal material would first gather to form the core, then the silicate material added on top of the core, and finally the gas and water to form the ocean and atmosphere sprinkled onto the surface. The second model is called *homogenous accretion* because the materials accreted to Earth were initially homogeneous and the separation of layers would happen after accretion, with the core sinking into Earth's center and the volatiles outgassing to form ocean and atmosphere. As for the crustal layer, we see that layer being created and recycled today, so heterogeneous accretion is not an option. It is, however, a potential model for the origin of core, mantle, and volatile-rich exterior.

Since planets accrete from solid objects in the solar nebula, heterogeneous accretion of the core would require a period when all solid objects in the early inner solar system were metal, and all the silicates remained in the gas. That would require that metals were more refractory (= less volatile) than the lithophile elements, so metal would be the first material to precipitate from the nebular gas, followed at lower temperatures by materials containing Si, Al, Ca, Ti, etc. Experiments, theoretical calculations, and observations of carbonaceous chondrites, however, all suggest that calcium and aluminum-rich silicates were the first materials to solidify (Table 7-3). Dating of meteorites also shows that the iron meteorites, thought to be remnants of cores of small solar system objects, formed slightly *after* the oldest materials in chondrites. Therefore, there is no evidence for a period in solar system history where metal was available to accrete in the absence of silicate. The metal cores of planetary bodies must have formed by separation of metal from rock after accretion.

Furthermore, astronomers studying interstellar clouds where star formation is occurring have found evidence for both metal and silicate in the clouds surrounding the star. This observation provides evidence for

Table 7-3
Condensation sequence

Temperature (K)	Element condensing	Form of condensate
3695	W (Tungsten)	Tungsten oxide – WO_3; wolframite – $FeWO_4/MnWO_4$
1760–1500	Al, Ti, Ca	Aluminium oxides: Al_2O_3, CaO, $MgAl_2O_4$
1400	Fe, Ni	Nickel-iron grains
1300	Si	Silicates: olivine $(Mg, Fe)_2SiO_4$, feldspars (Na, K) $AlSi_3O_8$
450–300	C	Carbonaceous compounds and hydrous minerals
<300	Ices	Ice particles: water, H_2O; ammonia, NH_3; methane, CH_4; argon-neon ices

the simultaneous presence of both materials during solar system formation. In its pure form, therefore, the heterogeneous accretion scenario is not well supported.

Since heterogeneous accretion is not viable for formation of core and mantle, core and mantle must have separated from one another after accretion. Is this physically reasonable? Two characteristics of metal and silicate account for this separation:

(1) The high density of metal relative to silicate means the metal would sink through the silicate under the force of gravity;
(2) this density separation is facilitated because metal and silicate are *immiscible*.

Immiscibility is a concept familiar to us from "kitchen science." Water is denser than alcohol, but addition of water to alcohol does not lead to the creation of a layer of pure water at the bottom of the glass. The two liquids are miscible; they mix to form a single homogeneous material. In contrast, oil and vinegar are immiscible; they remain separate phases when mixed together. Even after being vigorously stirred, the less dense oil floats to the top to form a pure layer, with the denser vinegar (which is largely water) underneath. Immiscibility and density differences lead to discrete layers with a sharp boundary between them. Metal and silicate are like oil and vinegar, they are immiscible. The combination of the large density difference and immiscibility makes downward separation of metal from silicate inevitable and irreversible.

TIMING OF CORE FORMATION

When did the core form and how long did core formation actually take? One possibility would be that melting and vigorous convection at the time of accretion would cause core formation to happen instantaneously with accretion. Alternatively, core formation could be progressive and even be happening in small amounts today.

Both geophysical arguments and geochemistry have a bearing on this question. Consideration of heat sources in the early Earth makes it hard to avoid the conclusion that substantial melting must have occurred during its formation. Heat sources in the early Earth include the heat released by impacts, heat created by the now extinct radionuclides, and heat from long-lived radionuclides. In addition, the process of core separation generated a great deal of heat as the metal "fell" to Earth's center. Calculations involving these heat sources show that so much heat was available at this point in Earth's history it is highly likely that substantial portions of Earth melted. If the moon formed from impact of a Mars-size protoplanet (discussed in Chapter 8), the impact alone would have generated enough heat to melt Earth's interior.

Good evidence also exists from studies of lunar rocks that the moon melted early in its history, as we will see in Chapter 8. Because of both heat sources and heat sinks, it is likely that if the moon melted, Earth melted also. The moon has smaller sources of heat than Earth. The smaller gravitational field causes heat generated by impacts to be less. The moon is volatile depleted, so the major heat producer ^{40}K was lower in abundance. The moon has no significant core, so heat generation by core formation would not contribute. Therefore, much more heat must have been available to melt Earth than to melt the moon.

Heat loss would also be more efficient on the moon because it is a smaller object. This can be understood experientially by considering the different time it takes a drop of hot water to cool compared to a cup of coffee. Small objects cool much more rapidly than large objects because heat contained in the entire volume is lost through the surface. The larger the ratio of surface area to volume, the more rapidly is heat lost. Since the moon has fewer sources of heat and more rapid dissipation of heat than Earth, if the moon melted, it is likely that Earth did also. Melting of substantial portions of Earth in the first few tens of millions of years of its history would have led to efficient segregation of core from mantle.

Quantitative estimates of the time to segregate the core come from our geochemical clocks—the radiogenic isotopes. In this case, the key is an extinct lithophile radioactive nuclide, ^{182}Hf, with a half-life of 9 million years, that decays to siderophile ^{182}W. ^{182}Hf accreted by Earth decayed completely to ^{182}W in the first 100 million years of Earth's history. If the core formed 100 million years or more after Earth accreted, then all the ^{182}Hf would have decayed away prior to core formation, and the tungsten isotopes of late-forming core would be the same as the mantle and bulk earth. Furthermore, the isotope ratios would also be the same as other solar system materials, such as chondrites, that never experienced metal-silicate separation (Fig. 7-8). On the other hand, if the core formed very quickly, then most of the tungsten would segregate into the core before ^{182}Hf had decayed away, and the mantle left behind would have had a very high Hf/W ratio while some radiogenic ^{182}Hf was still present. As the remaining ^{182}Hf decayed, the small amount of tungsten remaining in the mantle would become enriched in the daughter isotope, ^{182}W. Then the silicate Earth we can measure would have an excess of ^{182}W relative to undifferentiated objects such as chondrites. If Earth's rocks and chondrites have the same W isotope ratios, then core formation occurred $>$100 Ma after Earth accreted. If the W isotope ratios differ, core formation happened earlier. The earlier that core formation occurred, the more extreme would be the W isotope differences. Because Hf and W are rare elements, the measurements are technically difficult, and only in 2002 were definitive data obtained by several different laboratories. These measurements show enrichment of ^{182}W in the silicate Earth compared to chondrites. The magnitude of the anomaly suggests core formation took place in the first 30 Ma of Earth's history. Since the W isotope data show definitively that the core formed *after* Earth accreted, the data provide further evidence for homogeneous rather than heterogeneous accretion.

The various lines of evidence, then, suggest a dominantly homogeneous accretion of Earth, followed by widespread melting and rapid segregation of core from mantle by immiscibility and density separation in the first few tens of millions of years of Earth's history.

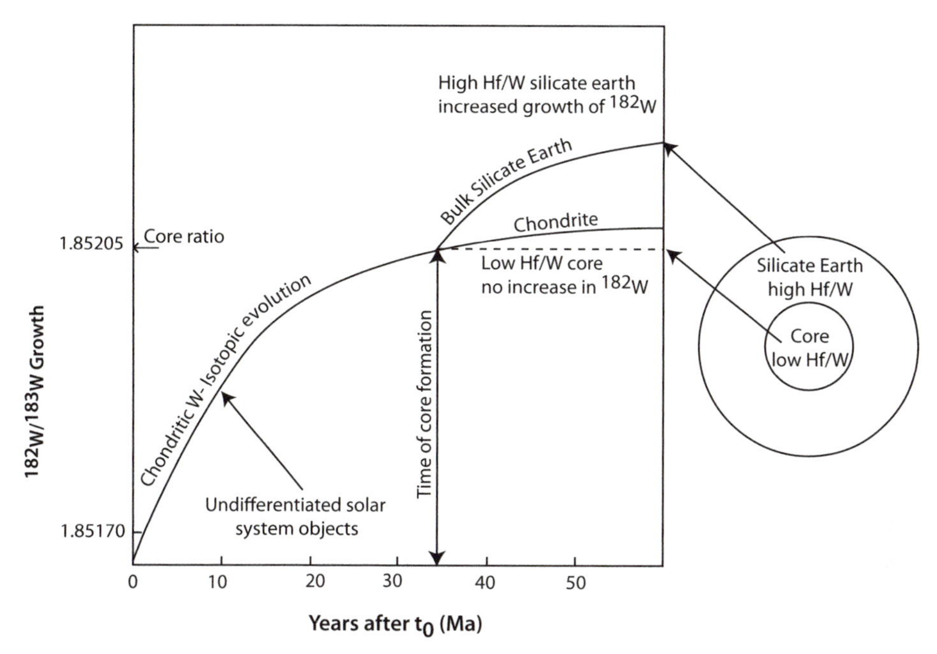

Fig. 7-8: Illustration of the effect of core formation on tungsten isotope evolution. T_0 is the time when Earth accreted. If the core formed >100 million years after T_0, all the radioactive ^{182}Hf would have decayed to ^{182}W, and core, mantle, and carbonaceous chondrites would all have the same tungsten isotope ratio. Because the silicate Earth has a different W isotope composition than chondrites, with enrichment in ^{182}W, the core must have formed before the completion of Hf isotope decay. (Data from Quing-zhu Yin et al. *Nature* 418 (2002): 949–52 and Schoenberg et al., *Geochim. Cosmochim. Acta* 66 (2002):3151)

This time frame fits well with evidence from other meteorites and isotope systems. These studies have shown that smaller planetary objects broken apart to form achondritic meteorites formed and differentiated into layers of metallic core, silicate mantle, and volcanic crust in as little as 5.0 million years. The heat source for such rapid differentiation is thought be ^{26}Al, which with its 0.7 million year half-life decays away fully in less than 7.0 Ma. Meteorites thought to be derived from Mars, which is about one-eighth the mass of Earth, give a timescale for formation of a Martian core of about 15.0 Ma. It appears that small objects formed and differentiated rapidly, while larger objects such as Mars and Earth took slightly longer.

From the perspective of the age of the solar system (4,565 million years), planetary formation and differentiation, powered by impacts and extinct radionuclides, occurred within the first 1% of solar system history. This is roughly equivalent in terms of percentages to the gestation periods of most mammals, when their primary physical structure becomes well defined.

ORIGIN OF THE CRUST

The crust is the uppermost part of the solid Earth above the seismically defined Mohorovicic discontinuity. The ocean and continental crust are distinct in their seismic properties and chemical composition. The ocean crust is made up largely of the mafic igneous rock *basalt* and its plutonic equivalent, *gabbro*. The continental crust is roughly granitic in composition (see Table 7-1), with a layer of sediments at the uppermost levels.

The origin of the crust is understood well in general terms, though many details remain to be elucidated. The major process that forms the crust is partial melting of Earth's interior. The magma produced is buoyant and either rises to the surface to erupt as volcanic lava or remains at slightly greater depths, where it cools more slowly and creates the coarsely crystalline *plutons*—gabbros in the ocean crust and granites in the continental crust. These processes can be observed today, and ancient rocks are sufficiently similar to modern ones to suggest that they formed by the same general process.

Processes of melting and crystallization near Earth's surface are subject to experimental investigation and well-constrained modeling, and geochemists over the last century have come to understand well the essential aspects of melting. Our normal human intuition is that melting is quite simple. If you heat something up, when it reaches its melting point, it melts. For example, H_2O has a single melting point of 0°C. Two factors contribute to more complex melting behavior for rocks: melting points depend on pressure (e.g., Figs. 4-3 and 7-4), and melting of complex mixtures like rocks melt over a range of temperatures, rather than a single temperature.

Unlike the constant pressure world we inhabit, Earth is subject to huge pressure variations owing to the vast amounts of rock that press down on the interior. For all substances for which the volume of the solid is less than the volume of the liquid (all rocks have this property), pressure and temperature have the opposite effect on melting of rocks. Increased pressure and decreased temperature enhance the stability of the solid. The high pressures of Earth's interior are the reason the mantle can be solid, even though the mantle is at higher temperatures than the melting temperature of mantle rock at the surface.

Since rocks are collections of minerals, and hence an assemblage of phases—not a single phase like water or ice—their melting behavior is different from pure compounds. For example, pure salt turns to liquid at a temperature of 350°C, and ice at 0°C, but an equal mixture of ice and salt melts at −20°C. We make use of these properties frequently. Small amounts of salt easily melt in water (we say "dissolve") at room temperature. In winter, adding salt to roads turns the ice to saltwater, even at temperatures below freezing. This general principle where adding one substance to another lowers the melting temperatures is called *freezing point depression* and can be represented graphically in a "phase diagram" as illustrated in the sidebar.

Careful experimental work also reveals that mixtures of minerals melt over a range of temperatures, not a single temperature. This range is bounded by the inception of melting, called the *solidus*, and the completion of melting, at the *liquidus*. Below the solidus, the mixture is solid. Between solidus and liquidus, the system is only partially melted. Above the liquidus, the mixture is entirely liquid. Because melting occurs over a range of temperatures, *partial melting* can occur, permitting segregation and separation of liquid from solid. We now arrive at the critical aspect for separation of Earth's crustal layers: *the composition of the partial melt is different from the bulk composition*. Partial melting to a different composition coupled with the buoyancy of the liquid relative to the solid permits segregation of the melt to form crustal layers that differ from the mantle that melted to produce them. These properties of melting mixtures can be viewed graphically and quantitatively using temperature-composition diagrams, called *phase diagrams* (see sidebar).

✷ SIDEBAR ✷

MELTING OF ROCKS

Our intuition about melting is based primarily on ice, which melts at a single temperature of 0°C. When there is more than one mineral present, however, melting takes place over a range of temperatures. Rocks are made up of many minerals, and therefore the details of the melting are quite complex. The important principles can be illustrated using mixtures of two minerals. In the first example, there are two distinct minerals that do not dissolve in one another in the solid state. In the liquid, however, all molecules mix together to form a single solution (analogous to salty water for ice/salt mixtures). In the second example, the molecules mix and form solutions in both liquid and liquid states (such as alloys), so there is only one solid phase present, but the solid can have varying proportions of its constituent molecules. Rocks are made up of both kinds of minerals—those of uniform composition and those which form solid solutions.

Equilibrium Melting in the Diopside-Anorthite Binary Eutectic Phase Diagram

The binary eutectic phase diagram (Fig. 7-9) illustrates melting where there are two distinct solid phases that do not form a solid solution, and a liquid phase where all the molecules are fully miscible. The horizontal axis represents the proportion of each mineral in the mixture. Each pure mineral plots at one end of the horizontal axis, and melts at a single temperature. All other compositions along the axis are mixtures of the two minerals. Temperature is plotted along the vertical axis, increasing upward.

Below the solidus, all mixtures are solid. Above the liquidus, all mixtures are a single liquid phase. What interests us is the transition between the two. To illustrate the transition, we explore the melting of one arbitrarily chosen mixture, which we call a bulk composition, indicated by a vertical line on the diagram. The vertical line in Figure 7-9 has a composition of 52% anorthite (An) and 48% diopside (Di). We could arbitrarily have chosen any bulk composition. Whatever the choice, when the bulk composition is either entirely solid or entirely liquid, it has exactly the proportions that it started with.

For systems of this type, there is a single minimum melting temperature for all mixtures, called the *eutectic* (*E* on the diagram), which then defines the solidus. As we heat our chosen bulk composition from an entirely solid state, the solid mixture encounters the solidus at point *S1*. Here the mixture will begin to melt. The first liquid, *L1*, necessarily appears at the "eutectic point," *E*, since it is the minimum melting tempera-

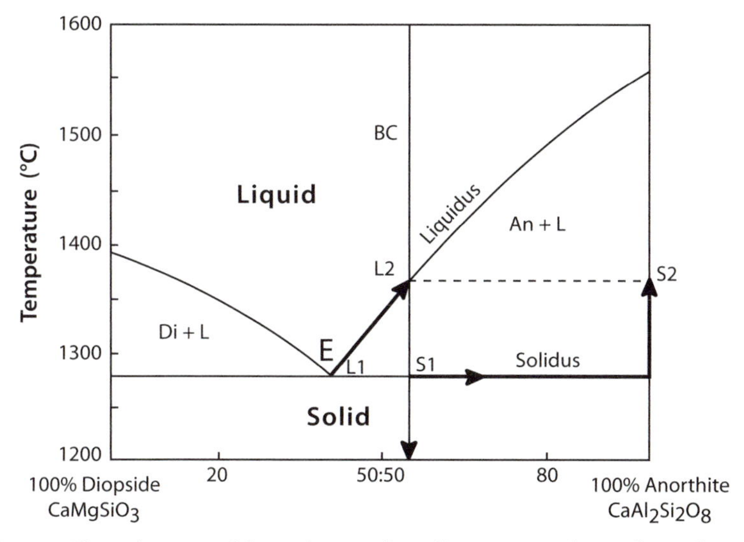

Fig. 7-9: Phase diagram of diopside-anorthite, illustrating melting of two phases that mix in the liquid state but remain separate in the solid state. The point labeled E is the minimum melting temperature, called the eutectic. BC stands for an arbitrarily selected bulk composition. Pressure is one atmosphere.

ture for all mixtures. Because this liquid is richer in Di than the bulk composition, as the liquid forms the solid composition becomes relatively depleted in Di, causing its composition to move along the horizontal axis directly away from *S1* to the anorthite axis. As long as both Di and An are present, the liquid is stuck at *L1* (= *E*) and the temperature remains constant. When the solid composition reaches the An axis, the last crystal of Di has melted out, and the system consists of 80% liquid of composition *E* and 20% An. Then the remaining anorthite begins to melt out as the temperature increases. Melting the anorthite causes the liquid to become more An rich, and it moves along the line from *L1* toward *L2*. As the temperature increases, the solid composition (pure anorthite) increases in temperature to *S2* as the liquid composition moves to *L2*. When the liquid composition arrives at the bulk composition, the system is entirely liquid, as the last crystal of An dissolves, at *S2*. Note that the liquid compositions (*L1* to *L2*) always differ from the bulk composition until the system is entirely liquid. Melting occurs over a range of temperature. Over this temperature range, partial melting occurs.

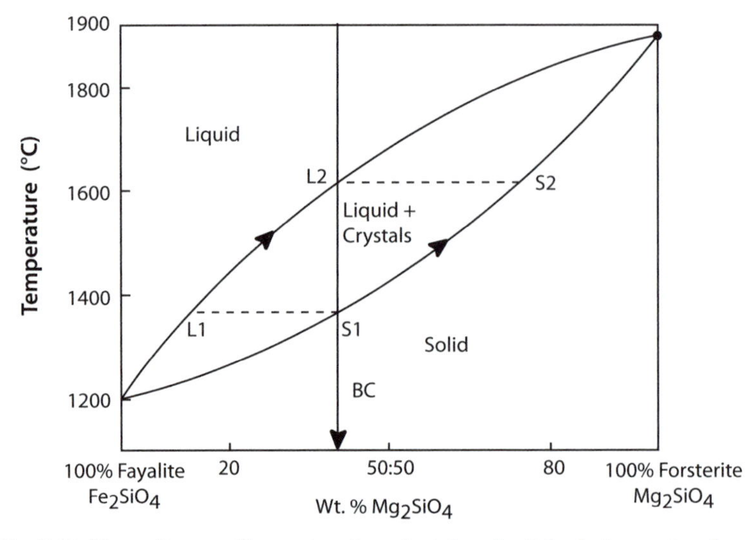

Fig. 7-10: Phase diagram illustrating the principles of solid solution, using the olivine solid solution. Olivine is the most abundant mineral in Earth's mantle. BC stands for an arbitrarily selected bulk composition. Pressure is one atmosphere.

Forsterite-Fayalite Solid Solution Phase Diagram

When there is solution in both solid and liquid, there is a single solid phase, but it can be of variable composition. The most common such solid solution in Earth's upper mantle is the mineral olivine, which is a solution between pure forsterite (Mg_2SiO_4) and pure fayalite (Fe_2SiO_4). Figure 7-10 is the olivine phase diagram. During melting, all solid compositions lie along the solidus and all liquid compositions lie along the liquidus. Co-existing solid and liquid always have the same temperature. Again we choose an arbitrary bulk composition, *BC*, indicated by the vertical line with arrow. At temperatures below the solidus, only solid olivine is present. As the temperature rises, the bulk composition arrives at the solidus and the first liquid appears at *S1*, with composition *L1*. As the temperature continues to rise, the solid follows the path *S1–S2* and the liquid follows the path *L1–L2*. When the liquid arrives at the bulk composition, the system is entirely molten, and with further heating the liquid rises above the liquidus to the liquid field. Notice that in this system as well, partial melting occurs over a range of temperatures, and during melting the liquid composition always differs from the bulk composition.

✶

The phase diagrams illustrated in the sidebar are for experimental data determined at one atmosphere pressure. Since pressure variations are very significant for Earth, we need to consider carefully the effects of pressure. Higher pressures cause the denser solid phases to become stabilized relative to the lighter liquids, and therefore it takes higher temperatures to cause melting to occur. For Earth, melting temperatures generally increase by 5°–10°C for every thousand bars (one kilobar, or 0.1 Gigapascals [GPa]) of pressure (equivalent to about 3 km of depth). At 120 km below the surface, the pressure is about 40 kbar (4 GPa), and the melting temperature of rocks is about 400°C higher than at the surface! A rock that is entirely solid at this depth could be well above its solidus at the low pressures of the surface, so if mantle material rises, it can melt because of the decrease in melting temperature with decreasing pressure. This process is called *pressure release melting*.

Pressure release occurs on Earth whenever mantle material rises. As we will learn in Chapter 11, the mantle convects slowly, and beneath ocean ridges and ocean islands the mantle ascends from depth toward the surface. As a given parcel of mantle rises, there is increasingly less weight of rock above it, and the pressure decreases. Ultimately the pressure decreases enough that the mantle crosses the solidus (the temperature at which melting begins) and starts to melt (Fig. 7-11). As it ascends, the mantle is farther and farther above its solidus and the extent of partial melting increases. The proportion of melt produced depends on the extent of rise above the depth where the solidus is crossed. This explanation of mantle melting contradicts our experience of melting by the application of heat and the raising of temperature, since we live in a constant pressure environment. Instead, the mantle cools down as it melts! Melting occurs by pressure decrease rather than temperature increase.

Now we come to the separation of chemically distinct crustal layers. Partial melts of the mantle have a different composition from the mantle itself. Instead of the 45% SiO_2 and 40% MgO of mantle peridotite, the partial melts have about 50% SiO_2 and about 15% MgO. Such liquids, which become basalts, have a density that is 10% less than the mantle and rise readily from the molten zone toward the surface. Partial melting of the mantle to produce basalts is the mechanism that forms the ocean crust, discussed at length in Chapter 12.

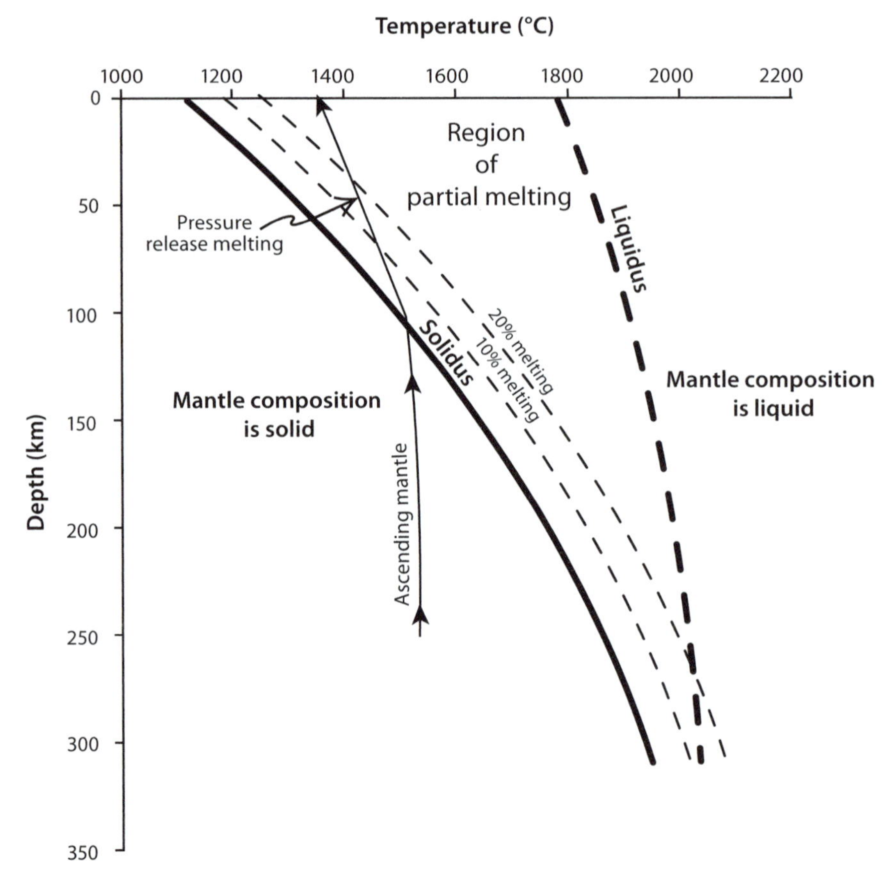

Fig. 7-11: Illustration of how the mantle melts on Earth by depressurization. Melting of mantle rock (peridotite) depends on temperature and pressure. The onset of melting is called the *solidus*; the temperature above which melting is complete is called the *liquidus*. Between the two is a partially molten region, as indicated by the contours of percent melting. The path for melting during mantle ascent changes after crossing of the solidus because melting takes energy, which lowers the temperature of the ascending mantle. (Solidus line after Hirschmann, *Geochem. Geophys. Geosyst.* 1(2000), paper no. 2000GC000070; liquidus temperature after Katz et al., *Geochem. Geophys. Geosyst.* 4 (2003), no. 9)

Continental crust formation follows many of the same principles, but there are multiple steps of partial melting. Multiple steps are necessary because the granite rocks and high Si and K contents of the continents cannot be produced by melting of the mantle. Instead, the continents represent the final product of sequential melting and cooling of magmas. When the mantle melts, basalts are formed. When these basalts melt in turn, granites form. Granites are also the end point of many other likely processes. When granites melt, granites are formed. When the mafic lower crust is melted, less dense granitic magmas are formed and rise to the upper crust. When granites or basalts are eroded to form sediments and these sediments melt, granites are formed. Therefore granites represent the logical endpoint of a sequence of many different melting and cooling events. The low densities of granitic rocks then ensure that they remain at the surface, floating on top of the mantle, well above the levels of the ocean crust.

Since continents are the ultimate endpoint of multiple melting processes, they have very effectively concentrated the magmaphile elements that prefer liquid to solid. Trace elements such as Th, U, Ba, Rb, K, and La have a substantial fraction of their total Earth budget in the continental crust. Perhaps as much as 70% of Earth's Rb, for example, has been concentrated in the outermost layer, effectively distilled to the surface by magmatic processes.

While this general conception of continent formation is sound, the specific physical processes involved in the initial creation of continental crust and the timing of when it happened are not well understood. If basalts come to the surface and then melt to produce granites, and if both granitic partial melt and mafic solid residue were to remain in the crust, the total composition of the crust would not change, it would simply be divided into two different layers. This occurs in part—the lower crust is more mafic than the upper crust—but the total crust is too high in SiO_2 and low in FeO and MgO to be a mantle melt. Furthermore, materials being added today to the continents by volcanic and plutonic rocks do not have the same composition as average continent—they are too mafic and do not have appropriate ratios of important trace elements. What we determine by experiments to be a mantle melt is dominantly basalt, what we observe being added to the continent is dominantly basalt, so how did this material become transformed to the granitic masses that form much of the continents we live on?

Three general models have been proposed to account for this puzzle. The first suggestion is that processes have changed though Earth's history, and in the ancient Earth the mantle was hotter and melted recycled basalt materials at depth to give rise to the continents. The residue remained in the mantle. The second idea is that continents form by a multistage process, where first a basaltic layer is formed, then the mafic residue that melted to produce granite "delaminates" and falls back into

Table 7-4
Compositions of crust and mantle

Chemical composition (wt%)	Continental crust*	Primitive mantle**	Ocean crust
SiO_2	60.6	45	50.39
TiO_2	0.92	0.2	1.72
Al_2O_3	15.9	4.45	14.93
FeO	6.71	8.05	10.2
MnO	0.1	0.14	0.18
MgO	4.66	37.8	7.34
CaO	6.4	3.55	11.29
Na_2O	3.1	0.36	2.86
K_2O	1.81	0.03	0.25
P_2O_5	0.13	0.02	0.35
(ppm) Li	16	1.6	
Sc	22	16.2	41
V	138	82	—
Cr	135	2625	—
Co	27	105	17
Ni	59	1960	150
Cu	29	30	74.4
Zn	72	55	—
Rb	49	0.6	1.35
Sr	320	19.9	124
Y	19	4.3	40.3
Zr	132	10.5	122.4
Nb	8	0.658	3.79
Cs	2	0.021	0.0141
Ba	456	6.6	14.8

Table 7-4 (Continued)

Chemical composition (wt%)	Continental crust	Primitive mantle	Ocean crust
La	20	0.648	4.36
Ce	43	1.675	13.4
Pr	4.9	0.254	—
Nd	20	1.25	12.3
Sm	3.9	0.406	4.1
Eu	1.1	0.154	1.46
Gd	3.7	0.544	5.67
Tb	0.6	0.099	0.99
Dy	3.6	0.674	6.56
Ho	0.77	0.149	1.42
Er	2.1	0.438	4.02
Yb	1.9	0.441	3.91
Lu	0.30	0.0675	0.59
Hf	3.7	0.283	3.12
Ta	0.7	0.037	—
Pb	11	0.15	0.59
Th	5.6	0.0795	0.2
U	1.3	0.0203	0.08

* *Continental crust composition from Rudnick and Gas Composition of the Continental Crust Treatise on Geochemistry #3, 1–64.*

** *Primitive mantle from Sun and McDonough Chemical and isotope systematics of ocean basalts: implications for mantle composition and process.* Geol. Soc. London Special Pas *42 (1989):313–345.*

the mantle (Fig. 7-12). A third intriguing possibility is that weathering is as important a process as igneous activity in controlling continental crust composition. Today we know that weathering of mafic minerals that make up basalt is a much more rapid process than weathering of the felsic minerals that make up granite. Weathering, therefore, might selectively remove the more mafic elements and deliver them to the ocean, leaving the felsic continent behind.

However they formed, granitic continents are an ancient feature of Earth. The oldest preserved continental rocks are very similar in composition to average continent today, suggesting that continent formation has been a repeatable process throughout Earth's history. In fact, the very oldest rocks are granites and sediments that were formed by weathering of granites. Both types of rocks are not "juvenile"—they require an extended history. This tells us that crustal, granite-forming processes must have been active in earliest Earth's history, even before the rock record.

ORIGIN OF THE ATMOSPHERE AND OCEAN

The scenario given above for the solid Earth also relates closely to the origin of Earth's atmosphere and ocean layers. The important volatiles H_2O and CO_2 were able to accrete in significant amounts because they can reside in the solid state in some of the minerals that make up rocks. Even minerals that have no volatile element in their chemical formula can contain small amounts of volatiles. Other minerals, such as amphibole or mica, contain substantial amounts of water. Limestone ($CaCO_3$) is by far the largest reservoir of CO_2 in Earth's crust. When these minerals break down and release gas, either by being heated or melted, the volatiles will rise to the surface, providing carbon, hydrogen, and nitrogen, the elements most essential for life. Degassing is an important process of atmosphere creation. When did the atmosphere form in Earth's history?

Once again radiogenic isotopes come to our aid with this question, making use of another exotic element, xenon (Xe). The noble gases such as xenon have an important place in the discussion of the atmosphere, because they are nonreactive and always remain in the gaseous state. Since they do not react with other elements to form minerals, a significant proportion of the noble gas budget of the Earth resides in the atmosphere. One of the xenon isotopes, ^{129}Xe, is the product of the short-lived radioactive isotope ^{129}I, with a half-life of 16 million years. While xenon is atmophile, iodine (I) is lithophile. Therefore just as Hf and W provide evidence on the separation of the core and mantle, I and Xe provide evidence for the separation of mantle and atmosphere.

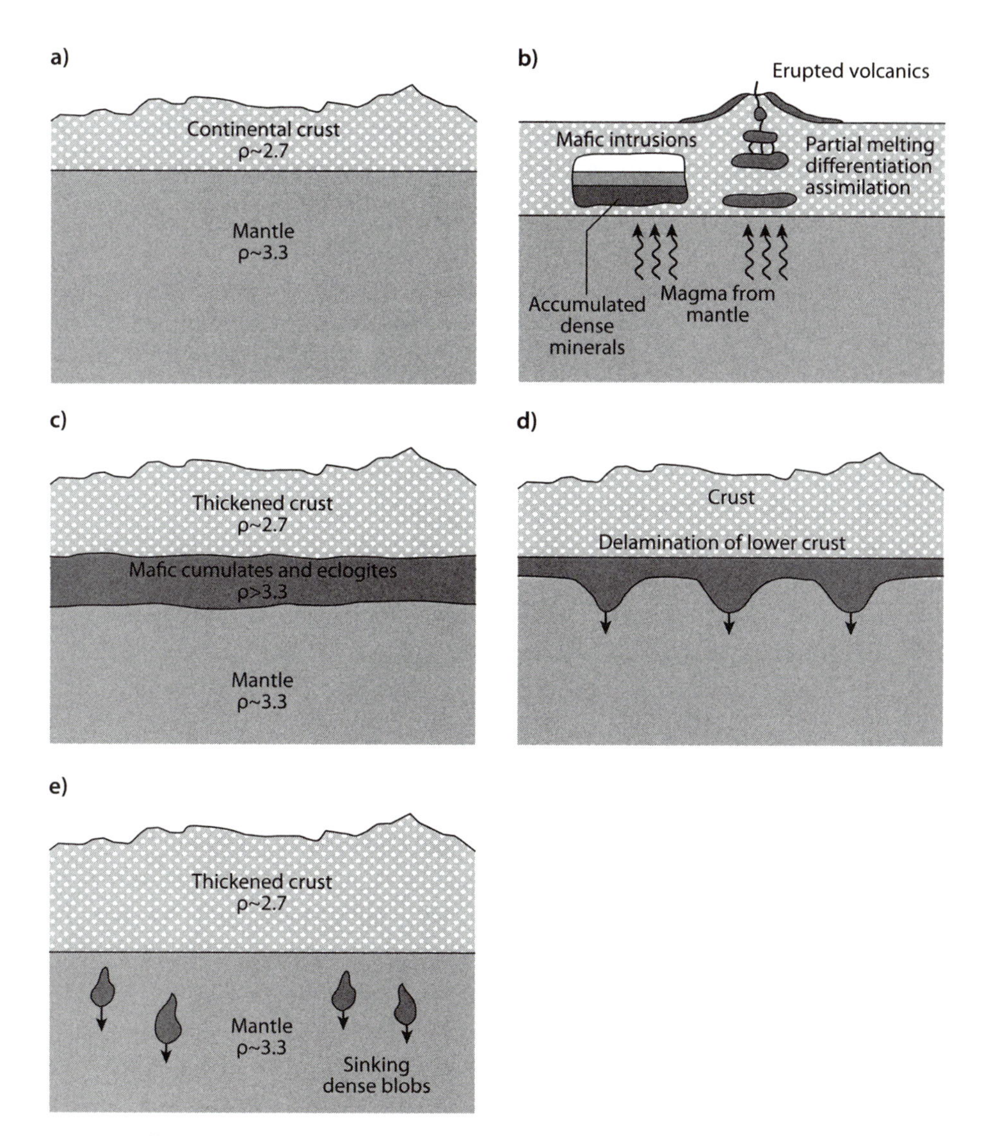

Fig. 7-12: Illustration of the role continental delamination may play in formation of the continents. Continental collision or arc magmatism thickens the crust and could lead to sufficient crustal thickening that melting takes place at depth. The granitic melts rise to the surface. The high temperature residue is sufficiently dense that it sinks into the mantle, leaving behind a high SiO_2 continental crust. Alternatively, magmas could crystallize dense minerals rich in Fe and Mg at depth in the crust, and these accumulated crystals could delaminate.

In 1983 Claude Allègre and coworkers determined the xenon isotope composition of volcanic rocks from ocean ridges and showed that they have an excess of ^{129}Xe compared to the atmosphere. Because ocean ridge basalts are derived by partial melting of the mantle, the inference is that the upper mantle has the ^{129}Xe anomaly. If the separation of mantle and atmosphere occurred 100 Ma or more after Earth accreted, then all the ^{129}I would have decayed away and all of Earth's reservoirs would have the same xenon isotopic composition. If on the other hand the atmosphere separated from the mantle very early in Earth's history, removing much of the xenon, then the remaining iodine would keep on producing ^{129}Xe. Since there was little xenon left in the mantle, an excess of ^{129}Xe relative to other xenon isotopes would result. The constraints from ^{129}I also suggest a period of about 30 Ma, similar to the timescale inferred from Hf-W for core/mantle differentiation. This evidence provides a consistent story where formation of the major layers of core, mantle, and atmosphere occurred over the first few tens of millions of years of Earth's history. The atmosphere and ocean would then be the result of homogenous accretion and mantle degassing.

For the atmosphere, however, the devil is in the details. Other noble gas isotope ratios are not readily explained either by degassing from Earth's interior or late addition of gases. Impacts of comets seem inevitable from models of solar system formation, but some of the measurements of isotope ratios of modern cometary material are not the same as those of the atmosphere and ocean. Giant impacts (see Chapter 8) or vigorous solar wind early in solar system history might have stripped the original atmosphere, so perhaps multiple atmospheres formed during the early Earth. A complex history that includes heterogeneous accretion of materials with different volatile budgets seems necessary to explain Earth's inner and outer volatile abundances. These remaining puzzles make the atmosphere the least understood aspect of formation of Earth's layers.

Summary

A family of different processes has led to the gradual differentiation of the Earth into layers sorted by density whose extent and composition are well constrained, with error bars and questions increasing with increasing depth. The innermost layer is a dense solid core of iron, nickel, and a small proportion of lighter elements. The liquid outer core is also metallic. Much lower in density is the mantle, consisting largely of solid solutions of ferro-magnesian silicates. Melting of the mantle leads to outgassing to create ocean and atmosphere and eruption of silicate magmas to form the crust. The denser, basaltic ocean crust is a result of mantle melting and resides at deeper levels than lighter, granitic continental crust, which must be derived by a sequence of melting processes.

Short-lived radioisotopes in the Hf-W and I-Xe systems both indicate that core and atmosphere formed early in Earth's history, and of course these reservoirs remain separated today. In contrast, crust formation and destruction are ongoing processes. Ocean crust formation can be observed today and is a direct result of mantle melting, but all of ocean crust is young, <150 Ma. Continental crust has a vast range of ages. Even the oldest continental rocks show evidence of a long previous history, and the earliest crustal differentiation events are hidden from view. Various theories are still being actively considered for the exact mechanisms by which continental crust has formed.

The diverse processes that have created Earth's layers have also led to a striking separation of elements. The siderophile elements reside largely in the core, and the large ion lithophile elements have been concentrated into mantle and crust. The magmaphile elements have been effectively concentrated into the crust, and the special classes of volatile magmaphile elements have been outgassed to form the least dense layers—ocean and atmosphere. This interior stratification sets the framework for life, which must rely on the molecules concentrated toward the surface. In particular, CO_2, H_2O, and N, which are so central to all living molecules, are placed at the surface by planetary differentiation, as are the magmaphile elements central to life—K, Na, Cl, and P. As we shall see in Chapter 9, CO_2 and H_2O and their interactions with the crust are also key players in establishing the long-lived climate stability upon which the origin and evolution of life depends.

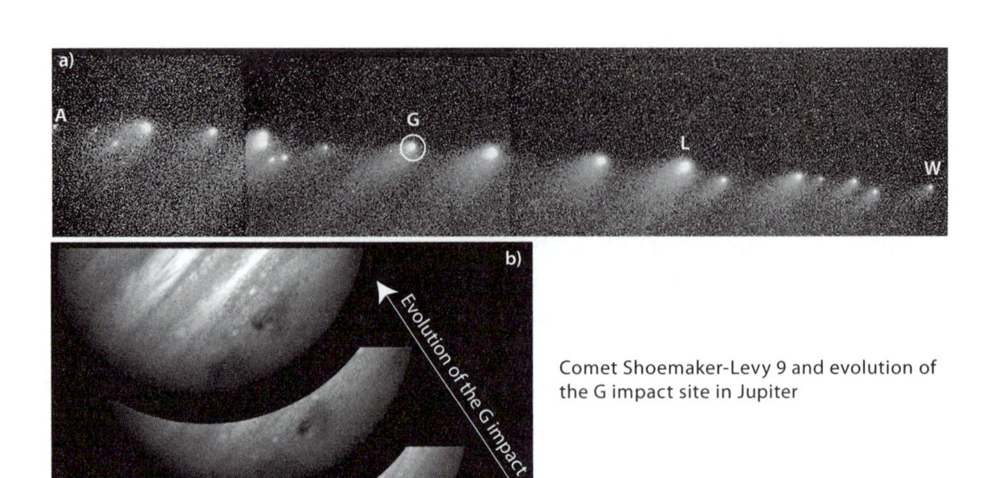

Fig. 8-0: (a) Hubble Space Telescope image of Comet Shoemaker-levy 9 in March 1994, four months before its collision with Jupiter. The comet had broken up into a "string of pearls" that ultimately impacted with Jupiter. (b) Mosaic of WFPC-2 images shows the evolution of the comet impact site on Jupiter. The comet broke into fragments, producing several impact sites. The change in one impact can be seen in the images from lower right to upper left, and the scar from a second impact appears in the third image. (Credit: (a) Courtesy of NASA; credit: H. Weaver (JHU), T. Smith (STScI). (b) R. Evans, J. Trauger, H. Hammel, and the HST Comet Science Team and NASA)

CHAPTER 8

Contending with the Neighbors

Moons, Asteroids, Comets, and Impacts

We are not alone in the Solar System. Since ancient times, it has been apparent that we have neighbors distinct from the stars, most obviously the moon and planets, with their bright presence and distinctive orbits. And occasionally one of the neighbors drops by for a permanent visit, seen in meteor showers or the occasional meteorite or comet that finds its way to Earth's surface. Particularly in the early history of the solar system, interactions with the neighbors had a major role in Earth's formation, and subsequent interactions have played a major role in the evolution of life and even today pose a threat of environmental catastrophe. From one point of view, Earth is simply made up of neighbors whose visits became permanent. If we could choose one planetary grain as the original Earth, then it ultimately grew to our current planet by the progressive accumulation of material through impacts, a process that still continues on a reduced level today. Therefore in a very real sense, we are *our former neighbors.*

Early Earth's history is closely tied to that of our nearest neighbor, the moon. Dates from lunar rocks are largely between 3.0 and 4.4 billion years. The absence on the moon of modern volcanism, of an atmosphere to produce weathering, and of any tectonic movement make the moon a kind of "planetary fossil" that records events in our vicinity early in solar system history, a period from which no rocks on Earth survive. Study of the lunar rocks not only tells us about the history of our nearest neighbor, it also provides key insights into Earth's early history that can be gained in no other way. The origin of the moon now appears to have been caused by the giant impact of a Mars-sized planet about 50 million years after the origin of the solar system. Owing to the heat generated from its rapid accretion, the moon likely underwent large-scale melting of its interior to produce an early "magma ocean." Floating crystals of plagioclase rose to form the predominant rock type of the light-colored lunar highlands. Separation of other crystals led to layering of the lunar interior. Melting several hundred million years later produced the younger black lava plains of the lunar maria. Thermal considerations suggest that Earth likely also underwent very large scale melting with formation of a magma ocean early in its history. Both Earth and moon later underwent a "late heavy bombardment" of meteorites when reorganization of the orbits of the outer planets destabilized the asteroid belt, leading to many large impacts in the inner solar system around 3.8 Ga. It may be for this reason that this age corresponds closely with that of the oldest surviving rocks on Earth, and only after this time was life able to establish a permanent foothold on Earth's surface.

Impacts that were the primary process of formation of planets and moons have continued through Earth's history, with marked consequences for life, including the extinction of the dinosaurs 65 million years ago. Historic impacts are evident from comets crashing into Jupiter (see frontispiece), as well as surviving young craters on Earth. Future impacts are inevitable, either of near Earth objects from the asteroid belt, or comets released from the vast cometary storehouses of the Kuiper Belt and Oort Cloud, from the outermost solar system beyond Neptune.

Introduction

Even a casual consideration of the world we observe around us shows that planetary evolution does not proceed in isolation. Instead, it involves close and diverse relationships with neighbors, large and small. Our size is negligible compared to the sun (Fig. 8-1); we depend on it for our orbit and our energy and light, and we are influenced by its varying magnetic field. The moon causes tides that influence all our shorelines

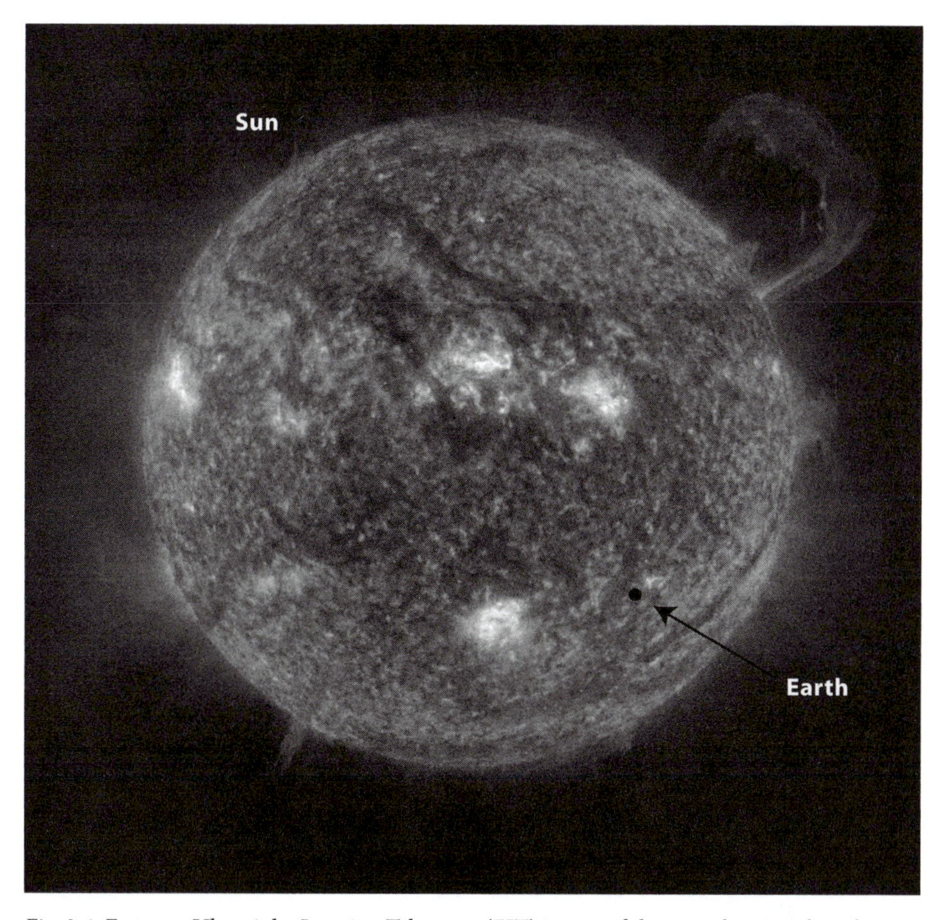

Fig. 8-1: Extreme Ultraviolet Imaging Telescope (EIT) image of the sun, showing the relative size of sun and Earth. More than a million earth-sized objects could fit inside the sun's volume. Earth is smaller than many sunspots. Prominences are relatively cool dense plasma suspended in the sun's hot, thin corona that escape the sun's atmosphere. Emission in this image shows the upper chromosphere at a temperature of about 60,000°K. Every feature in the image traces magnetic field structure. The hottest areas appear almost white, while the darker areas indicate cooler temperatures. (Information and image courtesy of NASA)

and ecosystems. Jupiter is also vast in size, and gravitational effects from the outer planets influence our orbit sufficiently to be a major cause of climate variations, as we will see in Chapter 18. The outer planets also influence the orbits of all the objects that traverse the solar system, and some of these disturbed orbits impact Earth and other planets today as meteorites and comets. Larger meteorite impacts in the past led to mass extinctions of life that contributed to biological evolution, as we will learn in Chapter 17. And in the future, an asteroid or comet impact may be the most significant catastrophe that human civilization will face. We are influenced by and depend upon these interactions with our neighbors. In terms of energy, climate, life, and matter, we are in relationship with the solar system, and Earth's habitability has been strongly influenced by these relationships.

We also have much to learn about Earth's history and habitability from the study of our solar system neighbors. Early Earth's history cannot be studied directly because the oldest rocks on Earth are about 4 billion years old, and less than 1% of Earth's surface contains rocks older than 3 billion years. Most of these rocks were not at the surface when they formed, and when they do become exposed at the surface, they are rapidly modified by erosion and biology. Earth's surface tells us very little about early solar system conditions. To put this 4.55 Ga–3.8 Ga data gap in perspective, it is a longer time period than the entire fossil record when multicellular life developed and evolved, and it is the time when Earth formed its primary layers, plate tectonics may have begun, and primitive life may have first appeared. How can we fill this important hole in Earth's history? Study of other solar system objects that preserve information from that time can tell us much about the earliest history of our habitable world.

The Diversity of Objects in the Solar System

In addition to the sun and the eight planets, the solar system contains a host of other objects. The most prominent of these are the more than 150 moons that circle the planets. The list does not end here. Circling the sun between the orbit of Mars and the orbit of Jupiter are about 100 billion asteroids. While most of them are 1 m or less in diameter, about 2,000 are greater than 10 km in diameter. Only one, Ceres, is more than 1,000 km in diameter. Jupiter and Saturn have rings made up of myriads of small objects. Finally, on the order of 1,000 billion comets are thought to ride in orbits more distant than Neptune's. The recent studies of comets, which are considered the leftovers from the formation of our solar system around 4.6 billion years ago, show that they consist of ices

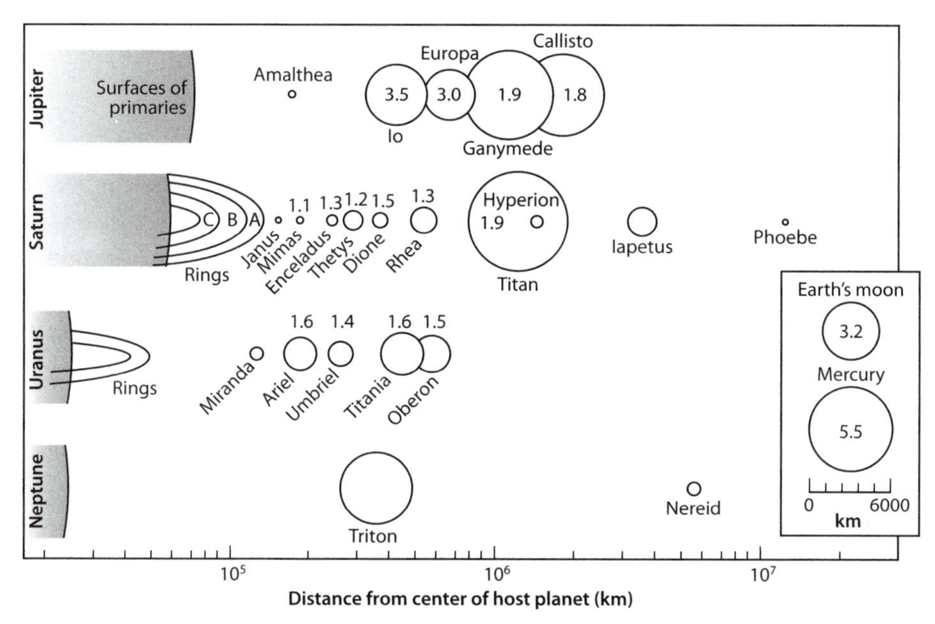

Fig. 8-2: Sizes of some of the moons of the four outer planets. The circles show the relative sizes of the moons, for which the scale is shown in the box on the lower right. Their distance from the center of the host planet is indicated on the horizontal axis. Note the scale is logarithmic. The moons appear to overlap because their size is greatly exaggerated relative to the distance scale. As can be seen, three of the moons are larger than Mercury, and five are larger than Earth's moon. The numbers within the circles give the bulk densities of the moons (in gm/cm^3). Io and Europa have densities close to that of our moon; the other moons have lower densities and must contain a substantial component of the ices.

wrapped around a rocky core. In this sense, comets are to the major planets (Jupiter, Saturn, Uranus, and Neptune) what asteroids are to the minor planets, namely, miniature versions. The famous "tails" of comets are the vapor created when in the course of their highly elliptical orbits they come close to the hot sun.

Six of the sun's eight planets have moons. As shown in Figure 8-2, all four outer planets have moons, and as our resolution of the outer solar system continues to increase and we can see smaller and smaller objects, the number of moons may increase still further. Jupiter has sixty-three, four of which are the large moons discovered by Galileo (Io, Europa, Ganymede, and Callisto) and two of which (Ganymede and Callisto) are larger than the planet Mercury! Saturn has fifty-three moons, one larger than Mercury. Uranus has twenty-seven, and Neptune has thirteen. Earth and Mars are the only terrestrial planets with moons. We have one large moon similar in size to the large moons of the outer planets. Mars has two very small moons.

The space probes sent from Earth to explore the solar system have photographed the moons of Mars, Jupiter, Saturn, Uranus, and Neptune, showing them to be solid objects of great diversity. Many of the moons as well as Mars and Mercury are dotted with craters made by the impacts of meteorites (Fig. 8-3). Others have no craters at all, suggesting an actively changing surface. Io has a smooth surface owing to its highly active volcanism (Fig. 8-4a). Europa (Fig. 8-4b) is entirely covered by ice that moves and deforms. The densest moons have densities similar to silicate rocks, but most of the moons of the outer planets have much lower densities, consistent with the cold environment in which the outer planets formed. Study of moons is a burgeoning field of planetary science, particularly as we try to understand the likely characteristics of solar systems other than our own and expand our views of the possible range of environments for life. For example, while the surface of Europa is ice, its density reveals a rocky interior, and there is evidence of a large liquid ocean beneath the snowball surface. It is natural to wonder whether liquid water and a rocky substrate led to the appearance of life at depths in the ocean there.

Some moons around the outer planets have characteristics that suggest they formed as mini solar systems around their parent planet; they have regular *prograde* orbits, circling the planet in the same direction as the planet is rotating, and they are aligned with one another around the planet's equator. Moreover, Jupiter's large moons show a regular decrease in density with distance from their planet, suggesting a possible gradient in the temperature of condensation caused by the formation and luminosity of the parent planet (Figs. 8-2 and 8-4). The moons with densities less than 3.0 must be considerably richer in volatile elements than are the terrestrial planets and contain substantial amounts of the ices. The miniature solar systems around Jupiter, Saturn, and Uranus have very small inner moons and large outer ones, similar to the organization of planets around the sun.

There is a second large class of moons that have higher inclinations (they do not rotate around the planet's equator) and often have *retrograde* orbits. These moons are thought to have been captured from other regions of the solar system. The two small moons of Mars are readily

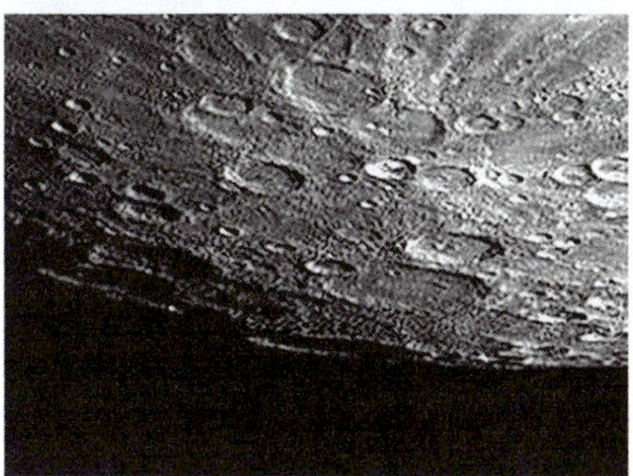

Fig. 8-3: Craters are a common characteristic of almost all bodies in the solar system. Here are two examples. The top image is a photograph of the surface of Saturn's moon Mimas. The main surface feature is the giant impact crater 130 km across in the upper right of the image; its walls are approximately 5 km high. A crater of similar relative dimensions on Earth would be as wide as Canada. Fractures on the opposite side of Mimas have been discovered and may have been created by shock waves from the impact traveling through the moon's body. The bottom image is a portion of the surface of Mercury near its south pole. Other impacts are visible in many other figures of this chapter. (Images courtesy of NASA)

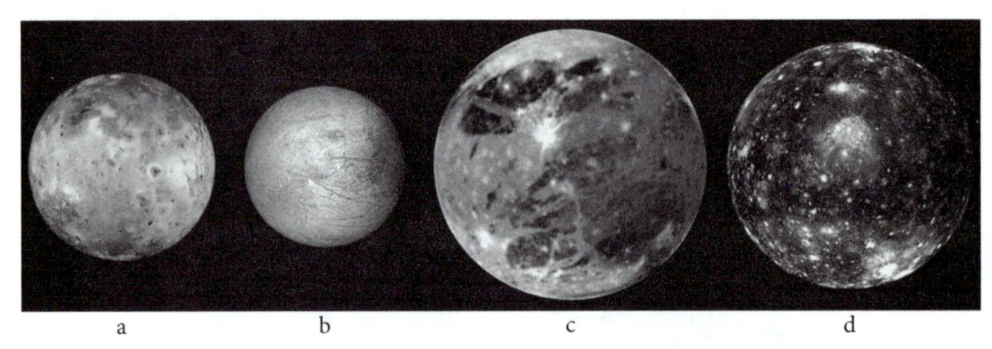

Fig. 8-4: The four "Galilean" moons of Jupiter, shown to scale. (a) Photograph of the surface of Io, the innermost of the four Galilean moons of Jupiter. Its bumpy surface is due to intense volcanic activity, which continually repaves the surface, destroying any craters that occur. (b) Photograph of the surface of Europa, another of the Galilean moons. Craters are mostly absent, because the ice-covered surface is in movement and is young in age. (c, d) The two outer moons of Ganymede and Callisto are made of rock and ice (see their densities in Fig. 8-2, and reveal partially cratered surfaces). Color images of Europa and Io can be seen in color plates 5 and 6. (Images courtesy of NASA)

explained by capture from the neighboring asteroid belt, but the asteroid belt is not a candidate for supplying the numerous moons of the outer planets. Instead, they are thought to come from a large region beyond the planets called the Kuiper Belt (Fig. 8-5a). Many hundreds of substantial objects have now been discovered beyond Neptune, and the Kuiper Belt is thought to include more than 70,000 objects larger than 100 km in size. Pluto is one of the largest of these objects, and it is now known to have a large companion named Charon. Neptune's largest moon, Triton, has a retrograde motion and is actually 18% larger than Pluto. Pluto's orbit actually crosses that of Neptune. Both Pluto and Triton are now thought be the largest representatives of Kuiper Belt objects. These considerations (among others such as its size and the inclination of its orbit) led to the removal of Pluto from the list of true planets.

The farthest reaches of the solar system are the Oort Cloud (Fig. 8-5), where billions of potential comets reside in vastly distant orbits extending much of the distance to the nearest star. Passing stars can perturb the orbits of Oort Cloud objects and send them rushing into the inner solar system as comets, many of which end up being accreted to one of the planets.

A striking aspect of the exciting new information coming from the moons and other objects of the outer solar system is their diversity. Io is

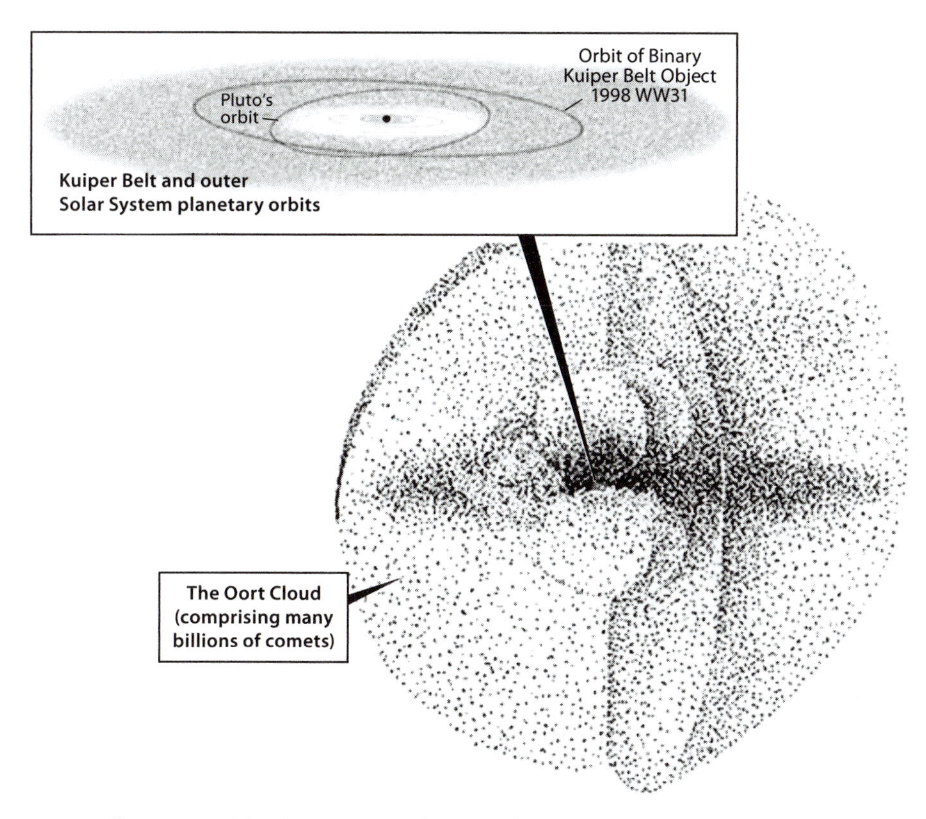

Fig. 8-5: Illustration of the the two major features of the outer solar system. The Kuiper Belt is a band of objects just beyond the orbit of Neptune. Note the tiny box representing the inner solar system. The former planet Pluto is the largest Kuiper Belt object. The Oort Cloud extends greatly beyond the Kuiper Belt and hosts many billions of objects, some of which are perturbed by the gravity from passing stars and enter into the inner solar system as visible comets. (Courtesy of NASA; http://www.nasaimages.org/luna/servlet/detail/NVA2~8~8~13317~113858:Hubble-Hunts-Down-Binary-Objects-at)

the most volcanically active body in the solar system, and liquid sulfur appears to play an important role in the eruptive style. Ice-covered Europa is a satellite where the rocky interior underlies a planetary scale ocean whose surface is completely ice covered. Saturn's Titan has an active surface climate driven by methane, which can be solid at the prevailing low temperatures and also forms rivers and lakes. Even greater diversity of environment and style must exist elsewhere in the galaxy, stretching our concepts of potential planetary environments and the diverse styles of life that might occupy them.

Origin of the Moon

The moon (Fig. 8-6) is of particular interest to us as we explore the origin of our habitable planet. It is our nearest neighbor, more than one hundred times closer then either Venus or Mars, and potentially has much to reveal about solar system events in our immediate neighborhood that are no longer preserved on the constantly changing terrestrial surface.

First-order consideration of the Earth-moon system reveals a number of curious features. Earth is the only inner planet with a moon of significant size. The moon is unusual even in the broader context of the solar system moons discussed above because it is the largest moon relative to the size of its planet. And unlike moons around the outer planets, it has a lower density (3.1 gm/cm^3) than its planet. Even more puzzling, the moon's density is lower than the density of *any* of the inner planets. Recalling our discussion of density in Chapter 5, the moon must be made up almost entirely of rock, without a core of any significant size. The evidence from other planets and meteorites suggests that planetary differentiation is commonly associated with separation of rock and metal, so how would it be possible to form such a large object in the inner solar system with no metallic core? The circular lunar orbit is also unusual. It is within 1% of being a perfect circle, although the general solution for satellites is an ellipse. For example, Jupiter's major moons have ellipticities of 4–15%.

Other puzzling features of the moon became apparent from the ages and chemical compositions of lunar rocks. The dates obtained on lunar rocks constrained the age of the moon to 4.43–4.52 billion years old. These numbers are 40 to 100 million years younger than the ages of chondrites discussed in Chapter 4. As we learned in Chapter 5, modeling of the solar nebular suggests that the major episodes of accretion of planetesimals would take less than 20 million years, so the age of the moon poses a bit of a puzzle. What happened to lead to its later formation?

Another puzzle is the low concentrations of siderophile elements in lunar rocks relative to chondrites. On Earth we explain this depletion by core formation. Since the low density of the moon precludes a signifi-

Fig. 8-6: Photograph of Earth's moon. The dark-colored portions (called *maria*) are areas of the surface that were flooded by lunar basalt flows. The light-colored portions (called *highlands*) represent the original anorthositic crust. The difference in cratering intensity shows that the maria are younger than the highlands. (Courtesy of NASA)

cant core, how could the moon become depleted in siderophiles? The moon is also very depleted in volatile elements, not only the obvious lack of water and atmospheric gases but also moderately volatile elements such as K, Na, and Cl. For Earth the ratio of the moderately volatile K to the refractory U is 12,000. For the moon, K/U is only 2,000. This difference suggests that the material that formed the moon was subjected to temperatures considerably higher than was the case for the material that formed Earth. The moon is also highly depleted in all elements more volatile than potassium relative to Earth.

There is one more piece of evidence bearing on the moon's origin coming from careful measurements of the oxygen isotopes. Careful measurement of the relative abundances of the three isotopes of oxygen, ^{16}O, ^{17}O and ^{18}O, showed slight differences between Earth and various classes of meteorites. The moon, however, is identical to Earth, suggesting a common parentage.

Hypotheses for the origin of the moon need to account for these diverse observations. The *capture hypothesis* has the moon accreted in an orbit similar to Earth, from which it was passively captured to enter orbit around Earth. This hypothesis does not explain the lack of a lunar core, and dynamically it is also very difficult to capture a large object like the moon and have it end up in a circular orbit.

The *fission hypothesis* is that the moon formed by fission from Earth after Earth's core had formed. This hypothesis has appeal because it accounts for most of the lunar puzzles. There would be siderophile element depletion despite the lack of core because Earth's core formation would occur prior to fission. Oxygen isotopes would be identical to Earth because moon and Earth were once combined. If the fission occurred at high temperatures, there might be a mechanism for loss of volatile elements. There are two difficulties with the fission hypothesis, however. The first is the young age of the moon. If early Earth were spinning fast enough to fission, why wouldn't it happen immediately after Earth formation and separation of the core, which we learned in Chapter 7 happened less than 30 Ma after Earth formed? The more serious difficulty is that fission would require the early Earth to spin with a two-hour day, causing some of the mantle to be ejected into orbit, where it could coalesce to form the moon. This hypothesis requires the ad hoc assumption that Earth had a much faster spin than observed elsewhere in the solar system. Even more seriously, the total angular momentum of Earth and moon today are not consistent with such high spin rates, unless a great deal of material was lost to space.

The *giant impact hypothesis* proposes that another planet about the size of Mars had a grazing impact with Earth, ejecting large amounts of material into space around the joined planets, which condensed to form the moon. A giant impact resolves most of the lunar puzzles. It would explain why Earth had a large moon and other inner planets did not. It would also explain the late age of the moon, since its formation would have to postdate the major accretion of the planets. The impact would happen after the formation of cores on the two impacting bodies. Modeling of the giant impact (Fig. 8-7) showed that the high density of the metallic cores would cause the two cores to fuse in Earth, leading to a hot cloud of silicate debris surrounding Earth that would be siderophile-element depleted. Condensation and accretion of this debris at high temperatures would lead to the formation of the moon, depleted in both siderophile and volatile elements.

A criticism of the giant impact hypothesis is that it requires a unique event. Advances in modeling planetary accretion, however, show that giant impacts are likely in early solar system history, and that most of the planets probably ultimately accreted from large protoplanets impacting one another. Large impacts have now been invoked to explain why Mercury has an oversize core—a large direct impact could have caused the removal of most of the silicate mantle. Recent results also suggest that the vast differences between the two hemispheres of Mars may result from a giant impact. A giant impact has also been proposed to account for the reverse rotation of Venus and the fact that Uranus has a horizontal spin axis compared to the other planets. These various lines of evidence, coupled with increasingly detailed modeling that provides compelling images of the giant impact, have led most planetary scientists to currently favor the giant impact hypothesis for the formation of the moon, and have led more broadly to a recognition of the importance of giant impacts to planetary accretion and early solar system history.

The hypothesis is by no means proven, however. A difficulty with the giant impact hypothesis is that modeling results suggest the moon would be formed largely from material from the impactor rather than Earth itself. To be consistent with the oxygen isotope evidence, the impactor would have to have had the same oxygen isotope fingerprint as Earth—and there is no way to verify whether it did. Since it is possible that this was the case, or that future models may find a way to make more of the moon from Earth itself, the giant impact hypothesis does not have major problems based on current data, and the fission and capture hypotheses

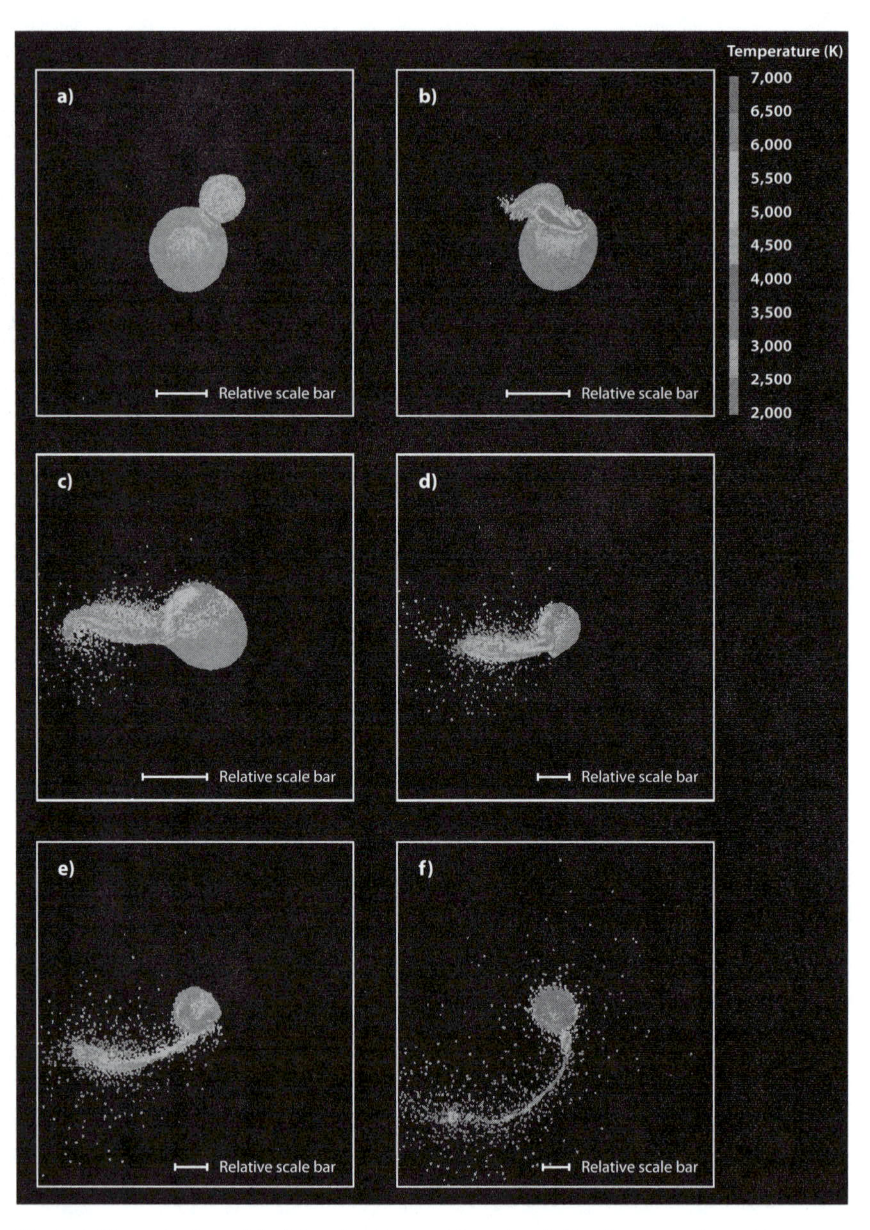

Fig. 8-7: Numerical model of the giant impact hypothesis for the moon formation. This model has the proto-Earth hit by a Mars-size body (called Theia) at a velocity of 40,000 km/hour and at an angle of 45°. The impact causes moon- forming material to be ejected into the Earth orbit, producing a hot silicate vapor; as it cools down, a solid particle disk is created, and through accretion of these particles, the moon is formed. For color version see color plate 7. (Courtesy of Robin M. Canup, Southwest Research Institute)

do have such problems. So the giant impact is currently the preferred model, but in view of the uncertainties, it ranks only a 5–6 on our theory scale.

Giant impact and fission share many features—both form the moon from Earthlike materials, after the formation of the core, from a hot cloud of debris surrounding early Earth. These common characteristics are likely to survive as our understanding of the moon's formation continues to evolve.

Using Impacts to Date Planetary Surfaces

After the giant impacts that marked the first 100 million years of the history of the solar system, impacts did not stop. When we look up at the moon, it is clear even to the naked eye that the lunar surface (Fig. 8-6) is pockmarked with craters of all sizes, some of them more than 1,000 km in diameter, and all of these must have happened after the moon had its primary differentiation into layers, since a rigid outer crust is a prerequisite for crater preservation. Impacts of significant size have been ongoing.

Today it is accepted that craters form as a result of impacting objects from space. This was not always the prevalent view. One of the difficulties in convincing early scientists that craters formed as a result of impacts is that most craters are circular. Impacts could come in at any angle, and if we carry out simple experiments firing projectiles in the laboratory, then low-angle impacts lead to elliptical craters, not circular ones. Furthermore, there was also often little evidence of the impacting object. Where was it? And there was often a lot of silicate melt around, suggesting that craters were caused by volcanic processes.

The breakthrough in understanding crater origin was the recognition that impacting objects travel at hypervelocities of 17–70 km/sec. At 70 km/sec, a meteoroid could travel from San Francisco to Paris in two minutes or travel from the moon to Earth in an hour and a half. The speed is about a hundred times faster than a bullet—Superman could not catch them. Energy increases as the square of the velocity, so a bullet at hypervelocities would be 10,000 times more damaging. At these high speeds, the pressures on impact are so great that a strong shock wave propagates from the point of impact. The shock caused by the impacting object rather than the object itself creates a circular crater some twenty times larger than the diameter of the impactor (Fig. 8-8). The impact also generates enough heat to largely vaporize the meteorite and melt the country rock, and creates high-pressure minerals never seen elsewhere on Earth's surface. The presence of the high-pressure form of quartz, named *stishovite*, is considered definitive evidence of an impact origin for numerous nonvolcanic, circular craters on Earth, and convincing evidence that the same morphological features elsewhere in the solar system resulted from impacts.

Earth's atmosphere plays an important role in impacts. Large meteoroids punch through the atmosphere and create a vacuum than can facilitate throwing some of the ejecta into space. Smaller meteoroids, however, are slowed down by the atmosphere. Many of them burn up in transit, making meteor showers, and others are drastically slowed by the friction of the atmosphere and have much less energetic impacts and better preservation of meteorite fragments. The moon is able to preserve a much more complete cratering record, because meteors of all sizes are able to arrive at the surface unchanged by interaction with an atmosphere.

Modern observations show that impacts remain important. On Earth today meteor showers are common, and these reflect a continuing inflow of solar system material. Larger recent impacts, since humans have inhabited the Earth, are also evident, such as the Barringer Crater in Arizona, where the Canyon Diablo meteorite, 50.0 m in diameter, impacted the Arizona desert to create a crater 1.2 km wide ~50,000 years ago (Fig. 8-9). In the early twentieth century, a likely comet exploded in the atmosphere over Tunguska, Siberia, leading to widespread devastation. And in 1994 the comet Shoemaker-Levy had a spectacular impact with Jupiter (see frontispiece). Another comet probably about 1 km in diameter impacted Jupiter in 2009.

Even larger terrestrial impacts are indicated by an unusual form of glassy rock called *tektites* (Fig. 8-10). Based on their aerodynamic shapes and surface textures, tektites are thought to form from liquids that crystallized to glass above Earth's atmosphere and then were modified by ablation as they fell back through the atmosphere to Earth's surface. Scientists are now convinced that these objects were formed during impacts that splashed material up from Earth's surface well above the atmosphere.

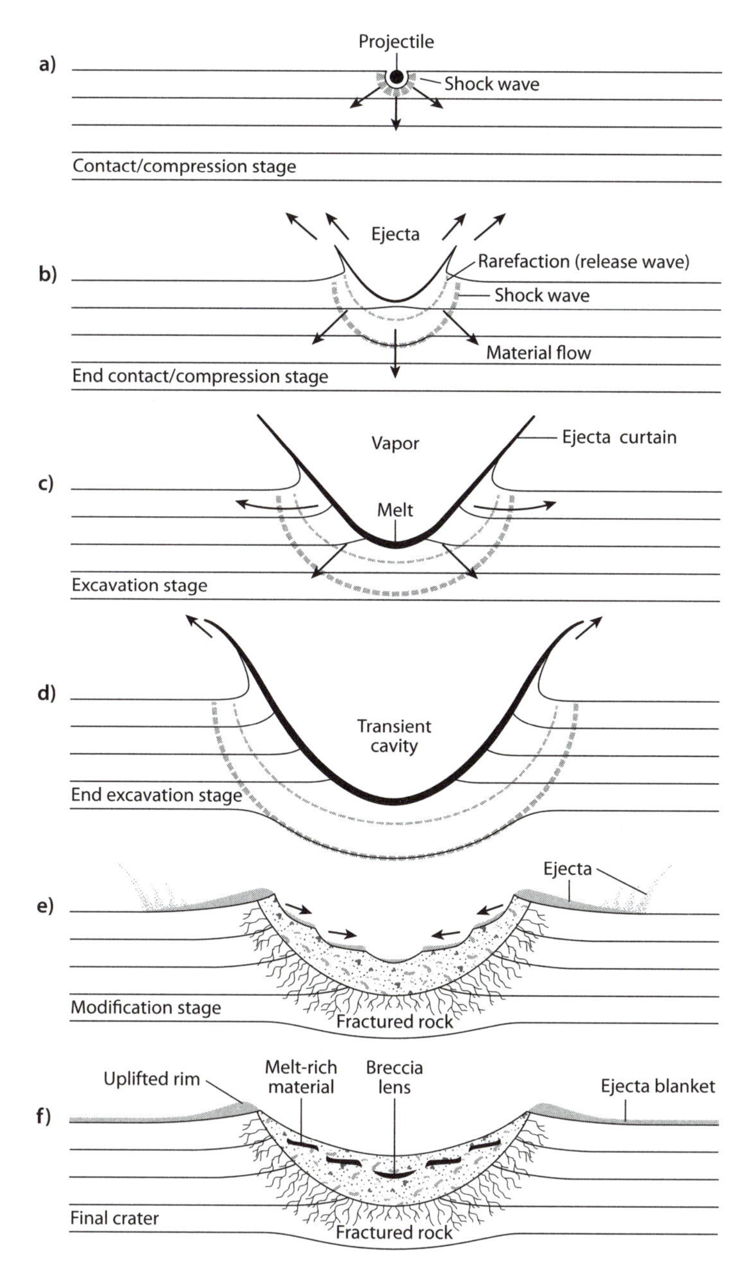

Fig. 8-8: Illustration of formation of an impact crater. Note that impacting objects travel at "hypervelocities" of 17–70 km/sec. At these high speeds, the pressures on impact are so great that a strong shock wave propagates from the point of impact. The shock caused by the impacting object rather than the object itself creates a circular crater some twenty times larger than the diameter of the impactor. The impact also generates enough heat to largely vaporize the meteorite and melt the country rock, and it creates high-pressure minerals never seen elsewhere on Earth's surface. (Modified from B. French (1998). Traces of Catastrophe, Lunar and Planetary Institute Contribution No. 954, with permission)

Fig. 8-9: Aerial photo of Barringer crater. The crater is about 1.2 km (400 ft) in diameter and 170 m (570 ft) deep . It was created by the Canyon Diablo meteorite, a meteor 50 m in diameter that impacted the Arizona desert 50,000 years ago. (Courtesy of U.S. Geological Survey)

The most abundant of the tektites are those found throughout Southeast Asia and Australia in soils and streambeds and in marine sediments adjacent to these landmasses. These tektites all have ages of 0.7 million years and likely formed from a single large impact. Other groups of tektites are found in North America (with an age of 30.0 million years), in central Europe (age 13.0 million years), and in the African Ivory Coast (age 1.1 million years).

In the case of the European tektites, the actual impact crater is thought to have been located in Germany. Even though it has been partially erased by erosion over the last 30 million years, the presence of stishovite in the sedimentary rocks deformed by this impact proves its origin. These various lines of evidence show that impacts of a wide range of objects continue historically and in the geologically recent past. Planetary accretion is ongoing.

Since impacts have persisted over billions of years, the extent of cratering of a surface reveals information about the age of the surface. Earth has few impact craters because its surface is continually reworked by erosion, mountain building, and volcanism. In contrast, the oldest planetary surfaces have received so many craters that their surfaces are "saturated" with them—the entire surface is cratered, so that new craters simply destroy older ones. Surfaces intermediate in age are moderately

Fig. 8-10: Photo of tektites, black glass objects whose shapes and textures bear witness of high-speed flight through the atmosphere. They are created from material melted and splashed high above the atmosphere by the impact of meteorites or comets. (Courtesy of Harvard Museum of Natural History)

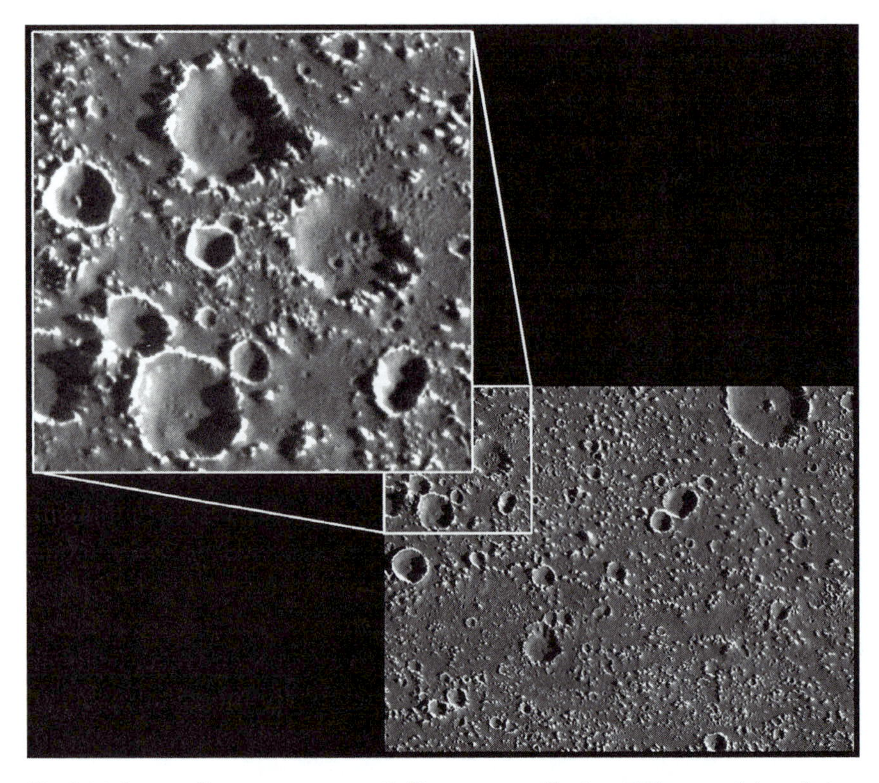

Fig. 8-11: Image of impact craters on Callisto, moon of Jupiter. Estimates of the relative ages of the surfaces of moons and planets can be obtained by the extent of cratering. Even for saturated surfaces, a relative age scale for craters can be constructed by looking carefully at their geological relationships. A crater that occurs within an existing basin, or destroys a preexisting crater rim, is younger, and "rays" of dust from impacts that overlie older craters also give relative ages. Craters with small numbers of impacts in their basins would also be younger than those with larger numbers of impacts. By dating impacts and combining the dates with the relative timescale, a cratering history can be constructed. (Courtesy of NASA)

cratered. This simple principle allows us to make estimates of the relative ages of the surfaces of moons and planets. Even for saturated surfaces, a relative age scale for craters can be constructed by looking carefully at their geological relationships. A crater that occurs within an existing basin, or destroys a preexisting crater rim, is younger, and "rays" of dust from impacts that overlie older craters also give relative ages. Craters with small numbers of impacts in their basins would also be younger than those with larger numbers of impacts (Fig. 8-11). By dat-

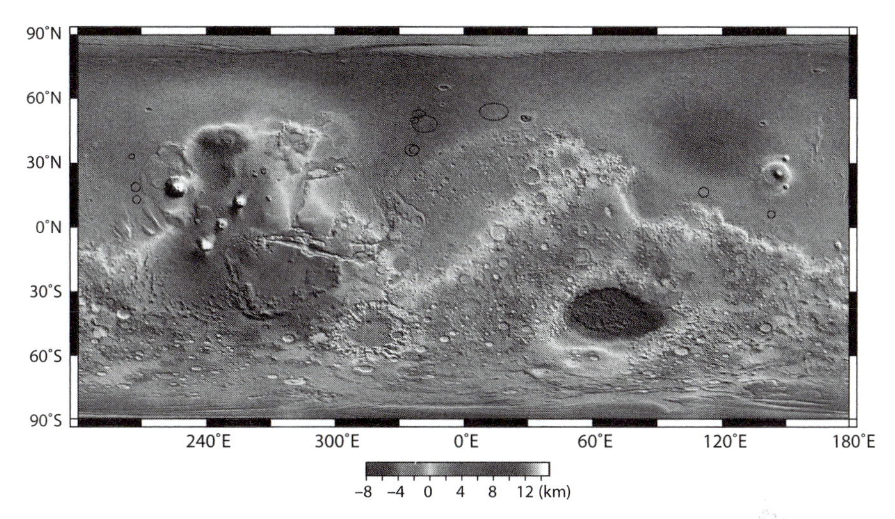

Fig. 8-12: Map of the global topography of Mars. The Mars Orbiter Laser Altimeter (MOLA), an instrument on the Mars Global Surveyor (MGS) spacecraft acquired the first globally distributed, high-resolution measurements of Mars topography. Topographic models have enabled quantitative characterization of global scale processes that have shaped the Martian surface, and as revealed on the image, Mars has two hemispheres with contrasting crater characteristics; the southern highlands have a large number of craters, suggesting an old surface, whereas the northern hemisphere plains are far less cratered, suggesting a much younger surface. See color plate 2. (From Smith et al., *Science* 284 (May 28, 1999):1495–1503; http://photojournal.jpl.NASA.gov/jpeg/PIA02031.jpg.

ing impacts and combining the dates with the relative timescale, a cratering history can be constructed.

The lunar surface is pockmarked with craters of all sizes, ranging from large craters 1,000 km in diameter to microscopic craters from impacting planetary dust. (These small craters are possible on the moon because it has no atmosphere to slow down and burn up incoming planetary debris, and the craters have the chance to be preserved owing to the absence of weathering.) This suggests that the lunar surface is very old. Mercury has a completely pockmarked surface like the moon, suggesting an ancient age for its surface. Mars has two hemispheres with contrasting crater characteristics (Fig. 8-12). The southern highlands have a large number of craters, suggesting an old surface. The northern hemisphere plains are far less cratered, suggesting a much younger surface, though craters are still far more abundant than they are on Earth. Venus has very few craters on its surface, showing that its exterior has been resurfaced during its history. Cratering intensity thus reveals much about the relative histories and activities of planetary objects.

Lunar Interior Modifications

After the moon formed, it underwent interior modifications that led to the formation of diverse layers controlled largely by density differences, just as took place on Earth. The moon provides the only direct evidence we have to permit a comparison of the interior modifications of two planetary bodies. As it turns out, there are important similarities and important contrasts both in the processes and final results of lunar and terrestrial planetary differentiation.

A similarity is that the moon appears to have a metallic core, but the contrast is that it is very small. The inference from the low lunar density was confirmed by instruments left on the moon by astronauts that radioed back seismograms generated by moonquakes. These results suggest that the moon has a small core (about 2% of its mass). The core likely formed by immiscibility, as on Earth, but the small amount of metal in the moon allowed only a tiny core to form.

The moon has no atmosphere or ocean. The reason is that the moon's gravity is so low that gaseous molecules can readily escape from the moon's surface. Here we encounter one of the prerequisites for habitability. If a planet is too small, it can retain neither atmosphere nor ocean.

The lunar crust carries a complex story that reveals much about the moon's evolution. Careful examination of the lunar surfaces shows that the density of impacts over the surface is not constant. The side of the moon that we see is divided into two distinct provinces of contrasting age. The whiter portions are more highly cratered than the plains that fill the giant craters. The smoothness of these surfaces led early observers to name them *lunar maria* (Latin for "seas"). The white regions of the moon are also at higher elevations, so they are called the *lunar highlands* (Fig. 8-6). The back side of the moon consists entirely of highlands terrain. The inference from the relative cratering age was confirmed by rocks returned from the highlands and maria by the Apollo and Luna missions in the 1960s. Study of these rocks then permitted a much more detailed assessment of how the two major terrains of the lunar crust

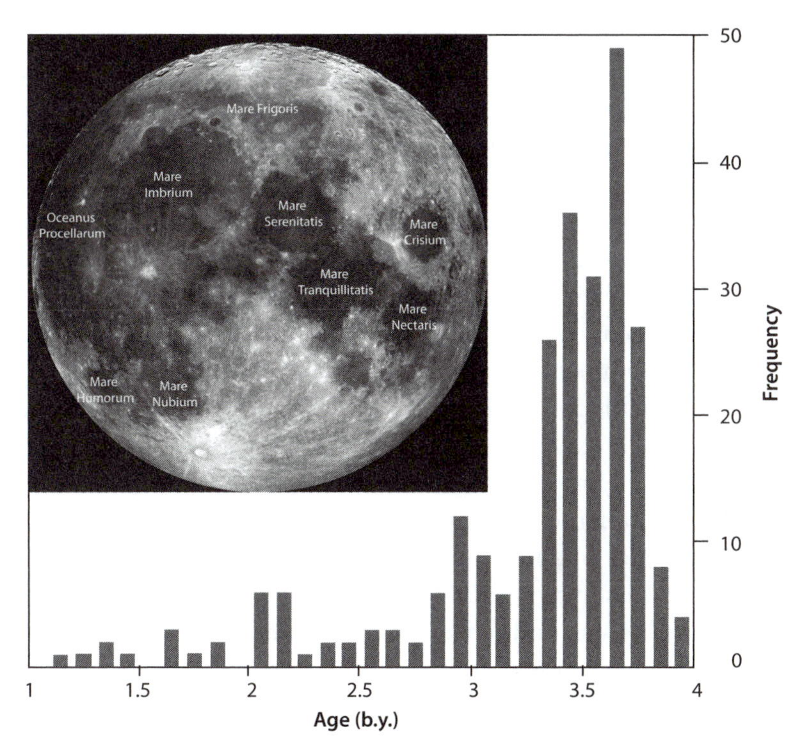

Fig. 8-13: Histogram showing the temporal distribution of ages for various Mare basalts. A map of the moon is shown as a reference for sample locations. Note that the Mare basalts show a peak in activity about 1 billion years after the formation of the moon. Recent high resolution photography of the lunar surface suggests there may be very small amounts of younger lava flows, inferred even to be as young as 1.3 billion years based on their cratering density. (Modified after Hiesinger et al., *J. Geophys. Res.* 105 (2000), no. E12: 29,239–75, and 108 (2003):1–27)

were formed, and a remarkably detailed model of the early history of the moon.

Dates from lunar rocks showed that the dark lunar maria consisted of basalts formed mainly during the period 3.1–3.9 billion years ago (Fig. 8-13). Rocks from the light-colored highlands are older, having formed as much as 4.4 billion years ago. Little or no volcanism has occurred on the moon during the last 3 billion years. Since then the moon has been a "dead" planet. No convection cells churn in its mantle. No plates collide on its surface. No volcanoes erupt. Why this big difference? Again, it is the moon's small size that is responsible. The small size of the moon leads to a large surface area/volume ratio, permitting heat to get out. And the low gravitational field of the moon means the pressure increases very slowly with depth, allowing melting to extend to great depths and extract the heat to the surface. Owing to its present cold and rigid internal state, great convection cells no longer carry heat from its interior to the top of its mantle.

The moon's maria are made of a rock superficially akin to Earth's basalt, and the highlands are rocks somewhat akin to Earth's granite; the highlands anorthosites consist primarily of the mineral plagioclase feldspar, also the most abundant mineral of Earth's crust, although the lunar feldspar is much more Ca rich because of depletion of the moon in the more volatile Na. Upon deeper inspection, however, the similarity between Earth's crust and the moon's crust breaks down.

As we saw in Chapter 7, continental crust on Earth is granite made up of multiple minerals that are the "minimum temperature melt" in the presence of water of a host of rocks—sediments, metamorphic rocks, basalts, or preexisting granites. These granitic melts crystallize to form quartz, two feldspars, and other minerals. Terrestrial granites have also concentrated the magmaphile elements to a hundred times higher levels than Earth's mantle. Not so the lunar highlands. These rocks consist largely of a single mineral, the Ca-rich plagioclase end member anorthite, and their concentrations of magmaphile elements are often very low. Examination of the binary phase diagram from the previous chapter shows that melts of *poly*mineralic substances do not lead to liquids that would have *mono*mineralic compositions. The lunar Highland crust does not have the composition of a partial melt of any planetary interior, and clearly formed by very different processes than Earth's continental crust.

The chemical compositions of the mare basalts also revealed bizarre compositions dissimilar to any terrestrial basalts. Whereas basalts on Earth generally have between 1 and 4 wt. % TiO_2 and lavas at the high end are rare, many lunar basalts had TiO_2 concentrations greater than 10.0 wt. %, while others had concentrations less than 0.5 wt.%. Clearly, processes that created both highlands and maria differed greatly from those that created crust on Earth. These new data challenged the creativity of terrestrial geoscientists. Could the well-understood principles of igneous petrology be used to understand the formation of the unusual compositions of the lunar crust?

Important additional clues came from trace element concentrations in the lunar rocks, particularly the *rare earth elements* (REE). The REE, also known as the lanthanide contraction series, occupy the lower interior of the periodic table (see Fig. 4-1) and have some very useful geochemical characteristics. Because they are all related by having additional electrons added to an inner rather than outer electron shell, all of them have a common outer electron shell structure. This leads to very similar geochemical behavior during igneous processes. As the number of protons in the nuclei increase, however, the ionic size of the REE decreases regularly over the entire series of seventeen elements. Because minerals discriminate for and against elements based on their charge and size, when only the size varies, the chemical differences from one REE to another tend to be gradual and smooth. This leads to "REE patterns" that are characteristic of different minerals and provide clues to what minerals have been involved in the genesis of different rocks.

All of the REE but one also have a common 3+ valence. The one element on the moon with a different valence is Europium (Eu) in the middle of the lanthanide contraction series. Europium can occupy two different valence states, 2+ and 3+, and this makes its behavior differ in important ways from the other REE. This difference in behavior is particularly marked for the mineral anorthite, because its mineral formula $CaAl_2Si_2O_8$ has Ca in a 2+ valence site of appropriate size where Eu^{2+} can substitute very easily, while the other 3+ REE are too large to substitute for the Al^{3+} in the mineral. For this reason, plagioclase feldspars take up much more Eu than other REE, creating a REE pattern with a marked positive concentration anomaly for Eu. When the REE pattern in a rock has a Eu anomaly, it shows that plagioclase has played an important role in the generation of the rock. Rocks that are formed from the accumulation of plagioclase have positive Eu anomalies, and those that have seen plagioclase removal have negative Eu anomalies.

REE patterns of the lunar highlands rocks have marked positive Eu anomalies (Fig. 8-14), showing that they formed by accumulation of plagioclase minerals. This corresponds with their monomineralic character—somehow plagioclase minerals accumulated preferentially to form the lunar highlands. Mare basalts, on the other hand, have strong negative Eu anomalies (see Fig. 8-14)! This sentence deserves the exclamation point because there was no plagioclase in these rocks, and

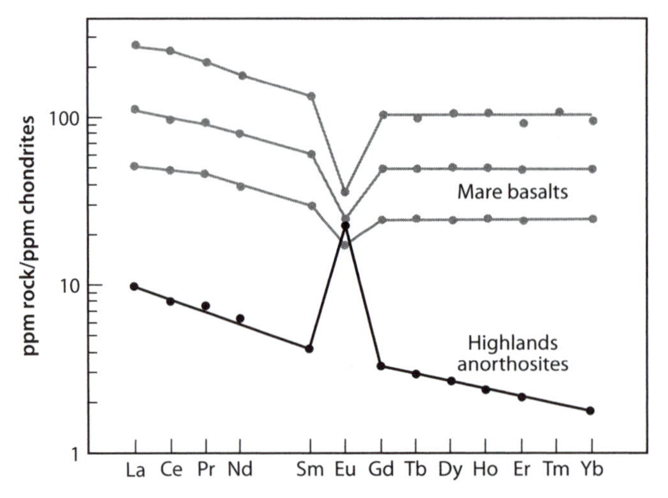

Fig. 8-14: Trace element patterns of Maria basalts from Apollo 17 and highlands anorthosites. Note that REE patterns of the Mare basalts have strong negative Eu anomalies, therefore the region that melted to form the rocks had already undergone prior separation of plagioclase; on the other hand, lunar highlands anorthosites have marked positive Eu-anomalies, showing that they formed by accumulation of plagioclase minerals. (Adapted from P. H. Warren, The Moon, in Andrew M. Davis, ed., *Meteorites, Comets, and Planets*, vol. 1 of *Treatise on Geochemistry* (Oxford: Elsevier Ltd., 2005))

experiments showed that their chemical compositions would not have crystallized plagioclase at any pressure. If no plagioclase could have been removed from the rocks, how could they have a Eu anomaly? The answer is that the source region that melted to form the rocks had already undergone prior separation of plagioclase. Then the source region would be both depleted in plagioclase and have a negative Eu anomaly. Subsequent melts would inherit the Eu anomaly and have so little plagioclase in the melt that they would not crystallize plagioclase during cooling.

To summarize the evidence, the lunar data showed old highlands crust up to 30km thick containing anorthositic rocks with strong positive Eu anomalies suggesting plagioclase accumulation. Mare basalts occurred hundreds of millions of years later, filling large impact basins, and had negative Eu anomalies despite the absence of evidence for plagioclase removal from the rocks. Moreover, Mare basalts included a wide diversity of compositions, from very high to very low TiO_2 contents.

These lines of evidence were creatively explained by a model where the moon had a large magma ocean early in its history. Accretion of the moon following the giant impact could have generated enough heat to cause most of the early moon to melt, creating a magma ocean. One of the first minerals to crystallize would have been plagioclase. Measurements of density showed that plagioclase solids were lighter than the magma of the magma ocean, because the magma ocean had a high FeO content (with 56 protons per iron atom), and all the elements making up plagioclase were of lower atomic number. The plagioclase crystallizing over hundreds of kilometers of thickness of magma ocean would rise to the surface, creating thick anorthositic crust. The plagioclase crystals would preferentially incorporate Eu from the magma ocean, leading to REE with positive Eu anomalies in the lunar crust and a negative Eu anomaly in the residual magma ocean liquid from which the plagioclase had separated. All the later minerals that crystallized would inherit the negative Eu anomaly produced by early separation of the anorthositic crust. Mafic minerals such as olivine and pyroxene would accumulate in other layers. Since these minerals contain little TiO_2, they would create Ti-poor source regions. Much later in the crystallization sequence, the dense Ti-rich mineral ilmenite ($FeTiO_3$) would crystallize, and consequently accumulation of this mineral would generate Ti-rich source regions. The solidification of the magma ocean would thus lead to a spectrum of source regions from Ti-rich to Ti-poor, all characterized by a negative Eu anomaly. After the solidification of the magma ocean, the heat generated by radioactive decay or other processes ultimately heated the lunar interior, permitting it to ascend and melt, generating the spectrum of mare basalt compositions up to a billion years later in lunar history. After these melting episodes, the moon became so cool that no further melting was possible.

This simple scenario (Fig. 8-15) explains the major features of the lunar crust and corresponds with the compositions and ages of the lunar rocks. The story is based, however, on a very sparse sampling of the moon. Only ~390 kg of lunar rocks are available for study, and they are from limited portions of the lunar surface. The entire back side of the moon and the lunar poles are not sampled at all. It is a tribute to geochemistry and the creativity of lunar scientists that such a complete

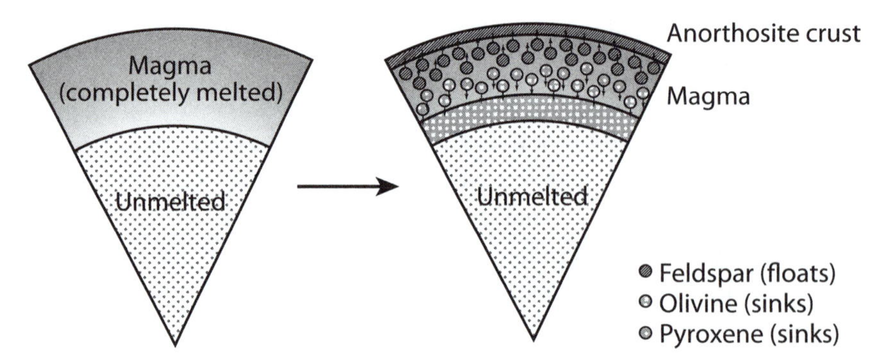

Fig. 8-15: Illustration of the lunar magma ocean hypothesis. Accretion of the moon following the giant impact could have generated enough heat to cause most of the early moon to melt, creating a "magma ocean." The plagioclase crystallizing over hundreds of kilometers of thickness of magma ocean would rise to the surface, creating the thick anorthositic crust. Mafic minerals such as olivine, pyroxene and the Ti-rich mineral ilmenite would accumulate in other layers.

model of lunar crustal formation has been able to be generated, but we should also recognize that more complete sampling would inevitably lead to important modifications of this story. The lunar magma ocean merits only a 4–5 on our theory scale—there will be exciting scientific discoveries when more lunar rocks are recovered.

History of Impacts in the Solar System

We know that impacts have occurred throughout the history of the solar system. Can we say anything about how the rate of impacts changes with time? One constraint comes from the current flux of impacts on Earth. About 40,000 tons per year of matter from space fall to Earth's surface each year, equivalent in total to a rock about the size of a college science building. Most of this matter is dust, scattered all over the Earth, but there is about one meteorite per year weighing more than 20 gm for every 10,000 square km (about the equivalent of a major metropolitan area). Most of these small fragments are never seen or recovered (Fig. 8-16). A handful of larger newly fallen meteorites are seen and collected

Fig. 8-16: Photo of the Carancas meteorite crater, Peru, which landed in 2007. The Carancas meteorite created an impact crater about 14 m wide in a rather remote area of that country. No casualties are known. If this insignificant impact had occurred in a populated area, there could have been hundreds of victims. Most such events occur in unpopulated regions (e.g., the ocean, high latitudes, etc.). (Image courtesy of Michael Farmer)

each year. Since meteorites are rapidly destroyed by weathering, *all* the meteorites collected by humans are relatively recent falls.

Could the present rate over billions of years lead to Earth's current size? No—current accretion rates over 4.5 billion years would produce less than one ten-millionth of Earth's mass. Impacts must have been vastly larger and more frequent in early solar system history. A natural preconception and first hypothesis for impact history would be a smooth exponential decline. Each time a planet or moon goes through its orbit it intersects a percentage of the objects with crossing orbits, gradually clearing out debris from the solar system like a giant gravitational vacuum cleaner. One could construct a "half-life," for example, for the number of asteroids that cross Earth's orbit, and over time this would lead to an exponential decline in numbers of asteroids, and associated impacts, similar to the decline of a radioactive isotope during its decay (Fig. 4-14c). The present impact rate would be one constraint, and we could use the cratering intensity on the moon to determine older points and specify the decay rate. This simple scenario, however, turns out to be only part of the story.

Constraints on a hypothesized exponential decline in cratering became possible after the return of lunar samples by the Apollo program. Early studies of the ages of *impact breccias* (broken-up rocks created by impacts) on the moon showed a clustering of ages between 3.9 and 3.8 billion years. Impact melts from lunar meteorites found on Earth gave the same age range. It appears that dozens of impact craters >300 km in size may have formed within this time interval. To account for these observations, a *terminal cataclysm* or *late heavy bombardment* (LHB) has been proposed, where cratering intensity increased markedly for a short period of time.

The LHB has been controversial because a mechanism for such late bombardment was hard to fathom. If planets gradually clean out their orbits, what would be the source of a large flux of impactors 700 Ma after the planets formed? What that would require would be some event that changed the numbers of objects in Earth- or moon-crossing orbits. If the solar system established its orbits very early on, how could that occur? As models of the early history of the solar system have become more sophisticated a mechanism has appeared, because the models suggest that the planets change their orbits in early solar system history. As the orbits change, different portions of the asteroid belt can become unstable, sending objects into the inner solar system. In particular, movement of the outer planets can greatly perturb the asteroid belt when Jupiter and Saturn pass through a mode where they have a 1:2 resonance, where Jupiter goes around the sun exactly twice as fast as Saturn. If this occurred at about 3.9 Ga, it would be a mechanism to generate a new flux of Earth- and moon-crossing objects. Recent studies of the distribution of asteroids in the asteroid belt provide strong evidence for outer planet migration early in solar system history. There is also evidence of a major heating event at the same time in the rare Martian meteorites. The confluence of evidence from the moon, lunar meteorites, Mars, the asteroid

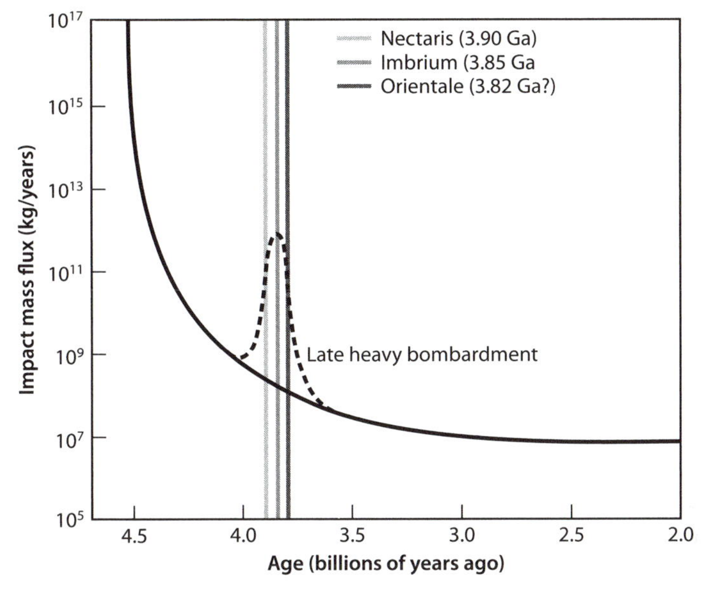

Fig. 8-17: Illustration of how impacts may have varied in solar system history. An overall exponential decrease in cratering intensity is punctuated by a massive bombardment episode at about 3.8 billion years, termed the "late heavy bombardment" (LHB). These results show the great importance of impacts in the early solar system, which would have had far-reaching consequences for habitability of early planetary surfaces. (Adapted from Koeberl, *Elements* 2 (2006), no. 4: 211–16)

belt, and solar system modeling is leading to much greater acceptance of the LHB hypothesis.

Figure 8-17 shows how impacts may have varied in solar system history. An overall exponential decrease in cratering intensity is punctuated by a massive bombardment episode at about 3.8 billion years. These results show the overwhelming importance of impacts in the early solar system, which would have had far-reaching consequences for habitability of early planetary surfaces.

Implications for the Earth

Study of the moon has revealed much about major events in the early solar system that were not evident from study of Earth alone, and these events have significant implications for Earth's early history.

If the giant impact hypothesis is correct, then shortly after its initial formation Earth underwent the immense catastrophe of giant, planet-size impact leading to ejection of material and formation of the moon. The energy of this impact was sufficient to lead to planet-scale melting, which upon solidification would have caused formation of an early crust and possible stratification of the mantle. Some have suggested that Earth got so hot that it formed a gaseous atmosphere of silicate gas. Depending on the atmospheric conditions, the magma ocean would have cooled rapidly, and one could construct models where it would become stratified. There is little evidence for a highly stratified mantle today, however, and it is possible that convection rehomogenized the mantle. That said, we do not have direct samples of the lowermost mantle, and some scientists argue for some layering of the mantle left over from the magma ocean event. Conclusions about the consequences for Earth of the giant impact rely almost exclusively on modeling, and models at that scale in the absence of known boundary conditions inevitably involve assumptions and creativity. The elegant calculations can be used to support plausible arguments for the consequences of the giant impact for Earth's early history and subsequent evolution, but the actual facts remain shrouded in uncertainty.

If early Earth did have a magma ocean, why didn't a large anorthositic crust form? If the lunar magma ocean could preserve 30 km of light anorthositic crust, wouldn't a terrestrial magma ocean have led to a similar crust hundreds of kilometers thick? No, Earth would not have generated a significant anorthositic crust because of the limited pressure stability of anorthite. All minerals have a limited range of temperature and pressure over which they are stable. The key difference in this regard between Earth and the moon is that the pressure change with depth is much less on the moon because of the moon's low gravity. That is why the astronauts could jump so high and so far when they landed on the lunar surface. Since rocks weigh much less on the moon than on Earth, pressure increases much more slowly with depth. Anorthite has a maximum pressure stability about 12 kilobars (1.2 GPa) (Fig. 8-18). The center of the moon, at a depth of 1,200 km, has a pressure of only 4.7 GPa (47 kbar), and anorthite is stable over 300 km of depth. Crystallization of 10% plagioclase over this interval would lead to a 30 km anorthositic crust. Earth, however, has an increase of 1 kbar (0.1 GPa) for every 3 km

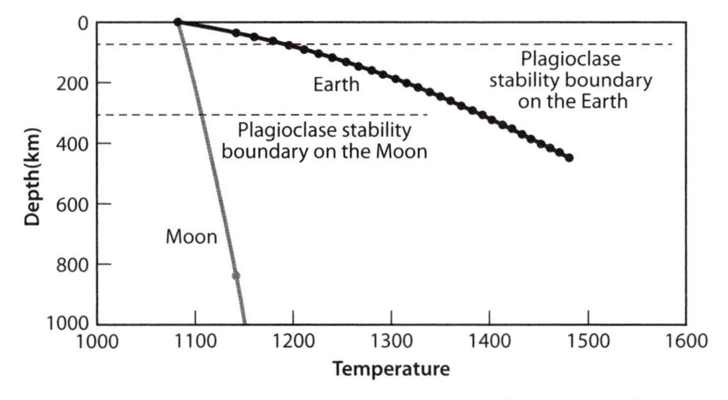

Fig. 8-18: Depth-temperature diagram with plagioclase stability for Earth and moon. Curves for Earth and moon differ because pressure increases more rapidly with depth on Earth. All minerals have a limited range of temperature and pressure over which they are stable. Anorthite is stable only up to about 1.2 GPa, or 12 kilobars. Because of its low total mass, the moon has a weak gravitational field and rocks weigh much less on the moon than on Earth. Anorthite is stable over 300 km of depth. Crystallization of 10% plagioclase over this interval would lead to a 30 km anorthositic crust. Earth, however, has an increase of 0.1 GPa (1 kbar) for every 3 km of depth, On Earth, the stability limit of anorthite of 1.2 GPa is equivalent only to 36 km, and plagioclase cannot crystallize below that depth. There is, therefore, no possibility of a thick anorthositic crust accumulating on the early Earth. It also takes much higher temperatures to melt Earth at depth.

of depth, so anorthite is stable only to a depth of 36km! The equivalent anorthositic crust would be only a few kilometers thick, which would easily be destroyed by impacts and subsequent magmatism. The rapid increase of pressure with depth on Earth also makes it more difficult to generate a terrestrial magma ocean, because melting temperature also increases rapidly with pressure. For the same temperature that would melt the moon to create a 600 km magma ocean, Earth would melt only to some 60 km, trivial in comparison to the 3,600 km depth of the terrestrial mantle.

The early Earth would also have experienced the extensive cratering for which there is evidence on the moon. If the moon suffered a terminal cataclysm, Earth would have received even more impacts owing to its larger radius and more extensive gravitational field. Based on radius alone, Earth would have received at least ten times more impacts than the moon. Adding in the Earth's additional gravitational pull could make the number far larger than that. David Kring and colleagues have estimated

Table 8-1
Hadean history: The first billion years of Earth's history

Year before present (Ma)	Time from the zero year	Event
4,566	0.00	Condensation of the first solid matter in the solar system
4,565	1 m.y.*	Formation of planetesimals
4,555	11 m.y.	Igneous activity in planetesimals
4,532	34 m.y.	Core's separation completed
4,500	66 m.y.	Moon forms from giant impact
4,450	116 m.y.	Atmospheric degassing largely completed
4,404	162 m.y.	Oldest zircon
3,980	586 m.y.	Oldest rock
3,800–3,900	~800 m.y.	Late heavy bombardment
3,500	1,066 m.y.	Evidence of life

**m.y. = million years.*

that a 20 km crater might have been formed on Earth every thousand years, and a 1,000 km basin every million years, sufficient to sterilize Earth's surface of any life that may have begun. The number of surviving rocks on Earth becomes vanishingly small earlier than 3.8 billion years, when the LHB would have ended. This could be a natural consequence of a terrestrial LHB. Only after that time was there sufficient stability of the surface that continental fragments had a chance of surviving. The LHB, then, reflects an important moment of passage for the history of the solar system, after which surface conditions would have become more stable, the rock record on Earth would have been able to be preserved, and the threat of surface sterilization by impacts would have eased. This overall framework provides us with a chronology for the early history of the Earth-moon system that we would not be able to obtain from Earth alone (Table 8-1).

No matter how the moon actually formed, scientists have long inferred that through a process called *tidal friction*, energy associated with Earth's spin is being gradually transferred to the moon. The extra energy gained by the moon speeds its movement around the Earth and lifts it to ever more distant orbits, as the tidal friction also slows down Earth's spin. This inference from calculation turned to proof, thanks to the Apollo program. One of the tasks given to the astronauts who visited the moon was to put in place reflectors that would provide precise points from which a laser beam shot from Earth could be bounced back to Earth. By measuring precisely the time required for a pulse of laser light to travel to the moon and back to Earth, it is possible to establish the distance from a point on Earth to a point on the moon with an accuracy of about 1 cm. These measurements have been repeated regularly over a period of decades. They confirm that the moon is moving away from Earth at the rate of 38 mm per year.

Proof of the complementary change in Earth's spin requires much longer timescales of measurement, which can be accessed only through the geological record. John Wells, a paleontologist at Cornell University, was aware that corals living in today's reefs are banded. While the most prominent of these bands have been shown to be annual, the porosity of the calcium carbonate deposited by these organisms changes subtly with the seasons. These changes can be seen on medical x-rays taken of slabs cut from a coral head (Fig. 8-19). In addition to seasonal bands, Wells also saw weaker banding, which he attributed to the monthly tidal cycle and to day-night cycles.

If the moon has retreated at a rate of 4 or so cm per year, then several hundred million years ago there must have been both more days and more months in a year. Wells found that in fossil corals about 360 million years old there were about 400 daily bands associated with each annual layer. As there is no reason to suspect that the time required for Earth to orbit the sun has changed significantly, these results suggest that the day was shorter in the past than it is today—i.e., that Earth was spinning faster. The changes found by Wells in the number of days tell us that the moon was 1.2×10^4 km closer to Earth at 360 Ma than it is today. If the record in fossil corals is being properly read, then the moon's retreat rate over the last 7% of Earth's history has averaged 4 cm per year, just as it has over recent decades. Estimates of the length of day at 900 Ma are about nineteen hours. If we extrapolate back to early periods of Earth's history, the day may have been as short as ten hours, leading to many more days in a year, and the moon would have been much closer

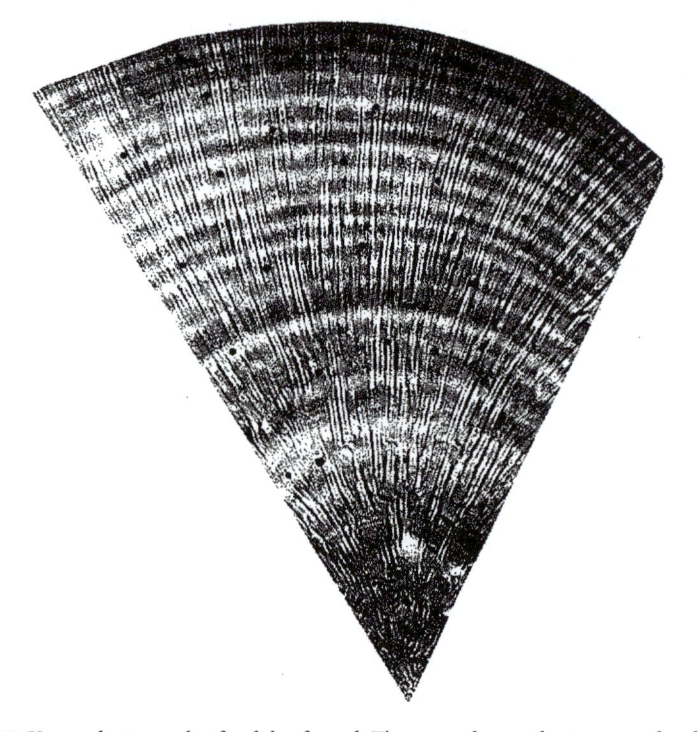

Fig. 8-19: X-ray photograph of a slab of coral. The annual growth rings are clearly visible. The dark bands represent growth during summer months. Proof that the most prominent growth bands in coral heads are caused by the seasonal changes in the character of the calcium carbonate they deposit was provided by studies of corals from the Eniwetok Atoll. The test of an early version of the hydrogen bomb conducted in this atoll in the 1954 produced an enormous amount of local radioactive fallout, so that the waters of the atoll became temporarily highly contaminated with the fission fragment ^{90}Sr. Since the element strontium substitutes readily for the element calcium in the $CaCO_3$ formed by coral, the 1954 growth band was 'marked' with radioactive strontium. When corals collected a decade or more after this event were analyzed, it was found that there was one growth band for each year that had elapsed since the 1954 test. Thus, the ^{90}Sr marking of the 1954 band allowed the annual growth-band hypothesis to be verified. (Courtesy of Richard Cember, Lamont-Doherty Earth Observatory)

and orbiting much faster, leading to a shorter month. This has important consequences for many aspects of the surface conditions of early Earth. Tides would have been much greater, which would have led to much more energetic shorelines and tidal environments, and the moon would have been twice as large in the night sky.

Future Impacts

The recognition of the importance of impacts throughout solar system history, along with the mass extinction of the dinosaurs that is accepted to be the result of a meteorite impact 65 million years ago, naturally leads to the question of the likelihood and danger of impacts today and their consequences for Earth and human civilization.

We know that a major impact occurred in Siberia in the early twentieth century, and the energy released was equivalent to a large atomic bomb. Widespread destruction and loss of life could result from such an impact today, if it occurred either at sea or in populated regions, whether directly on land or by the creation of an enormous tsunami.

Impacts require Earth-crossing orbits, and such orbits can result from three different sources in the solar system. The first are the asteroids with Earth-crossing orbits, referred to as *near-Earth objects* (NEO). A systematic mapping program has identified the largest 1,000 of these, and it appears that none of them are likely to impact Earth in the near future. Note that many of the orbits are chaotic, and uncertainties increase considerably moving further out in time.

Two other sources of impacts come from the outer reaches of the solar system (see Fig. 8-5). Beyond the orbit of Neptune lies the Kuiper Belt, of which Pluto is the largest current representative. More than 1,000 Kuiper Belt objects have now been observed, and there are estimates of more than 70,000 of them larger than 100 km in size. Much further from the Sun is the Oort Cloud, where a massive amount of solar system debris shot into highly elliptical orbits rotates around the sun at distances up to one light year, or 50,000 times the distance of Earth from the sun. The Oort Cloud is the outermost portion of our solar system. It is so far away that objects there can have their orbits perturbed by neighboring stars or by the Milky Way galaxy itself. These perturbed orbits can then come rushing into the inner solar system. Halley's comet, for example, spends most of its life in the Oort Cloud. The Shoemaker-Levy comet was not previously known and impacted Jupiter in 1994 with spectacular results (see frontispiece). A more recent impact of Jupiter by an unknown comet was accidentally observed by an amateur astronomer in July 2009.

It may appear surprising that impacts continue with such frequency despite the fact that materials in planet-crossing orbits have been largely removed from the solar system over its long history. This occurs because most current impacts arise from materials whose orbits have been recently perturbed. The billions of objects in the Kuiper Belt and Oort Cloud are in cold storage awaiting such perturbations. Comets are objects from these regions whose orbits were recently perturbed to send them zooming into the inner solar system. Some are quickly captured through impact. Others are ejected from the inner solar system. Still others, like Halley's comet, make regular visits but lose a little bit of their mass with each orbit, so their total lifetime is only a few million years. That means that current comets are not remnants of earlier objects with Earth-crossing orbits. They are simply the most recent crop of perturbed orbits. Since there are billions of potential comets beyond the orbit of Neptune, and orbit perturbations happen periodically in the outer solar system, a steady supply of potential impactors is assured.

One might think from Hollywood that we could simply send up a rocket with a nuclear warhead to destroy the incoming object. Our missiles fly up to about 1,000 km from Earth's surface. At 50 km/sec, an incoming comet passes that point 20 seconds before impact, and the comet is moving faster than anything our missiles are designed to intercept. Even if a missile were able to explode nearby, the impactor would simply break up into fragments and disperse damage over a wider area. While NEOs can be mapped and there would be many years of warning prior to impact, most Kuiper Belt and Oort Cloud objects cannot be mapped, and there would be little warning. Shoemaker-Levy was discovered only months before its impact with Jupiter. There is no effective defense against impacts of comets, and these are sure to happen at some future point in Earth's history.

Summary

Not all of the materials from the early solar nebula were accreted by planets. More than a hundred moons, almost exclusively around the outer planets, reveal a great diversity of planetary accretion and style and also show the important role of capture through solar system history.

Even after the major steps of formation of planets and planetesimals occurred, impacts continued to play a central role in early solar system history. Earth's moon is an exceptional body in the solar system—the only significant moon of the inner planets, of very large size and low density relative to its parent planet, depleted in siderophile elements while lacking a significant core. A giant impact was the likely cause of the origin of the moon, and this event would have largely melted the early Earth some 50–100 million years after its formation. Study of the moon shows the importance and scale of early planetary differentiation, likely also to have influenced earliest Earth's history, though the remaining evidence for a terrestrial magma ocean is far from clear. Impacts then progressively decreased until migration of the outer planets destabilized the asteroid belt and led to a "late heavy bombardment" during 3.9–3.8 Ga. This is also the age of the oldest rocks on Earth and suggests that from this time forward the terrestrial environment became more stabilized with a surface environment where an equable climate could be permanently established and life had the potential to flourish.

Vast amounts of debris remain in the solar system, ranging from the rocky materials of the asteroid belt to the billions of objects in cold storage in the Kuiper Belt and Oort Cloud of the outer solar system. Some of these objects become modified by the inevitable gravitational perturbations created by the outer planets and passing stars. Some of the perturbed orbits intersect the inner solar system, where ultimately they become captured by planets and moons to form modern impacts. Impacting objects from the solar system had an important influence on the evolution of life on Earth, and the process continues today, with potentially catastrophic consequences for human civilization at some unknown future time.

Supplementary Readings

Neil McBride and Iain Gilmour, eds. 2004. *An Introduction to the Solar System*. Cambridge: Cambridge University Press,.

William K. Hartmann. 2005. *Moons and Planets*, 5th ed. Pacific Grove, CA: Thomson Brooks/Cole.

Robin M. Canup and Kevin Righter, eds. 2000. *Origin of the Earth and Moon*. Tucson: University of Arizona Press.

Fig. 9-0: Seascape photograph by Ansel Adams of Point Sur during a storm. Earth has had a persistent liquid ocean since at least 3.8 Ga. (Photo by Ansel Adams. Collection Center for Creative Photography, University of Arizona, © The Ansel Adams Publishing Rights Trust, with permission)

CHAPTER 9

Making It Comfortable

Running Water, Temperature Control, and Sun Protection

The solar system is marked by temperature extremes, from the millions of degrees of the solar corona to the near absolute zero of interplanetary space. The moon has temperature variations of ~300°C between lunar night and lunar day. Although Venus is nearly equal in size and bulk composition to Earth, the Venusian ground surface is 450°C warmer. Mars is 80°C colder. None of these environments are able to host liquid water, which is a critical need for life as we know it. Earth, in contrast, has a "goldilocks" temperature, with evidence for continuous liquid water at the surface for all of Earth's history recorded in the rock record. During this time the luminosity of the sun has brightened by about 35% as the sun has consumed its hydrogen fuel. How did this equable climate come about? When did it begin? How has it been maintained?

Earth's climate stability is dependent on its volatiles. The volatile budget of a planet depends on its accretion history and bulk composition, the amount of volatile loss to space from impacts and the solar wind, and the cycles of the volatile elements between planetary interior and exterior. Once the core formed a few tens of millions of years after accretion, Earth would have developed a magnetic field that deflected the solar wind, minimizing its effects on the atmosphere, preventing volatile loss, and protecting the surface from ionizing radiation harmful to life. Liquid water was present very early in Earth's history. Sediments, formed by deposition from water, and pillow basalts, solidified lavas that formed in contact with water, are present in the oldest rocks at 3.8 Ga. There is evidence for even earlier liquid water from tiny crystals of zircon that have ages as old as 4.4 Ga. Climate stability is a long-standing characteristic of Earth's evolution.

A planet's surface temperature depends on the luminosity of its star and on the planet's distance from the star. It also depends on the reflectivity of its surface and on the "greenhouse power" of its atmosphere, caused by molecules consisting of more than three atoms, such as CO_2*,* H_2O *and* CH_4*. Earth and Venus exemplify the importance of the greenhouse effect. Most Venusian carbon is in the atmosphere as* CO_2*, creating a powerful thermal blanket. By contrast, Earth's carbon is nearly all stored in sediments as carbonate minerals and organic residues. Feedbacks must have modulated greenhouse gases in the atmosphere to maintain long-term climate stability. The most likely feedback is a "tectonic thermostat" that relates subduction and volcanic outgassing of* CO_2 *to changes in weathering. High* CO_2 *or high temperatures enhance weathering that releases* Ca^{2+} *to the oceans. This then leads to cooling caused by removal of* CO_2 *as* $CaCO_3$*. Low* CO_2 *or low temperatures permit greater buildup of* CO_2 *from volcanoes, causing warming. Weathering is itself influenced by plate movements and mountain building. Hence, Earth's climate reflects linkages among the sun, plate tectonics, and surface biogeochemical cycles, providing climate stability consistent with liquid water. The tectonic thermostat depends on the coexistence of ocean and continent. Earth's surface has just the right amount of water for this balance. Whether this is a happy accident or the result of feedbacks in early planetary history remains an unsolved puzzle.*

Introduction

We have yet to consider the attribute of our planet most important to living organisms—a stable climate that permits water on the surface. What fixes its water supply? What sets the temperature of its surface? What permits the lucky coexistence of continents and oceans? In a nutshell, what makes our planet habitable?

There are, of course, no simple answers to these questions. In previous chapters we have seen that the habitability of our planet is in part determined by its nebular heritage that sets its size, orbit, spin, and bulk chemical composition. It is in part determined by the evolution of its interior and crust. As we shall see in this chapter, it also critically depends on what happened to its volatiles after planetary accretion, and how they are cycled through planetary processes.

The Planetary Volatile Budget

For life of any complexity to develop on a planet, abundant liquid water must be available. Water is essential for all of life as we know it. Water is a fundamental medium of transport and chemical communication that make cellular processes possible. Living cells are about 70% water by weight. The average human being is 60% water (watermelon is >90% water). Water's centrality to life is reflected in the stark differences we see from region to region related to water availability. Given abundant rainfall, we have lush forests teeming with all manner of living organisms. Where there is little rain, we have deserts sparse in life. Where only snow falls, we have barren ice caps. These contrasts are found on a planet whose surface is 70% covered by liquid water!

Carbon is also essential for habitability, since it is the central element for all the organic molecules (those with C-H bonds) of which life is made. As we will see later in this chapter, C as CO_2 is also a pivotal molecule for climate stability, and its exact concentration in the atmosphere is a fundamental control on surface temperature. The *carbon cycle* links the carbon in organic molecules, the atmosphere, the oceans, the mantle and limestone ($CaCO_3$) in a balance that supports both life and the climate that is essential for life. For a habitable planet, the right amounts of H_2O and CO_2 are both critical.

Given the centrality of H_2O and CO_2, the first requirement for habitability is that the planet must have captured enough volatiles, including sufficient water to make a sizable ocean. The silicate earth as a whole contains only small amounts of H_2O and CO_2—about 700 ppm H_2O (0.07 wt.%) and 200 ppm CO_2 (0.02 wt.%). For H_2O these small amounts mean that Earth ended up with only one water molecule out of every three million that were in the pool of matter from which Earth formed. Most of the carbon in the nebula was in the form of methane gas. Earth, however, somehow managed to capture about one in every 3,000 carbon atoms. These numbers further illustrate that Earth is very volatile depleted relative to the solar nebula. A curious fact is that the ratio of H_2O/CO_2 in the silicate earth (~3.5) is significantly higher than the ratios in chondrites (H_2O/CO_2) <1.5. One solution would be that much of Earth's volatile budget comes from comets, which have a high H_2O/CO_2 ratio. Or carbon may be one of the light elements in the core. Water and CO_2 are further pieces to the puzzle of the origin of Earth's atmosphere and ocean prior to the beginning of the rock record.

The small amounts of volatiles in Earth as a whole lead to the second habitability requirement—given a modest volatile budget, the volatiles must be concentrated at the surface. This occurred on Earth, since at the surface a low volatile budget is not at all evident. To the contrary, the proportion of H_2O and CO_2 in the combined atmosphere, ocean, and crust is rather high—7.2 wt. % H_2O and 1.5 wt. % CO_2. These numbers reflect ~100 times enrichment relative to the total planet and are just the right amounts for a large liquid ocean and climate stability conducive to life. At the high temperatures that must have prevailed during the period of the formation of our iron core, extensive melting and active convection would have cycled mantle rocks to the surface, where their H_2O and CO_2 would have been transported to the atmosphere in gaseous form. As the iron migrated to the core, H_2O and CO_2 would have migrated to the surface. It is also possible that volatiles degassed during impacts of incoming planetesimals, so that volatiles were preferentially accreted at the surface.

Sufficient volatiles at the surface, however, are not all that is required for habitability. The water at the surface must be in liquid form. Can we somehow ascertain when liquid water, essential for the formation of life, might have first appeared? The answer to this question is critical for the origin of life, discussed in Chapter 13, because it constrains the time interval over which the processes leading to life had time to operate. For example, if life and liquid water appeared at the same time, then life's origin would be geologically instantaneous. If water were present from the beginning, life could have a billion years or more to develop. Which is it?

Evidence for Liquid Water before 4.0 Ga

Some of the oldest rocks are sedimentary rocks, and most sediments require weathering, transport ,and deposition by liquid water. The evidence for the earliest fossils comes from rocks that are 3.5 billion years old. The oldest sediments are those of the 3.8 Ga Isua formation in Greenland. These rocks include cherts, carbonates, and banded iron formations. All of these rocks require liquid water to form, and the same rock types occur in much younger rocks when we know liquid water was present. The rocks show that water was present at least as early as 3.8 Ga.

We are able to push back the presence of liquid water even earlier by using evidence from a surprising source, the mineral zircon ($ZrSiO_4$)—one of the highest-temperature and most stable minerals found in rocks. Zircon is very low in abundance (usually less than 0.02% of the rock), but also very common. Virtually all granites and sandstones contain some grains of zircon. Zircons are also very stable minerals and are difficult to alter or dissolve. The high chemical fidelity and chemical resistance of zircons allows them to survive during weathering and sedimentary transport; they are so robust they often survive even multiple melting

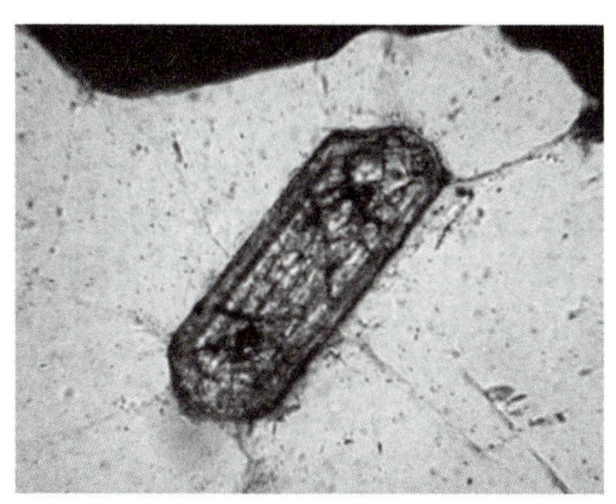
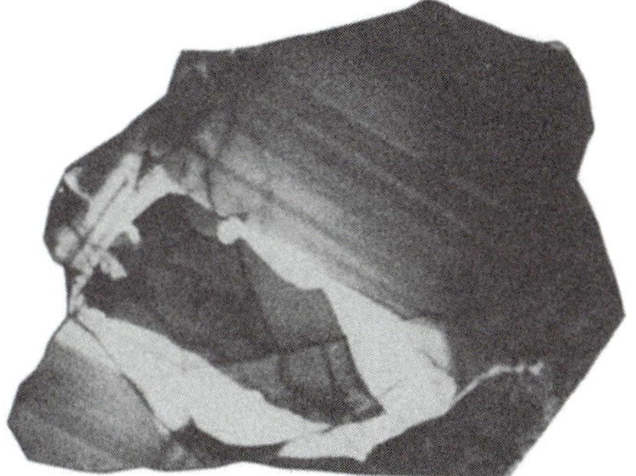

Fig. 9-1: Images of zircons. *On the left:* A single zircon crystal in a biotite. The crystal is about 100 microns (1/10 mm) in length. *On the right:* A zircon viewed under cathode luminescence that reveals its zoning and growth history. Each point in a zircon can give a precise date using the U-Pb system, (see Fig. 9-2). The zircon has a complex history, with an old inner core that was partially corroded, surrounded by younger generations of mineral that grew around it. The old core has an age of 4.4 Ga, the oldest age found for any terrestrial material. (Photograph courtesy of John Valley)

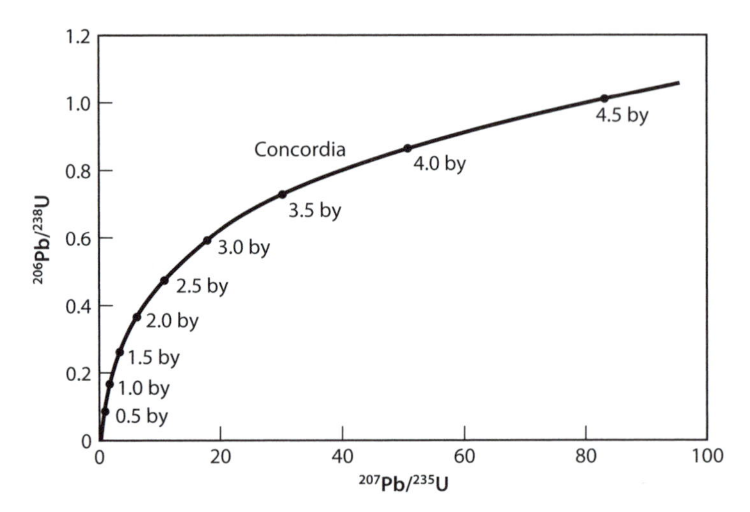

Fig. 9-2: The "Concordia" diagram used to illustrate U-Pb dating that is possible with zircons. Zircons contain no Pb initially, so all the Pb is created by radioactive decay of the two isotopes of U. Because the half-life of ^{235}U is much shorter, much more ^{207}Pb is created in older rocks than in younger rocks. Samples that have not lost any Pb since they formed will plot on the Concordia, and their age is confirmed by two independent methods. Pb loss causes the data to move directly toward the origin at the time the Pb loss occurred. This can also be used to constrain ages, an original age and a metamorphic age when the Pb loss occurred.

events, creating zoned minerals where each part of the mineral formed at a different time (Fig. 9-1).

Zircon has one other key characteristic—mineral grains can be individually dated. Zircon concentrates the parent element U and excludes the ultimate daughter element Pb. These are the ideal initial conditions for radiometric dating. Furthermore, ^{238}U and ^{235}U have different decay constants and decay to ^{206}Pb and ^{207}Pb, respectively, so that two ages can be determined independently. Because of the different decay constants, relatively more ^{207}Pb is formed at the beginning and more ^{206}Pb more recently. If Pb is lost during subsequent geological processes, all the Pb isotopes are lost proportionally, and the ages no longer agree. When they do agree, the age is robust and the zircon is said to be *concordant* (Fig. 9-2). Zircons are thus very effective messengers from the past, because they record the time of their formation and they resist subsequent chemical change.

These characteristics make ancient zircons the oldest Earth materials that have not been modified since earliest Earth's history. The most ancient zircons are preserved in Archean sedimentary rocks. These rocks are as old as the igneous rocks that surround them, but the sedimentary rocks, particularly sandstones, preserve zircons that were created by a previous generation of igneous events and then were preserved during weathering to survive in the younger sediment. The most famous locality for such ancient zircons is an apparently nondescript sedimentary rock in Australia, called the *Jack Hills formation*. This formation has been exhaustively studied, and zircons have been recovered with concordant ages as old as 4.4 billion years. This is much older than the oldest known rock with a reliable date associated with it—the Acasta gneiss, near 4.0 Ga. So how do these little minerals tell us about ancient water?

Zircons provide evidence for the early presence of water through two fairly detailed lines of reasoning. Water has a strong influence on freezing point depression and permits melting of Earth materials at low temperatures. The lowest temperature silicate melt is granitic magma, which can be produced by hydrous melting of basalt, sediment, or other granites. Higher-temperature magmas also differentiate toward granitic compositions, as they crystallize in the presence of water. Granites imply water. While zircons can be found in anhydrous mafic rocks, they are rare. In hydrous granitic rocks they are ubiquitous. The common existence of zircons in the Jack Hills formation suggests granites, which requires water. This is not quite proof, however, since some zircons also occur in high-temperature, anhydrous magma.

Further evidence comes from the trace element contents of zircons. Titanium is incorporated as a trace element in zircon because it has the same 4+ valence state as Zr. Bruce Watson and Mark Harrison showed that the amount of Ti incorporated in zircon is very sensitive to temperature. Measured Ti contents of the Jack Hills zircons show that they formed at temperatures of about 750°C, a temperature associated with hydrous, granitic magmas and not with anhydrous mantle-derived magmas. The zircons come from low temperature granites, which requires water. Granites are also the characteristic rock type of continents, so the data suggest continents were present as well.

STABLE ISOTOPE FRACTIONATION

Here we need a brief interlude to introduce the concept of *stable isotope fractionation*, which is a final line of evidence from zircons. In previous chapters we have discussed variations in isotope ratios of elements resulting from radioactive decay. Oxygen is not a radioactive decay product, and all isotopes of oxygen have the same electron shell structure, so how could the ratios of oxygen isotopes vary? It turns out that at low temperatures, processes discriminate slightly among isotopes of the same element based on their mass. The variations are so small that they are reported as parts per thousand (or "per mil") relative to a seawater standard. One consequence of this isotopic fractionation is that rainwater that falls on continents is isotopically heavy—i.e., has a slightly higher ratio of $^{18}O/^{16}O$. In order to make the numbers a bit more intuitive, stable isotope variations are always referred to as "per mil" variations relative to a well-known standard, with the heavier isotope in the numerator and the light isotope in the denominator. For oxygen the standard is mean seawater. The notation used is $\delta^{18}O$, so seawater has a $\delta^{18}O$ value of 0 per mil. "Heavy" oxygen with higher $^{18}O/^{16}O$ than seawater has positive values of $\delta^{18}O$ (e.g. 10 per mil would be 1% heavier than seawater).

The mantle has $d^{18}O$ of about 5 per mil, as indicated by the oxygen isotope measurements of mantle materials reported in Figure 9-3a and 9-3b.

Rocks that have been influenced by a low-temperature water cycle involving evaporation and precipitation (i.e., weather with a liquid ocean) have heavier oxygen, with $\delta^{18}O$ greater than +5. This interaction can involve either sediments in the source regions that melted to produce the rocks or interactions with migrating fluids in the crust that are derived from rain. In either case, a low-temperature water cycle is indicated. We can see the evidence of this water cycle by examining data from rocks when we know a water cycle was present. For example, some igneous rocks are known to be formed from sediments that require a water cycle. These can be seen in Figure 9-3c as the "metasedimentary province plutons" with d180 of 6 to 7. So are the Jack Hills zircons from sedimentary or mantle-derived materials? The Jack Hills zircons are isotopically heavy, very similar to continental zircons formed in modern times (Fig. 9-3). The heavy oxygen suggests an active low-temperature water cycle was present during the formation of the rocks whose zircons survived in the Jack Hills formation. Otherwise, the oxygen isotopes would not be fractionated.

The zircon evidence suggests that as far back as 4.4 billion years there was a robust water cycle on Earth, and liquid water was present. It is a beautiful piece of geological detective work that some of the smallest

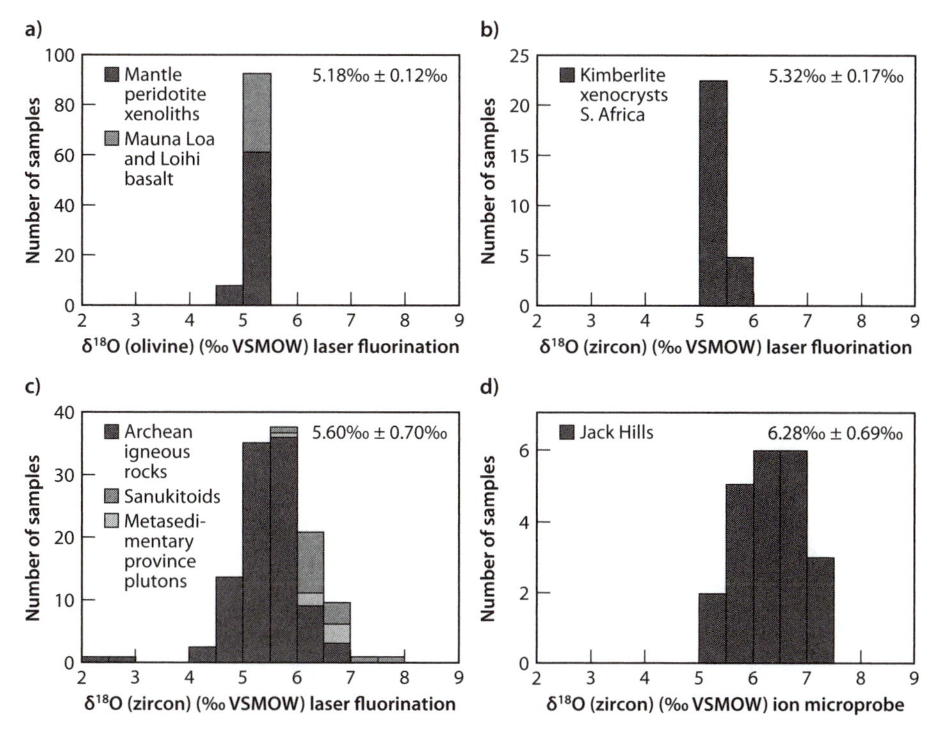

Fig. 9-3: Oxygen isotope data from various rocks compared to the data from the ancient zircons collected in the Jack Hills formation. The top two panels, a and b, show the data for mantle derived magmas which have had no interaction with the water cycle, with values of 5.2–5.3. The lower left panel c shows Archean igneous rocks are similar to mantle values, but sediments that have interacted with the water cycle have elevated values. The higher values of the Jack Hills zircons in panel d require that their source rocks had seen the influence of a low temperature water cycle. (Modified from J. Valley, *Reviews in Minerology and Geochemistry,* v. 53, no. 1, 343–385)

and rarest mineral grains found in rocks hold the essential clues to water's presence, providing crucial evidence for appropriate conditions for life's origin.

Controls on Volatiles at the Surface

For habitability over Earth's history, we need to consider the long-term controls on the total amount of water and carbon at the surface—not how the volatiles cycle through surface reservoirs over a few thousand years or less, but how they cycle between the interior and exterior over hundreds of millions of years.

If impacts caused degassing of incoming planetesimals, or core formation led to massive early degassing of Earth's interior, volatiles would have been concentrated to the surface in earliest Earth's history, consistent with the evidence cited above for a substantial ocean at that time. If early degassing were so efficient, we might expect much of Earth's volatile budget to be initially in the atmosphere and ocean. Continued volcanism over time would have led to further addition of volatiles to the surface, causing most of Earth's volatile budget to be in the atmosphere and ocean. Surprisingly, however, substantial H_2O and CO_2 remain in Earth's interior. Volcanoes today are still emitting volatiles, and these permit estimates of the concentrations of volatiles in the mantle. While the concentrations are low, the volume of the mantle is so large relative to the crust that about half of Earth's CO_2 and H_2O still reside in the interior. Another entire ocean volume remains trapped in the solid Earth. Furthermore, current volcanic emissions over time would generate an ocean volume of water in only 2–3 Ga, and yet as we saw above there is evidence for an ocean prior to 4 Ga. Has the ocean then increased substantially in size?

Studies of rocks from the continental crust can be used to show that sea level has remained remarkably constant through Earth's history, implying near constancy of the volume of the oceans and hence water at the surface. Since volatiles (including H_2O) are steadily being supplied from Earth's interior, how can the volume of H_2O at the surface have remained in a narrow range? And since degassing is an inevitable result of early Earth's history and continued volcanism, how can so many volatiles remain in Earth's interior?

These questions require a consideration of the surface volatile budget, not just as a progressive degassing but also as a dynamic process involving fluxes among Earth's reservoirs and space, causing volatile addition and removal. Water and CO_2 removal could occur by volatile loss to space or by returning H_2O and CO_2 to the interior during recycling of Earth's plates.

ATMOSPHERIC LOSS TO SPACE

Once at the surface, any gaseous substance has the opportunity to escape into space. The most well known mechanism is called thermal escape.

Just as a space vehicle can escape Earth's gravitational field with sufficient velocity, individual atoms or molecules can escape when their velocity is high enough. Velocities increase with temperature and decreasing atomic mass. For Earth the temperature of the outer regions of the upper atmosphere is about 1500°K, while that at the surface is only about 300°K. The high temperatures of the upper atmosphere greatly increase the probability for molecular escape.

The velocity required for escape depends on the planet's gravitational field and on the mass of the molecule itself. These dependencies are extremely strong. The escape velocity from Jupiter is 60 km/sec, from Earth it is 11.2 km/sec, and from the moon only 2.4 km/sec. For smaller planets, the escape velocity is much lower, and gases can escape more easily. Escape also depends critically on the mass of the gas molecule. A factor-of-2 difference in mass changes the likelihood of escape by several orders of magnitude. While Earth and Venus are massive enough to prevent thermal escape of all but the lightest gases, the moon has insufficient gravity to hold even the heaviest gas. Thus, Earth and Venus have substantial atmospheres while the moon has none at all. Jupiter has such a high escape velocity that it has been able to retain even the lightest gases, hydrogen and helium. Planet size and the ability to retain an atmosphere are critical for habitability.

There are other mechanisms of atmospheric loss. Particles of the solar wind can have very high velocities that can strip gases from the outer atmosphere. Impacts can also cause stripping of the atmosphere by accelerating molecules to escape velocities. And if planets are too close to their star, the very high temperatures can produce various effects that create atmospheric loss. The diversity of processes of atmospheric loss and how they change over planetary history may help to account for much of the diversity of planetary atmospheres and compositions in the solar system.

For our present purposes, however, we need to estimate to what extent H_2O and CO_2 might be lost from Earth. By turning to the evidence from helium, we are able to estimate the extent these heavier gases have been lost. This method involves a comparison between the total number of helium atoms in the atmosphere and the number of helium atoms that leak into the atmosphere from Earth's interior each year (Fig. 9-4). The number in the air is obtained from the mass of the atmosphere and its

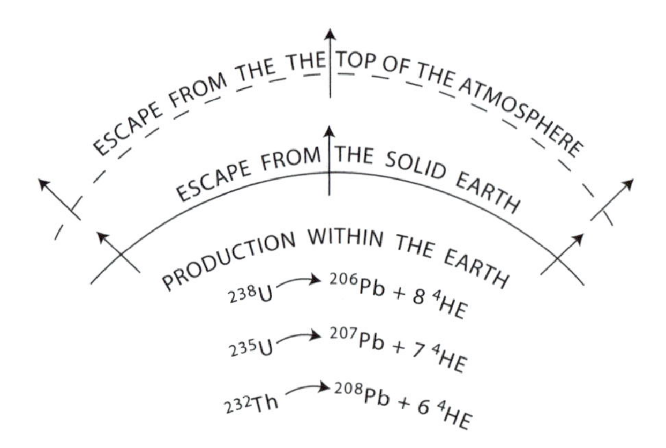

Fig. 9-4: Earth history of helium atoms. ^{4}He atoms are generated within the Earth's crust and mantle by the decay of uranium and thorium. By measuring the amount of heat escaping from Earth, we have a fairly good idea of how much uranium and thorium are present in Earth. Thus we know the rate at which ^{4}He generation occurs. After a billion or so years of entrapment in solid Earth, the average helium atom manages to reach the surface, where it resides an average of 1 million years before escaping from the top of the atmosphere. All the helium atoms manufactured by radio decay within Earth are eventually lost to space.

helium content. Helium production at ocean ridges can be determined by measuring the helium concentrations in ocean ridge basalts, in the hydrothermal fluids at deep sea vents (discussed at length in Chapter 12), and in seawater. Since helium is produced by the radioactive decay of ^{238}U, ^{235}U, and ^{232}Th, estimates of these concentrations in continental rocks, combined with measuring heat flow on continents that is caused by radioactive decay, permit estimates of how much helium is escaping from continents each year. The number of atoms being added to the atmosphere in a year is about one-millionth the number of helium atoms currently residing in the atmosphere. This suggests that helium atoms reside in the atmosphere for an average of 1 million years before they escape from the atmosphere to space. Using the helium escape time, the escape time for the other gases can be calculated from molecular theory. With masses of 20 for neon, 28 for N_2, 32 for O_2, and 44 for CO_2, escape times are so long that atmospheric loss is negligible over Earth's history.

With its atomic weight of 18, water has a mass similar to neon, and therefore water molecules do not escape from the atmosphere. But the hydrogen atoms that are essential parts of the water molecule are a dif-

Table 9-1
Composition of Earth's atmosphere today*

Gas Name	Gas Formula	Percent by Volume
Nitrogen	N_2	78.08
Oxygen	O_2	20.95
Argon	Ar	0.93
Carbon dioxide**	CO_2	0.039
Neon	Ne	0.0018
Helium	He	0.00052
Krypton	Kr	0.00011
Xenon	Xe	0.00009
Hydrogen	H_2	0.00005
Methane**	CH_4	0.0002
Nitrous oxide**	N_2O	0.00005

**In addition, the atmosphere contains water vapor (H_2O) in variable amounts (up to 2% for warm air and down to a few parts per million for very cold stratospheric air). Water is also a greenhouse gas.*

***Greenhouse gases.*

ferent story. As they are only half as massive as helium, hydrogen molecules have an escape time far less than 1 million years. Fortunately, H_2 is a very rare gas in our atmosphere (see Table 9-1).

Today any H_2 molecules generated by bacteria living in soils survive only a few years in the atmosphere before being converted to water ($2H_2 + O_2 \rightarrow 2H_2O$). There is another route, however, by which a planet can lose its hydrogen and hence its water. High in the atmosphere, ultraviolet light from the sun breaks H_2O molecules apart, creating free H atoms. These free H atoms are exceedingly vulnerable to escape. The O atoms left behind eventually react with either iron, sulfur, or carbon. This process is the likely mechanism by which Venus lost most of its water.

The reason why this process has not decimated Earth's water reserves is that our atmosphere has a "water trap" that keeps almost all of Earth's water in the lower atmosphere, preventing it from being carried to higher

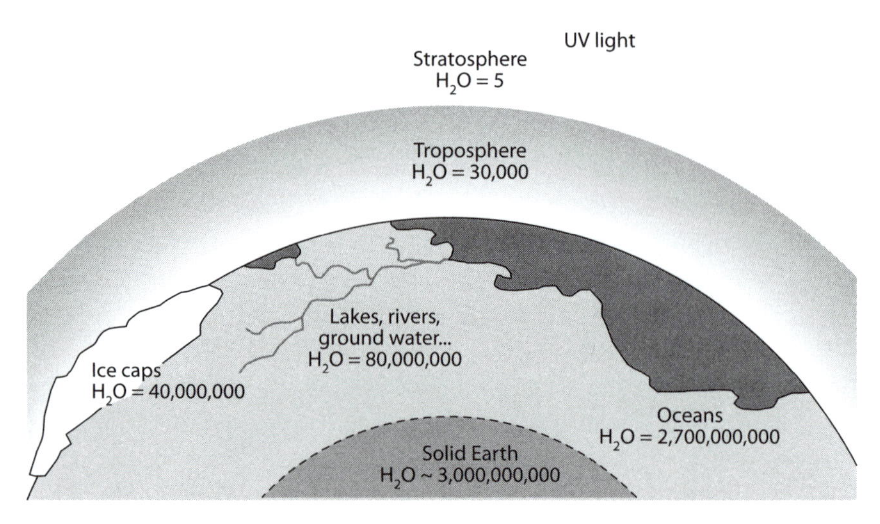

Fig. 9-5: Most of Earth's hydrogen is in the form of water, about half of which resides in the ocean. The majority of the remaining water is trapped in solids making up the mantle and crust. Earth's freshwaters (lakes, rivers, groundwater, etc.) make up only about 3.0% of the total. Its ice caps make up only 1.5% of the total. Only a very small fraction of our water is at any given time stored in the atmosphere as vapor, which is confined almost entirely to the lower, well-stirred part of the atmosphere (referred to by meteorologists as the troposphere). Only two of every billion water molecules reside in the stratosphere. This is important, because only that water present in the stratosphere has a chance to be dissociated by ultraviolet light. Dashed line separates Earth's interior from exterior.

levels from which it could escape. As tallied in Figure 9-5, the bulk of our water resides in the ocean, sediments, and ice. At any given time only about one H_2O molecule in 100,000 is in the atmosphere. The lower atmosphere is called the *troposphere*, and as we all experience, the temperature in the troposphere declines rapidly with increasing height above sea level (Fig. 9-6). This temperature decrease with altitude causes water vapor to precipitate as dense ice crystals to form the clouds and snow. It can be a warm, sunny day in Los Angeles, California, or Nice, France, while snowing a short distance away in the mountains at heights of a few thousand meters. Airplanes flying at 10,000 m are at –60°C even on summer days. At the top of the troposphere the temperature is so cold (–60°C) (Fig. 9-6) that virtually no water vapor can exist, and only extremely dry air migrates from the troposphere to the next highest atmospheric level, called the *stratosphere*. These attributes of Earth make

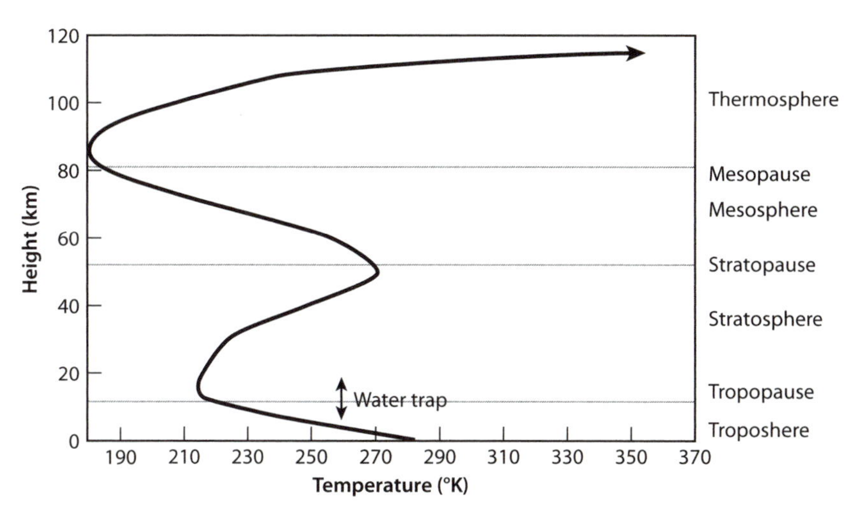

Fig. 9-6: Temperature profile through Earth's atmosphere. All of weather occurs in the troposphere. At the top of the troposphere, temperatures are so cold that all water has precipitated as ice, and none migrates to the upper atmosphere where it can be broken apart, which would allow H atoms to escape.

its hydrogen escape hatch very small. Over the last 4.0 billion years where there is a rock record and evidence for liquid water, only a small fraction of our hydrogen has been lost!

Of course, what may have happened in earliest Earth's history is not so clear. In the early Earth there was no O_2 and bacteria likely converted hydrogen to methane ($4H_2 + CO_2 \rightarrow CH_4 + 2H_2O$). While methane is in very low concentration in the atmosphere today, it might have been more abundant on early Earth, and there is no cold trap as there is for water, so methane would be able to rise to the upper atmosphere and be broken apart with loss of H. Massive impacts can also cause atmospheric loss even of heavier molecules, and very high atmospheric temperatures after the moon-forming impact would have had major effects on the atmosphere. Is there another way we can determine whether Earth lost a lot of its hydrogen?

Here once again stable isotope variations provide important evidence. Hydrogen has two isotopes, ^{1}H (called H) and ^{2}H (deuterium, or D), which is a factor-of-2 mass difference. This leads to *much* easier escape for ^{1}H relative to D. But Earth's D/H is about the same as chondrites, suggesting little hydrogen loss. Earth was able to retain its water.

All the evidence, therefore, suggests that Earth's atmosphere was stable with respect to loss of H_2O out the top, certainly by the time of the advent of a stable climate and the end of the largest impacts some 3.8 billion years ago. We thus need to look downward as well as upward to understand the controls on the long-term volatile budgets at the surface.

CYCLING OF VOLATILES BETWEEN THE SURFACE AND EARTH'S INTERIOR

Since volatile loss to space cannot explain the relatively constant water budget at the surface over time, the solution must lie in recycling of volatiles from the surface to the interior. Return of volatiles to the interior would also account for the significant amounts of volatiles that reside today in the mantle.

The process of volatile recycling is the subduction of Earth's tectonic plates, which will be discussed at length in Chapters 10 and 12. The new ocean crust formed at ocean ridges leads to extensive circulation of seawater through cracks in the crust. Interactions between the rocks and the water cause alteration of the crust to form new minerals that contain water and CO_2 in the solid state (e.g., sheet silicates and calcium carbonate.

Further alteration occurs as the spreading plate moves the crust away from the ocean ridge, and at the same time sediments fall through the ocean to create a progressively thickening sedimentary layers, including clays and carbonates made up of volatile-rich minerals. When the plate reaches the subduction zone, this volatile-rich package is "subducted" down into the mantle, returning volatiles to the interior. By this mechanism a flux from the surface to the interior balances the volcanic flux to the atmosphere.

Precise estimates of the down-going flux are difficult to obtain because no single drill hole has yet been able to penetrate the entire ocean crust, and only a few holes have been able to penetrate a kilometer or more. Considering the vast expanses of the oceans, and the diversity of ocean crust environments, our knowledge of the average composition of altered ocean crust is very limited. The data that do exist suggest that the ocean crust contains sufficient H_2O and CO_2 to easily balance the volcanic outgassing and supply sufficient CO_2 and H_2O to the interior to explain the substantial amounts that reside there. In fact, so much H_2O goes down the subduction zone that if it all remained at depth the oceans would diminish with time!

It remains curious that the amount of water at the surface appears to have remained within a tight range throughout Earth's history. This implies a balance between outgassed water and subducted water. A possible solution to this puzzle is that most of the subducted water is efficiently processed at subduction zones and returned to the surface through the volcanism that occurs there. As we will see in Chapter 12, the volatile-bearing minerals of the subducting plate break down and release their volatiles at the high pressures and temperatures of Earth's interior. The released volatiles then trigger volcanism that transfers water to the surface. If more or less water were subducted, more or less outgassing would occur. The net flux of water in or out of the Earth could then be small enough that there would be little variation in total amount of water at the surface.

For CO_2, the budget is much less clear. Unlike water, where changes in sea level provide information about the surface water budget, there is no clear geological indicator of the amount of CO_2 at the surface over Earth's history. Carbonate minerals are also more stable than hydrous minerals in the subducting plate, and substantial amounts of CO_2 in carbonates could get past the subduction zone filter and returned to the mantle. Much remains to be understood concerning the overall CO_2 budget related to outgassing and subduction.

Surface Temperature

For the H_2O at the surface to be in the essential liquid state, the Earth's surface temperature must have been maintained within a narrow range. Our water would be of little use to life were it tied up in massive ice sheets. Nor would it be of use if it were all in the atmosphere as steam. And if temperature changed from near freezing to near boiling on a regular basis, life could also not survive. How has Earth's surface temperature maintained a narrow range consistent with liquid water over billions of years?

A planet's temperature depends not only on the amount of sunlight it receives but also on the reflectivity of its surface and on the content of so-called greenhouse gases in its atmosphere. If a planet had the surface properties similar to those of a *blackbody*, its temperature would be fixed solely by the amount of sunlight reaching its surface. To qualify as a blackbody, the surface of an object must be nonreflecting—i.e., all the sunlight reaching it must be absorbed and reradiated as infrared light (Fig. 9-7). Also, there can be no gases in its atmosphere that absorb outgoing infrared light. Were Earth a blackbody it would have a mean surface temperature of about 5°C (see Table 9-2).

No planet we know of, however, is a perfect blackbody. All have an *albedo* that indicates what proportion of the light is reflected. The higher the albedo, the more light is reflected and the cooler will be the planet. The reflectivity of a planet has much to do with the amount and state of its water. Ocean water has a low reflectivity; ice and clouds have a high reflectivity. As plant leaves absorb nearly all the light they receive, there is little reflection from a forest. By contrast, about half the light reaching bare soil is reflected. As we have already pointed out, the extent of plant cover depends on the distribution of rainfall. On Earth, clouds, ice caps, and soils reflect back to space a sizable portion of the sunlight impinging on Earth, leading to an albedo of about 0.3. Thirty percent of the sunlight is reflected back into space and plays no role in heating Earth's surface. Were this the only deviation from the ideal blackbody, Earth's surface temperature would average –20°C and all water would be frozen. High albedo lowers planetary temperature.

The factor that counterbalances albedo and warms a planet is the greenhouse effect caused by particular molecules of gas in the planetary atmosphere. All molecules with three or more molecules are greenhouse gases. Such molecules absorb energy of packets of infrared light through the vibrations of their molecular bonds (Fig. 9-8). Incoming solar radiation has short wavelengths that are not absorbed by greenhouse gases, so the energy makes it through the atmosphere to the surface. When the light is reflected from the surface, however, the wavelengths of light correspond with planetary temperatures of about 300°C and are largely in the infrared. These wavelengths are effectively absorbed by greenhouse gases. Important greenhouse gases in Earth's atmosphere are water vapor (H_2O), carbon dioxide (CO_2), methane (CH_4), nitrous oxide (N_2O), and

Table 9-2

Summary of the factors influencing the surface temperatures of the terrestrial planets

	Mass of atmosphere kg/cm^2	Distance from Sun 10^6 km	Solar energy received 10^6 $watts/m^2$	Black-body temperature °C	Fraction sunlight reflected	Reflective cooling °C	Green-house warming	Actual surface temperature °C
Mercury	0	58	9126	175	.068	−8	0	167
Venus	115*	108	2614	55	.90	−144	+553	464
Earth today	1.03**	150	1368	5	.30	−25	+35	15
Early Earth		150	958	−26	.30 (?)	−21	62(?)	15 (?)
Mars	0.016*	228	589	−47	.25	−16	+3	−60
Moon	0	150	1368	5	.11	−7	0	−160 – +130

**Mostly CO_2.*
***Mostly $N_2 + O_2$.*

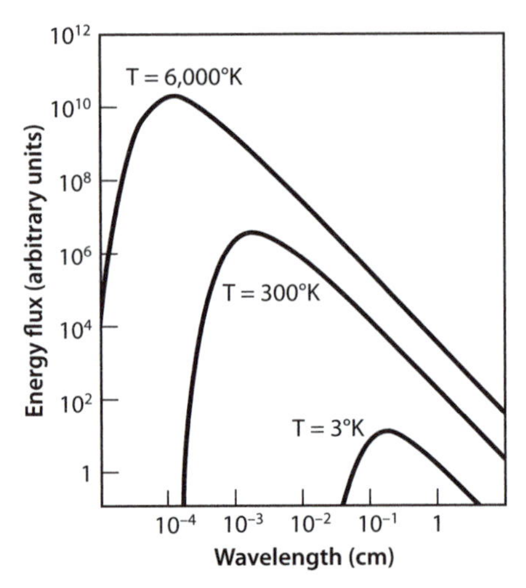

Fig. 9-7: Light emitted from black bodies of three quite different temperatures: The hotter the body, the more energy it emits (the 6,000°K body gives off about 10,000 times more energy per unit area than the 300°K body and 100 million times more per unit area than the 3°K body). The wavelength representing the peak of energy emission for a star whose surface temperature is 6,000°K lies in the visible range, that for a planet whose surface temperature is 300°K lies in the infrared range, and that for a universe whose background glow temperature is 3°K lies in the microwave range. Earth's atmosphere is transparent to the short wavelengths of the sun's radiation but absorbs some of the energy of Earth's reflected radiation.

ozone (O_3). The capture of outgoing Earth light by these gases serves as a thermal blanket that keeps the planet warm. For Earth, the greenhouse warming more than compensates for the sunlight lost through reflection. Earth's mean surface temperature (i.e., 15°C) is 10°C warmer than if it were a perfect blackbody (see Table 12-2).

To assess another planet's surface temperature, we also need to know how much sunlight it receives, the reflective properties of its surface and the amounts of infrared-absorbing gases contained in its atmosphere. Solar radiation received is a straightforward function of distance from the sun. Albedo can be measured from space. Greenhouse effect can be determined where the composition of the atmosphere is known or can be adequately estimated. Table 9-2 shows how the combination of solar radiation, albedo, and greenhouse effect lead to a planet's surface temperature. On Earth, it is clear that habitability is strongly influenced by

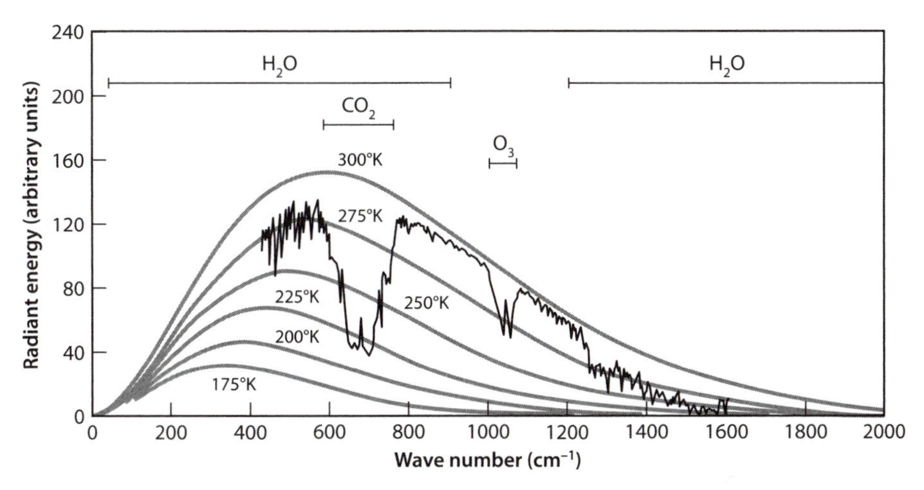

Fig. 9-8: Absorption of outgoing Earthlight. The jagged curve shows an actual spectrum of Earthlight leaving the top of the atmosphere over the island of Guam. For comparison, the smooth curves are blackbody curves that show the expected spectra if there were no greenhouse gases. These curves are drawn for a series of temperatures; that appropriate for Guam would be just less then 300°K. The wiggles and large dips are the result of absorption of the radiation by water, carbon dioxide, and ozone in the atmosphere. Water creates a broad reduction in the region 400–600 cm^{-1}As can be seen, the dip created by CO_2 is especially prominent. The wave number is a measure of the frequency of the radiation.

the fact that the greenhouse warming is large enough to more than offset reflective cooling.

Table 9-2 also shows the great importance of solar energy for a planet's blackbody temperature, which is the baseline upon which planetary temperature depends. Has solar energy been constant through time? While we have no direct measurements of long-term changes in energy coming from the sun, there is evidence from other stars in the galaxy at diverse stages of evolution that reveals how stellar energy production changes with time. Stars similar to the sun increase their energy output with time, as the hydrogen fuel in the interior converts to helium and increased temperatures are needed to balance the contracting forces of gravity. Astronomers estimate that the sun was producing about 39% less heat in Hadean times than it is emitting today. Using this luminosity, we can calculate the blackbody surface temperature of Earth for its early history, shown in Table 9-2 and illustrated in Figure 9-9.

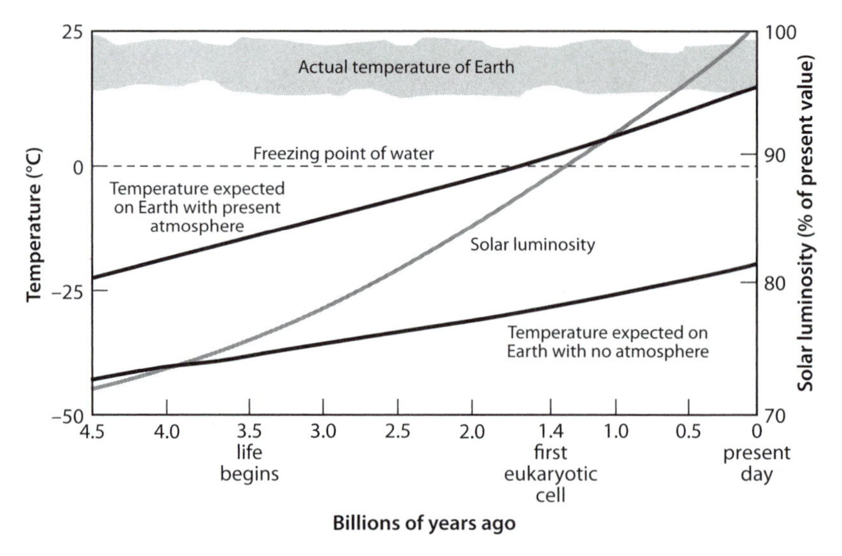

Fig. 9-9: Illustration of how the sun's luminosity has changed over Earth history and its consequences. The sun today is emitting 39% more energy than it was in the Hadean. In the absence of any greenhouse effect, Earth would have been below freezing throughout its history. With today's atmosphere, Earth would have been below freezing for all times older than 2 billion years ago. The fact that Earth's rocks provide evidence for liquid water since at least 4 Ga requires some combination of enhanced greenhouse effect or lower albedo on the early Earth. Figure modified from Kasting et al. 1988, *Sci. Amer.* 256:90–97.

Calculation of early Earth's black-body temperature for a sun with 61% of its present luminosity leads to a baseline (blackbody) temperature of only −26°C. If Earth's albedo were the same as it is today, the temperature absent a greenhouse effect would have been a frigid −47°C, similar to Mars's temperature today! With a greenhouse similar to today, temperatures on Earth would still have been below freezing. This conflicts with evidence for abundant liquid water throughout Earth's history. This conflict, referred to as the *faint young sun paradox*, can be resolved either by having substantially less reflective cooling with an albedo more akin to Mercury or the moon, or a much enhanced greenhouse effect that could provide some 55°C of greenhouse warming. This enhanced warming would require higher concentrations of greenhouse gases. Proposals include much higher CO_2 contents or an atmosphere with significant methane, which would be possible in an atmosphere with no O_2.

Plate 1. See also figure 6-1.

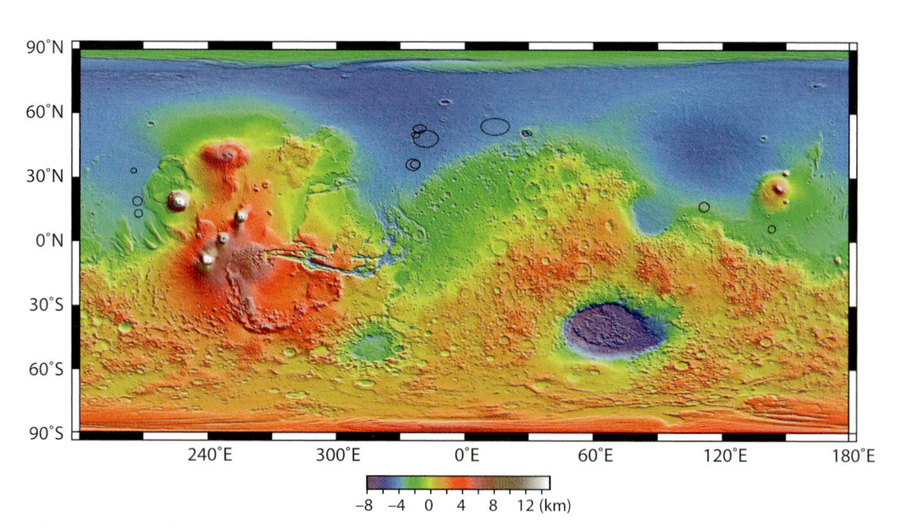

Plate 2. See also figure 8-12.

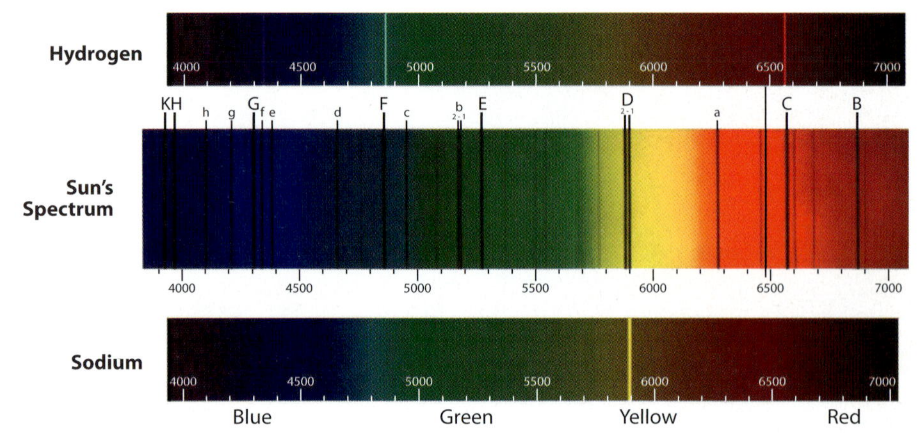

Plate 3. See also figure 2-3.

Plate 4. See also figure 5-0.

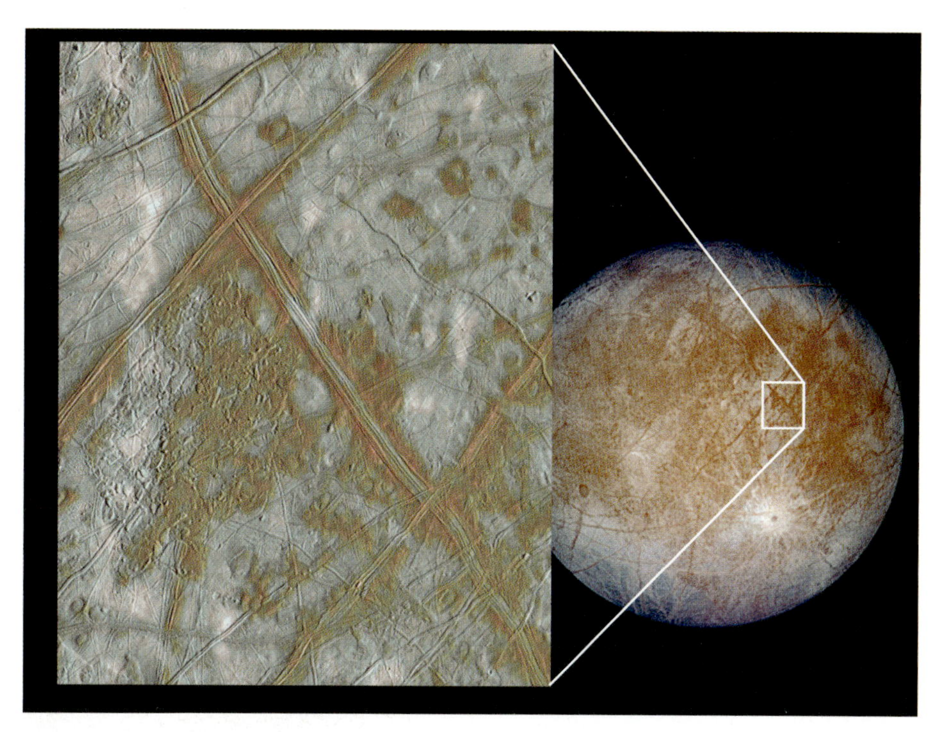

Plate 5. See also figure 8-4.

Plate 6. See also figure 8-4.

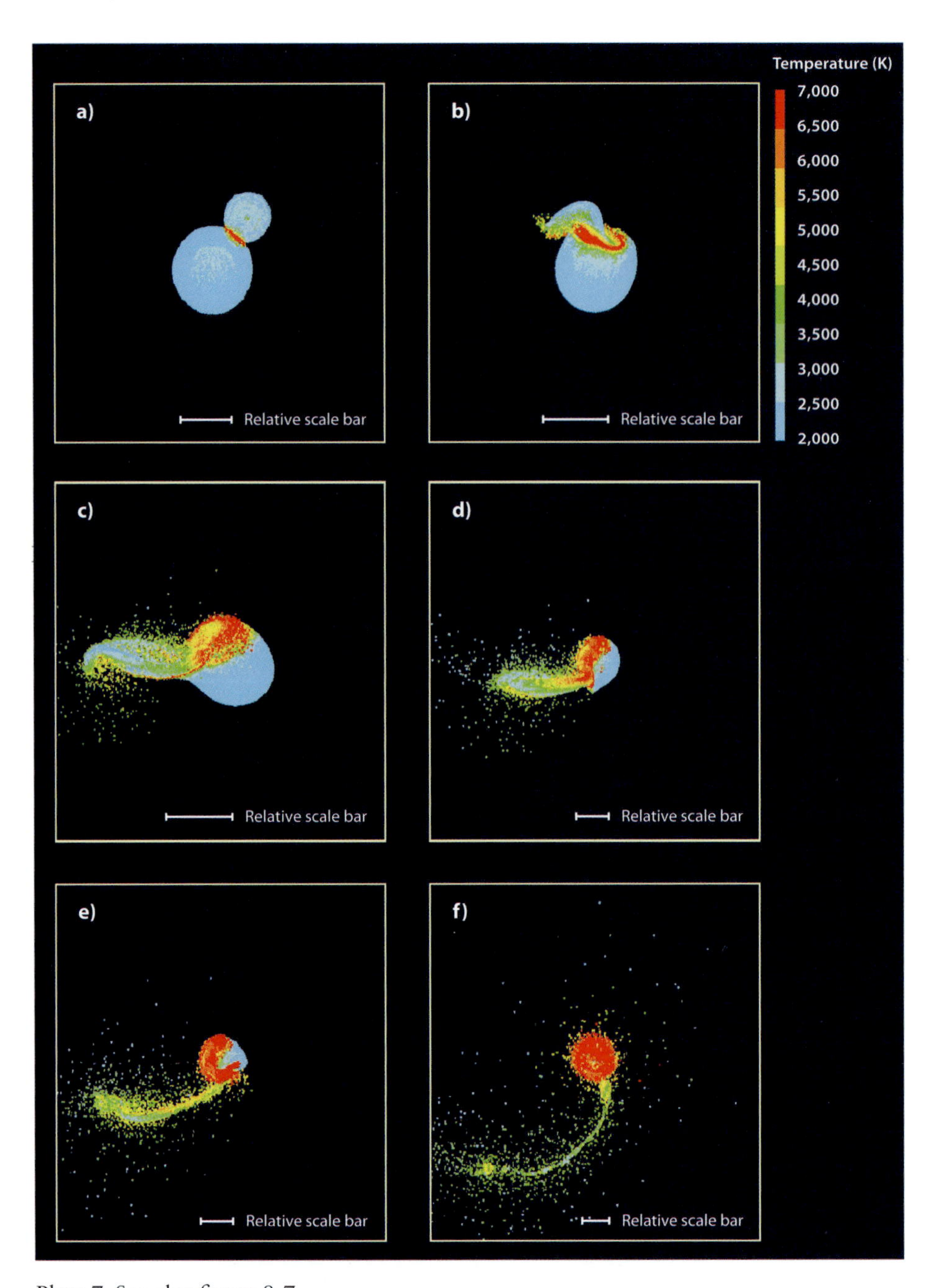

Plate 7. See also figure 8-7.

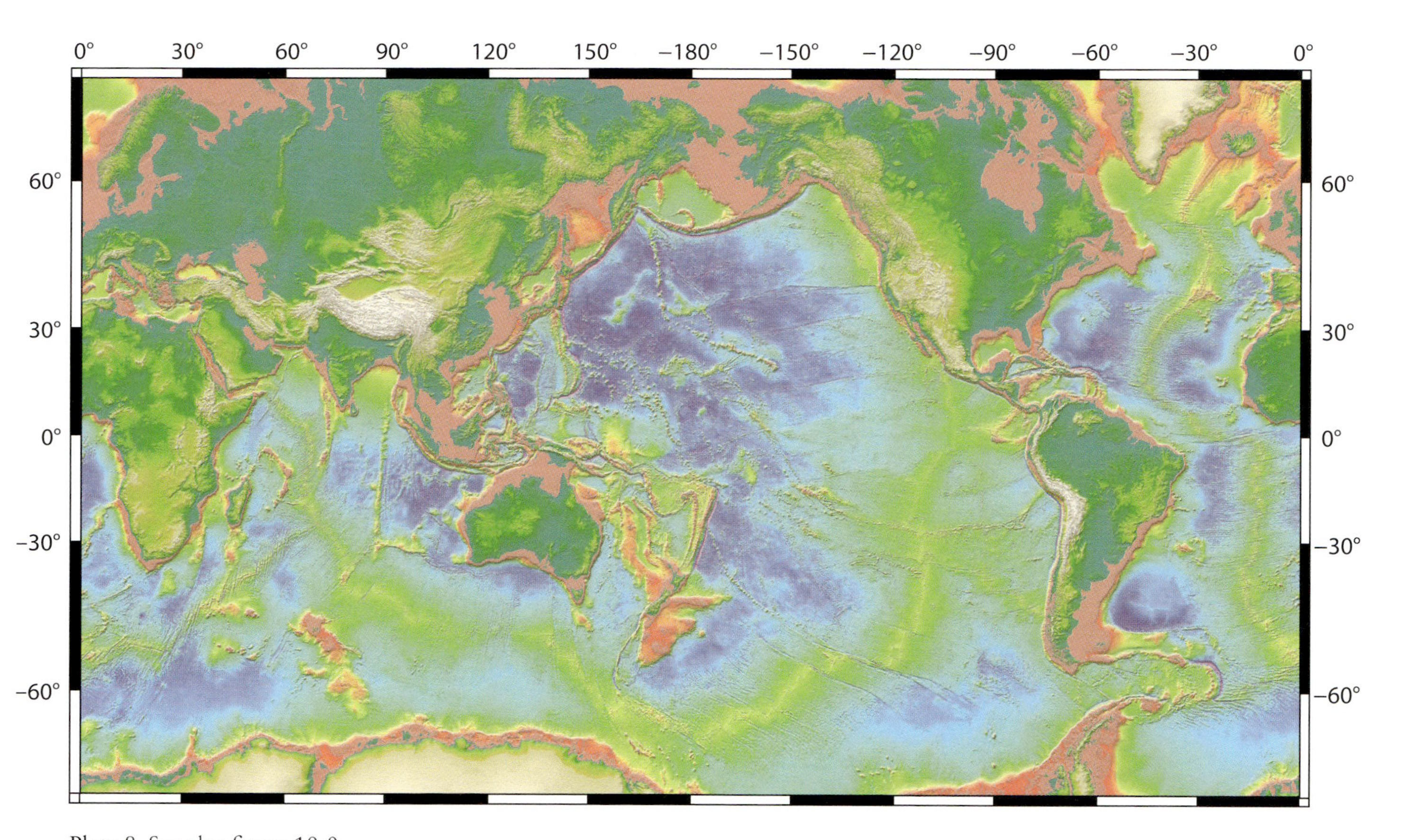

Plate 8. See also figure 10-0.

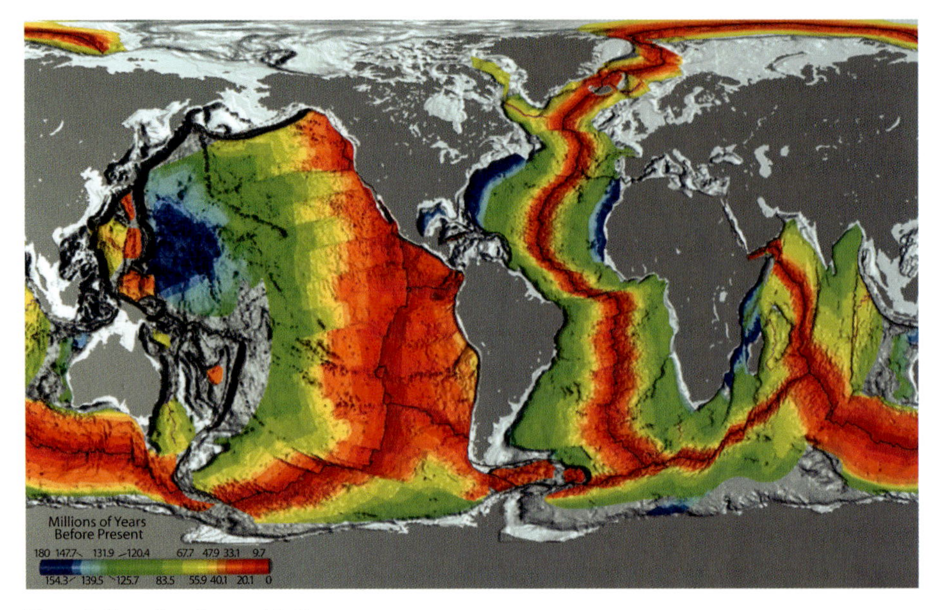

Plate 9. See also figure 10-5.

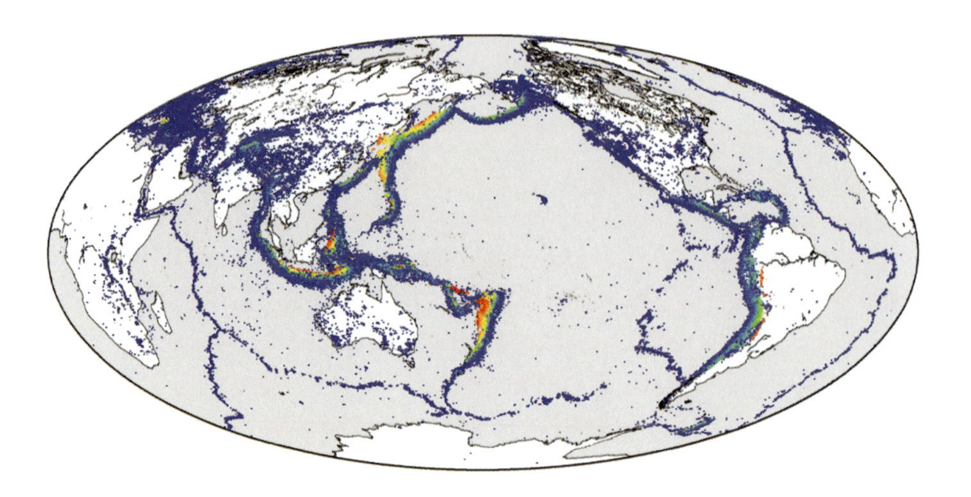

Plate 10. See also figure 10-7.

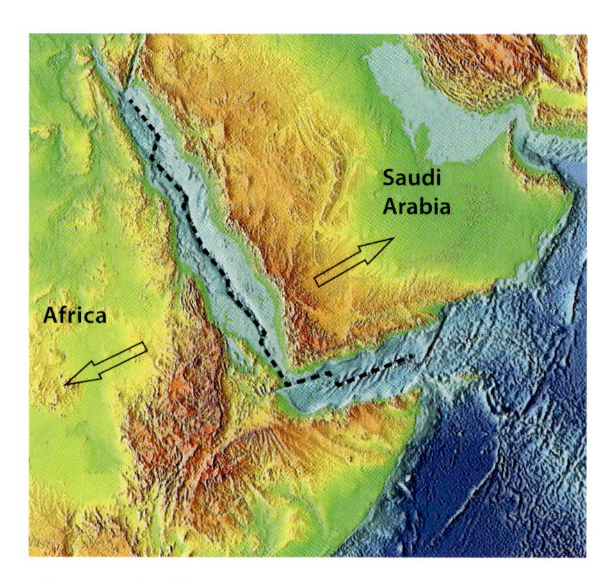

Plate 11. See figure 10-14.

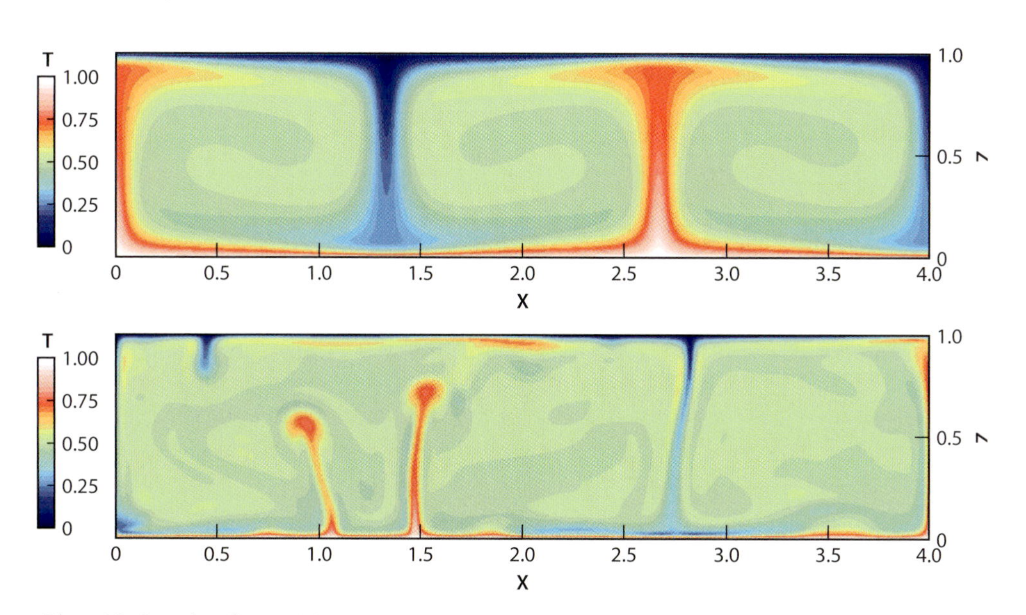

Plate 12. See also figure 11-5.

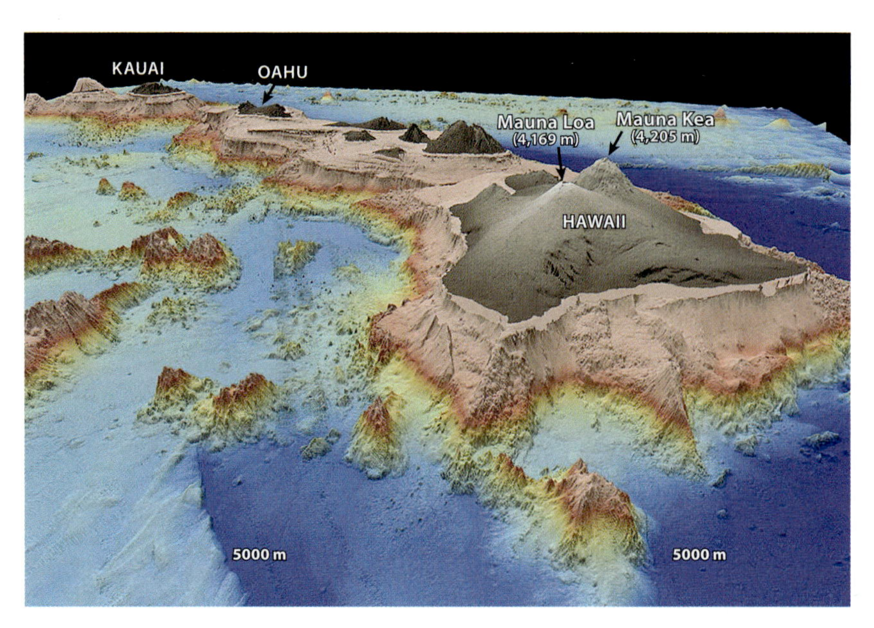

Plate 13. See also figure 11-0.

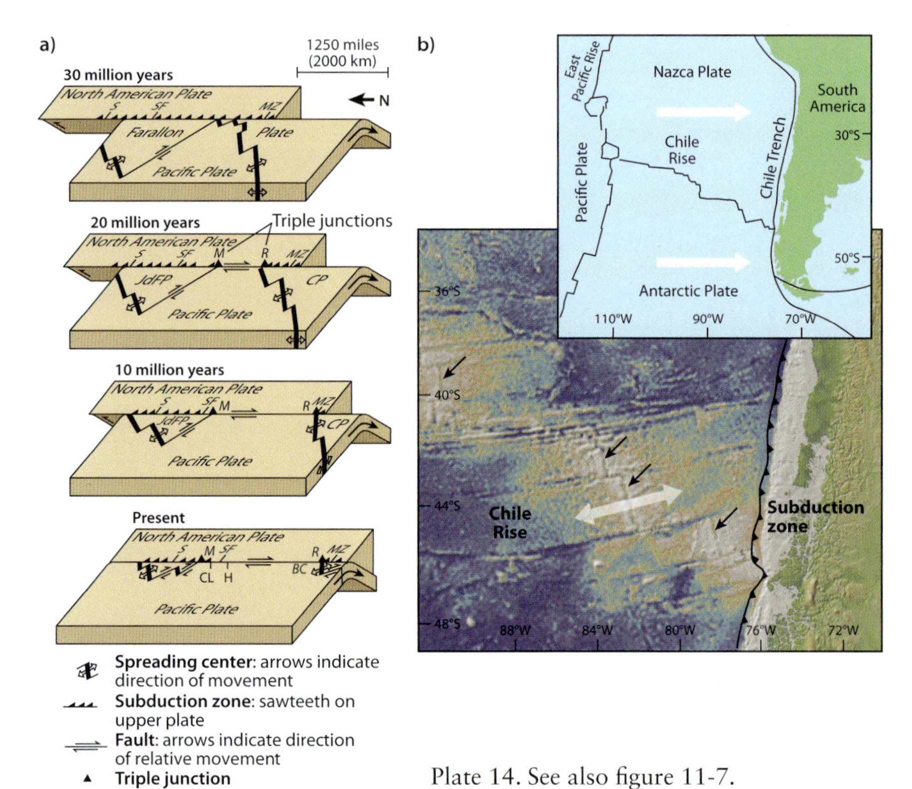

Plate 14. See also figure 11-7.

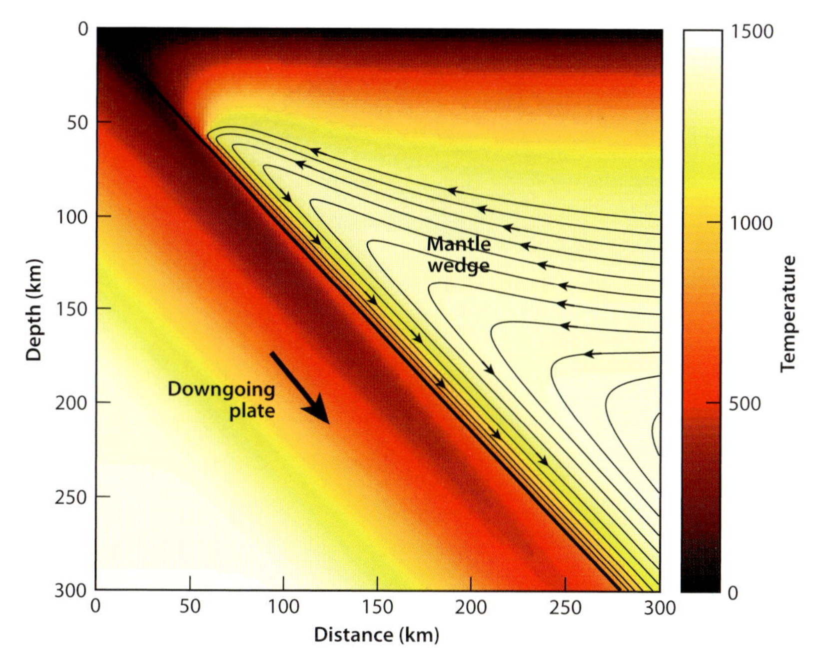

Plate 15. See also figure 11-9.

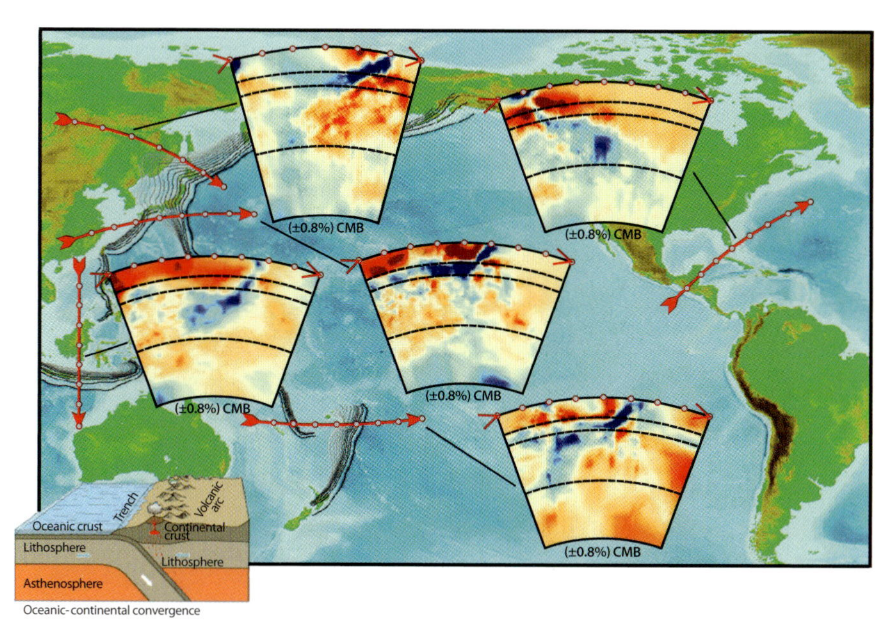

Plate 16. See also figure 11-10.

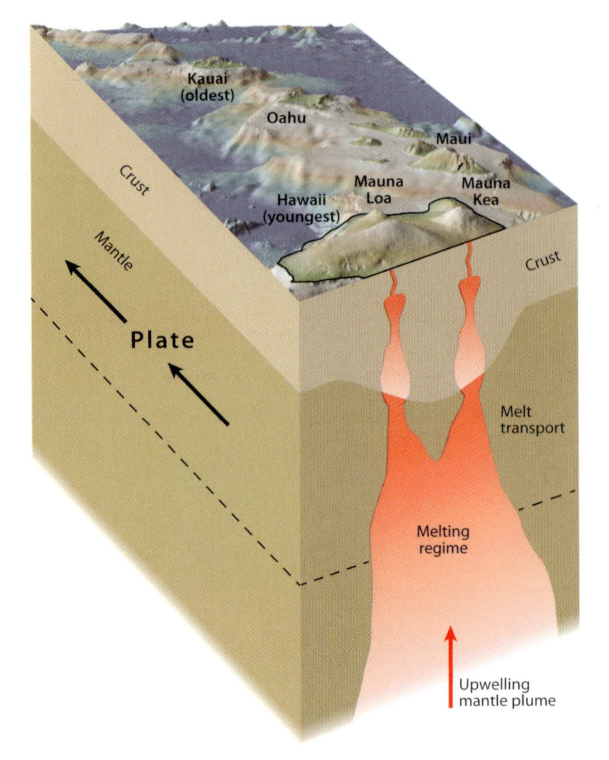

Plate 17. See also figure 11-13.

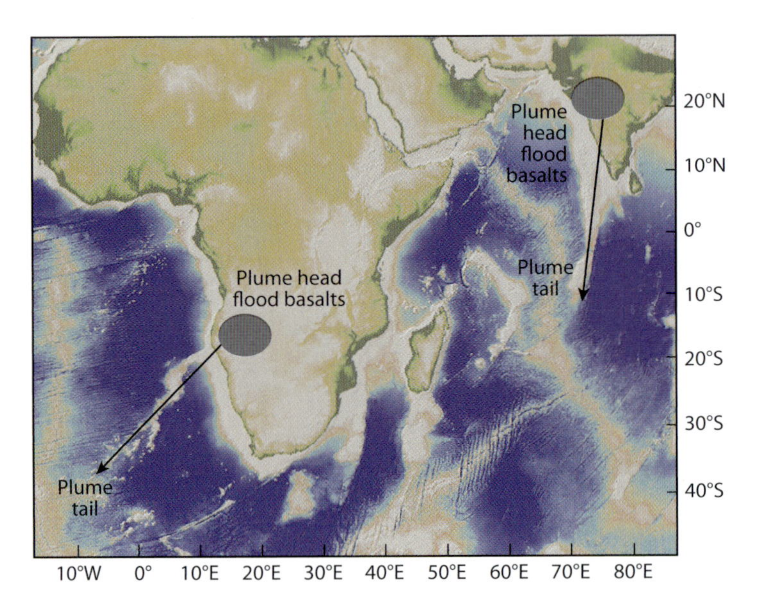

Plate 18. See also figure 11-15.

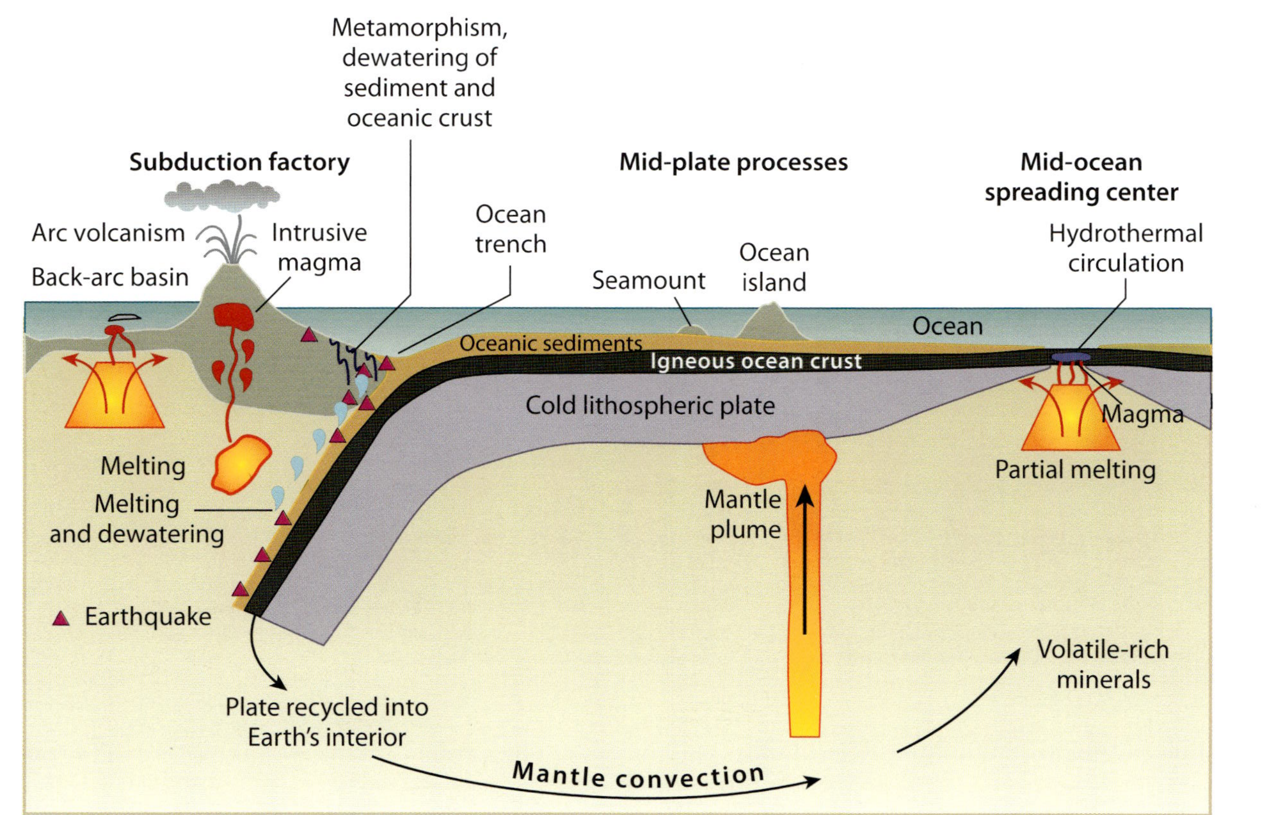

Plate 19. See also figure 12-1.

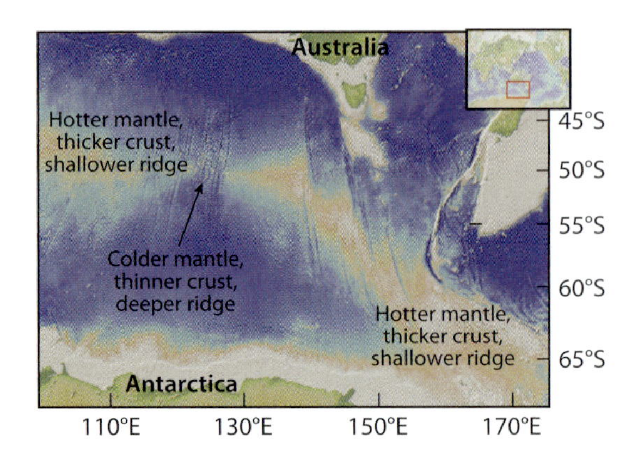

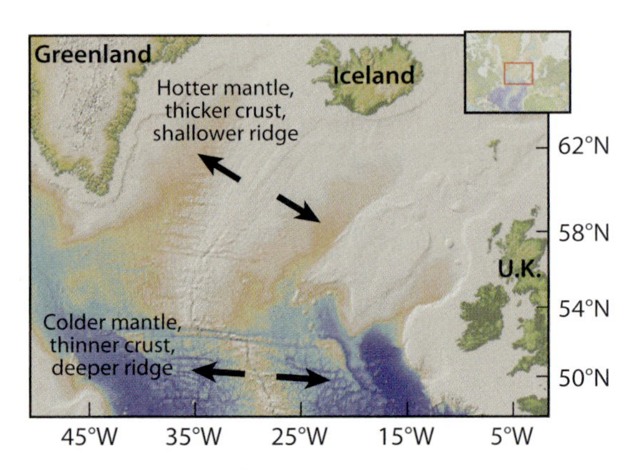

Plate 20. See also figure 11-19.

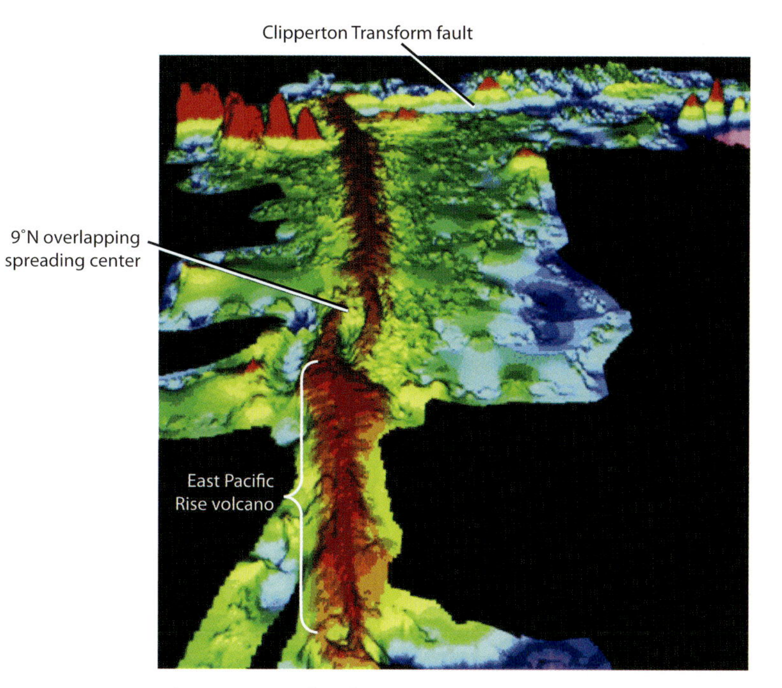

Plate 21. See also figure 12-2.

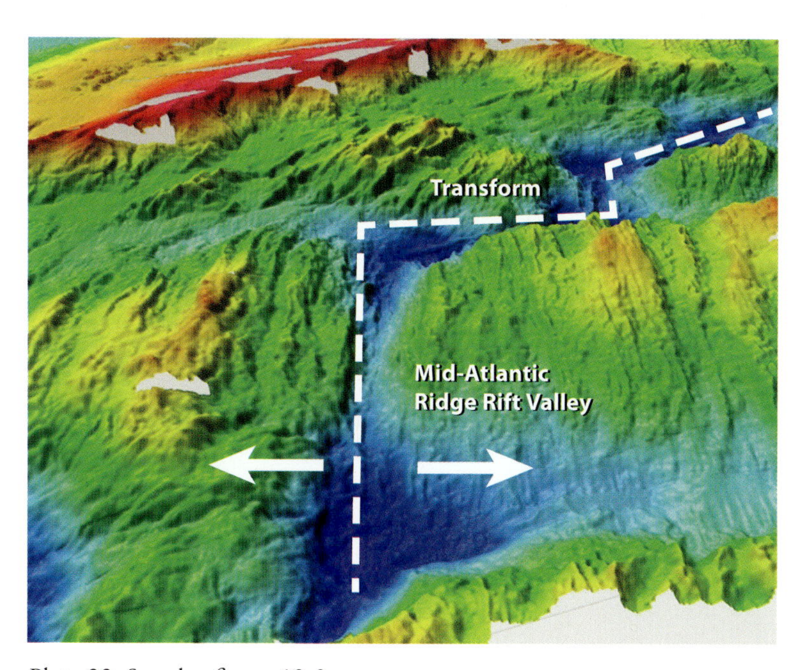

Plate 22. See also figure 12-3.

Plate 23. See also figure 12-0.

Plate 24. See also figure 12-6.

Plate 25. See also figure 16-0.

Plate 26. See also figure 16-5.

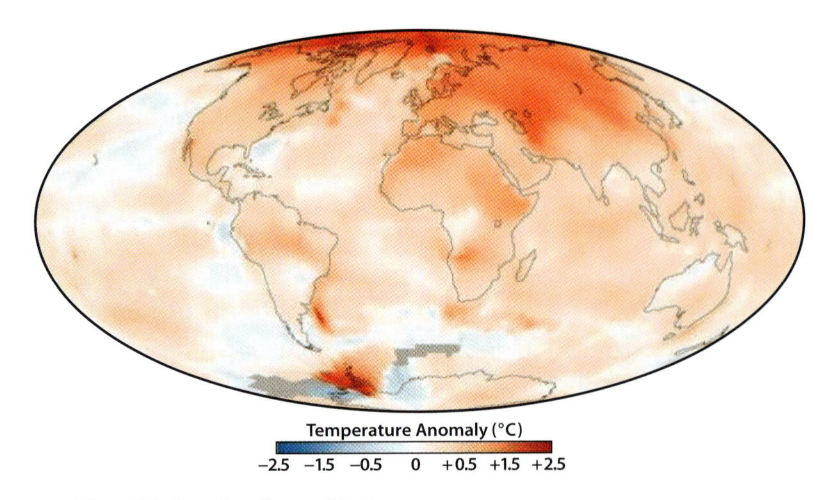

Plate 27. See also figure 20-6.

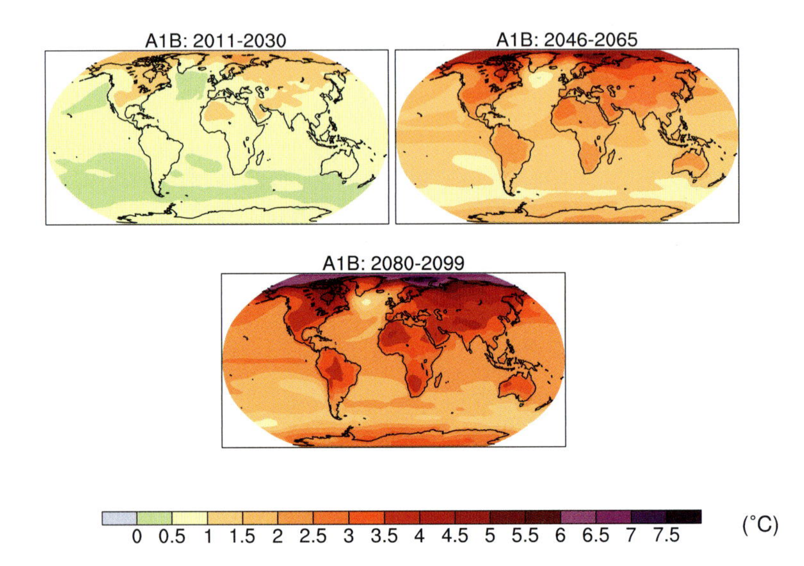

Plate 28. See also figure 20-17.

Even more notable is that Earth's surface temperature has stayed in a narrow range over billions of years of progressive change in solar luminosity (Fig. 9-9). As the sun's luminosity increased over time, Earth's atmosphere precisely accommodated to maintain a stable surface temperature. This fact suggests a sensitive feedback mechanism, a kind of terrestrial thermostat that is able to maintain a constant surface temperature in the face of changes in external forcing. What could be the detailed mechanisms leading to such stability? Why has this mechanism not acted on other planets?

Earth's Long-Term Thermostat

Despite changes in solar luminosity, moving continents, periods of ice ages when glaciers covered much of the globe, warm periods when reptiles thrived at the poles, vast changes in life, and the amount of O_2 in the atmosphere, Earth's temperature appears to have remained comfortably within the range of 0° to 100°C for most of geologic time. We now explore the likely mechanism that has permitted this steady-state condition in the face of large changes in conditions.

Carbon dioxide plays an extremely important role in establishing a planet's surface temperature. In Earth's atmosphere CO_2 is second only to water in its greenhouse capacity. Nonetheless the amount of carbon in the atmosphere is trivial compared to the vast quantity of carbon at the surface. Most of this carbon resides in sediments, part as calcium carbonate (called *limestone* by geologists) and part as organic residues (called *kerogen* by geologists). Fortunately, only a tiny fraction (about sixty atoms out of every million) is currently in the atmosphere as CO_2. Were all Earth's carbon in the form of gaseous CO_2, its amount would exceed that of N_2 and O_2 by a factor of about 100. The pressure exerted by this CO_2 atmosphere would be a staggering 100 atmospheres (similar to the pressure experienced by the hull of a nuclear submarine submerged to a depth of 1 km). Since CO_2 plays such a central role in greenhouse warming, the partitioning between solid $CaCO_3$ and gaseous CO_2 must play a key role in climate stability.

A very interesting argument can be constructed to show how Earth's climate may have been maintained in conditions where liquid water is stable by a carbon cycle involving atmospheric CO_2 and $CaCO_3$. The long-term climate thermostat has to do with the connections between solid Earth geochemical cycles, weathering, composition of the atmosphere, and seawater composition. Climate control is not just the weather!

As rain falls on rocks and soils, there are chemical reactions that lead to weathering and release of chemicals. River water then carries these elements to the oceans. The Fe is immediately precipitated; the Mg and much of the Na are taken up by the ocean crust. Al is kept in clay minerals and is relatively inert. This leaves Ca and Si as important elements. Not surprisingly, both these elements are deeply involved in modern biogeochemical cycles and make up the shells of diverse organisms that form in the ocean and whose precipitation leads to cherts (in the case of SiO_2) and carbonates (in the case of CaO). Most of the Ca and Si come from the breakdown of feldspars, pyroxenes, and other mafic minerals that are common in the continental crust. Because the major players in the climate context are Ca and Si, we can simplify the discussion to breakdown of the Ca-silicate component using the mineral wollastonite, with the formula $CaSiO_3$. This mineral is broken down by interaction with water and CO_2 dissolved in soils to form dissolved ions calcium, bicarbonate, and neutral silicate:

$$3\,H_2O + 2\,CO_2 + CaSiO_3 \rightarrow Ca^{2+} + 2HCO_3^- + H_4SiO_4^o$$

These ions percolate through the soil to a nearby stream and eventually to the sea. In the modern ocean, organisms use these constituents to manufacture their shells. Prior to the evolution of shell-forming organisms, calcium carbonate could precipitate directly from seawater inorganically. In both cases the reaction can be written as:

$$Ca^{2+} + 2\,HCO_3^- \rightarrow CaCO\mathbf{3} + H_2O + CO_2$$

Silica can also be precipitated as opal by the reaction:

$$H_4SiO_4^o \rightarrow SiO_2 + 2\,H_2O$$

The calcite ($CaCO_3$) and opal (SiO_2) hard parts fall to the seafloor to contribute to the sediment that accumulates on the oceanic plates as they move to the convergent margin where it is subducted. As will be described in Chapter 12, the high temperatures and pressures of the interior cause the minerals to break down in a process called *metamorphism*. Restricting ourselves to the Ca and Si components of those reactions, the calcite reacts with the opal to yield wollastonite and carbon dioxide gas:

$$SiO_2 + CaCO_3 \rightarrow CaSiO_3 + CO_2$$

The calcium silicate is returned to the mantle, compensated by the Ca and Si that are included in the magmas rising to the surface from melting of the mantle. The CO_2 dissolves in the magma that rises to the surface, where the solubility of CO_2 is so low that it is released out the convergent margin volcano to the atmosphere (Fig. 9-10).

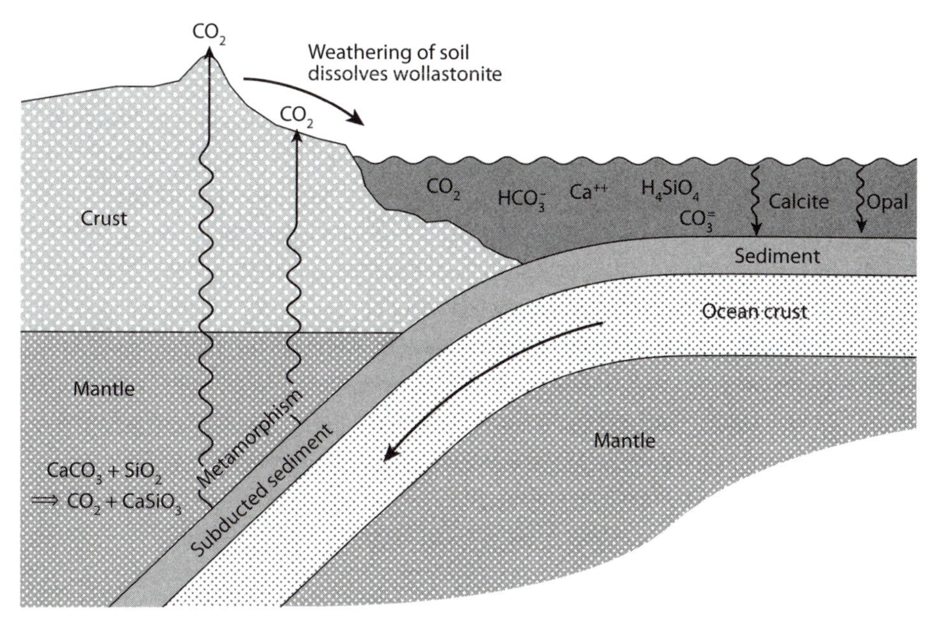

Fig. 9-10: The subduction of oceanic crust beneath the continents carries part of the blanket of sediment deep into Earth's mantle: Here they are heated and metamorphosed. During this process, some of the carbonate minerals contained in the sediment are broken down, releasing CO_2. This CO_2 migrates back to Earth's surface and rejoins the ocean-atmosphere reservoir. Eventually it recombines with calcium in the mineral calcite. This calcite is buried on the seafloor and starts another trip toward a subduction zone.

So much for the background. The interesting aspect of this cycle is how it interacts with the CO_2 content of the atmosphere. The basic driving mechanism for the cycle is the plate motions that carry sediment from Earth's surface to the interior, releasing CO_2 gas. Along with the amount of carbonate sediment on the down-going plate, this sets the rate at which CO_2 is returned to the atmosphere-ocean system. If at any given time CO_2 is not removed from the ocean through calcite burial in the sediments as rapidly as it is added to the atmosphere-ocean reservoir, then the CO_2 content of the atmosphere would steadily increase. If, on the other hand, organisms were to remove calcite from the ocean too rapidly, the CO_2 content of the ocean-atmosphere system would steadily decline. Somehow a balance between the supply of CO_2 to the ocean-atmosphere reservoir and removal of CO_2 must be achieved. Key to this balance is the requirement that in order to form calcite, organisms need calcium as well as CO_2. The CO_2 that leaks from Earth's interior must mate with CaO dissolved from its crust to form CaO CO_2 (i.e., $CaCO_3$). Thus, calcite can accumulate in marine sediments no faster than calcium is being made available through the chemical reactions taking place in continental soils. The rate of these chemical reactions depends on the temperature of the soil (all chemical reactions go faster when the reactants are heated), the acidity of the water (which causes minerals to break down more rapidly, and the rainfall (the more water that runs through the soils, the more that can be carried away).

Now come the feedbacks to this cycle of events (Fig. 9-11). As stated above, if CO_2 were to be added to the atmosphere faster than it was removed by calcite deposited in deep-sea sediments, then the CO_2 content of the atmosphere would increase. This would make the planet even warmer (because of the increased greenhouse blanketing) and make the planet even wetter (warmer air holds more water vapor and hence makes more rain). The higher CO_2 content also makes the water more acidic. Thus, a rise in atmospheric CO_2 content would increase the rate at which calcium dissolves from the continents and thereby permit calcite to accumulate more rapidly in marine sediment. Eventually the calcite production rate would become great enough that CO_2 could be removed from the atmosphere-ocean system as fast as it was added. The CO_2 buildup would be stemmed.

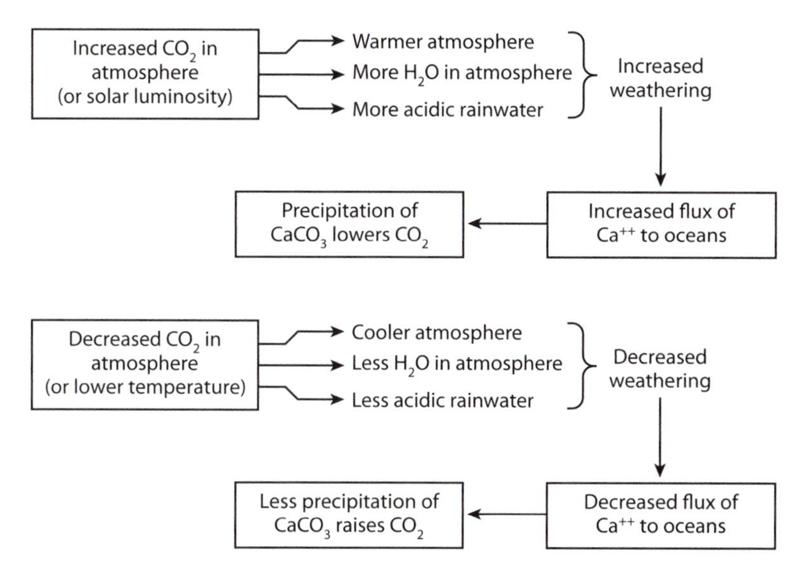

Fig. 9-11: Illustration of feedbacks that control atmospheric CO_2 and surface temperature on Earth, as described in the text.

The feedback also works in the other direction. If for any reason temperature in the atmosphere gets particularly low, weathering will slow, diminishing Ca supplies and carbonate precipitation, causing CO_2 to rise. Or if CO_2 becomes too low in the atmosphere, temperature, rainfall and acidity will decline, limiting CO_2 removal from the system by cutting off the supply of Ca. This process requires long periods of time to take place, because weathering, sediment precipitation, and subduction are slow processes. It is Earth's *tectonic thermostat*, operating on tectonic timescales of $>10^5$–10^7 years.

The hypothesis of a tectonic thermostat has great appeal because some strong feedback mechanism is necessary to preserve climate stability over billions of years of planetary history where the radiation from the sun has changed substantially and where events such as great volcanic outpourings, meteorite impacts, or "snowball Earth" episodes (discussed below) might lead to catastrophic long-term effects. This hypothesis lacks direct tests, however. Quantitative estimates of the changes in weathering rates with atmospheric CO_2 are hard to come by, the fate of carbonates during subduction is a subject of debate, the chemical compositions of atmosphere and oceans have changed over Earth's history in ways that are not well defined, and thermal evolution of Earth's interior may have influenced metamorphic reactions in subducting plates. Absent a viable competing hypothesis, the idea is widely accepted, but on our scale of certainty it still merits only a 6.

A LESSON FROM VENUS

We have a dramatic reminder that the situation with regard to CO_2 could well be different. The reminder is the planet Venus, which has a whopping atmosphere made almost entirely of CO_2, with ninety times the surface pressure on Earth. The greenhouse effect of this CO_2 atmosphere gives Venus the scalding surface temperature of 464°C. As Venus and Earth have nearly the same size and nearly the same density, it seems reasonable that they also started with a similar inventory of volatiles. Indeed, the fact that the amount of carbon in the CO_2 of the Venusian atmosphere is about the same as the amount of carbon locked up in limestone and kerogen on Earth's surface provides evidence that this is true.[1] Thus, Venus has the conditions that would prevail on Earth if all the CO_2 locked up in limestone and kerogen were to be released as CO_2 to the atmosphere.

However, when comparing Earth and Venus, a problem arises in connection with water. If Venus started with the same component of volatiles as Earth, it should have a sizable ocean (or rather, at its high temperature, an atmosphere dominated by steam).[2] Not only is the atmosphere of Venus not dominated by steam; water vapor is barely detectable.

Most scientists believe that the hydrogen initially present on Venus as water escaped to space. In the very hot Venusian atmosphere, water vapor would be effectively transported to the "top." Here it could be disassociated by ultraviolet light to form hydrogen atoms that would then escape. The "left behind" oxygen atoms would be stirred back down through the atmosphere to the surface of the planet, where they would

[1] Because Venus is so hot, it surely has no life and hence no kerogen. Also, $CaCO_3$ would decompose under these conditions, releasing its carbon as CO_2 gas. Hence, it is likely that nearly all carbon on the Venusian surface resides as CO_2 gas in its atmosphere.

[2] If the Earth were heated to the point where its ocean was converted entirely into steam, this steam would exert a pressure about 270 times that of the present Earth atmosphere.

gradually convert the FeO in the hot Venusian crust to Fe_2O_3. Evidence in support of this hypothesis was obtained when an unmanned American space probe was dropped into the atmosphere of Venus. Before this probe was rendered inoperative by the high temperatures, it measured and radioed back to Earth the isotopic composition of the trace amount of water present in the Venusian atmosphere. The astounding finding was that the amount of deuterium (^2H) relative to ^1H in Venusian water is >100 times higher than the $^2H/^1H$ ratio in Earth water. Because of their twofold larger mass, deuterium atoms have a much lower escape probability than do hydrogen atoms. Hence, the escape of hydrogen from Venus would tend to enrich the residual water in deuterium. While the observed hundredfold enrichment of deuterium does not prove that Venus once had as much water as Earth, it can only be explained if Venus once had at least a thousand times more water than it does now!

Thus, it is entirely possible that Venus and Earth started with roughly the same volatile ingredients. Earth for some reason evolved along a path that kept its carbon safely locked up in sediments and hence avoided the disastrous consequences of a so-called runaway greenhouse warming. Venus, on the other hand, at some point slipped and let CO_2 build up in its atmosphere. This buildup led to high temperatures, which would have terminated life (if indeed it ever achieved a foothold on Venus). It is hard to imagine how a planet once in this very hot state could ever cool off.

We have little knowledge of the history of Venus. It is difficult to imagine that astronauts will ever roam about its surface as they did on the moon. While the Russians and Americans have managed to land several unmanned space probes on the hot surface of Venus, these vehicles survived the hostile conditions only long enough to radio back information about the temperature, pressure, and composition of the Venusian atmosphere and (as we learned in Chapter 4) about the potassium to uranium ratio in the rock surface upon which the probe landed. Radar beams bounced off Venus tell us that its surface has large topographic features and a young surface lacking in large-impact craters. Because of the young surface, we do not know whether Venus might have had an earlier history more like Earth's. Certainly in the early solar system with the luminosity of the sun 30% less than present, Venus would seem to have been in a favorable position for life within the solar system.

In any case, after the runaway greenhouse, conditions permitting the development of life were never reestablished. We can speculate that Venus's runaway climate occurred because Venus is closer to the sun than Earth and was not able to manage the increase in solar luminosity. Was it also that Venus spins hundreds of times more slowly than Earth? Was it because life never got started on Venus? Was it because the initial component of water on Venus was much smaller than that on Earth? In any case, Venus's presence reminds us that climate stability is not assured, sending a message that planetary climate can go catastrophically off course.

SNOWBALL EARTH

With this background in mind we can try to envision what would happen if Earth's ocean were to freeze. There would be no marine organisms to make calcite, nor would there be water from which calcite could precipitate inorganically. There would also be no chemical erosion. Under these conditions, the CO_2 released from Earth's hot interior would build up and up in the atmosphere until the temperatures became warm enough to melt the ice. The secret of this escape hatch is that CO_2 outgassing is driven by Earth's internal heat and is therefore insensitive to the surface temperature. A part of Earth's history where such a scenario took place would be a useful test of the tectonic thermostat.

Until very recently it was believed that Earth had never been entirely frozen. But then two Harvard geologists, Paul Hoffman and Dan Schrag, picked up on a suggestion made in 1992 by CalTech's Joseph Kirschvink that during the episodes of glaciations that occurred during the Neoproterozoic period from 580 to 750 million years ago, Earth did indeed become totally frozen. Kirschvink referred to these episodes as *snowball catastrophes.*

The key observation is as follows. Deposits formed at this time by glaciers were intermingled with marine sediment. In other words, the glaciers must have reached sea level. Further, paleomagnetic measurements demonstrate that these deposits were widespread over a range in latitude. More important, some of these glaciers were located near the equator at sea level, giving rise to the possibility that the entire Earth was frozen. Such a state cold arise more easily if oceans covered the

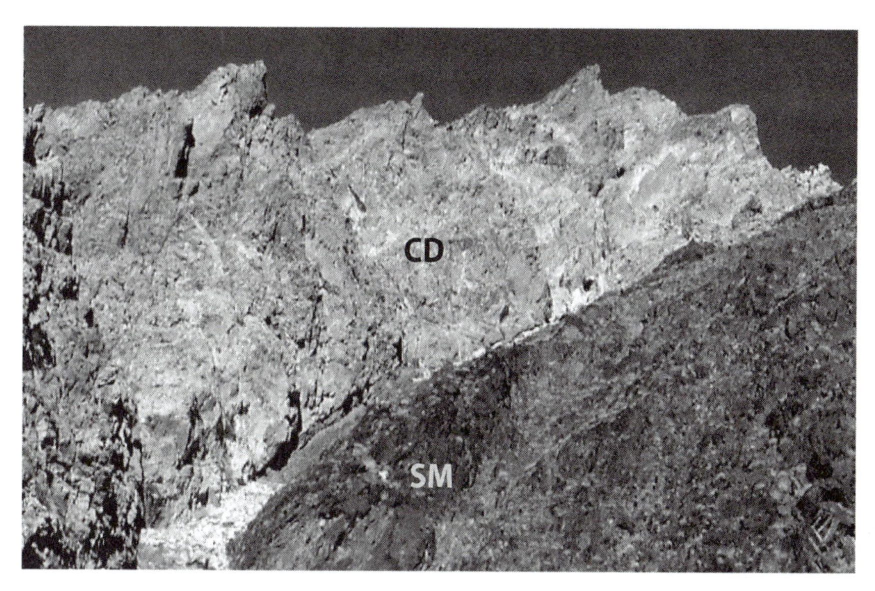

Fig. 9-12: The Noonday cap carbonate (labeled CD) lies above the upper Kingston Peak (Surprise Mb) (labeled SM) glacial deposit called a *diamictite*, The outcrop is in the Panamint Range, Death Valley area, California. Height of cliff ~300m (~1,000,ft). Note the abrupt transition between thick glacial deposits, indicating an ice age, and the carbonate sequence that indicates warm conditions. (Courtesy of Paul Hoffman; www.snowballearth.org)

poles, since growth of oceanic ice cover is far easier than growth of continental ice sheets. If polar ice begins to grow, it increases Earth's albedo, causing more light to be reflected back into space. If the ice passes a critical surface area, then the albedo lowers the atmospheric temperature, and there is a positive feedback leading to more sea ice that can extend all the way to equatorial latitudes.

Key to the thinking of Hoffman and Schrag was the observation that these glacial deposits are overlain by thick sequences of calcium carbonate (Fig. 9-12). Carbonates are generally deposited in warm seas, so explaining why they would occur immediately above a glacial deposit was quite a puzzle. These so-called *cap carbonates* have textures unlike any other limestone in the geologic record. Not only are their textures different but so also is the isotopic composition of their carbon. Its composition lies well away from that of "normal" limestones and instead closely matches that of average Earth carbon.

Hoffman and Schrag created a scenario based on the tectonic thermostat that is an intriguing hypothesis to explain these observations. After the ocean froze over and continental glaciers descended to the shores of the sea, chemical erosion ceased. No more Ca was delivered to the oceans by rivers (some would still be delivered by hydrothermal vents), and the deposition of calcium carbonate and organic residues would drastically decrease. Plate tectonics continued to operate, however, so CO_2 continued to escape from the interior through volcanoes that would melt their way through the ice. As no mechanism existed for the removal of this CO_2, its concentration in the ocean and atmosphere went up and up. As it rose, Earth gradually warmed despite the very high albedo of the 100% ice cover. After 10 or so million years, the CO_2 greenhouse effect was large enough that the ice began to melt. The melting led to a runaway warming, for as the highly reflective ice and snow was replaced by far less reflective sea and land, less sunlight would be reflected, leading to a positive feedback toward increased melting. Erosion would then have reinitiated with a vengeance, supplying calcium to the CO_2-rich ocean. This, of course, led to the precipitation of massive calcium carbonate deposits. As the CO_2 was used up, our planet's surface cooled down toward its ambient state.

Sun Protection

One final ingredient is needed to make the planetary surface conducive to the origin and long-term evolution of life. The sun emits both ultraviolet radiation and a solar wind of charged particles with very high velocities. Particularly for early Earth, the solar wind could have contributed to stripping of the early atmosphere and possibly led to loss of the volatiles that are critical for climate stability and life. At the same time, galactic cosmic rays coming from distant stars also have the potential to deliver doses of radiation detrimental to life. Once life began, this radiation would have created a high dose of radioactivity that would have been hazardous to life as we know it. While the sun is the ultimate source of habitability, some protection against cosmic radiation is necessary.

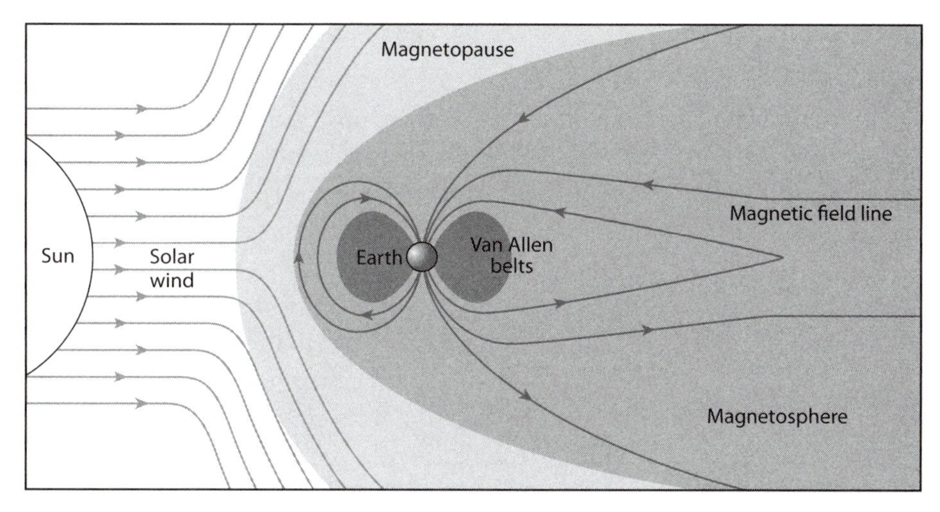

Fig. 9-13: Graphical illustration of the protection from the Sun's solar wind provided by Earth's magnetic field. The solar wind is deflected away from Earth, providing protection for the atmosphere and surface.

Earth's sun protection is its atmosphere and magnetic field. In the modern atmosphere ozone absorbs much of the Sun's ultraviolet radiation, protecting life on land from its effects. It may be no accident that macroscopic life was able to evolve and take over the continental surface only after there was sufficient oxygen in the atmosphere to give rise to an effective ozone shield.

For cosmic rays and the solar wind, the principal protection is Earth's magnetic field. We are all familiar with Earth's magnetic field through compasses and their importance for navigation. Magnetic fields also exert a force on charged particles such as those in cosmic radiation. Earth's magnetic field causes most of the particle radiation from the sun to be diverted around the planet (Fig. 9-13).

Earth has the largest magnetic field of any of the terrestrial planets. The magnetic field is generated by convection in the liquid outer core. Since Earth has been cooling since its early history of accretion, a liquid core would also have been present in the early Earth, and the sun protection provided by the magnetic field likely contributed to early habitability and the possibilities of an origin of life.

Summary

Earth's habitability is critically dependent on a surface environment with adequate volatiles, oceans and continents, liquid water, and a temperature that remains within a restricted range over billions of years. Because Earth as a whole has few volatiles, early concentration of volatiles to the surface is essential to provide an ocean and atmosphere. A sufficient planetary mass, protection from atmospheric loss, and a "water trap" in the lower atmosphere enabled Earth to retain all its volatiles with the exception of helium. The surface temperature of a planet depends on the amount of solar radiation it receives, its albedo, and its greenhouse effect. Evidence from ancient zircons suggests an active water cycle at the surface, including liquid water prior to 4 billion years, and the record of sedimentary rocks shows that water has been present at Earth's surface since that time. It is remarkable that Earth has managed a stable climate over billions of years, despite a 39% change in the solar luminosity. This fact suggests a robust feedback mechanism that would enable greater greenhouse warming in the early Earth. A tectonic thermostat involving the carbon cycle is the most likely mechanism for climate control. A warmer atmosphere with high CO_2 would lead to greater weathering, causing more CO_2 to be sequestered in carbonate rocks. The efficacy of this mechanism is apparent by comparison with Venus, where CO_2 was not sequestered into carbonates, leading to massive greenhouse warming and loss of planetary water, making the planet unsuitable for life. A planetary surface suitable for life is also greatly enhanced by the presence of a magnetic field. Earth's liquid outer core provides the largest magnetic field of any of the terrestrial planets. The magnetic field would have contributed to prevention of atmospheric loss in early Earth's history, and it protects the planet from deadly cosmic rays, providing Earth with billions of years of homegrown sun protection.

Supplementary Readings

Kevin J. Zahnle and David C. Catling. 2009. Our planet's leaky atmosphere. *Scientific American*, May 11:29.

James Callan Gray Walker. 1977. *Evolution of the Atmosphere*. New York: Macmillan.

James F. Kasting and David Catling. 2003. Evolution of a habitable planet. *Annu. Rev. Astron. Astrophys.* 41:429–63.

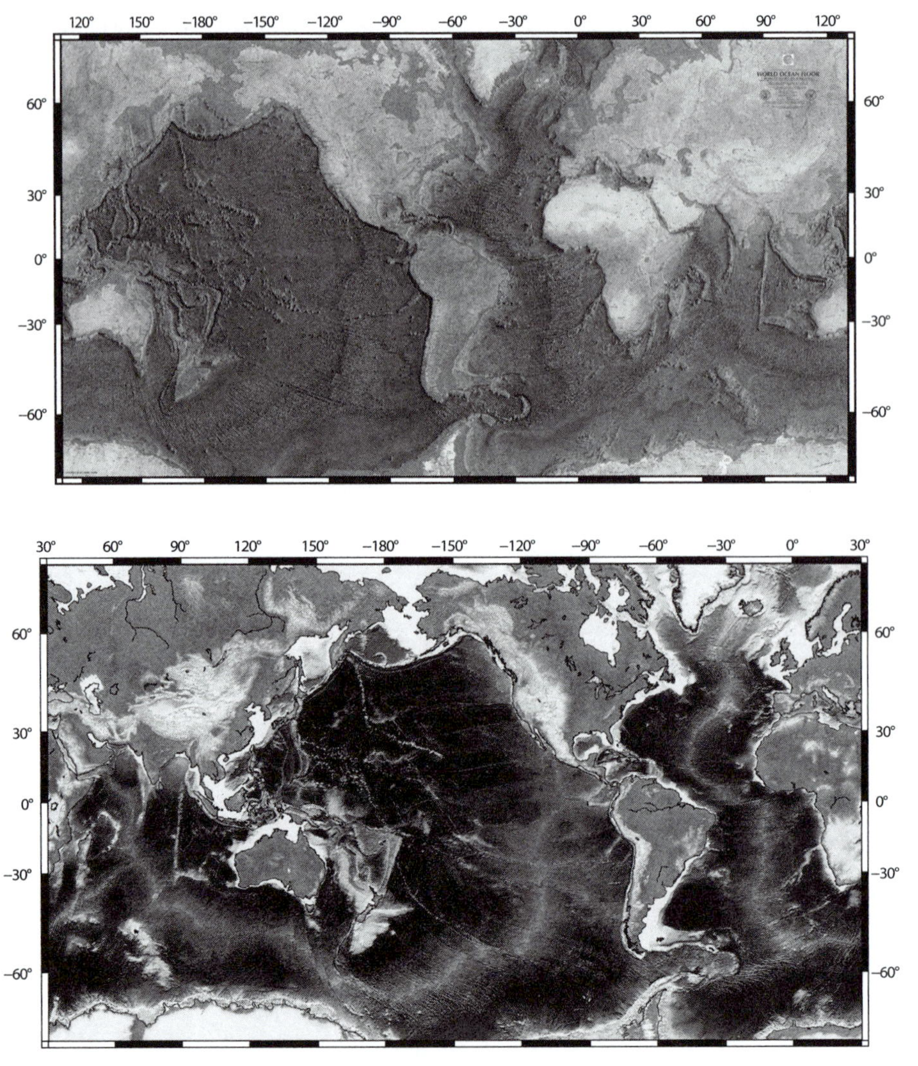

Fig. 10-0: Two renditions of a map of the ocean floor. *Top:* Map of ocean ridge system produced by Bruce Heezen and Marie Tharp. While the map appears to be detailed, it is in fact largely schematic. Heezen and Tharp had data in a few places and interpolated and guessed to make a continuous image. (Base map reprinted by permission from *World Ocean Floor* by Bruce C. Heezen and Marie Tharp, copyright 1977) *Bottom:* Modern and much more accurate map of the sea floor based on both bathymetry and global gravity data obtained by satellite. See color plate 8. (Map provided by David Sandwell, Scripps Institution of Oceanography)

CHAPTER 10

Establishing the Circulation

Plate Tectonics

Planetary differentiation created the large-scale layers of the terrestrial planets billions of years ago, and these layers still exist today. Core, mantle, crust, and volatile rich exterior were a common heritage for Earth, Venus, and Mars. Our conception of these layers as fixed and static is deeply ingrained in our daily experience. Rocks are solid, and while they can be broken, they do not flow. Missouri is far from the ocean and Ireland is an island. Such facts are both concrete observations and common sense. Therefore, when Alfred Wegener proposed early in the twentieth century that Africa and South America were once joined together and that continents move across Earth's surface, the idea was met with skepticism, harsh criticism, and even derision by much of the geological community. After World War II, a new generation of Earth scientists began to explore the oceans. The gradual acquisition of new data revealed a remarkable symmetry to the Atlantic Ocean. Down the center ran the Mid-Atlantic Ridge; depth of the ocean and thickness of sediment increased regularly from the ridge toward the continents. The symmetry extended to the pattern of magnetic anomalies that were related to the periodic reversals in Earth's magnetic field that had been discovered on land. All these data could be explained by seafloor spreading, where new ocean crust was created at the ridge and aged progressively as it spread away to both sides. Sampling confirmed this idea by showing that young volcanic rocks were recovered at the ridge axis and the oldest sediments were near the margins of the continents, distant from ridges. Global seismology then showed the complement to seafloor creation at ocean ridges—the ocean crust is recycled into the mantle at ocean trenches, and its return to the mantle can be mapped precisely by the dipping planes of earthquakes called Benioff zones *in locations such as Japan. The new theory of plate tectonics accounted for these observations by proposing that Earth's surface is made up of fixed plates that are continually in motion—created at ridges, destroyed at subduction zones. The plates consist of brittle* lithosphere, *and they float on top of the mobile* asthenosphere. *Continents drift, not by plowing through the oceans as Wegener implausibly suggested, but as light rafts floating on top of the plates that are created and destroyed in the surrounding oceans. The continents are too light to be recycled, hence continents preserve a much longer record of Earth's history than the continually recycled ocean floor. Mountain belts occur where plates collide. Earthquakes and volcanoes occur at the margins of plates where they spread apart, converge toward each other, or slide by one another. In a period of a few years during the mid-1960s, our view of Earth changed, from a static surface with fixed and isolated continents and oceans to a surface in constant motion, with plates moving as fast as 20 cm/year. Long-standing geological questions such as the origin of the ocean floor, earthquakes, volcanoes, and mountain belts became simple consequences of plate motion. More recent measurements of plate movement using the global positioning system match the speeds inferred from magnetic anomalies and are striking confirmation of plate tectonics, changing its standing from theory to factual observation.*

Introduction

Early in its history, Earth differentiated into layers of increasing densities toward the core, and this process is likely to be a general phenomenon for the terrestrial planets (see Chapter 7). This primary differentiation, occurring more than 4 billion years ago, could give the impression of a static planet, with little movement and effective isolation between the layers. This is the case for the moon. On Earth, it is obvious that the ocean and atmosphere move vigorously. And as we shall see in the next two chapters, Earth's surface and interior are in constant movement as well. Indeed, these movements permit circulation, recycling, and exchanges between layers and appear to be critical requirements for habitability. What is the evidence that the surface of the Earth moves? What are the characteristics, speeds, and driving forces of this movement? How is this movement important for the operation of Earth's cycles and the interpretation of its history?

The Static Earth Viewpoint

The continual movement of all Earth's parts is not an easy concept for us as human beings to grasp and has only become known to specialists in Earth science in the last few decades. While we see the devastating local influence of earthquakes or volcanic eruptions, we rely on Earth's major features to be static. San Francisco is a port, and waterfront property bought a hundred years ago is still waterfront property today. Antarctica is at the South Pole, covered by ice. These are concrete observations verified over and over again. Therefore, one would not be wrong to accept these kinds of generalizations as established truths. It is no wonder that an understanding of Earth as a continually changing and dynamic environment has been slow to develop—and is still developing.

We saw in Chapter 4 the difficulty human beings had to fully comprehend the timescale pertinent to Earth. It turns out that comprehending a mobile Earth was even more difficult. Even with the emerging comprehension of Earth's geologic timescale in the nineteenth century, the solid layers of Earth were viewed as relatively static and unchanging. Geologists in the twentieth century were cognizant of the billions of years of Earth's history. Volcanic eruptions would gradually build huge edifices. Erosion would wear down high mountains and modify coastlines, and the climate in the past included ice ages. Observable processes acting over time would create all the physical features of Earth. Since there were no observations of moving continents, the continents and oceans were viewed as fixed. Geologists were not able to provide good answers, however, to some of the most obvious questions that anyone would ask about the planet:

- Why are there mountain belts, and why are they located on some continental margins but not others? Most but not all mountain ranges

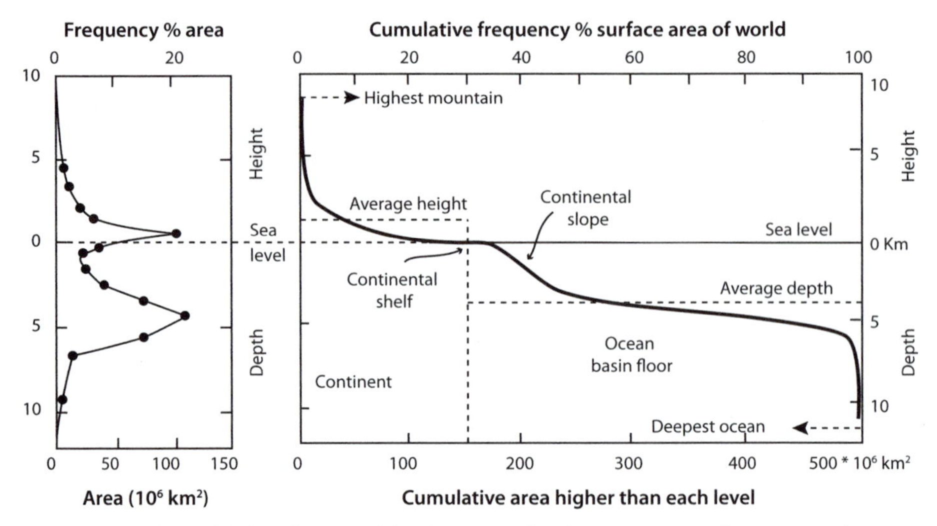

Fig. 10-1: The bimodal distribution of depths on Earth. There are two peaks, one just above sea level that corresponds to continents and the other at a depth of 4,000 m below sea level that corresponds to the oceans. (Adapted from Wylie (1972), *The Dynamic Earth,* John Wiley & Son).

occur on the edges of continents. On the other hand, the Alps and Himalayas are steep and high, but both are in the interior of their respective continents. The Appalachians are low and rolling, and on the eastern edge of North America; the east coast of South America has no mountains, while the west coast has the high Andes.

- Why is Earth's surface divided into continents and oceans with an interesting distribution of elevations? The oceans are deep, with a mean depth of 5,000 m below sea level. They are made of the volcanic rock basalt, covered by sediment. Continents reside mostly above sea level, with a mean elevation of less than 1,000 m. Very little of Earth's surface exists at intermediate levels (Fig. 10-1).
- Why do earthquakes and volcanoes occur where they do? California and Alaska have frequent large earthquakes, but New York and Florida have almost none. Western Europe is earthquake free, while Japan suffers great damage every decade.
- Why do the continents of Africa and South America fit together like pieces of a giant jigsaw puzzle, and yet North and South America do not?
- If oceans and continents are fixed features, why are the sediments in the oceans so thin? Accumulation rates over billions of years of conti-

nental erosion should by the principle of uniformitarianism have made them vastly thicker.

- Why are oceans such as the Atlantic shallowest at the center and progressively deeper toward the continental margins?
- Why are animals on some continents relatively similar to one another, while other continents, such as Australia, have very different species?

These important questions, which should have been answered by a thorough understanding of Earth, had no clear answers through the first half of the twentieth century. Before plate tectonics, the static view of the Earth made these issues impossible to consider appropriately and completely.

Continental Drift Theory

The apparent fit between the continents on both sides of the Atlantic was noted by Francis Bacon as early as 1620. In the early twentieth century the meteorologist Alfred Wegener considered the fit more carefully by looking at the submerged continental margins rather than just the coastlines. The fit was even better. Wegener noted that some distinctive formations in South Africa were very similar to rocks in Brazil and that fossils typical of tropical areas, such as ferns, are found far north of the tropics today in Spitsbergen. Glacial deposits found in Africa and South America provided additional clues; if the continents were brought back together, a single continental glacier could explain them. He became convinced that the continents had moved, and in 1912 he marshaled a wide variety of evidence in support of the theory of *continental drift* (Fig. 10-2).

These ideas from someone outside the field of geology ran into stiff resistance and were harshly ridiculed, particularly by North American geologists. Ultimately a conference in 1928 produced a book that summarized the state of the debate at that time. Critics emphasized a fundamental flaw in Wegener's theory—the lack of a mechanism by which the continents could plow through the crust of the seafloor. The ocean floor should have been strongly deformed by such movement, but adjacent to continents the ocean floor is smooth. Furthermore, no known forces

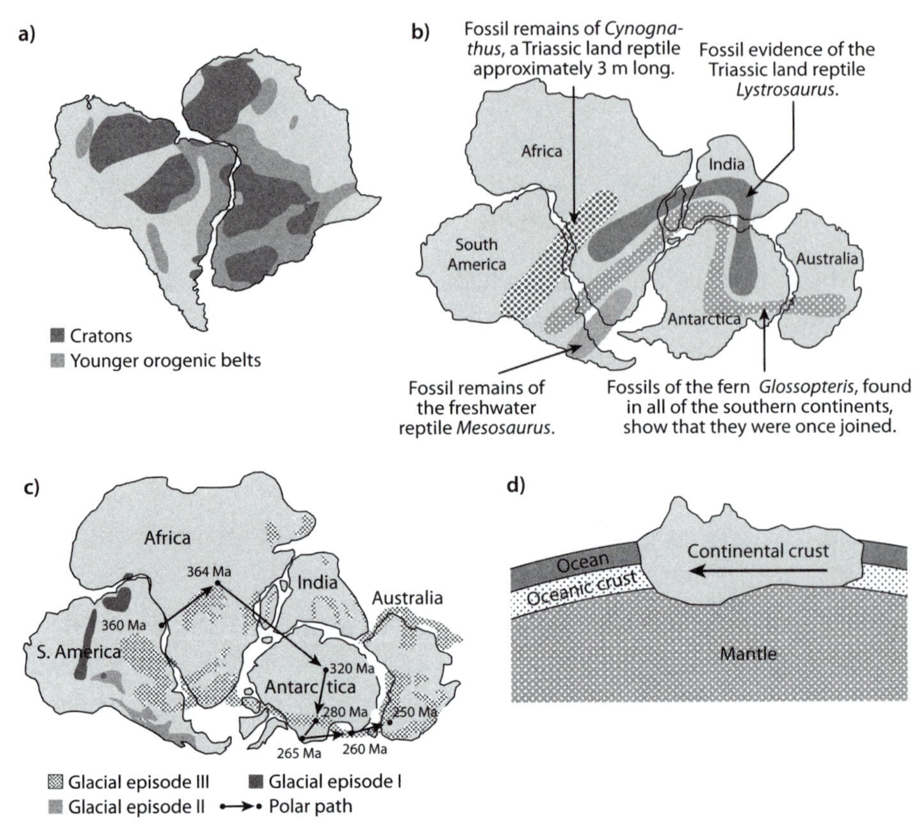

Fig. 10-2: (a) The fit of continents and correspondence of rock formations that led Alfred Wegener to propose the continental drift theory; (b) correspondence of fossils across continents that are currently separated (credit: Image from U.S. Geological Survey's *This Dynamic Earth*, http://pubs.usgs.gov/gip/dynamic/dynamic.html); (c) correspondence of ancient glacial deposits with former positions of the South Pole. Glacial deposits now near the equator originally formed at high southern latitudes (from *Late Paleozoic Glacial Events and Postglacial Transgressions in Gondwana*. Boulder, CO: Geological Society of America, 2010); (d) The major criticism of Wegener's theory—how could thick continents plow through the ocean crust and mantle?

could cause the huge masses of the continents to move about Earth's surface, and the forces that Wegener proposed were insufficient. Since Earth's interior was viewed as solid rock, moving the continents through it was not physically possible (Fig. 10-2d). While drifting continents could solve one of the great problems in geology—the formation of mountains such as the Alps and Himalayas by one continent ramming into another, this explanation engendered an even greater problem—what could cause the continents to move? Wegener's untimely death in 1930 at the age of fifty, while carrying out a scientific expedition in Greenland, left the theory without a persistent champion. Continental drift was dismissed as a crackpot idea by most geologists in the northern hemisphere, deserving of only a footnote in many geological textbooks. Even introductory textbooks published in the early 1960s included only brief discussions of the possibility:

> Many pages would be required to sketch the points of argument involved in controversy over the hypothesis of continental drift. Those who reject the hypothesis....—and these are perhaps the majority of geologists of the English-speaking countries—do so on two counts: first, that the points of similarity between parts of the continents can be explained without requiring their once having been in contact; second, that drifting of a rigid continental plate across the basaltic oceanic layers is a physical impossibility.[1]

The idea was dramatically resuscitated, however, by the force of new observations coming from the exploration of the ocean floor in the 1950s. By the late 1960s the theory of plate tectonics was accepted by almost all geologists.

New Data from the Ocean Floor

Knowledge of the seafloor is difficult to obtain because rocks at the bottom are hidden by several kilometers of water and the deepest parts of the ocean are at pressures of several hundred atmospheres. We cannot see through the water, can dive with our bodies only to some 60 m, and even military submarines can only descend to a few hundred meters. Therefore the seafloor remains more technologically challenging to explore than outer space. While a small satellite can circle Mars and return a high-resolution map of its entire surface, seawater acts as a shield and prevents such a broad perspective on the oceans. In early exploration of the seafloor, the only way to obtain data was for a ship to make measurements as it traveled over the ocean surface. There were no computers to

[1] Arthur N. Strahler, *The Earth Sciences*, Harper's Geoscience Series, Carey Croneis, ed. (New York: Harper & Row, 1963), pp. 420–21 (QE 26 S87).

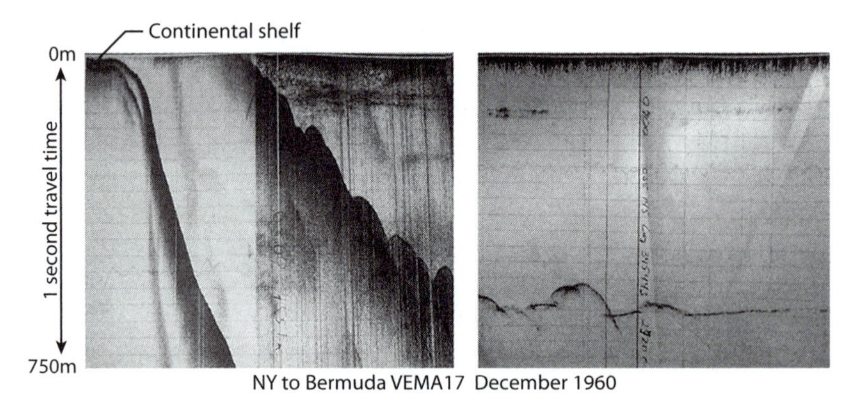

Fig. 10-3: An example of a depth profile collected by the Research Vessel *Vema*. Thousands of such profiles collected at a speed of about 10 knots gradually revealed the bathymetry of the ocean floor. (Research Vessel Vema expedition, Lamont-Doherty Earth Observatory)

automate the process or store the data, so printed records on chart recorders were the stock in trade. Sharing this information between institutions was difficult, and the collection and organization of a large body of data at a single place became critical. Scientists at Lamont-Doherty Geological Observatory of Columbia University began an almost random walk exploration of the world's oceans in the years following World War II. The Lamont ship, the *Vema*, would traverse the surface at some ten knots, sending down sound waves every few seconds and measuring how long they took to come back in order to determine the depth at a single point (Fig. 10-3). The ship also towed a magnetometer to measure the variations in the magnetic field, dropped explosives to look at the reflection of seismic waves, and stopped every eighteen hours to lower a cable to the bottom and recover a sample. A month at sea could lead to a track a few thousand kilometers in length along which these data would be collected. If one imagines even a local pond or river, it is remarkable how little we know of the bottom, and how much work it takes to determine what is down there. For the two-thirds of Earth covered by the ocean, the problem is of such a grand scale that every scientific cruise today is still a voyage of discovery, where new features are seen for the first time.

As the data accumulated over a period of many years, two scientists at Lamont, Marie Tharp and Bruce Heezen, gathered together all the data on depths of the ocean that were being recorded. They confirmed the initial suggestion from the pioneering nineteenth-century expedition of HMS *Challenger* of a huge ridge in the center of the Atlantic Ocean, running down its entire length. In the exact middle of this ridge was a cleft, the Mid-Atlantic Ridge rift valley. Extensions of the ridge, while less pronounced, could be followed all around the globe, traversing every ocean basin (see frontispiece to this chapter). This discovery of the global system of ocean ridges, based as it was on tireless work at sea—twenty-four hours a day, seven days a week, by hundreds of investigators—remains one of the most stunning geographical accomplishments. How often in human history has a new geographical feature of global scale been discovered?

The discovery of the ocean ridges made the entire topography of the ocean floor systematically symmetric. Moving away from the ridge, the depth of the oceans progressively increases. The increase in depth is largest nearest the ridge, and gradual far from the ridge, until one reaches an ocean trench, where the depth plunges to the greatest depths in the seafloor, often just adjacent to continents. Depth variations are not the only systematic feature of the ocean floor. Dredges of the center of the rift valley recovered mostly fresh volcanic rocks. Away from the ridge, only sediment was seen on the ocean floor. Seismic records showed that with increasing distance from the ridge there was progressive thickening of sediment with harder rock beneath it. Such systematic patterns were crying out for explanation!

Evidence from Paleomagnetism

One of the more perplexing aspects of the newly acquired data from the oceans was that variations in the intensity of the magnetic field were also symmetrically aligned with respect to the ocean ridges. If the ship steamed perpendicular to the ridge, the magnetic fluctuations on the two sides were almost a mirror image. The symmetrical bands of higher and lower magnetic intensity were common to all the world's ridge axes.

Other scientists working on land had been examining the variations in Earth's magnetic field through time as it was recorded in continental rocks. Almost all rocks contain magnetic minerals, notably magnetite, and these minerals act as small compasses that record the direction of Earth's magnetic field. Measurements in sequences of sedimentary and volcanic rocks of known age showed that minerals from the youngest rocks pointed north, but those older than 750,000 years pointed south! Working further back in time it turned out that the direction switched back and forth every million years or so. These "magnetic reversals" could only be explained if Earth's magnetic field switched in polarity. Such reversals could be accounted for by quantitative models of flow in the liquid outer core, whence Earth's magnetic field is derived. The variation was not entirely regular—some intervals were as long as 2 millions years, others less than 100,000 years. If plotted in a row with time on the horizontal axis, and white and black colors used to represent normal and reversed polarity, the pattern of magnetic reversals led to a kind of "bar code" defined by the differing intervals of time when the field was "normally" or "reversely" magnetized.

In a series of papers published in the 1960s, the connection between the oceanic data and magnetic reversals became the "smoking gun" for the new theory of seafloor spreading, which would also cause drifting continents. When the magnetic pattern of higher and lower magnetic intensity on the seafloor was plotted alongside the bar code of magnetic reversals, there was a perfect match (Fig. 10-4). The natural explanation of the mysterious magnetic "wiggles" recorded initially with no particular purpose by oceanographic research vessels was that they resulted from reversals of Earth's magnetic field.

The variations in the intensity of the magnetic field work like this. The youngest rocks, at the center of the rift valley, have all their magnetic

Fig. 10-4: Evidence for seafloor spreading from the pattern of magnetic variations (*magnetic anomalies*) measured at sea. The alternating bands of black and white are the high and low intensities of the magnetic profile. Black represents high intensity, above the horizontal line in panel (c) (called "normal") when the magnetic North Pole was in the Northern Hemisphere as it is today. White represents low intensity (called "reverse"), when the intensity was below the line in panel (c), and the magnetic pole was in the Southern Hemisphere. The ages in the bottom panel (e) are determined from paleomagnetic studies carried out on land, where the reversal timescale can be well calibrated. The profile in panel (d) is calculated intensity for reversals in the magnetic field. The relative widths of "stripes" of high and low intensity are the same as the magnetic reversal intervals determined on land. (Modified after Vine, *Science* 154 (1966), no. 3755: 1405–15, 1966)

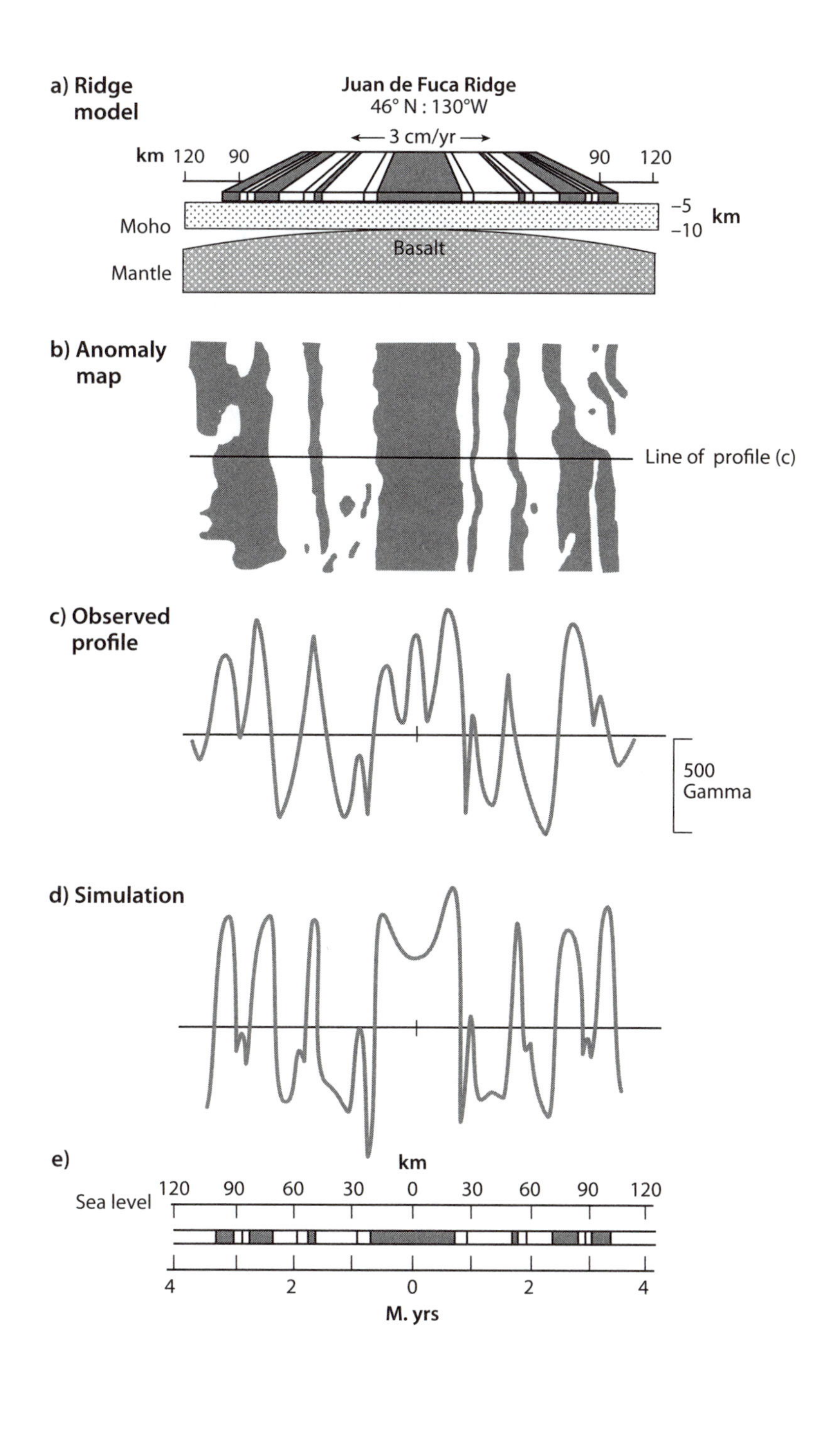

a) Ridge model
Juan de Fuca Ridge
46° N : 130°W
3 cm/yr
km 120 90
90 120
–5
–10
km
Moho
Basalt
Mantle
b) Anomaly map
Line of profile (c)
c) Observed profile
500 Gamma
d) Simulation
e)
km
Sea level
120 90 60 30 0 30 60 90 120
4 2 0 2 4
M. yrs

minerals pointing north. The magnetized rocks then add to the intensity of the current magnetic field because they also point in the current direction, making the total intensity relatively large. Slightly older rocks formed when the magnetic field was pointing south, and the magnetic minerals point in the opposite direction from the current field, canceling part of that field out and leading to lower magnetic intensities. The "smoking gun" occurred when the relative widths of the periods of greater and lesser magnetization around ocean ridges was fit exactly by the relative duration of the magnetic reversals determined from rocks on land. This regularity was gradually extended further and further backward through time, with each new location and extension in time a confirmation of the idea. New crust is being created today exactly at the cleft of the Mid-Atlantic Ridge rift valley. The crust just off the eastern coast of North America and off the western coast of Africa was created at an ancient ridge some 140 million years ago. The ocean floor in the Atlantic has a regular age sequence that increases from the center of the ridge to the continental margins. These data could be explained by the new model of seafloor spreading, where the ocean basins were formed as giant conveyor belts spreading symmetrically outward from the ocean ridge. The ocean ridges were *spreading centers*, with new ocean crust being formed by volcanism from Earth's interior. The calibration of magnetic anomalies back through time enabled the interpretation of all the magnetic data collected from the oceans and the production of a map showing the age of the ocean floor, symmetrically distributed around all the ocean ridges (Fig. 10-5).

These ideas were soon tested further by the advent of the ocean-drilling program, whereby a specially designed ship could drill all the way through the sedimentary layers to the basaltic basement underneath. The age of the sediment could be determined from its fossils. Drilling at different distances from the ridge axis showed that the farther one went from the ridge, the older the age of the sediment immediately overlying the basalts (Fig. 10-6). At the ridge axis, there was no sediment. Close to the axis the sediment was always thin and the oldest sediment very young. With increasing distance the thickness and oldest age of the sediments progressively increased, all the way to the continental margins, where the oldest sediment was recovered. These observations using classical geological techniques confirmed the more esoteric

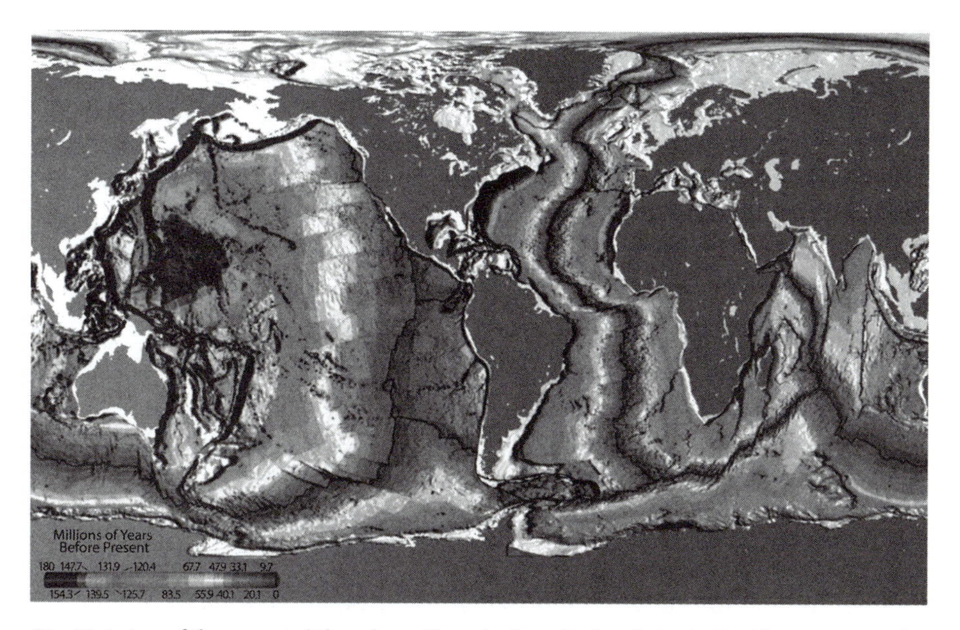

Fig. 10-5: Age of the oceanic lithosphere. Note the broader bands in the Pacific as compared to the Atlantic because of faster Pacific spreading rates. See color plate 9. (From Müller et al., *Geophys. Res.* 102 (1997), no. 82: 3211–14)

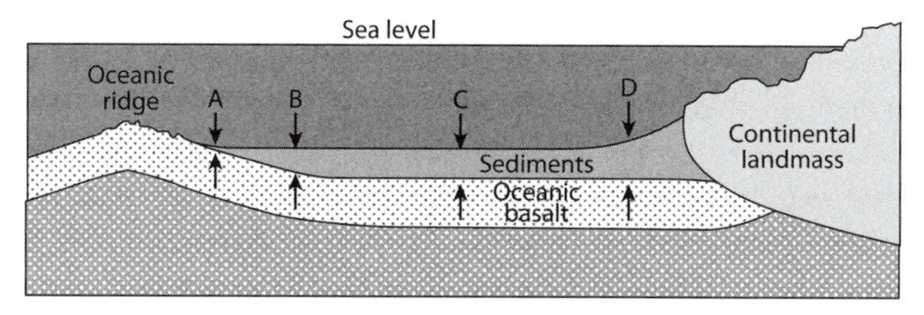

Location	Thickness	Sediment age	
		Surface	Bottom
A	1–5 m	recent	10^6 years
B	10–100 m	recent	10×10^6 years
C	500 m –1 km	recent	75×10^6 years
D	1–3 km	recent	130×10^6 years

Fig. 10-6: Illustration of how sediments were used to test the hypothesis of seafloor spreading. Near the ridge (location *A*) the sediment is very thin and the age of the oldest sediment overlying the igneous ocean crust is only 1 million years. Drill holes through sediment at increasing distance from the ridge revealed a thicker sedimentary column and progressively older sediments at the base.

inferences from paleomagnetism, and demonstrated that the ocean ridge was indeed the active location where new crust was being created.

The rate of spreading for different ridges around the world could be determined by dividing the distance from the ridge by the age determined from the magnetic anomalies. The *spreading rate* was found to vary from 1 to 20 cm per year. While such speeds cannot be seen with our eyes, they are not trivial. They are similar to the growth rates of our hair and fingernails. We may not see it happening the moment we look, but over short intervals of time, there is visible movement.

Global Distribution of Seismicity

Further confirming evidence came from the global distribution of seismicity. Earthquakes do not occur randomly over Earth's surface, they are restricted to very well defined belts (Fig. 10-7). Thousands of small, shallow earthquakes occur in a long line that runs like a seam around the globe and corresponds with the global system of ocean ridges. These earthquakes are caused by the active volcanism and tectonism associated with the spreading that is taking place there (Fig. 10-8a). The second predominant zone of earthquakes forms an inclined plane that extends hundreds of kilometers down into the mantle, starting at the very deep oceanic trenches that occur around the edges of the ocean basins. These inclined planes of seismicity are called *Wadati-Benioff zones*, usually referred to as *Benioff zones*, named after the two seismologists who first discovered them (Fig. 10-8).

The small, shallow earthquakes along ridges are consistent with faults that offset the ocean ridge system as well as small faults associated with active volcanism or creation of the rift valley seen on the new bathymetric maps. The Benioff zones required a different explanation, and the simplest one tied in neatly with the discovery of seafloor spreading. For the overall surface area of the Earth to be preserved, the creation and spreading of crust at ocean ridges has to be balanced by destruction of an equal area of crust. The crustal recycling occurs at *convergent margins*, also called *subduction zones*, where the crust sinks back down into the mantle, creating an ocean trench and descending into the mantle below. The Benioff zones mark the fault between the descending ocean

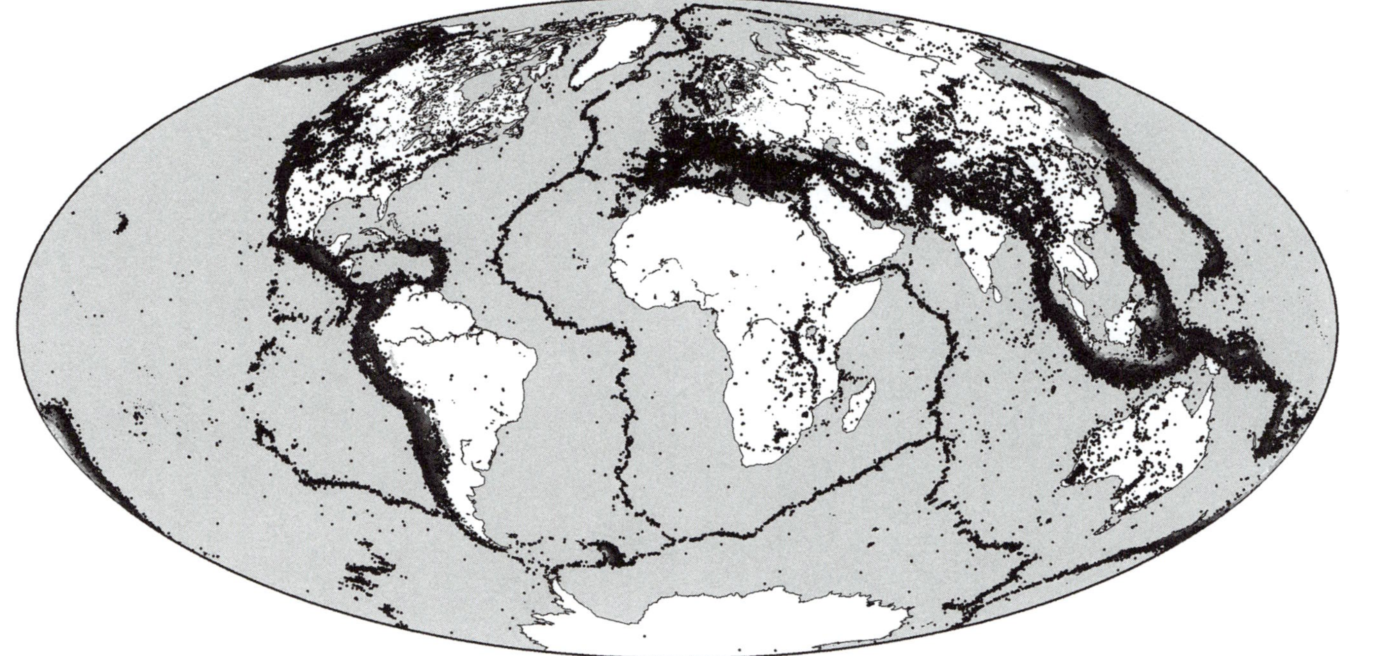

Fig. 10-7: Global distribution of seismicity. Dark circles are shallow earthquakes, gray circles are intermediate-depth earthquakes, and light gray circles are the deepest earthquakes. Note the band of shallow earthquakes that define the ocean ridge system. Earthquakes at convergent margins are shallow at the trench and become progressively deeper with distance from the trench, defining the Benioff zone that is the trace of the plate as it subducts into the mantle (see Fig. 10-9b). Note that almost all earthquakes are confined to plate margins. See color plate 10. (Courtesy of Miaki Ishii, Harvard University)

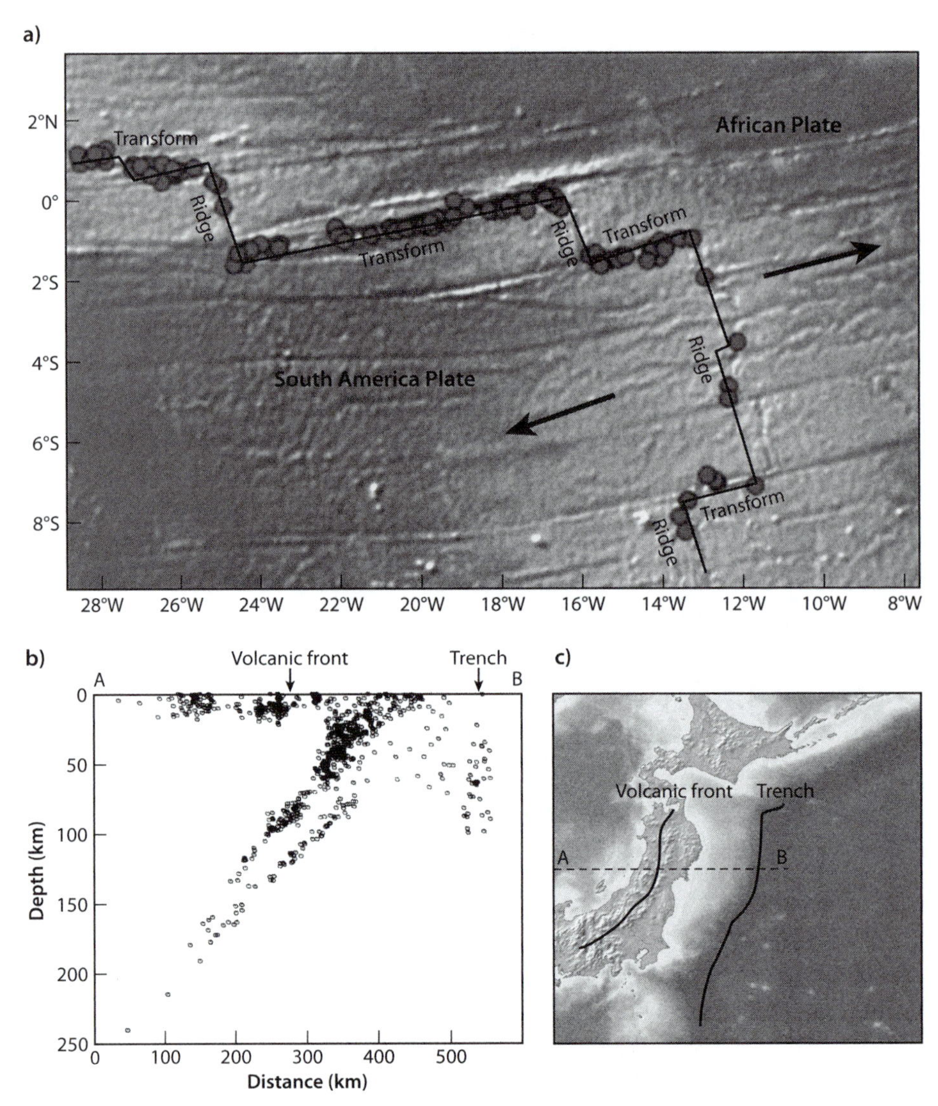

Fig. 10-8: (a) A portion of the Mid-Atlantic ridge where spreading separates the African and South American plates. The spreading center consists of volcanic ridge segments where plates spread directly apart, offset by transform faults where two plates slide by one another. Circles are earthquake locations, showing that tectonic activity occurs only along the active plate margin and particularly at transform faults (Image from GeoMapApp (www.geomapapp.org)); (b) depth profile across Japan showing the depth distribution of earthquakes (Hasegawa et al., *Tectonophysics* 47 (1978):43–58). The top of the band of earthquakes is showing the top of the plate being subducted into the mantle. *A–B* indicates the profile being shown in the bottom right panel (c).

crust and the overlying mantle. The ocean crust is involved in a long-term cycle. Continually created out of the mantle at ocean ridges, the crust traverses the ocean floor, to be recycled as it returns to the mantle at subduction zones.

Not all earthquake locations occur at ocean ridges and subduction zones. Earthquakes also occurred along large faults that cross cut the ocean basins. These faults, called *transform faults*, are apparent in the large *fracture zones* that in the Atlantic can extend all the way from Europe or Africa to the Americas (see frontispiece). The seismically active portions of the fracture zones host much larger earthquakes than occur along the spreading axes, and the earthquakes occur only along limited portions of the fracture zones, not their entire length (see Figure 10-8a).

The ideas of seafloor spreading led to a simple explanation for this third class of earthquakes. In a paper entitled "A New Class of Faults and Their Significance," J. Tuzo Wilson showed how these long linear faults could be explained as connection zones between ridge segments. Seafloor spreading made a prediction of the sense of movement along the fault. If seafloor spreading occurred, then crust should be moving in opposite directions only along limited portions of these faults. The fracture zones would then be fossil remnants of the transform fault, with no present active slip along them. The simple prediction would then be that earthquakes should occur only along the transform fault part of the fracture zone. Beyond the connection between the fracture zone and spreading axis, the adjoining crust would be moving in the same direction, and hence no earthquakes would occur. Precise determination of earthquake locations along the fracture zones showed that this was indeed the case (Fig. 10-8a), and that the sense of motion along the fault conformed to the predicted motion from seafloor spreading (Fig. 10-9). Some transform faults even cut through continents, forming zones of very high seismicity. For example, the San Andreas Fault in California connects spreading centers in the Gulf of California and off the coast of Oregon.

The Theory of Plate Tectonics

All of these ideas and observations came together with the theory of *plate tectonics*. This theory proposed that Earth's surface is made up of

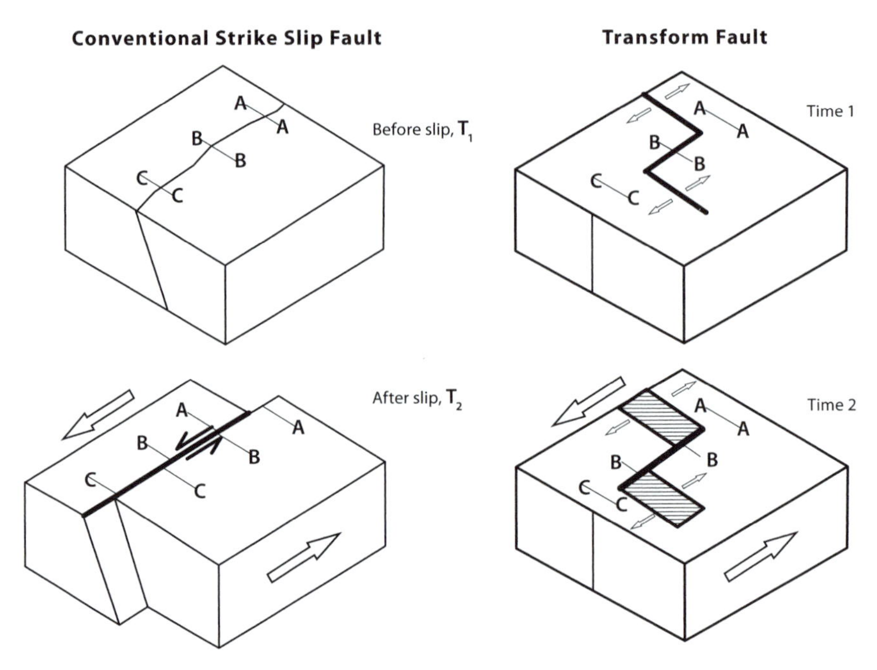

Fig. 10-9: Illustration of the contrast between a conventional fault, as envisioned prior to plate tectonics, and a transform fault. The top diagrams show an initial point in time, and the bottom panels a later point in time. The left-hand block diagrams show a conventional "strike-slip" fault, which displaces along its entire length, with identical offsets for all features on both sides of the fault. The right-hand block diagrams show the movement of a transform fault. The ridge stays at a constant offset along the fault, and only features between the two ridge segments are offset by fault movement. Note, for example, that A—A is not offset by the transform fault. Stippled areas represent new seafloor created in a time interval between the top and bottom panels.

"plates" that glide over a mobile mantle. The plates are defined as the rigid part of the Earth, called the *lithosphere*, which can crack, making it subject to earthquakes. The mobile interior, called the *asthenosphere*, is not rigid. Instead, at the high temperatures and pressures of Earth's interior, the solid mantle flows as if it were a very viscous fluid. The plates form and diverge at ocean ridges, converge and are destroyed at subduction zones, and slide past one another at transform faults. Since the plates are defined by their physical properties, they are not the same as either oceanic or continental crust. The plate includes both the crust and the cold and brittle portion of the upper mantle. The bottom of the plate is a transition where flow changes from brittle to ductile (Fig. 10-10).

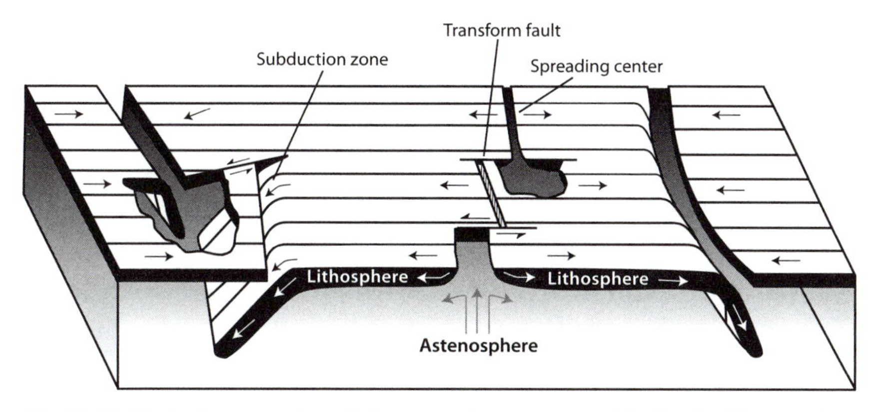

Fig. 10-10: The basic conception of plate tectonics as presented by Isacks, Oliver and Sykes. The earth's surface consists of rigid plates, defined by a lithosphere that is sufficiently brittle that it cracks and makes earthquakes. The underlying asthenosphere flows and does not crack. The plates are created at spreading centers, and consumed at trenches. They can also slide by one another at transform faults. (Modified after Isacks, Oliver, and Sykes, *J. Geophys. Res.* 73 (1968), no. 18: 5855-900)

The temperature boundary between these two regimes is ~1,300°C, and the plane underneath the plate where that temperature occurs, called an *isotherm*, is used to represent the lithosphere/asthenosphere boundary. In actuality, of course, the boundary is a gradual change in physical properties. The plates of the lithosphere then can glide over the more mobile interior or be moved along by it. The plates are large pieces of a spherical jigsaw puzzle that entirely covers the earth, moving continually about the surface as they are created and destroyed.

Since the plates are rigid and inert, all the important action takes place at the plate boundaries. At divergent boundaries, mantle upwells to create the ocean ridges and the volcanism and earthquakes associated with them. At convergent margins, subduction causes huge earthquakes and leads to volcanoes that occur with great regularity approximately 110 km above the Benioff zone. An important feature of plate tectonics is that these two types of plate boundaries are also the bathymetric extremes of the oceans. With the exception of ocean islands and plateaus, which will be discussed in the Chapter 11, the ridges are the shallowest large features in the ocean basins, while the subduction zones are associated with the deepest regions defined by the ocean trenches. The third type of plate boundary, transform faults, has no associated volcanism, but large earthquakes occur, as the huge plates have to slide by one another. Earth's entire surface can then be divided into different plates (Fig. 10-11), and the relative motions of all the plates conserve the total surface area. A simple and conclusive test of the model was that the plate velocities and directions determined by magnetic anomalies on the seafloor should add together to conserve Earth's surface area, and they did!

A central aspect of plate tectonics that distinguishes it from continental drift is that the continents are not the plates. Because of their low density, the continents float on top of the mantle lithosphere and are carried along as passengers sitting on top of the plates. Some plates, like the Pacific, are entirely made up of ocean crust. Many others, like the North American, contain both ocean and continent. Because of their low density, the continents cannot be subducted. They stay on Earth's surface, bobbing like corks as the plates are created and destroyed in ocean basins that surround them. When the plate movement brings two continents together, the continents have a massive collision that leads to the creation of huge mountain ranges. This occurred when India collided with Asia to create the Himalayas, and when Italy collided with Europe to create the Alps. The Appalachians on the East Coast of the United States were formed when Europe and North America collided prior to the opening of the Atlantic Ocean.

This combined information provided a revolutionary new perspective on the ideas of continental drift proposed by Wegener. All his conclusions about the continents wandering over the surface of the earth, coming together and breaking apart, were essentially correct. His physical conception, however, was flat wrong. He proposed that the continents moved through and across Earth's surface, and the oceans were simply "in the way" of continental movement. In his theory the continents moved through the ocean crust like icebergs moving through thick sea ice. This concept is a physical impossibility, and this flaw seemed then to be a refutation of his theory. The new perspective was that continents are the passive passengers floating on the top of plates. The ocean basins were where the physical processes could be observed, where plates were created and destroyed. Rather than acting as passive resistance to continental movement, the oceans were the dynamic expression of forces coming from Earth's interior.

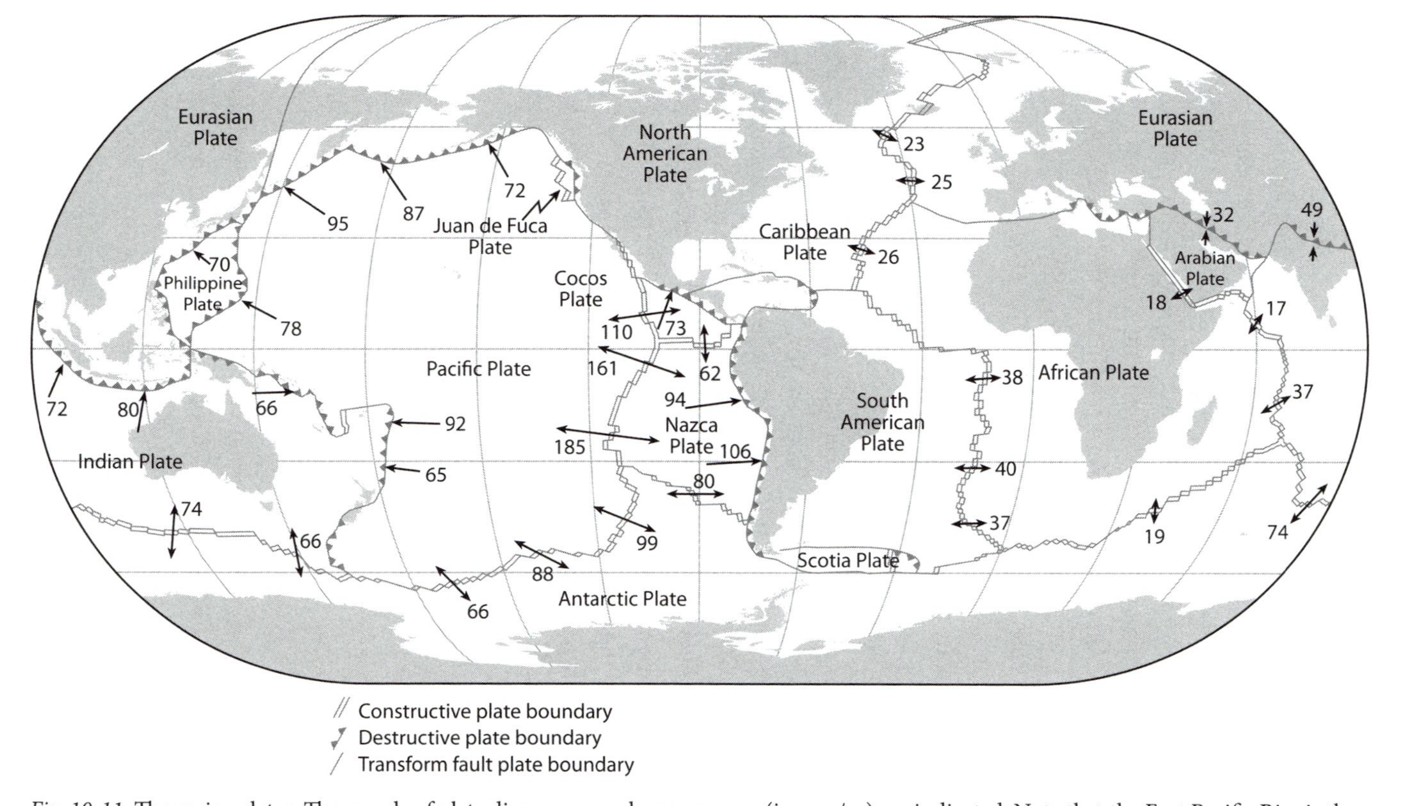

Fig. 10-11: The major plates. The speeds of plate divergence and convergence (in mm/yr) are indicated. Note that the East Pacific Rise is the fast-spreading ocean ridge, which speeds up to 185 mm/yr, while the Southwest Indian Ridge, south of Africa, spreads at only 19 mm/yr. (Courtesy of U.S. Geological Survey)

A further test of plate tectonics comes from the depth of the oceans. The plates are formed at the ocean-spreading center and become progressively older with distance from the ridge. When they form, the plates arise from cooling of magma that comes to the surface near 1,200°C, and once solidified the plates have a thickness that is similar to that of the ocean crust itself, ~6–10 km. The lithosphere is very thin. The 1,300°C isotherm that separates lithosphere from asthenosphere is very close to the surface, separated from the very cold water of the oceans by only ~10 km. This is a very steep thermal boundary layer. As the plate moves away from the ridge, the cool blanket of seawater gradually cools off the plate as it ages, and the 1,300°C isotherm retreats deeper into the mantle, causing the plate to thicken. As the plate thickens through cooling, its density increases and it sinks gradually into the mobile asthenosphere. How much should it sink as it ages?

The evolution of a thermal gradient across a boundary layer is a classic problem in physics for which there are straightforward mathematical solutions. In fact, Lord Kelvin used these equations in his discussions of Earth's cooling. The mathematical results have a very robust characteristic—the thickness of the cold layer (i.e., the lithosphere) increases as the square root of age. As the cold, dense layer thickens, it should then sink gently into the mantle. A test of the plate model would be that the amount of sinking, recorded as the ocean depth, should change linearly with the square root of age, and that the sinking rate with distance should vary depending on the spreading rate. This regular behavior is what is observed in the worldwide oceans (Fig. 10-12) and was a compelling confirmation of the plate tectonic theory.

The Plate Tectonic Revolution

In a very small number of years plate tectonics revolutionized our understanding of Earth's surface and brought together in a single framework data from all the fields of Earth science. Earthquakes, volcanoes, variations in the magnetic field over the oceans, seafloor bathymetry, the distribution of fossils on different continents, the distribution of ancient glacial deposits, the ages of mountain belts—formerly all appearing to be isolated from one another, were now diverse aspects of a single

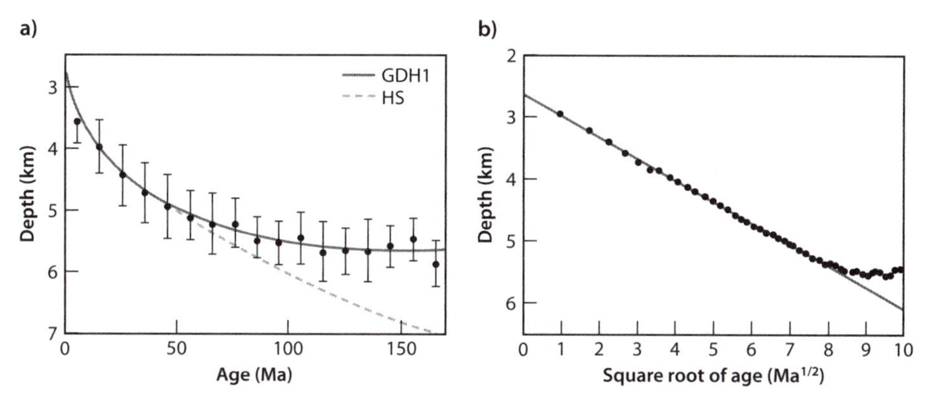

Fig. 10-12: Left: Seafloor depth plotted vs. of the age of the crust. Note the rapid subsidence at young ages compared to older ages. *Right:* The same data plotted vs. the square root of age (i.e., 64 Ma plots as 8 and 100 Ma plots as 10). A linear relationship with the square root of age is consistent with plate tectonics where a plate cools progressively from above as it spreads away from the ridge. As the plate thickens, it takes longer and longer for the same amount of heat to be released. As the plate cools, its density increases, causing the plate to sink into the mantle. An intriguing and as yet not fully explained aspect of the data is the deviation from the linear relationship as ages approach 100 Ma. (Modified after Stein and Stein, *Nature* 359 (1992):123)

process. Geologists working in Africa could relate to South America as well. Scientists studying ice ages in northern Europe could look at rocks near the equator formed by ice ages from previous epochs. Paleontologists needed to understand tectonics, and continental geologists needed to understand what was happening in the oceans. Earth became an interconnected whole, and fields that were previously separated by data type and scientific culture could no longer be understood and studied in isolation. The power of this new theory could be seen in the fact that over a period of only a few years major questions and long-standing puzzles became simple consequences of plate tectonics.

- Mountain belts on continents are built by processes at convergent margins. Interior mountains such as the Alps and Himalayas are the consequence of collision of continents. Volcanic mountain ranges such as the Andes and the Cascades are caused by subduction. Continental margins that are not plate boundaries have no young mountains at all.
- Oceans are young because they are being continually repaved by crust formation at spreading centers and recycled at subduction zones.

Continents are old because their low density prevents them from being subducted.
- Earthquakes and volcanoes are the result of the movements of the plates and the associated convection of the mantle.
- The continents around the Atlantic fit together perfectly because they were once joined together and broke apart, separated by seafloor spreading along the Mid-Atlantic Ridge.
- Sediments in the deep ocean are thin because oceanic plates have a limited lifetime. They are continually recycled back into the mantle.
- The oceans deepen away from ocean spreading centers because of the progressive thickening of the plate as it is cooled from above.
- Animals differ from one continent to another in relationship to the time the continents have been separated. Over longer periods of time, gradual evolution leads to greater differences than over shorter periods of time. Biological differences between continents depend on their history during continental drift.

The new observations from the ocean floor, the existence of a clear mechanism through seafloor spreading and plate tectonics, and the vast number of observations that became simple to understand led to almost universal acceptance of plate tectonics over a period of about five years in the mid-1960s. As with all new ideas, a few scientists, some very prominent, resisted plate tectonics, since it challenged their fixed views.

So is plate tectonics "just a theory" or is it a fact? The clear test would be to measure the plate velocities in real time and see if they are moving. In the 1960s this was very difficult, because accurate surveying of continents to within a few centimeters is no small task! The advent of the global positioning system (GPS) has made such measurements possible. Very accurate benchmarks set up on the continents combined with measurements over periods of many years have allowed direct determination of the relative movement of the continents. This ability to measure the plates permitted obvious questions to be addressed by direct measurement. Do the plates move? Is their movement steady and continuous, or in fits and starts? How do the plates move relative to each other? Measured plate movements have been found to correspond precisely with the rates inferred from paleomagnetism. The plates move continuously, and this movement places stress on faults, which ultimately slip and slide abruptly to create earthquakes. The somewhat surprising, but beautiful, result is that the average rates of spreading over hundreds of thousands of years, inferred by the somewhat arcane method of looking at small wiggles in the magnetic field over the ocean, agree very well with the instantaneous rates measured by GPS. Hence plate movements reflect steady and continuous flow over Earth's surface. The direct measurement of plate movement makes plate tectonics no longer just a theory—it is an established observation of how Earth works. In our calibration of theories, it is a perfect 10!

Movements through Time

While plate motions are not directly perceptible to us, they are actually amazingly rapid. One percent of Earth's history is 40 million years, and in this time the Pacific plate has moved 4,000 km! The ocean floor never gets much older than 150 million years before it is subducted, so the ocean crust provides a record only of the last 4% of Earth's history. The corollary of this is that ocean basins open and close on about the same timescale, and many generations of ocean basin formation and destruction must have occurred throughout Earth's history, hidden forever from modern investigation. Continental rocks contain a far longer record, extending back more than 3 billion years. Since most of the action happens in the ocean, however, the continental record is more difficult to interpret. After all, plate tectonics became obvious after a couple of decades of exploration of the seafloor but was not clear after 150 years of modern geology on the continents! Careful mapping of the continents, however, combined with studies of paleomagnetism that reveal the latitude of rocks when they form, can provide information on movement of the continents through Earth's history. These reconstructions show that continental collisions lead to the assembly of all continents together in supercontinents that can persist for substantial periods of time. Rifting events then cause the continents to break up into fragments and journey across Earth's surface, until they are then reassembled into another supercontinent. The most recent supercontinent was Pangaea, which broke up beginning 225 Ma. Figure 10-13 shows continental movement since the breakup of Pangaea. The continents have indeed split apart and come together, wandering from pole to pole throughout Earth's history. These movements then provide an entirely new context for interpreting all of

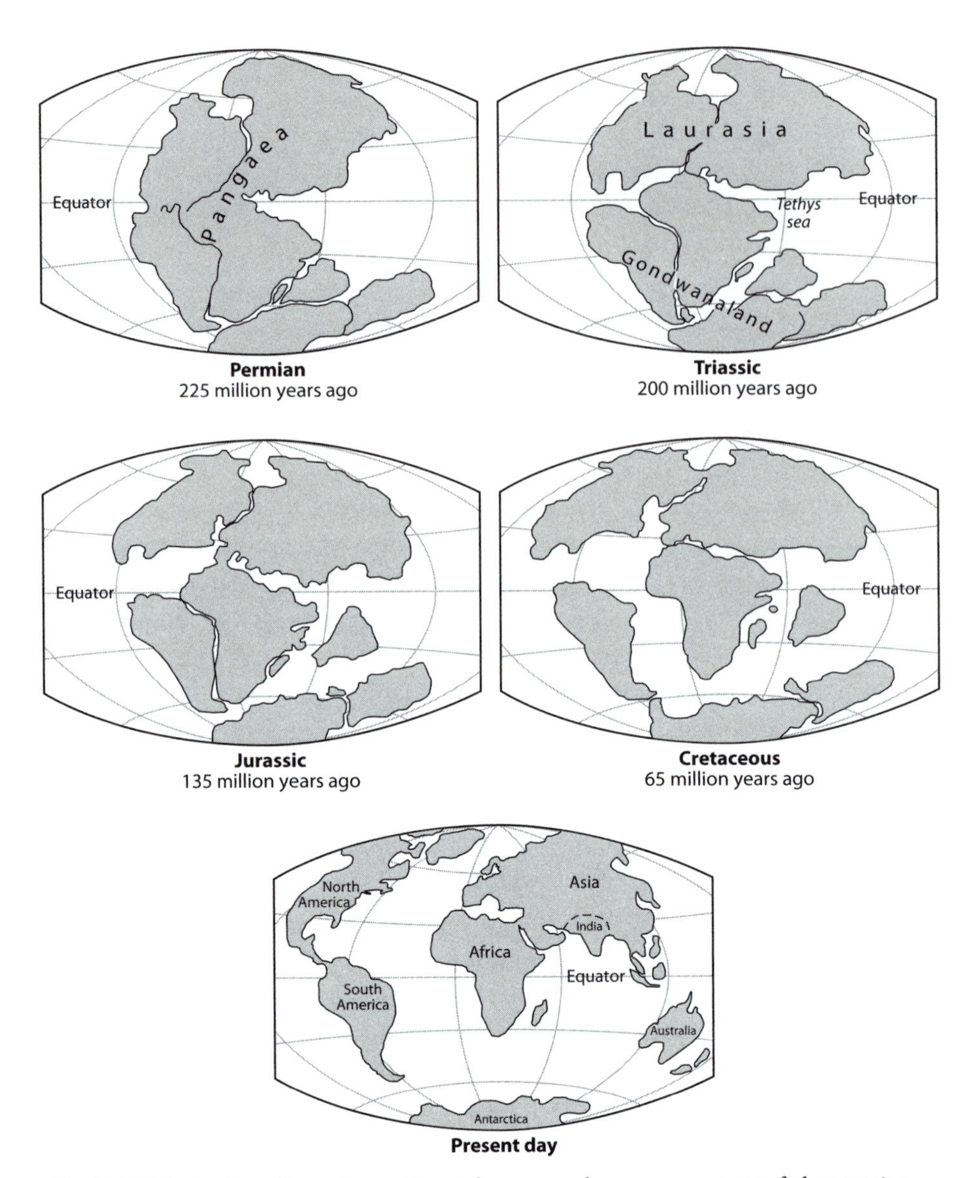

Fig. 10-13: Illustration of how the continents have moved as a consequence of plate motion since the Permian period 225 million years ago. The supercontinent Pangaea breaks up into two halves, Laurasia and Gondwanaland. These then break up into smaller fragments as the Atlantic opens and India and Australia move north. Africa, India, Australia, and Eurasia are now reassembling into a supercontinent, which would become fully formed once subduction commenced in the Atlantic, permitting it to close. (Modified after C. R. Scotese, *Atlas of Earth History*, vol. 1, *Paleogeography* (Arlington, Tex.: PALEOMAP Project, 2001), 52 pp)

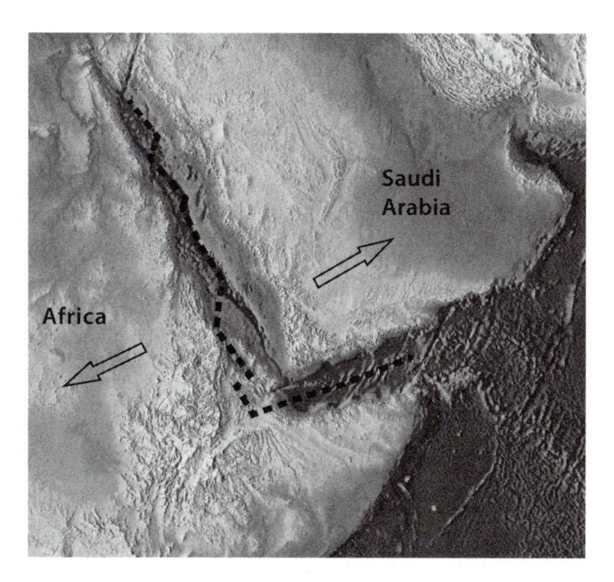

Fig. 10-14: Satellite image of the Red Sea and Gulf of Aden that is the passage to the Indian Ocean. The Red Sea is a recently opened rift that is gradually separating Saudi Arabia from Africa. Should spreading continue, a new ocean basin will eventually form that would be joined with the Indian Ocean. See color plate 11. (Background bathymetry from GeoMapApp; www.geomapapp.org)

geological history, and the classic science of geology has been revolutionized by plate tectonics.

Can we predict the future positions of the plates? We see evidence of ongoing continental assembly by the collision of India with Asia, and the impinging collision of Africa with Europe, closing up the Mediterranean Sea. Ultimately subduction zones likely will form in the Atlantic, the Atlantic Basin will close, and North and South America may join Africa and Eurasia in the next supercontinent. At the same time, we see incipient rifting in the Red Sea and East African Rift, which are the initial stages of formation of a new ocean basin (Fig. 10-14).

Summary

Earth's surface is continually in movement. The movement can be accurately described by plates that cover the surface of the globe and move relative to one another. The plates are stiff and can crack, and this rigid layer is called the lithosphere. Beneath the plates is the mobile asthenosphere, which moves steadily and progressively by mantle convection (to be discussed in more detail in the next chapter). The lithosphere forms the thermal boundary layer between the cool temperatures of Earth's surface and the temperatures of more than 1,300°C in Earth's interior. Because the external shell is brittle, it interacts with the interior only along the margins of the plates. The plates form by volcanism at oceanic spreading centers. Here the plates can be only 10 km thick. As the plates age and move away from the spreading center axis, they are cooled by interaction with the surface. Cooling increases density and causes the plate to thicken linearly with the square root of age. The increased thickness of dense, cold material causes the oceanic plates to subside, deepening the oceans away from ridges. Oceanic trenches mark the locations of subduction zones where plates are recycled into the mantle, and two plates converge. Volcanism occurs on the overriding plate, producing linear chains of volcanoes that also demark the locations of plate convergence. At some locations the plates slide by one another. These are the transform faults, such as the San Andreas Fault in California. Continents are light and buoyant, and float like rafts on top of the plates. They cannot be subducted, and when they collide as a result of plate motion they make huge mountain ranges such as the Alps and Himalayas. Plates move so rapidly that all the ocean basins are young, and the oldest oceanic rocks record only the last few percent of Earth's history. The rapidity of movement means the continents skate across the surface, continually colliding, breaking up, and changing their positions. Because they remain on the surface and are permanent residents of the lithosphere, continents contain much older rocks than the oceans. Erosion and mountain building are continually reprocessing the continental rocks, however, making older rocks increasingly rare. The ocean basins show the present and recent history of Earth's dynamic interior; the continents contain the complex record of the ancient past.

Supplementary Readings

Alfred Wegener. 1966. *The Origin of Continents and Oceans*. Biram John, trans. New York: Dover Publications.

Naomi Oreskes, ed. 2003. *Plate Tectonics: An Insider's History of the Modern Theory of the Earth*. Boulder, CO: Westview Press.

Henry Frankel. 1987. "The Continental Drift Debate." In H. Tristram Engelhardt Jr. and Arthur L. Caplan, *Scientific Controversies: Case Solutions in the Resolution and Closure of Disputes in Science and Technology*. Cambridge: Cambridge University Press.

Philip Kearey, Keith A. Klepeis, and Frederick J. Vine . 2009. *Global Tectonics*. New York: John Wiley & Sons.

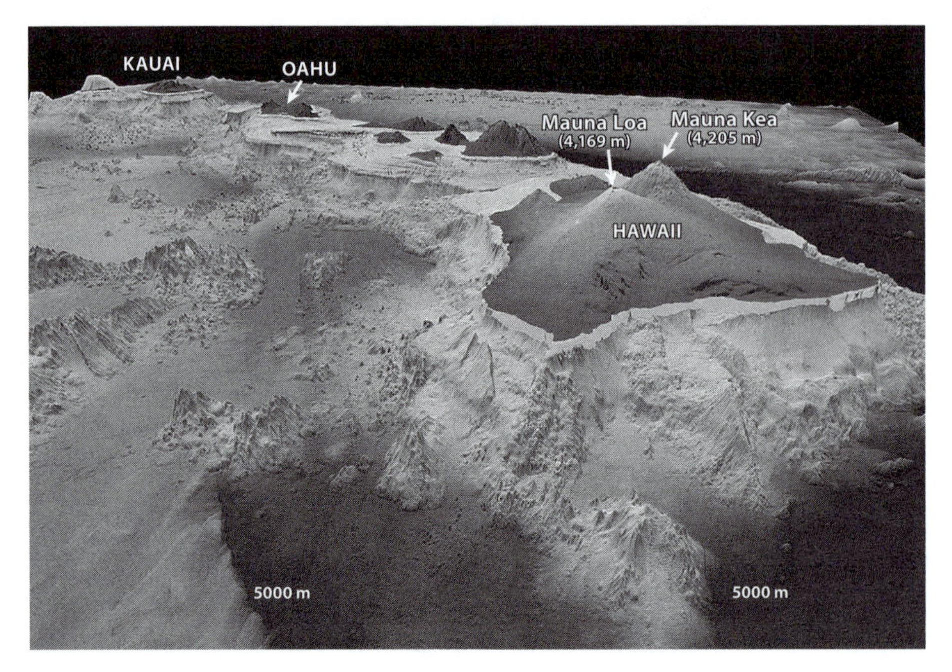

Fig. 11-0: Image of Mauna Loa volcano, the largest volcano on Earth. It reaches almost 14,000 feet above sea level and has a base that is more than 150 km in diameter. Much of the volume of the volcano is below sea level. With a total volume of about 70,000 cubic kilometers, this volcano is more than one hundred times the size of most convergent margin volcanoes and is, most likely, the surface manifestation of deep convective mantle plumes that quite probably rise from the core/mantle boundary. See color plate 13. (University of Hawaii School of Ocean and Earth Science and Technology and U.S. Geological Survey; http://oregonstate.edu/dept/ncs/photos/mauna.jpg)

CHAPTER 11

Internal Circulation

Mantle Convection and Its Relationship to the Surface

Plate tectonics describes the movement of plates around Earth's surface but not how this movement relates to deeper layers. Subducting plates must displace the mantle as they descend, and ocean ridge melting requires solid mantle to ascend in order to depressurize and melt. Both observations require the mantle to flow in the solid state. Study of how Earth's topography responds to the loading and unloading of ice sheets shows that continents move up and down as mass is added and removed according to Archimedes' principle, called isostatic adjustment. Isostatic adjustment can only occur if the mantle flows.

What is the pattern of mantle flow, and how does it relate to surface observations? Flow in response to density differences is called convection. Whether or not convection occurs depends on how large the density differences are, the viscosity of the material, and other parameters that all combine to give the Rayleigh number, which is a kind of convection index. High Rayleigh numbers mean convection occurs, and the Rayleigh number of the mantle is so high that mantle convection is inevitable. Early concepts related spreading centers to convective upwelling and subduction zones to downwelling of simple, large-scale convection cells within the mantle, but such a simple view clashes with observations from Earth's surface. The ocean ridges move rapidly over Earth's surface, and occasionally are subducted at trenches, indicating they are superficial phenomena not associated with deep convection. Downward flow at subduction zones, on the other hand, can be traced to greater depths. These aspects of mantle convection are passive, driven by the plates themselves.

At high Rayleigh numbers, convection cells break down and there can be multiple rising jets of hot material. Evidence for active convection by rising plumes within the mantle comes from mantle hot spots that create ocean island chains such as the Hawaiian Islands and that can occur in the middle of plates. Plumes are probably produced by heating from below at the core/mantle boundary.

The various forms of convection lead to significant temperature contrasts in the upper mantle. Because the ridges migrate freely over Earth's surface, they provide an instantaneous sample of mantle temperature variations. Where the mantle is hotter, it melts more beneath the ridge, creating a thicker crust that floats higher owing to isostatic adjustment, creating shallow seafloor, such as at Iceland. Low temperatures occur distant from hot spots, or where the ridge crosses a former subduction zone. Narrow hot rising jets producing island chains and plateaus, long downwelling zones associated with subduction, and ocean ridges skating over the surface are the diverse connections between plate movement and mantle flow. These movements produce chemical fluxes between surface and interior that are an essential aspect of planetary habitability.

Introduction

Plate tectonics revolutionized our understanding of all geological processes and provided a framework that united the Earth sciences. The limitation of plate tectonics, however, is that it is primarily descriptive, documenting the facts of the movements of the surface—what are called *plate kinematics*—without providing an understanding of the underlying causes of plate motion. Why do the plates move? Why does Earth have plate tectonics and Venus and Mars do not? Why are there volcanoes at spreading centers and convergent margins? What are Earth's inner movements and how do these relate to the exterior movements of the plates? For our comprehension of Earth as a total system, the plate tectonic revolution provided only the skeleton. What we begin to probe in this chapter are the connections among Earth's diverse layers and, in particular, how the movement of plates at the surface is connected to circulation of the mantle beneath the plates.

At first glance this question may seem far removed from planetary habitability. As we shall see in subsequent chapters, however, the fluxes of materials to and from Earth's solid interior are central to a stable climate, to the existence and chemical compositions of ocean and atmosphere, and to the origin and evolution of life.

Movement of Earth's Interior

Simple consideration of plate tectonics leads to the conclusion that movement of plates at Earth's surface must have a corresponding movement of the mantle beneath. At both spreading centers and subduction zones, mantle flow is required. Material injected into the mantle at subduction zones must displace the mantle below. At ocean ridges, mantle melting requires the upward flow of solid mantle for melting to occur (see Chapter 7).

Flow of rocks is not an intuitive concept for us. Near the surface, within the lithosphere, rocks do not flow—they crack and displace along faults. Fault movements create the earthquakes that are familiar to us. Earthquakes, however, only occur in the outermost portions of Earth—the top 10–15 km at spreading centers, and down to a few hundred kilometers at Benioff zones. Apart from the seismic zones, movement must also be occurring to compensate for spreading and subduction. The boundary between brittle movement along faults and ductile flow without faults divides the mantle into the brittle lithosphere and the ductile asthenosphere. Seismic waves tell us that the mantle is solid; therefore, flow in the asthenosphere must be solid flow.

Although flow of solids has a ring to it that is not familiar to us, deformation of solids is within our experience. For example, each time we bend a paper clip, or when iron or copper are heated and forged into new shapes in the solid state, solids are deforming. In glaciers there is flow of solid ice that is visible and observed on human timescales, and the folds observed in rocks that are uplifted to the surface also clearly show deformation and flow (Fig. 11-1a). Solids can flow when they are at temperatures that are close to their melting temperatures. Therefore

Fig. 11-1: Illustration of flow in the solid state. (a) Aerial photograph of the Barnard Glacier that shows glacial flow (courtesy of Robert Sharp, California Institute of Technology and the University of Oregon Press, Eugene, Oregon); (b*)* Photograph of a folded rock, Gneiss, Cabonga Réservoir, Québec (courtesy of J. P. Burg, ETH Zurich).

ice flows on the surface, but rocks do not. At depth in the mantle, and at some locations in the deep continental crust, temperatures are high enough that flow in the solid state must be possible.

Proof of mantle flow in the solid state also comes from studies of the origins of Earth's topography: why are mountains high? Why are the oceans deep? Why are the continents rising today in northern Canada and Scandinavia?

EARTH'S TOPOGRAPHY AND MANTLE FLOW

Our common sense perception of topography is that it simply reflects the thickness of material that is added to or removed from a certain spot. If we dig a big hole in the ground, the elevation is lower at the bottom of the hole and higher where the excavated dirt is piled up. Such conclusions apply, however, only when the underlying material is rigid and does not respond to the change in weight. If the underlying material can flow, it adjusts to the weight of the object placed upon it.

For example, we can't dig a hole in a liquid because liquid flows to accommodate forces. Wood placed in water sinks to a certain level, pushing the water underneath aside. The depth that the wood sinks follows a simple principle—the water flows from high pressures to low pressures until the pressure is equalized everywhere at the same depth. Therefore the pressure at the interface of the bottom of the wood with the water beneath it must be identical to the pressure of a column of water next to the wood at the same depth. If it weren't, the water would flow and adjust until the pressures equalized. This depth where the pressures are the same is referred to as the *depth of compensation*; when material floats so that pressures are all the same at depth, the material is said to be compensated. This principle was discovered originally by Archimedes, who recognized that floating objects must displace a mass of liquid equal to their weight. For Earth, this concept is referred to as *isostasy*—iso (equal) stasis (standstill). *Isostatic adjustment* means that Earth's surface rises or falls in response to a change in load.

To visualize how this works in practice, consider two examples, one where the mass is compensated and one where it is not. What happens if we place a sheet of plywood 1 cm thick and a beam 20 cm thick with the same density in a tub of water? The beam sinks deeper than the plywood and also sticks out of the water a few cm higher than the thin sheet, but not the full difference in thickness. Looking underwater, we would see the beam protruding deeper than the plywood. At the depth of compensation, below both pieces of wood, the pressure is the same everywhere and the masses are compensated. If instead we place the beam and plywood on a rigid table, the top of the beam rises fully 19 cm higher than the plywood, the bottoms are at the same level, and there is no spot below them where the pressures are equal. The pressure beneath the beam is much greater than beneath the plywood. The water flows; the table does not.

Which of these possibilities reflects how Earth works? If the inner Earth were rigid, a mountain range 2 km high could simply reflect two extra kilometers of rock (Fig. 11-2a). Or, if the supporting material beneath the mountains can flow over long periods of time, then the mountains must have a deep root to compensate for the higher elevation (Fig. 11-2b).

If we could measure the pressures at depth beneath the mountain range and beneath flatter regions, it would be straightforward to tell these models apart, but that would require impossible measurements deep in Earth's interior, some 100 km or so below the surface. Geophysicists found a clever way to answer this question, however, by making use of the slight variations in Earth's gravity field. The gravity field reflects the total mass of a column extending down from Earth's surface to the core. If the mountains are extra mass piled on top of the crust, then their additional mass should cause the gravitational field to be greater. If the topography is compensated, then the total mass of any column is the same, and the gravity field should be similar over mountainous regions and low elevation plains. Even the first crude measurements made in the nineteenth century showed that mountains do not add substantially to the gravity field, and therefore vertical columns through the Earth all have about the same mass. Conclusion: the mantle must flow in response to the weight of the crust that is above it (Fig. 11-2).

One additional complexity comes from density differences. As we learned in Chapter 4, different materials can have different densities, and varying density can also lead to topographic differences. Imagine a block of styrofoam and a block of the *same thickness* of dense hardwood floating in the tub of water. The styrofoam sticks up much higher, because its light weight is compensated by a very small volume of water. Both thickness and density contribute to topographic differences (Figs. 11-2b and

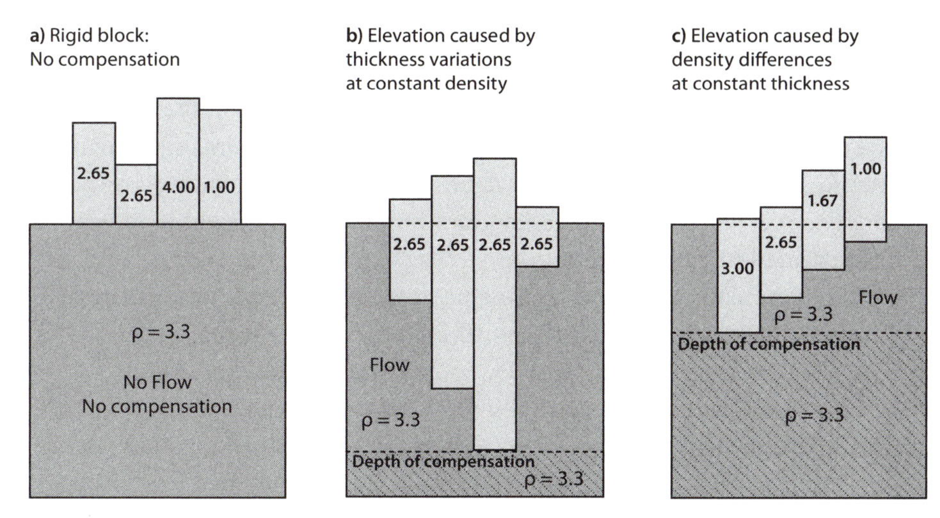

Fig. 11-2: Three different models of variation in elevation. (a) differences in elevation result from accumulation of material without compensation, leading to pressure differences at depth. The rigid block underneath the material does not allow flow; (b) differences in elevation result from differences in thickness of the lighter material because the underlying material can flow; (c) differences in elevation results from differences in density. Model *b* applies to continents; model *c* applies to the increase in ocean depth with age discussed in Chapter 9. At any one age on the seafloor, model *b* can apply where there are differences in crustal thickness. Model *a* is not how topography on large scales is accommodated on Earth.

11-2c, respectively). So an alternative compensation model could be that mountains are high because they are made up of light rocks. But are they? Inspection of the rocks on continents—generally granites and metamorphosed sediments—shows that their densities are similar in all continental regions, suggesting mountain ranges are regions of thick crust rather than light rocks. Seismic results have confirmed these inferences from the gravity field; mountain ranges have deep roots and continental topography largely reflects crustal thickness (Fig. 11-2b).

Isostasy then gives us the tools to consider the bimodal distribution of elevations on Earth seen in Figure 10-1. Why are the oceans deep? The difference between oceans and continents has to do with both thickness and density. The ocean crust is both thinner (~6k m vs. 35k m) and 10% denser than the continental crust, causing it to exist at much lower elevations.

Isostasy also explains the variations of depth within the ocean basins. The increasing depth of the seafloor with age in the oceans discussed in Chapter 10 is caused by subsidence, as the density of crust and mantle increases as the plate ages and cools. There can also be depth variations at constant age. Iceland, at zero age on the Mid-Atlantic Ridge, is above sea level! Depth variations at constant age depend on the thickness of the ocean crust, as influenced by the temperature of the mantle during melting. Figure 11-2b applies to depth variations of the ocean floor at any one age, while depth variations as a function of age are akin to what is illustrated in Figure 11-2c .

The fact that oceans and continents both reflect isostatic adjustment requires that the mantle can flow, like water but much more slowly. If Earth's interior were rigid, isostatic compensation would not occur.

How rapidly can the mantle flow? To provide us with an answer to this question, Earth has provided a natural experiment that allows observation of isostatic compensation in action. During the last glacial period, several kilometers of ice covered Canada and Scandinavia. The weight of the ice was an added load and caused the continents to sink as the mantle flowed away in response to the load. When the ice melted, the load was removed, and the continents were able to rise (Fig. 11-3). Because isostatic adjustment takes many thousands of years, the continents in these regions still rise at measurable rates today (Fig. 11-3b), a phenomenon called *postglacial rebound*. The stiffness of the mantle recorded by postglacial rebound provides crucial information as we evaluate another larger scale form of mantle flow, mantle convection.

Mantle Convection

All flow is a response to forces. As we have just seen, isostatic adjustment occurs because of changes in load at the surface. Force can also be exerted within the mantle whenever light material underlies denser material. For example, a hot thermal boundary layer between the outer core and the deep mantle causes the mantle to be heated from below; thousands of km above, cold boundary layers at the top of the mantle cause the mantle to be cooled from the surface. Heating causes expansion and lowered density, while cooling causes contraction and increased density, so these boundary layers create light material underlying denser material. If the mantle is not too stiff, then it will circulate through a process called *thermal convection*.

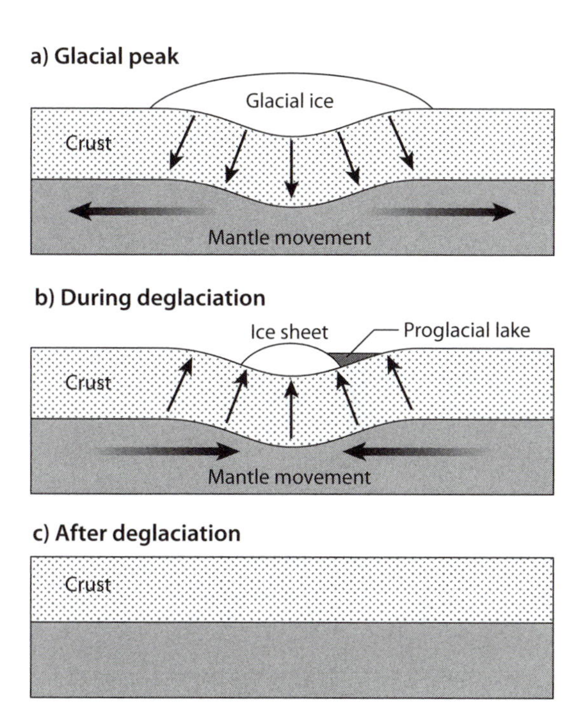

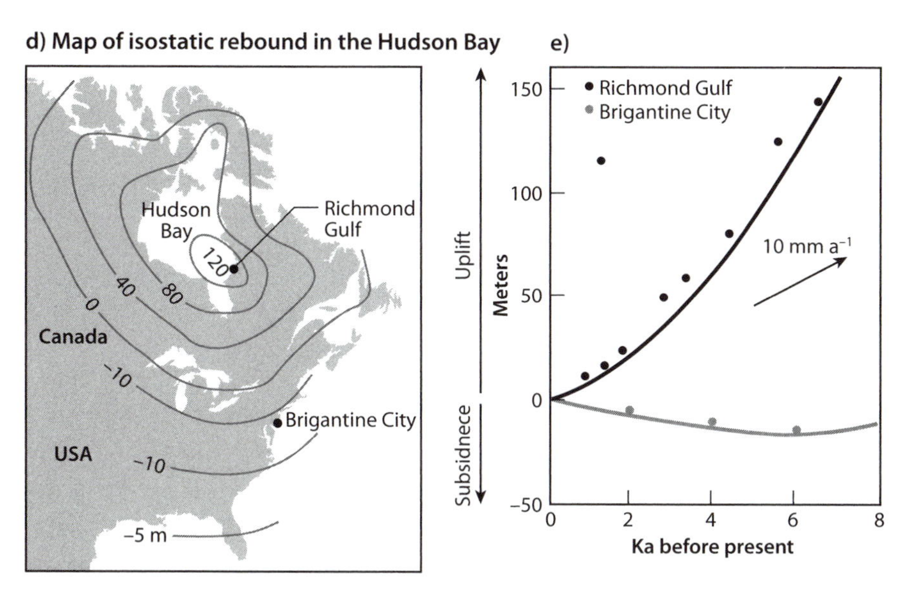

Fig. 11-3: (a)–(c) Illustration of isostatic adjustment caused by glacial loading and unloading. Glaciation adds mass to the surface and causes the continent to subside as the mantle flows away. When the ice melts, postglacial rebound occurs. (d) A map of postglacial rebound of the Hudson Bay area and part of North America, a process that continues today. Numbers within the contours are uplift or subsidence in meters. Note the major uplift centered on Hudson Bay. (e) Uplift over time of two cities in (d). Figure from R. Walcott (1973), *Ann. Rev. Earth and Plan. Sci.,* 15.

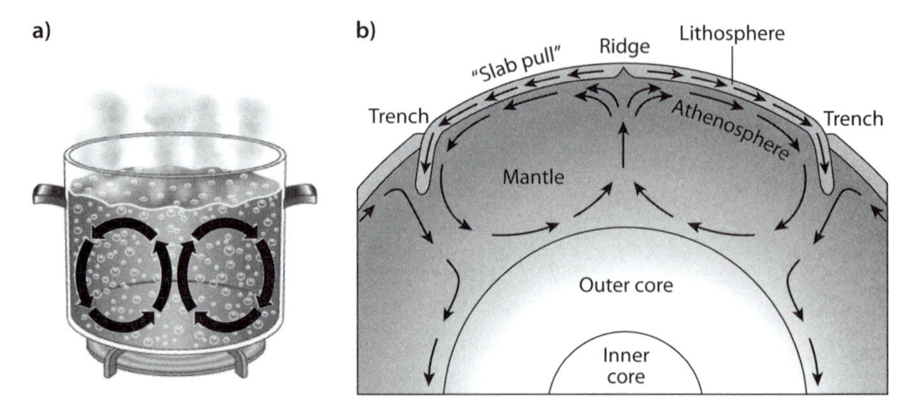

Fig. 11-4: (a) Comparison of the simplest form of convection illustrated with water in a heated stove, leading to a circular convection cell; (b) schematic view of a hypothetical (and partially wrong) relationship between subducting plates being associated with the downwelling limb of a convection cell and ridges being caused by convective upwelling of the cell. (Note that this simple framework does not generally apply to Earth's plates.)

The simplest kind of convection leads to *convection cells*, where hot material ascends at one location, flows laterally, cools off, and descends. The zones of ascent and descent define the cell (Fig. 11-4a). We see convection in action when liquid is heated on a stove, when hot air rises above a heater, or when cold air flows downward off a cold drink or ice tray. Solids as well as liquids can convect, provided the solid material can flow. The convection cell concept has a superficially compelling parallel with plate tectonics—plates form above ascending mantle at ocean ridges, cool off as they spread away from the ridge, and descend as cold slabs at subduction zones. Many drawings of plate tectonics (e.g., Fig. 11-4b) show just such a relationship between mantle flow and plate movement. If this were true, it would suggest direct and simple connections between the interior flow of the mantle, the creation and destruction of crust and plates, and the lateral movement of plates across the surface. The plates would be driven by the pattern of mantle convection.

This scenario requires that mantle convection occurs. Not all materials convect—most solids we are familiar with at the surface do not convect, and that includes the large masses of rock that make up mountains. Isostasy requires flow, but not necessarily convection—it all depends on the forces. So we have two fundamental questions:

(1) Does the mantle have the properties that make convection likely at conditions within the Earth?
(2) If so, does the pattern of convection conform to plate boundaries at the surface?

MUST THE MANTLE CONVECT?

While convection is the response of matter to variations in density, even in the face of density variations convection does not always occur. It depends on the magnitude of the density anomalies, the viscosity of the material, and the distances involved. Large density anomalies, low viscosities, and large distances all enhance convection.

Density differences are often driven by changes in temperature. The critical factor driving convection is when convection would lead to more rapid dissipation of temperature differences than escape of heat through conduction. The mantle has large temperature differences, but the rocks of the mantle are very stiff, inhibiting convection. A battle exists between the forces driving and inhibiting convection. How do we tell what wins? Careful investigation of convection both in theory and in the laboratory has given rise to a parameter that indicates whether or not convection will occur. The parameter is known as the *Rayleigh number*, developed by Lord Rayleigh in 1915. The Rayleigh number makes use of parameters that are familiar to us, such as distance (h), temperature (T) and viscosity (η) and two other parameters that may be less familiar. One is the coefficient of thermal expansion (α), which is the measure of how much an object expands in response to increasing temperature. The second is the thermal conductivity (κ), which is a measure of how quickly heat diffuses through a material. Metals, for example, have high thermal conductivity, which is why they are used for pots, because the metal transfers the heat rapidly. Rocks have lower thermal conductivity, which is why rocky materials are used on the outside of blast furnaces—they do not let much heat out. Very hot stones are used for tabletop cooking in some restaurants because the stones retain their heat for a long time.

Convection is *enhanced* by factors that cause large density differences to exist and persist. The product of temperature difference and coefficient of thermal expansion (α) determines the density differences, and the forces they cause are also enhanced by a larger gravitational field (i.e., the same mass of rock on the moon weighs less than on Earth). Greater distances (h) make it more difficult for heat differences to diffuse away, hence the density contrasts that drive convection remain in place, and make convection more likely. Convection is *inhibited* by increased thermal conductivity (κ), which allows the temperature differences to diffuse away rapidly, and by high viscosity (η) that makes it difficult to flow.

The Rayleigh number puts the terms that enhance convection (temperature differences, coefficient of thermal expansion, distance, gravitational acceleration) in the numerator and those that inhibit it (thermal diffusivity and viscosity) in the denominator. The Rayleigh number has the form:

$$R_A = \frac{\alpha g \Delta T h^3}{\eta \kappa} \tag{11-1}$$

where g is the gravitational constant and T is the temperature contrast between top and bottom. When the Rayleigh number is greater than about 2,000, convection is inevitable. As the Rayleigh number increases, the convection becomes more and more active, and ultimately turbulent and chaotic (Fig. 11-5).

The Rayleigh number for the mantle can be calculated. Distances are known, and the coefficient of thermal expansion and thermal conductivity of mantle peridotite have been measured in the laboratory. Temperature is constrained by the compositions of volcanic rocks (see later in this chapter) and the heat that flows from Earth's rocks. The most difficult parameter to constrain is viscosity. The best estimate comes from the isostatic response of the continents to glacial loading and unloading, illustrated in Figure 11-3. The rate of rise depends on the viscosity of the mantle. In a bathtub, a cork will pop up very quickly if a weight is removed from it because of the low viscosity of water. The cork would pop up much less quickly in peanut butter because peanut butter has higher viscosity and would flow into the space under the cork more slowly. By measuring how quickly North America and Scandinavia rose after the last glaciation, the mantle viscosity can be calculated to be about 10^{21} Pascal seconds—which is 10^{24} (a trillion-trillion) times more viscous than water! (To put that in perspective, peanut butter is about 200,000 times more viscous than water.)

While the mantle viscosity is large, the distances and temperatures are large as well, and diffusive removal of temperature differences is very slow. Once all the numbers are plugged into the Rayleigh number equation, it turns out that despite the very high viscosities, the Rayleigh number for the mantle is a million or more! This number vastly exceeds the value of 2,000 that is the threshold for convection, and mantle convection is inevitable. The high Rayleigh number of the mantle led the great British geologist Arthur Holmes to be the first to suggest a possible link between mantle convection and continental drift in the 1950s.

At very high Rayleigh numbers, however, convection is no longer a simple pattern of symmetrical convection cells (e.g., Fig. 11-5a) and can have multiple ascending hot jets and a much less organized form (e.g., Fig. 11-5b). The uncertainties in the mantle Rayleigh number are sufficient that we cannot know a priori what form of convection the mantle would undertake. And of course, Earth is not a simple box like many convection experiments. Are there simple convection cells corresponding to plate margins? Or is the convection pattern more consistent with upwelling plumes that come and go in a complex pattern? What

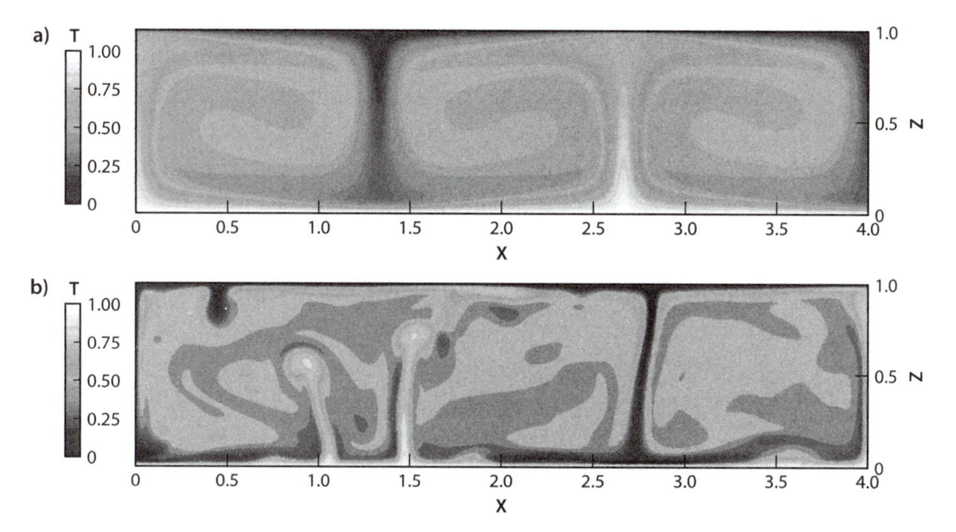

Fig. 11-5: Numerical models of convection illustrating the effect of increased Rayleigh number: (a) at a modest Rayleigh number of 10^5 (*top panel*) a simple convection cell forms; (b) as the Rayleigh number increases ($Ra = 10^7$), convection becomes much less organized and more turbulent, with vigorous ascending and descending plumes. In the figure you can see some plumes extending from the base to the surface, and some that are just beginning to rise. See color plate 12. (Figure provided by Thorsten Becker)

does Earth have to tell us about convection of its interior? For answers to those questions, we need to turn to the evidence that Earth provides.

Does Plate Geometry Correspond to Mantle Convection Cells?

A more careful consideration of the details of plate boundaries shows there cannot be a simple relationship between plate margins and convection cells. While cartoons of plate movement often show nice, regular convection cells, the actual geography on Earth is far more complicated. For example, a reinspection of Figure 10-11 shows that the distance from the East Pacific Rise to the western Pacific subduction zones is some 10,000 km, while off the northwestern United States the Juan de Fuca Ridge is only a few hundred kilometers from the Cascadia subduction zone. How could convection cells be of such different sizes? Even more problematically, the African and Antarctic plates are mostly surrounded by ridges (Fig. 11-6) with no associated subduction zone that could be a downwelling limb of a convection cell. Clearly, the mantle cannot consist of regularly sized convection cells.

The final nail in the coffin for the simple convection cell concept is the fact that ridges can be subducted. The East Pacific Rise used to be continuous from the South Pacific all the way up to Washington State, and subduction off the coast of California led to the creation of the Sierran volcanoes, now eroded down to their granitic roots. Because of the expansion of the Atlantic Ocean basin, the Pacific is contracting, and this caused the North American subduction zone to swallow up the ridge. Once this ridge was subducted, California became cut by the San Andreas Fault (Fig. 11-7a). Today ridge subduction is occurring at several places, including the subduction of the Chile Rise beneath southern Chile (Fig. 11-7b). Upwelling and downwelling convection limbs cannot be in the same place, so spreading centers cannot correspond with upwelling limbs of major convection cells.

So then, what are ocean ridges and how do they relate to mantle flow? This question is also amenable to evidence from observations. Because Africa has ridges on the east, west, and south, the continent is approxi-

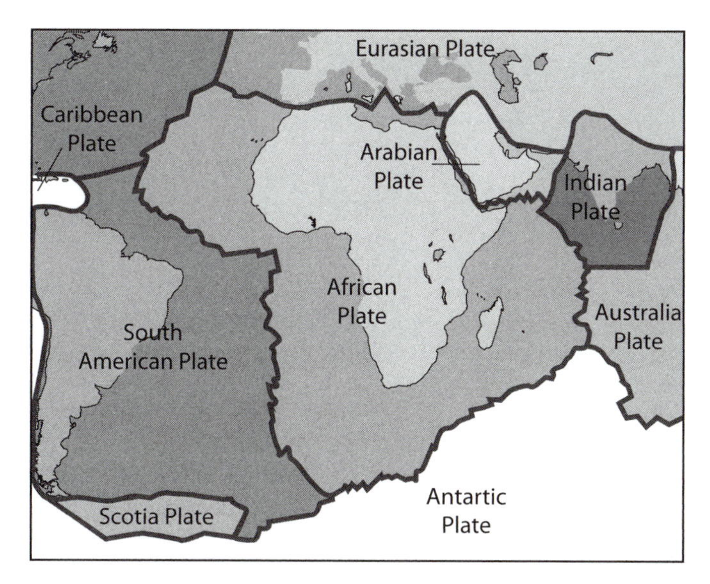

Fig. 11-6: Map of some of Earth's plates. Note that the African plate is largely surrounded by ridges with no associated subduction zones and therefore is growing to south, east, and west, as the ridges become increasingly distant from the continent through time. There can be no regular pattern of "convection cells" associated with plate margins. Instead, the growth of the Atlantic is compensated by excess subduction in the Pacific as the Pacific Ocean basin shrinks in size. (Image from U.S. Geological Survey's *This Dynamic Earth;* http://pubs.usgs.gov/gip/dynamic/dynamic.html)

mately stationary. As spreading occurs, the ridges are moving away from the continent as they produce more and more ocean crust and the African plate grows in size. The Mid-Atlantic Ridge, for example, began adjacent to the African continent when South America and Africa split apart and has steadily migrated westward as the South Atlantic has opened. Its westward migration is half the speed of the total separation of South America from Africa. At the same time, the Central Indian Ridge on the eastern side of the continent has migrated eastward. The two ridges steadily move across the mantle below and increase their separation. Limbs of convection cells in something as large as the mantle will be quite stable for long periods of time, and if ridges were above upwelling limbs, they should stay put. Instead, ridges glide rapidly over the mantle, at speeds that can be similar to spreading rates! Careful measurements around the globe show that virtually all the ridges are migrating over the surface.

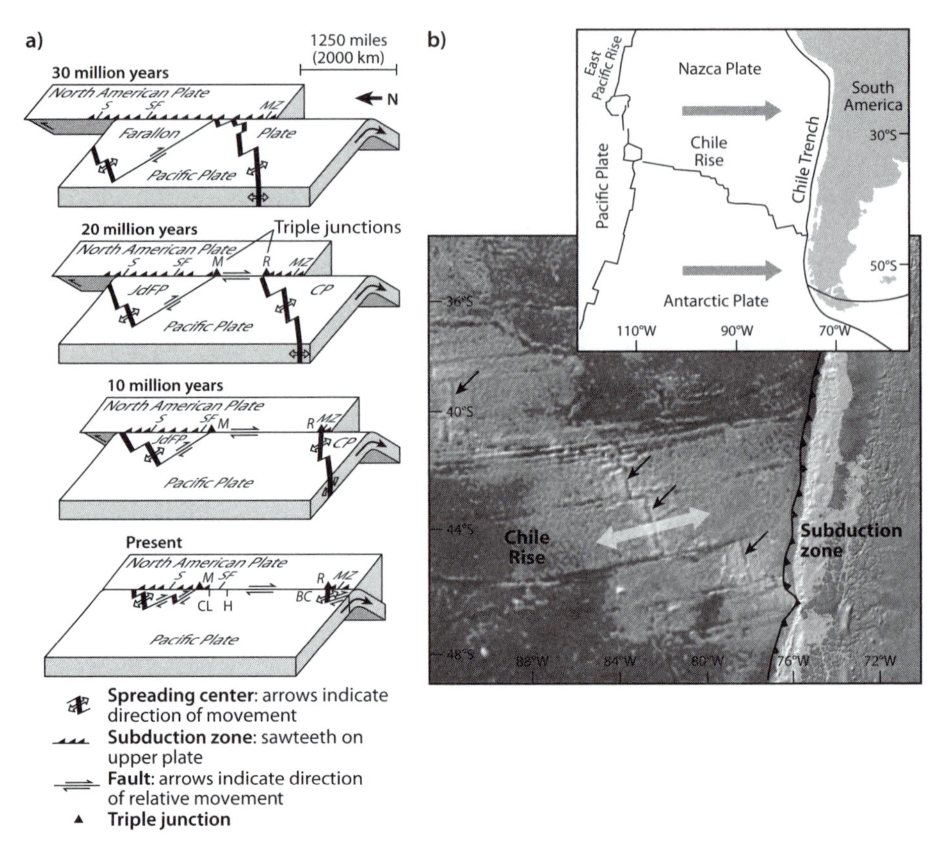

Fig. 11-7: (a) History of the former Farallon plate that used to exist west of California. The contraction of the Pacific Ocean basin caused the plate and its associated ridge to be progressively swallowed by the subduction zone. At some later time the Juan de Fuca Ridge will also be subducted (courtesy of U.S. Geological Survey). (b) The Chile Rise spreading center is currently being subducted beneath southern Chile (base map is from GeoMapApp; http://www.geomapapp.org). See also color plate 14.

The solution to all these observations is to recognize that most spreading centers (we will see the exceptions in a moment) generate their own mantle flow as they spread, rather than reflecting the active forces associated with deep convective upwelling. This is referred to as *passive upwelling* in contrast to dynamically driven active upwelling associated with large-scale mantle convection. An illustration of how this works is shown in Figure 11-8. Because plates separate from each other at the ocean ridge and thicken with age, each increment of spreading opens up a vertical gap in the mantle beneath, and the mantle must rise to fill

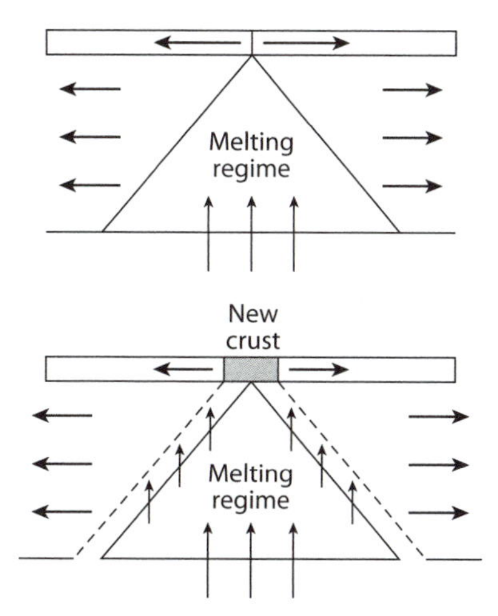

Fig. 11-8: Illustration of mantle upwelling driven by plate movement. The two panels show the sequence of spreading. Top panel shows a steady state of melting regime beneath the ridge. Bottom panel shows one increment of spreading. As the plates move laterally, the movement opens up a "gap" that the mantle rises to fill. The ascent depressurizes the mantle, causing it to melt. The melt rises to create new ocean crust. Of course, the process is not incremental but continuous. Continuous spreading of the plates causes localized upwelling beneath the ridge, independent of whatever else is happening in the mantle beneath. Because the plates extend only to about 100 km depth, the upwelling is primarily in the uppermost mantle.

the gap. As the ridge spreads, the shallow mantle rises beneath it. The upwelling is a shallow feature caused by the spreading itself—local flow imposed from the kinematics of spreading, rather than convective flow driven by the Rayleigh number. This flow is localized in the uppermost mantle, rather than reflecting deep-seated mantle convection. In this case ridges can skate across Earth's surface like water spiders, and when the ridges reach subduction zones they just get carried into the mantle and disappear. When a plate cracks in a new location and spreading begins, the spreading causes local upwelling to begin directly beneath it. Spreading centers cause local upwelling generally not associated with deep-seated mantle convection.

Subduction zones are a different story, because the down-going plate is thick and cold, and we know from the Benioff zones that the plates extend down to 700 km in the mantle. They cannot be associated only with shallow processes. The deep downward movement of the plate also causes the adjacent mantle to move in the same direction, causing a wide region

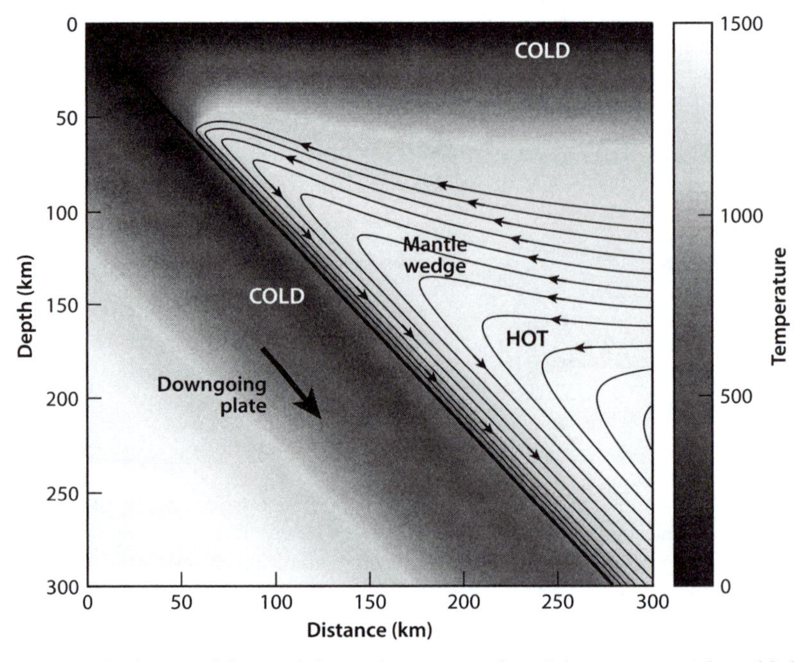

Fig. 11-9: Calculation of flow and thermal structure of a subduction zone. The cold slab dips down into the mantle, causing cold temperatures in the overlying mantle wedge, and "corner flow" of the mantle where the mantle on the left moves to the right, makes a sharp turn downward, and is dragged down by the movement of the down-going plate. This leads to widespread mantle downwelling associated with subduction zones. See color plate 15. (Figure courtesy of Richard Katz)

of cold downwelling (Fig. 11-9). Sophisticated seismic imaging techniques of recent years show "slabs" of old subducting plates extending deeper than the seismic zone. In certain regions they extend down 1,500 km or more into the mantle (Fig. 11-10). Subduction zones are therefore associated with deep downwelling convection, although the flow is by no means vertical. They are the cold downwelling half of mantle convection.

Brad Hager and Rick O'Connell at Harvard expressed this behavior as "mantle convection driven by the plates." Plate movements cause local upwelling at ridges and major downwelling at subduction zones. The upper mantle flows in part in response to the movements of the overlying plates. Rather than being driven by mantle convection, it appears that the plates are themselves important drivers!

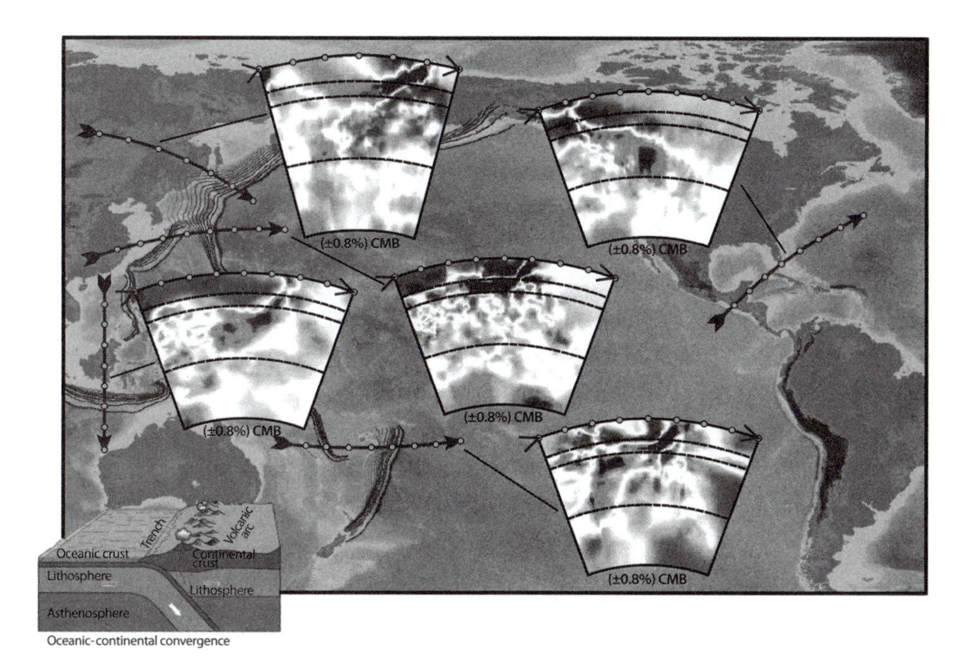

Fig. 11-10: Mantle images created by seismic tomography, showing the diversity of slab subduction. Heavy arrows on the maps show the cross sections. Insets are vertical profiles into the Earth showing seismic velocity variations. Slabs can be seen as darker colors at depth. The top two dashed lines in each inset show the phase transitions of the upper mantle. The third dashed line is at depths of 1600 km. Note that for Central America on the far right and Sudan on the far left the slab (dark colors) appears to extend to great depth. In other cross sections, the slab seems to be detained near the bottom of the upper mantle. See color plate 16. (Modified after Li, C., et al. A new global model for P-wavespeed variations in Earth's mantle, *Geochemistry, Geophysics, Geosystems,* vol. 9, Q05018, doi:10.1029/2007GC001806. Figure courtesy of Rob Van der Hilst)

If plates are not riding along on top of mantle convection cells, why do the plates move? One driving force appears to be the change in mineralogy that takes place at subduction zones. As we learned in Chapter 4, the stability of minerals depends on pressure and temperature; at higher pressures mineral assemblages become denser. Such a transformation occurs for the basaltic rocks of the subducting crust. At the surface they have a density of about 3.0gm/cm^3, but at depths of about 50 km basalts undergo a transformation to a garnet-bearing rock called *eclogite* with a density of 3.35 gm/cm^3. This high-density material is a large weight on the deep end of the subducting plate and tends to drag it downward (Fig. 11-11). The other end of the plate, at the spreading center, is much

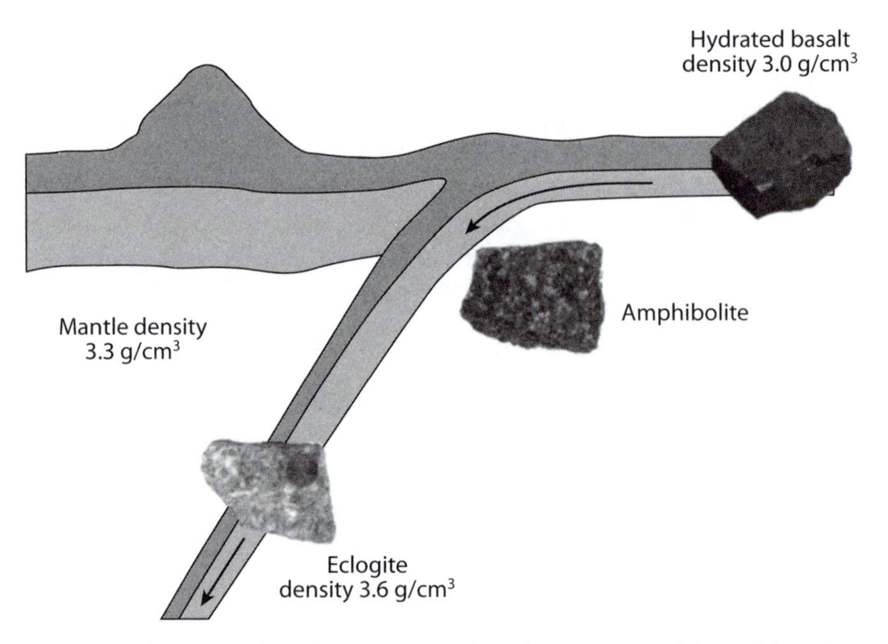

Fig. 11-11: Illustration of the change in mineralogy that occurs as a slab is subducted. At high pressures, the basalt converts to eclogite, which has a higher density than the surrounding mantle, which makes the slab dense and leads to "slab pull," one of the important driving forces of plate motion.

higher than the surrounding ocean floor. The elevation of the ridges tends to push the plates away from the ridge. One can think of it as a gravity slide, sliding downhill from the ridges and being pulled by the extra weight at the bottom of the subduction zone. Both "ridge push" and "slab pull" tend to make the plates slide across the surface.

But wait—what about the high Rayleigh number of the mantle that we saw earlier in the chapter would drive active mantle convection, including active upwelling? If the ridges are shallow and passive, where is this inevitable active upwelling component of mantle convection? And if the subduction zones move deep into the lower mantle, where is the compensating movement that takes material from the lower mantle to make room for them? Because of its high Rayleigh number, the mantle must also have a convective life of its own, with temperature variations and complex circulation. This deeper aspect of mantle convection is revealed from observations from the chains of ocean islands that occur throughout the ocean basins.

Active Mantle Upwelling: Plume Heads and Tails

While ~90% of Earth's current volcanism is associated with plate boundaries, there are a significant number of volcanoes that occur in the middle of plates, a phenomenon called *intraplate* or *hot spot* volcanism. And in the past there have been vast volcanic outpourings that made continental flood basalts and oceanic plateaus. Many of these also occur on continents or in the middle of oceanic plates. Some of the most famous volcanoes in the world—e.g., on the island of Hawaii (Fig. 11-12a) or at Yellowstone National Park—are intraplate volcanoes, in some cases thousands of kilometers distant from any plate margin. This volcanism must reflect active upwelling because it is even able to penetrate old, thick plates and light continental crust. Here is where we find the evidence for the active upward movement of mantle convective flow and where we can address the questions of whether the flow is diverse hot rising jets or stable upwelling limbs.

Much intraplate oceanic volcanism is associated with island chains, which are prominent features in maps of the seafloor (Fig. 11-12a). The chains might reflect a long rift associated with upwelling, or they could be produced by a point source that was fixed in location as the plate migrated over it. Which is it?

The answer to this question comes from both geochemistry and geophysics. The Hawaiian chain begins with the active volcano of Hawaii at the eastern end and then progresses to lower and lower islands toward the west, followed by a large number of submerged volcanoes, called the Emperor Seamounts (Fig. 11-12a). The active volcanoes are on the youngest island. Volcanoes on other islands and submerged seamounts are extinct. Careful dating of rocks from these young and old volcanoes shows there is a simple age progression all the way along the chain (Fig. 11-12b), and the age corresponds closely with the independently determined spreading rate of the Pacific plate.

Such chains of volcanism were a mystery until the discovery of plate tectonics suggested a simple origin. Jason Morgan, one of the founders of plate tectonics, noticed that the chains followed the direction of absolute plate motion and could be explained by a fixed "hot spot" in the underlying mantle. This concept is illustrated in Figure 11-13. As the

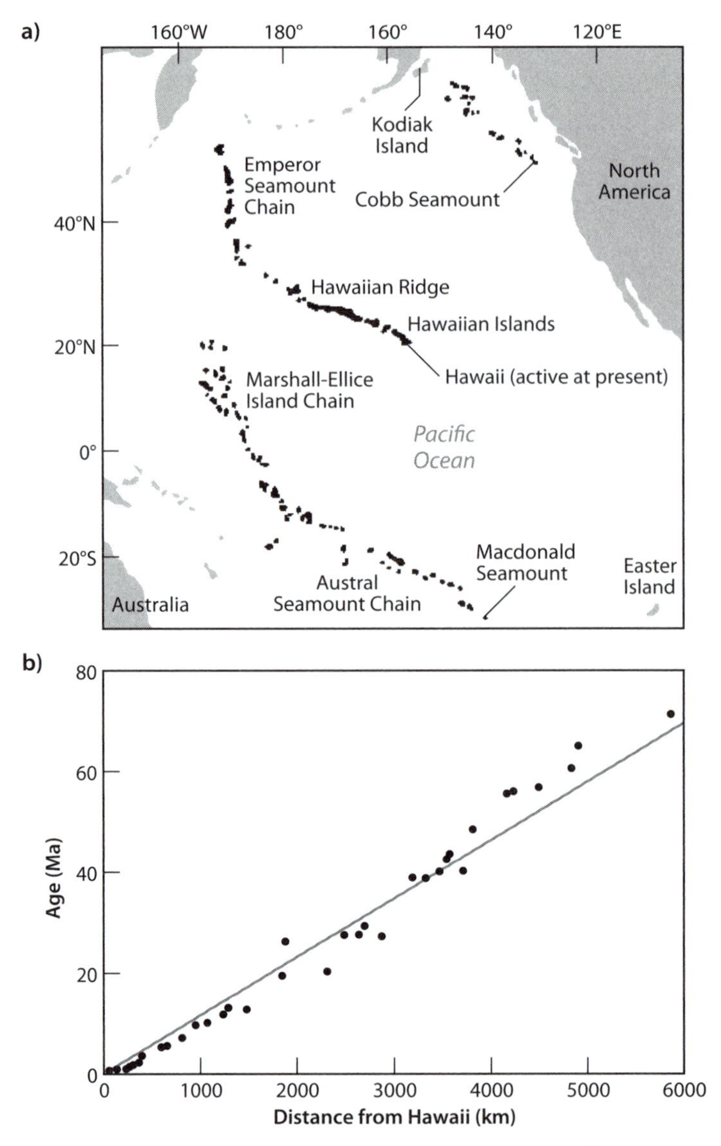

Fig. 11-12: (a) Map of the Pacific Ocean showing the linear chain of islands and seamounts associated with Hawaii. Active volcanoes on the island of Hawaii form the currently active end of the chain, where island growth is taking place today. Older islands were formed in the past above the same spot but have been carried away by motion of the Pacific plate. As they age, the islands are eroded and subside, ultimately leading to completely submerged structures that make up the Emperor Seamount Chain. (b) Graph showing how the ages of Hawaiian islands and seamounts vary regularly with age. The slope corresponds with the spreading rate of the Pacific plate. (Modified after K. C. Condie, *Mantle Plumes and Their Record in the Earth History* (Cambridge: Cambridge University Press, 2001))

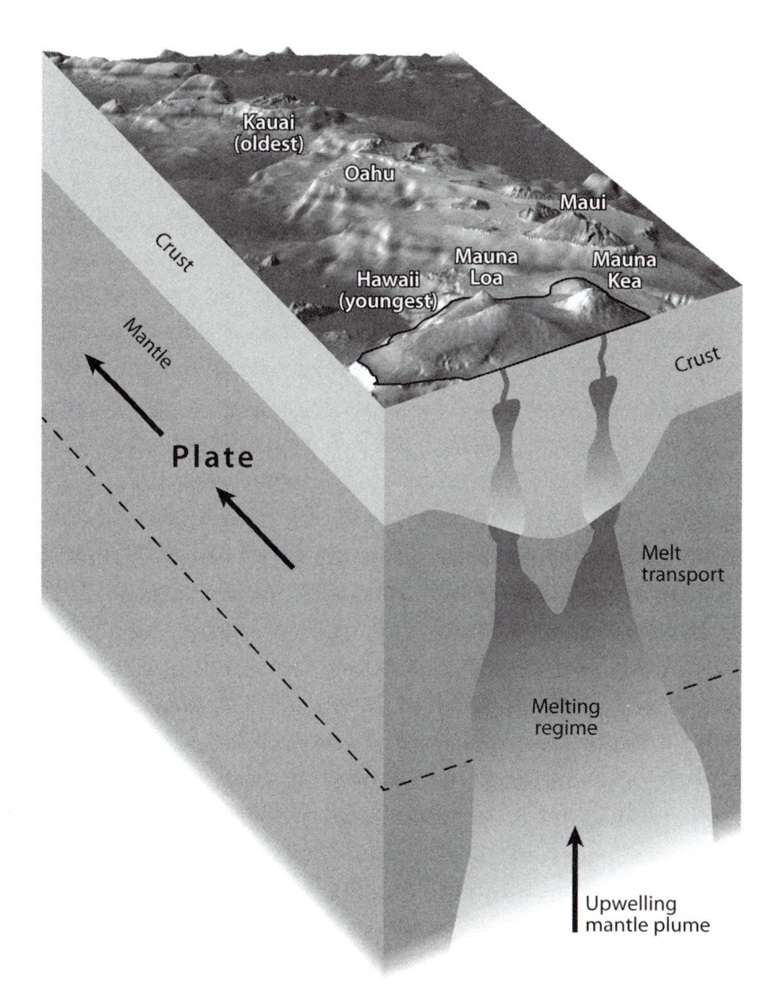

Fig. 11-13: Illustration of the model for the age variation in the Hawaiian islands and seamounts, where there is a fixed hot spot in the upper mantle, likely fed by a stationary mantle plume. As the Pacific plate migrates over the hot spot, islands are formed and then carried away by the movement of the plate. See color plate 17.

plate moves across a fixed point in the upper mantle, the spot on the plate currently over the fixed hot spot has an active volcano. The next volcano down the line was recently at the hot spot, but has been moved away from it by plate motion. Over time, this creates an island chain of progressively older volcanoes; only the youngest volcano directly above the hot spot is active. The "bend" in the Hawaiian chain possibly indicates a major change in plate motion that occurred at 43 Ma, when a change in the configuration of subduction zones led to an abrupt change in direction of the movement of the plate. The young volcanoes rise above sea level because the excess volcanism creates thick crust. Then they subside as the plate ages. The hot spots are *mantle plumes* rising beneath the plate at a single point, akin to the pattern of upwelling shown in Figure 11-5b rather than 11-5a. Active mantle convection consists of rising jets of hot material that occur independent of spreading center location and are usually far from downwelling limbs of cold material associated with subduction.

The mantle plume hypothesis received additional evidence from high temperatures of plume-derived magmas and the close association of "plume heads and tails" that linked dynamic understanding of plumes to evidence from Earth's surface. We now explore this in greater detail.

We know that melting must occur by pressure release at hot spots, and this requires upwelling of solid mantle. For plumes the upwelling is not produced by the plate motion. Instead, the plate serves as a cold cap that actually can inhibit melting. The only way to produce plume volcanism in the middle of a plate is through active upwelling, that is, a plume of buoyant material rising through the mantle. Since the physical properties of the mantle do not vary much, the main creator of buoyancy is high temperatures, and hence ocean islands are believed to be created by hot mantle plumes that are fixed in position in the mantle.

If this is true, then mantle temperatures of plumes must be hotter than their surroundings. In this case, the pressures and temperatures of hot spot magmas should be greater than ambient mantle, as reflected, for example, by normal ocean ridges. Both predictions correspond with the detailed compositions observed at hot spot volcanoes and also explain the excess crustal thickness that always occurs where plumes reach the surface.

The fixed positions of hot spots also suggest they are derived from a deep and hot thermal boundary layer. The upper mantle is continually being moved about and stirred by plate motion, and contains an important component of horizontal movement. If plumes were generated from within this convective flow, they should be transported by the convective flow and move across the surface. Instead, the hot plumes at fixed position imply that the plumes come from much deeper levels, where their position is independent of upper mantle convection. This can occur if plumes arise from hot boundary layers, as seen in Figure

11-5b. Just as in a lava lamp, where material is heated from below, it gradually becomes buoyant and rises as a plume to the top of the lamp, mantle heated from below can produce plumes that would rapidly ascend to the surface. The existence of the fixed mantle plumes suggests a hot boundary somewhere deep in Earth's interior.

Great debate has taken place concerning whether this boundary layer exists at intermediate depths in the mantle or occurs at the core/mantle boundary. Recent high-resolution seismic imaging suggests that at least some of the mantle plumes originate in the deepest portions of the mantle and therefore may result from heating at the core/mantle boundary. Thermal modeling of the temperatures of the core suggests it is much hotter than the lower mantle. With its low viscosity from its liquid state, the outer core must convect vigorously, removing any cooled material that forms at the core/mantle boundary and maintaining a very thin and extreme thermal boundary layer with the mantle above. The hot thermal boundary layer beneath the mantle would be a natural, perhaps inevitable, source of rising convective plumes.

Modeling of plumes from a hot thermal boundary layer shows that they begin with a "plume head," a massive rising hot ball of material that is followed by a "plume tail," which is a narrow rising jet following the same track (Fig. 11-14). The plume heads would lead to an initial massive volcanic episode at plume initiation, followed by a long period of smaller amounts of volcanism from the same spot.

These features predicted from the hypothesis of hot plumes from a deep boundary layer correspond with the plume tracks that begin with a massive volcanic outpouring and continue for tens of millions of years of hot material of lesser volume. Examination of many of the age-progressive chains of hot spot volcanism shows that they begin with flood basalt if it occurs on land, or an oceanic plateau if it occurs in the ocean. These are massive outpourings of lava over very short periods of time. Several of these are apparent in the South Atlantic and Indian Oceans (Fig. 11-15). For example, the Deccan flood basalt province in India is the manifestation of a plume head and is contiguous with a long ridge of progressively younger basalts heading to the south (Fig. 11-15). In North America, the Columbia River Basalts appear to be the earlier flood basalts associated with the hot spot that is currently beneath Yellowstone National Park. The age of the flood basalts corresponds with the age progression measured along the track of the hot spot. Many hot

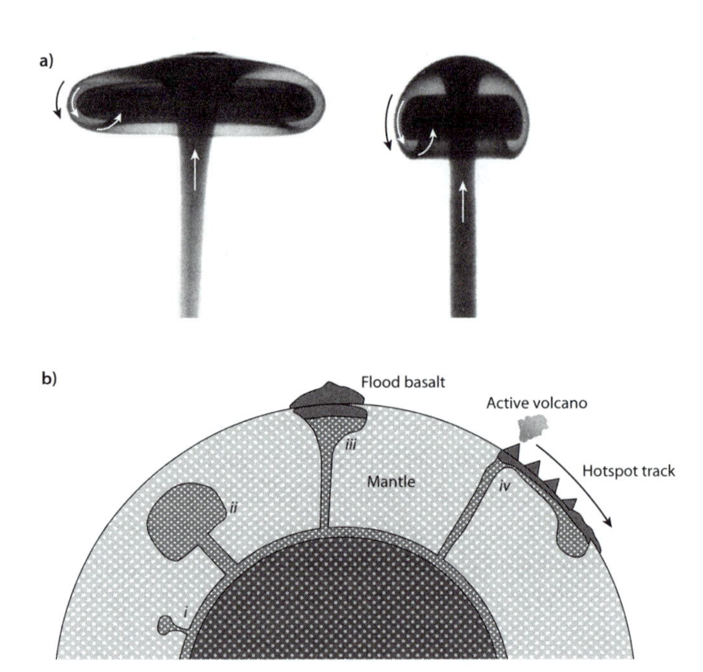

Fig. 11-14: (a) Model of plume head and tail based on work by Ross Griffiths, Ian Campbell and others. The first arrival of the plume at the surface is associated with a large plume head that can give rise to a massive volcanic outpouring. Subsequent volcanism would come from the "tail" of the plume, producing steady volcanism but of lesser volumes. (After Griffiths and Campbell, *Earth Planet. Sci. Lett.* 99 (1990):79–93; http://www.mantleplumes.org/WebDocuments/Campbell_Elements.pdf) (b) Cartoon of the manifestation of plume head and tail in Earth's interior: (*i*) plume forms at the core mantle boundary; (*ii*) plume head expands because of rapid transport within the conduit; (*iii*) plume head reaches surface to create a flood basalt province; (*iv*) after the flood basalt, plume tail creates a hot-spot track of smaller volcanoes as the plate at the surface displaces over the plume. (Modified from Humphreys and Schmandt, *Physics Today* 64 (2011), no. 8: 34).

spot tracks appear to have a flood basalt initiation, followed by tens of millions of years of intraplate volcanism along a well-defined spatial trajectory. (Hawaii is an exception in having no flood basalt province, possibly because its distal end is being subducted). The topographic expression at the surface is almost like a plume "head and tail" laid on its side on Earth's surface (Fig. 11-15).

There are many hot spots on Earth. Those associated with intraplate volcanism are concentrated in two large regions approximately one

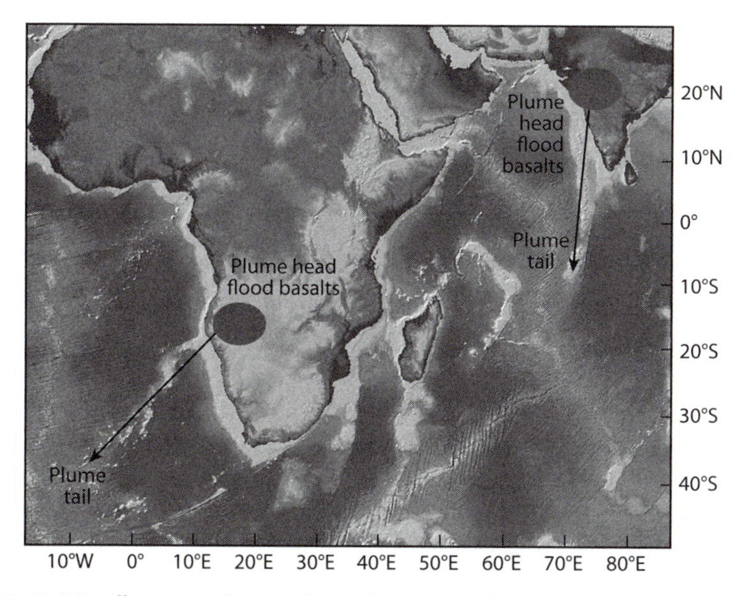

Fig. 11-15: Map illustrating the actual manifestation at the surface of a plume head and tail. The plume head would lead to a *flood basalt province* of massive volcanism. As the plume remained stationery, movement of the plate over the surface would create a long ridge of volcanism with a steady progression of ages. See color plate 18. (Map modified from image created by GeoMapApp; www.geomapapp.org)

hemisphere apart from each other. One is the central Pacific, where there are many ocean islands such as Tahiti, Samoa, Mauritius, and Hawaii. The second is in the vicinity of Africa. The Pacific is generally surrounded by subduction zones. The African plate is surrounded by ridges, as we have seen in Figure 11-6. Characteristic of both regions where hot spots are concentrated is that they are far from plate boundaries—they occur in the middle of large plates and are probably associated with a general component of plume-driven mantle upwelling in these regions.

These results suggest that mantle convection on a real planet such as Earth is far more complicated and interesting than tank experiments and calculations. Flow in the upper mantle is driven by the movements of the cold boundary layer (the plates) at the surface. The component of hot, active rising flow is manifested by plumes rising from hot thermal boundary layers at depth. Of course, these two aspects of convection interact with one another, creating a complex flow and temperature field for the mantle as a whole that is only beginning to be known.

The ocean ridges provide a potential window into lateral temperature variations in the mantle. As the ocean ridges traverse the surface, they passively sample the mantle underneath. The temperature of the mantle controls the variations in depth and thickness of the ocean crust, creating observable features on the seafloor that relate to conditions in the mantle. Ocean ridges provide a potential "window" into the temperature structure of the mantle, which should correspond with the convective structure of the mantle. To understand how this works, we need to examine more carefully the melting processes that take place to form ocean crust at spreading centers.

Formation of the Ocean Crust at Spreading Centers

We learned in Chapter 7 how partial melting of the mantle by pressure release (rather than temperature increase) is the fundamental mechanism by which melting occurs at ocean ridges. As we saw in Figure 11-8, the spreading of the plate creates an inevitable upward movement of the mantle underneath. This movement can be related to pressure release melting. Figure 11-16 shows how the path of mantle flow beneath the ridge creates a *melting regime* that is more or less triangular in shape. The amount of melting increases from 0% at the base of the triangle, where the solidus is crossed, to a maximum extent of melting for the central column of melting that ascends to just beneath the ocean crust. If melting increases linearly with decreasing depth, the mean extent of melting is half the maximum.

Since the ocean crust is formed from the melt extracted from the mantle, we can relate the thickness of crust to the volume of melt extracted from the melting regime. This is most simply visualized by considered the vertical column through the mantle and crust that is the end product of each increment of spreading. Each level in the residual mantle column comes from a different final depth in the melting regime. Experimental data show that the mantle melts about 1% for each 3 km of upwelling. So if the mantle ascends 60 km, it melts to a maximum of 20%. The mean extent of melting is then 10%, and 10% of 60 km is 6 km, which is the average thickness of ocean crust today.

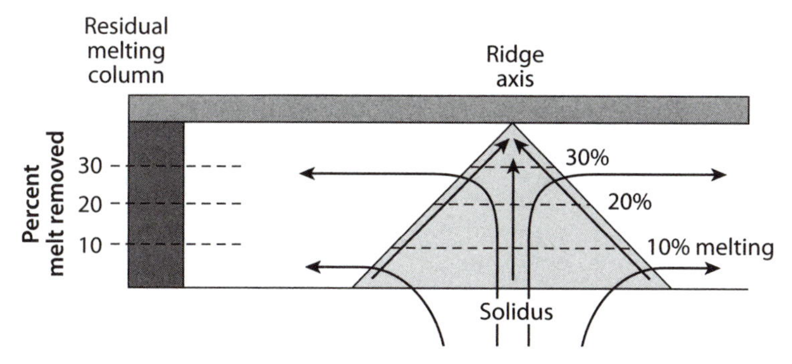

Fig. 11-16: Illustration of ridge melting regime showing the formation of a *residual mantle column* created by each increment of spreading. The crustal thickness is the total melt removed from the column, which is the mean percentage of melting times the column length. Most ocean crust forms by about 10% melting of a 60 km column, leading to 6 km of ocean crust.

Now let's consider what happens if the temperature of the mantle varies, as it must as the ridges journey across the upper mantle convective system (see Fig. 11-17). At higher temperatures, the mantle solidus would be crossed at greater depth beneath the ridge, and the size of the melting regime and length of the residual mantle column both increase. If melting were to begin at 90 km, then the maximum extent of melting would be 30% (90 km/3), the mean extent of melting 15%, and the crustal thickness would be 15% of 90, or 13.5 km. If on the other hand melting were to begin at 30 km, then the maximum extent of melting would be 10%, the mean extent of melting only 5%, and only 1.5 km of crust would be produced. These changes in crustal thickness have direct consequences for the depths of ocean ridges. The ocean crust is buoyant relative to the mantle and is subject to the principles of isostatic compensation. Ridges with thick crust float higher than thin crust, and therefore shallow ocean ridges should be associated with thicker crust and with higher temperatures in the underlying mantle.

The conclusion is that changes in the depth at which melting begins, which are a direct response to the temperature of the underlying mantle, control the thickness of the ocean crust and the depth of the ocean ridge. Hot mantle melts more and creates thicker crust and shallower ridges. The temperature differences can be quantified by knowing that the temperature at which melting begins increases by about 4°C for every kilometer of depth. Changing the initial depth of melting from 100 km

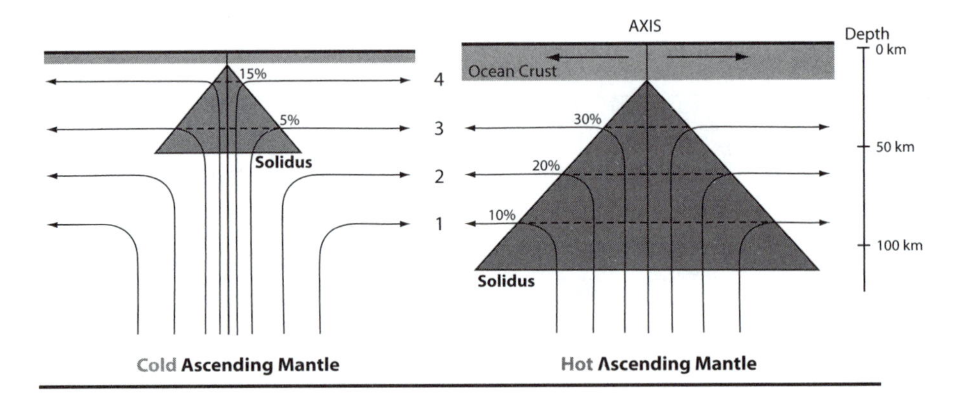

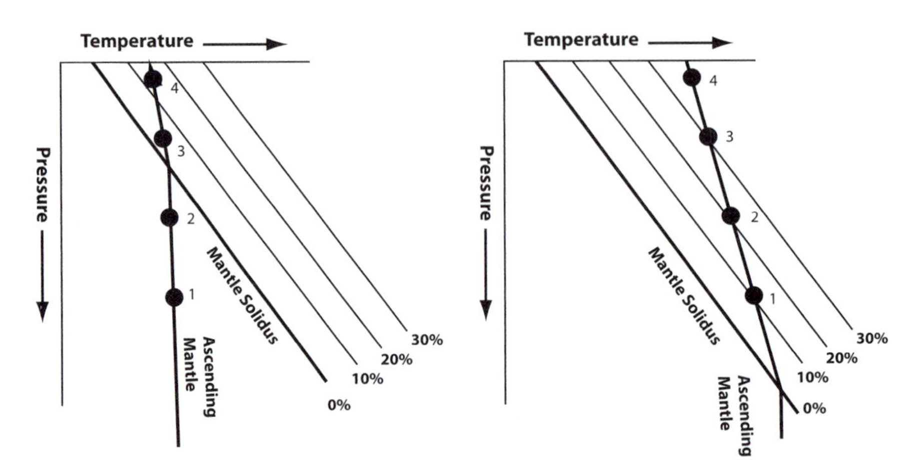

Fig. 11-17: Illustration of how variations in the temperature of the mantle beneath the ridge influence the depth of melting, the size of the melting regime, the mean extent of melting, and the crust thickness. Hot mantle melts more, leading to thicker crust, which has shallower depths below sea level in response to isostatic compensation.

to 60 km, for example, would reflect a temperature difference of about one hundred degrees.

There are also chemical consequences of varying the extent of melting of the mantle. As we learned in our investigation of simple phase diagrams, liquid compositions change progressively with the proportions of melt produced. Therefore we would expect ocean crust derived by larger extents of melting to differ in composition from crust created by smaller extents of melting. One of the simplest chemical consequences relates to the magmaphile elements—those elements that are strongly concentrated into the liquid fraction no matter what proportion of liquid is present. The concentrations of these elements are inversely proportional to the fraction of liquid. If all of the sodium, for example, goes into the liquid, and the liquid forms only 1% of the total, then the sodium would be one hundred times enriched in the liquid relative to the whole. If the liquid proportion increases to 2%, then the sodium would be fifty times enriched, and at 10% melting, the sodium would be only ten times enriched. The same numbers of Na atoms are diluted by the increased melt volume as the percentage of melt increases. In practice, not all of the sodium goes into the liquid, but most of it does. For this reason, the same mantle composition that melts to greater extents owing to higher mantle temperatures should have lower Na contents that go along with the thicker crust and shallower depths.

This line of reasoning shows that the depths of ocean ridges, the thickness of the ocean crust, and the chemical composition of the ocean crust all should be related to one another. This relationship can be seen in Figure 11-18, where the sodium concentration in the ocean crust relates simply to the depth of the ocean ridge, a result that is a natural consequence of variations in mantle temperature.

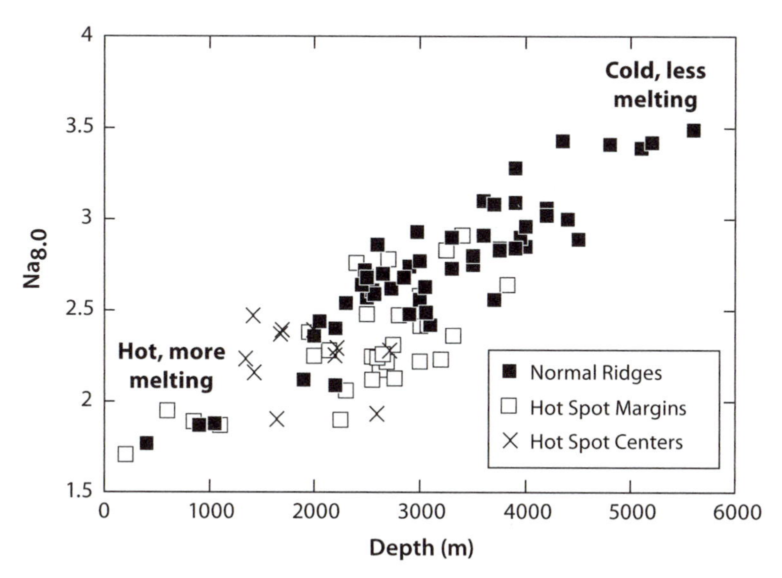

Fig. 11-18: Variations in the composition of the ocean crust vary with the regional depth of the ridge axis. Each point represents an average crustal composition over distances of ~100 km. Low Na contents and shallow depths are both consequences of greater extents of melting of the mantle owing to higher mantle temperatures.

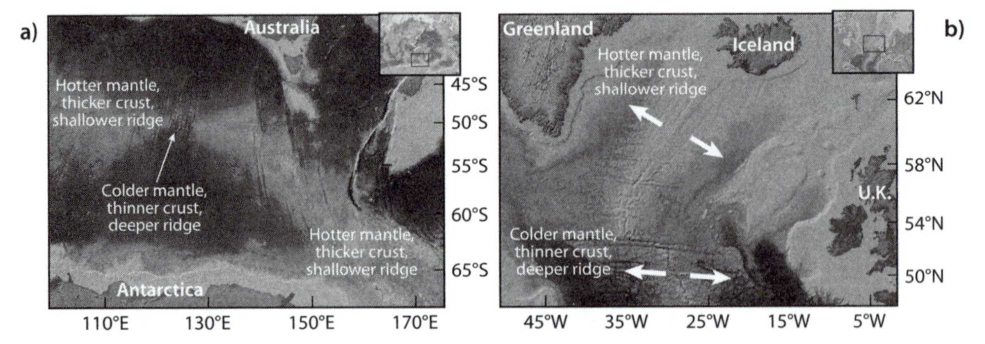

Fig. 11-19: (a) Map of the Southeast Indian Ridge between Australia and Antarctica, a region called the Australian/Antarctic Discordance (AAD). Light colors along the ridge axis midway between Australia and Antarctica are shallow and indicate a ridge created by hotter mantle, and dark colors are deeper, where the crust is thinner, produced by mantle at a lower temperature. The AAD is a cold region that separates distinct chemical provinces associated with the mantle beneath the Indian and Pacific basins (b) Map of the Iceland region. Iceland is a mantle plume where temperatures are inferred to be the highest of the entire ocean ridge system, leading to shallow depths and thick crust. Temperature apparently decreases both north and south of Iceland, leading to greater depths and thinner crust. See color plate 20. (Background maps from GeoMapApp; www.geomapapp.org)

While relating the depth of the ocean ridge to the temperature of the mantle requires a long chain of reasoning, the inferences are quite straightforward—the depth of the ocean ridge reflects the temperature structure of the underlying mantle. A global map of the ocean ridge system then gives a snapshot into the temperature of the mantle underneath! The ridges that are shallowest are those associated with hot spots that occur near ocean ridges, providing additional evidence that hot spots are indeed hot. The most obvious example is Iceland, where temperatures are high enough that the crust is five times thicker than normal ocean crust, leading to the creation of an island along a spreading ridge.

Temperature variations inferred from ridge depths then provide us with additional information about the pattern of mantle convection. The shallow ridges associated with hot spots such as Iceland are consistent with hot spots derived from mantle plumes from a thermal boundary layer in the deep mantle (Fig. 11-19b). The deepest ocean ridges generally occur at the margins of the great ocean basins—e.g., the separation between the Indian and Pacific oceans south of Australia and the equatorial Atlantic. The Australian-Antarctic Discordance (Fig. 11-19a) is a boundary between Indian and Pacific mantle that is consistent with cooler mantle temperatures in that region. In general, ridges show an undulation over distances of 5,000–10,000 km of highs and lows that must reflect the overall mantle convective circulation, with upflow zones associated with hot spots and return flow associated with the edges of ocean basins far from hot spots. The large variation in the depths of ocean ridges confirms that ridges migrate over the Earth's surface and simply respond to the mantle temperature above which the spreading occurs.

Summary

The movements and surface topography of the oceans are closely linked to processes occurring in Earth's interior. Isostasy and plate movement show that the mantle must flow. The high Rayleigh number of the mantle shows that active convection is inevitable. Mantle convection is the essential link between the thermal energy of the interior and the external functioning of the planet. Exterior movements of the plates cause circulation of the mantle and influence its flow. Other forms of mantle convection, represented by plumes, also transport material and energy from the core and mantle to the surface and influence the depths of ocean ridges and the detailed position of plate boundaries. The plumes are most abundant far from subduction zones, suggesting feedback between the different modes of convection of the mantle. Mantle convection links a hot thermal boundary at the core/mantle boundary to the cold thermal boundary between the mantle and the surface. The core-mantle and mantle-plate connections produced by interior modification in Earth's early history exchange and transfer energy that keeps the planet in motion. We shall understand from subsequent chapters how this continual circulation of the solid Earth and core is an essential aspect of Earth's habitability.

Supplementary Readings

Goeffrey F. Davies. 2001. *Mantle Convection for Geologists*. New York: Cambridge University Press.

Gerald Schubert, Donald L. Turcotte, and Peter Olson. 2001. *Mantle Convection in the Earth and Planets*, 2 vols. Cambridge: Cambridge University Press.

Fig. 12-0: Pictures of black smoking chimneys on the East Pacific Rise. The mineral-rich fluids are clear at depth in the crust and precipitate sulfides and oxides upon encountering seawater, creating the "smoke," which is not carbon! Exit temperatures of the fluids are as high as 400°C. See color plate 23. (Photograph courtesy of K. H. Rubin)

CHAPTER 12

Linking the Layers

Solid Earth, Liquid Ocean, and Gaseous Atmosphere

The physical description of plate tectonics is only the beginning of an understanding of its importance for Earth. The chemical circulation that takes place during that process, the plate tectonic geochemical cycle, is central to Earth's habitability. It is an important control on the steady-state composition of seawater, produces the oceanic and continental crust, transfers the elements necessary for life to Earth's surface, and lowers the viscosity of the mantle to permit active convection.

The cycle begins at ocean ridges where magma is transferred from the mantle to create the ocean crust. At the spreading center magma at 1,200°C is in close proximity to the ocean at a few degrees C. Seawater circulating through cracks in the crust becomes strongly heated, leading to intense hydrothermal activity that manifests as fields of black smokers, high-temperature chimneys that can reach heights of ten-story buildings and belch fluids at temperatures as hot as 400°C. These reactive fluids modify the igneous rocks of the ocean crust to create new minerals that contain H_2O *and* CO_2 *as solid constituents. These minerals persist as the plate ages, and the bulk composition of the solid crust, beginning with essentially no* H_2O*, ends up with an* H_2O *content of about 2% and substantial* CO_2*. As seawater modifies the compositions of the rocks, the composition of the ocean is changed by the exchange with the rocks. The fluids coming out of hydrothermal vents have a very different composition from seawater itself, balancing the input of river water in ways that maintain the steady-state composition of the ocean.*

As the plate journeys across the seafloor, diverse sediments accumulate, leading to a complex sandwich of materials rich in water and other elements that were not initially present at the ocean ridge. As the plate subducts, these materials undergo processes of mineralogical change called metamorphism *that release the mineral-bound* H_2O *and* CO_2*. The released water lowers the melting temperature of the slab and the mantle wedge overlying the slab, producing wet magmas that rise to become the hydrous, explosive magmas that build volcanoes at convergent margins. The hydrous magmas of the convergent margins differentiate upon cooling to compositions that are rich in silica and low in Mg and Fe, creating low-density materials that are added to the continents and permit the continental crust to float above sea level. The material remaining in the slab is recycled back into the mantle, where it contributes to the creation of a mantle of heterogeneous volatile and trace element composition. The recycling of* CO_2 *at convergent margins leads to long-term climate stability on Earth. Recycling of* H_2O *maintains the steady state volume of the oceans and lowers the viscosity of the mantle, enhancing convection. Further transfers to the surface come from the volcanism from mantle plumes that likely originate at the core/mantle boundary. The solid Earth geochemical cycle then relates the mantle, ocean ridges, oceans, volcanism at convergent margins, creation of the continental crust, mantle convection, and long-term climate stability. The cycles of the solid Earth and their interactions with surface reservoirs are essential for the long-term habitability of the planet.*

Introduction

The basic structure of plate tectonics is plate formation by mantle melting at ocean ridges, movement of the plates across surface, and their return to the mantle at subduction zones. Through this process, plate tectonics recycles material from the interior to the surface and back again in a grand cycle of the solid Earth. This process, however, is only the physical framework of a comprehensive system of exchanges, circulations and feedbacks among Earth's inner and outer layers, including mantle, crust, ocean, atmosphere, life, and even the core. This total process, which we will call the *plate tectonic geochemical cycle* (Fig. 12-1), is not simply the physical recirculation of rock through the plates. It is an essential aspect of planetary habitability that maintains climate stability during Earth's history, keeps seawater volume and composition near steady state, and permits the formation and persistence of the continental crust where life on land has had the chance to flourish. The cycle also provides ecosystems in the sunless deep sea, which as we shall learn may be important for the origin and maintenance of early life. Our aim in this chapter is to understand the overall chemical exchanges involved with plate tectonics, and the processes that contribute to the steady state characteristics far from equilibrium that characterize Earth's reservoirs. This aim requires an exploration in some detail of the processes of plate creation and destruction at spreading centers and convergent margins. In this chapter we explore this cycle from the creation of the plate at ocean ridges with the attendant interactions with the ocean, the transport of the plate across the seafloor, and the complex processing that occurs as the plate is returned to the mantle at convergent margins, where arc volcanism builds continents and releases volatiles to the atmosphere. These processes relate mantle convection and plate tectonics to the ocean and atmosphere.

The Global System of Ocean Ridges

In early presentations of plate tectonics, ocean ridges were drawn as a line on a map with transform offsets. Investigation of the ridge reveals it instead to be a dynamic system extending from mantle to microbe, linking mantle, crust, ocean, and life. The chemical fluxes associated with ocean ridges create new environments for life, permit volcanism thousands of kilometers away at convergent margins when the crust is recycled, and contribute to the steady-state chemical composition of seawater. All of these have significance for global habitability. The suitability of Earth for advanced life on continents depends on the hidden operation of the ocean ridge system, hidden from view and virtually unknown even to specialists only decades ago.

Simple volumetric considerations show that *most* of Earth's volcanism occurs at ocean ridges. The amount of new magma delivered from the

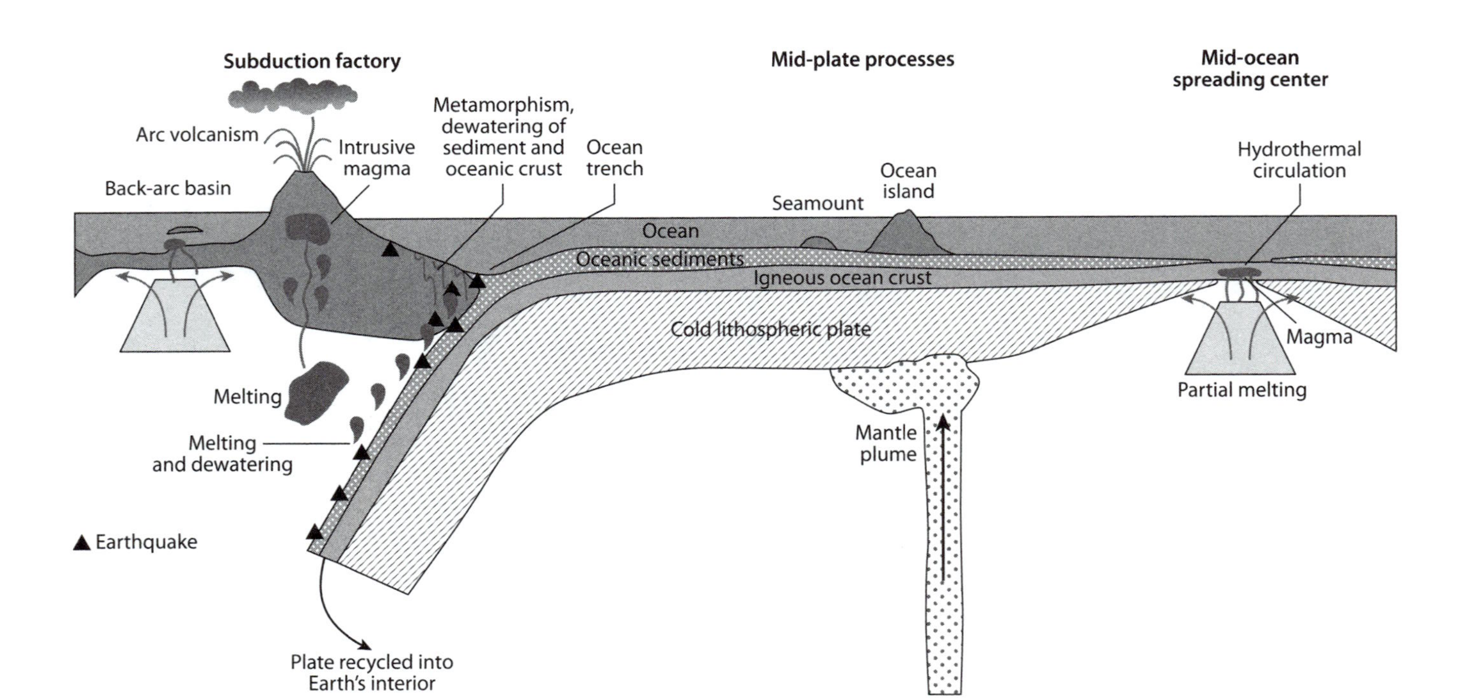

Fig. 12-1: Cartoon of the plate tectonic geochemical cycle discussed in this chapter. Mantle upwelling creates the crust at ocean ridges and drives hydrothermal circulation. The hydrothermal activity modifies seawater chemistry and creates hydrous minerals in the crust. Sediments accumulate as the crust ages. The crust may also be influenced by hot spots. The complex package of materials is subducted. Metamorphism of the crust releases water and causes melting and formation of island arcs and ultimately continental crust. The slab returns to the deep mantle and creates mantle heterogeneity, later sampled by volcanism. See color plate 19.

mantle to ocean ridges can be estimated from the product of crustal thickness (6 km) times mean spreading rate (about 5 cm/yr) times the ridge length of about 70,000 km. This amount is about 21 km^3 per year. Calculations for convergent margin volcanoes yield estimates of about 2–3 km^3/yr, only a tenth as much. Estimates for intraplate volcanism yield numbers similar to those of convergent margins. Hence, ocean ridge volcanoes are responsible for 80% of Earth's volcanic output.

Careful mapping of the seafloor in the last decades of the twentieth century gradually revealed that ocean ridge volcanoes are not the symmetrical cones that come to mind with the word "volcano" but are long linear features where magma rises to fill the crack created by the separation of the plates. The ridge is also "segmented" by transform faults that offset the volcanic crack, and therefore ridge volcanoes are not referred to by names of volcanoes but by names of segments, often associated with their bounding transform faults.

The detailed morphology of ocean ridge volcanoes depends on the spreading rate. At the fast spreading rates (>10 cm/yr) of the East Pacific Rise (Fig. 12-2), volcanoes are very long and narrow and do not rise very high off the seafloor. While a volcano like Mount Fuji might be a circular feature some 20 km across that rises 3,000 m above its surroundings, East Pacific Rise volcanoes can be 100 km long, 2–3 km wide, with relief of only a few hundred meters. Their shape can be understood from the fact that at fast spreading rates the hot mantle rises so close to the surface that the magma oozes up all along the opening crack of the spreading center, making a long linear feature—what Jeff Fox has described as "the wound that never heals." At a spreading rate of 10 cm/yr, the crust moves like a geological race car, 100 km per million years. Since the volcanoes are only a few kilometers wide, old lavas are rapidly spread away, so there is not the opportunity to build up a high undersea cone.

At slow-spreading ridges (< 4 cm/yr), the spreading center is still a long crack, but the hot mantle upwells slowly enough that the hottest temperatures are kept at bay. The lithosphere thickens rapidly close to the spreading axis and toward the transform faults where the ridge abuts older lithosphere. Shallow magma chambers are either ephemeral or absent. The lithosphere forms a kind of cold tent that causes deeper faulting and also causes mantle-derived magmas to be focused along the ridge axis (see Fig. 12-3). The deep faults lead to a prominent rift valley,

Fig. 12-2: Bathymetric map of the East Pacific Rise off the coast of South America from south of the Siqueiros transform to north of the Clipperton transform. Notice that near 9°N there is also an *overlapping spreading center*, which is a ridge offset smaller than a transform fault where two ridge volcanoes overshoot and overlap. In order to help visualize the topography, the bathymetric maps (*bathymetry* is the word used for undersea topography) of the seafloor are usually presented with different shades corresponding to different depths. Of course, these shades are not the colors of the actual seafloor but are simply a visual aid to see shapes. The long linear features between offsets are the characteristic form of ocean ridge volcanoes at fast spreading rates (>8cm/yr). The vertical scale of the figure is ~200 km. The total depth change across-axis is a fewer hundreds of meters. See color plate 21. (Image courtesy of Stacey Tighe, University of Rhode Island)

and the focused magmatism leads to large depth changes along the ridge—shallower toward the centers of segments and deepening toward the transform faults.

HYDROTHERMAL CIRCULATION AT SPREADING CENTERS

Ridge volcanoes also differ from those on land in that all of them are covered by a 2–3 km blanket of ocean. This makes the ridge an interface

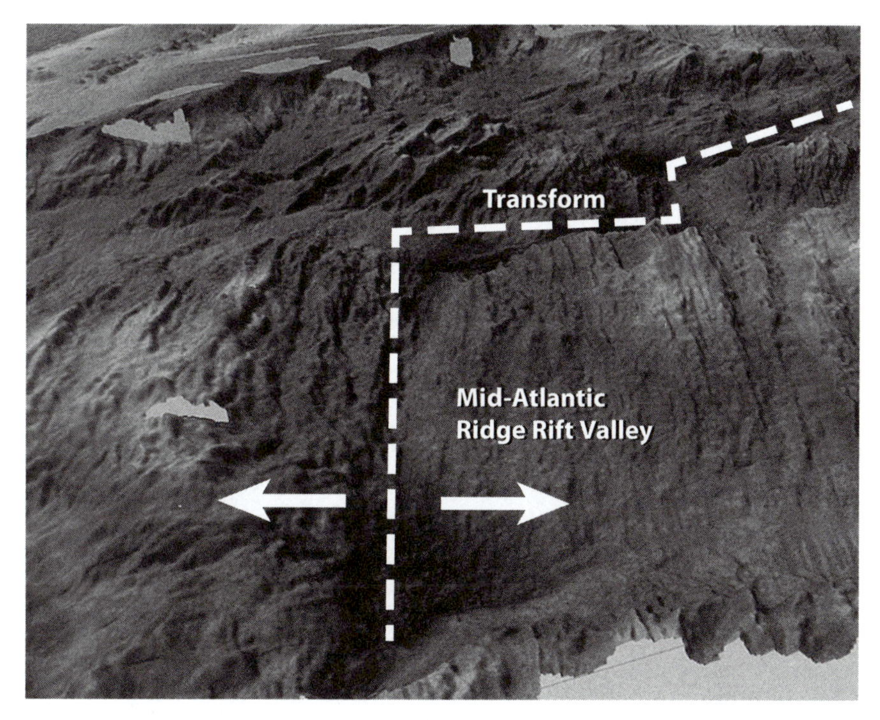

Fig. 12-3: Bathymetric map of a portion of the northern Mid-Atlantic ridge, looking north. The region, near 36°N, is known as the FAMOUS (French American Undersea Study) area. The segment is about 55 km in length, bounded by transform offsets on the north and south. Notice that the spreading axis is located within a rift valley with steep walls. The magnitude of bathymetric variations is much greater than the fast-spreading ridge, with the rift valley being 1,000 m deeper than the walls of the rift mountains that bound it . The gray scale is very different from Fig. 12-2, since the amount of relief is about five times greater. See color plate 22. (Figure courtesy of Javier Escartin, Institut de Physique du Globe, Paris, using data from Cannat et al., *Earth Planet. Sci. Lett.* 173 (2001):257–69, and Escartin et al., *J. Geophys. Res.* 106 (2001):21,719–35)

between cold seawater near 0°C and mantle-derived magma at 1,200°C. The eruptions and the faults created by spreading create pathways for fluids to pass through the crust and come in very close contact with molten magma and very hot rock. The interaction leads to vigorous hydrothermal systems with chemical exchanges that modify both seawater composition and the composition of the ocean crust.

The vigor of ridge hydrothermal systems caught most marine geologists by surprise. Exploring the northern portions of the East Pacific Rise using a submersible, spires were discovered with billowing clouds of black "smoke" coming out the top (see frontispiece). Arriving at the smoking "chimneys," the manipulator arm stuck a temperature sensor into the fluid flow, and the sensor went off scale and then appeared to fail. Upon return to the surface it was found that the temperature sensor had melted! When the temperatures of black smoker vents were finally able to be measured, the temperatures were found to be as hot as 400°C—far hotter than the 100°C of boiling water at Earth's surface.

How could such high temperatures be possible? The key is the influence of pressure on boiling points. If we camp at high altitudes, the pressure is lower and therefore water boils at a lower temperature and cooking takes longer. Just the opposite occurs when the pressure is higher—the boiling point goes up. This is apparent from the pressure-temperature phase diagram for H_2O seen in Figure 4-3. The weight of water causes the pressure to increase greatly with depth so water can be heated to much higher temperatures before boiling. The molten magma rising from the mantle has a temperature of 1,200°C and heats seawater all the way to its boiling point. Therefore, rather than being a cold blanket that suppresses hydrothermal activity, the high pressures of the ocean allow far higher temperatures and more vigorous activity than is possible on land.

The hot, saline water is chemically very reactive. Magma coming from the mantle contains almost no water, and there is disequilibrium between the rock and hot seawater. The subsequent chemical reactions cause both rock and water to change composition. Rocks that originally consisted of the minerals pyroxene, plagioclase, and olivine react to form hydrous minerals such as chlorite, amphibole, and epidote, discussed further below. The water changes its composition at the same time, losing virtually all of its Mg, some Na, and dissolving Fe, Mn, Cu, Zn, Pb, and other metals as its oxidized sulfate transforms to reduced sulfide. At high temperatures the density of this mineral-laden solution decreases, and it becomes buoyant and rises turbulently to the surface. As it rises it deposits sulfide minerals along veins in the crust and then flows out of surficial vents at high velocities. Here it contacts deep seawater with a temperature of only a few degrees C. The rapid cooling of the reduced and acidic hydrothermal fluid causes the metals dissolved in the fluid to precipitate, creating the dramatic "black smoke" and leading to the construction of chimneys that rise from the seafloor.

While the study of ocean ridge hydrothermal systems is still in its infancy, certain generalizations are emerging about their type and distribution. The ocean ridge hydrothermal systems relate to magmatism and faulting. At fast spreading rates, a shallow magma chamber 2–3 km below the surface exists in the ocean crust. Above the magma chamber are shallow faults that serve as pathways for fluid motion. Seawater penetrates the cracks, is heated by the underlying magmatic heat, and buoyancy drives it forcefully upward to the surface. The hydrothermal system is an active convective system driven by a shallow layer with very large thermal contrast at its base. The convection is by way of flow of water through cracks and porous rock, rather than convection of the rock itself. The Rayleigh number for the fluid is very high, because the viscosity of water is low and the temperature contrast large. Since the spreading rate is fast, magma supply is large and frequent eruptions reset the hydrothermal system. Hydrothermal activity manifests as a rapidly changing assembly of small groups of "black smokers" distributed quite extensively along the ridge segment (Fig. 12-4, right panel).

At slow-spreading ridges, the magma supply is smaller, lithosphere thicker, and faulting deeper (see Fig. 12-4, left panel). Shallow magma chambers, when they exist, are intermittent. Locations of faults and access to the heat source that they provide are important controls on the locations of hydrothermal systems. These tectonic contrasts influence the hydrothermal system and create a variety of hydrothermal environments. Where water penetrates more deeply along faults, heat is extracted from large volumes of hot rock rather than the boundary layer of a magma chamber. This leads to large systems that may be separated from the zone of active volcanism and persist for relatively long periods of time. In periods when magma approaches the surface to form a shallow magma system, usually at the center of a ridge segment, the systems that develop are more similar to fast-spreading ridges. Slow-spreading ridges can also host a very different form of hydrothermal system, at lower temperatures, that arises when hot mantle peridotite is raised by faults to be close to the surface, and is altered by its interaction with seawater. This type of system can release reduced species of gases such as CH_4 and H_2 that we shall see may be important for the origin of life.

The influence of hydrothermal systems does not stop at the seafloor. The rising plumes from the hydrothermal vents contain reactive particles

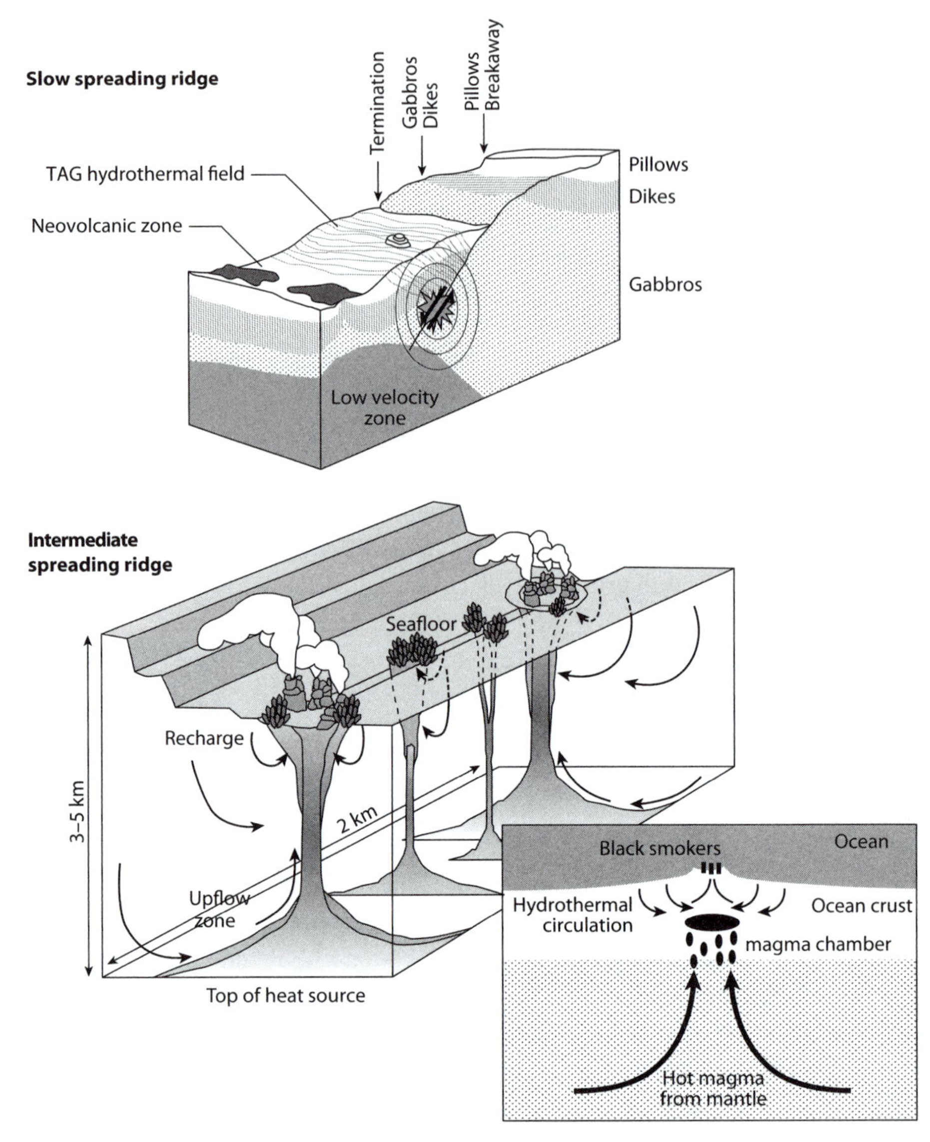

Fig. 12-4: Cartoons illustrating hydrothermal flow at fast- and slow-spreading ridges. At intermediate and fast spreading ridges (*bottom panel*), a shallow magma chamber is common, and hydrothermal circulation is near the center of the ridge above the chamber, supported by shallow faults. (Figure courtesy of D. Kelly and J. Delaney) At slow-spreading ridges (*top panel*), such as the TAG area near 20°N on the Mid-Atlantic Ridge, magma reservoirs are often absent and hydrothermal circulation is often associated with deep faults slightly off axis.

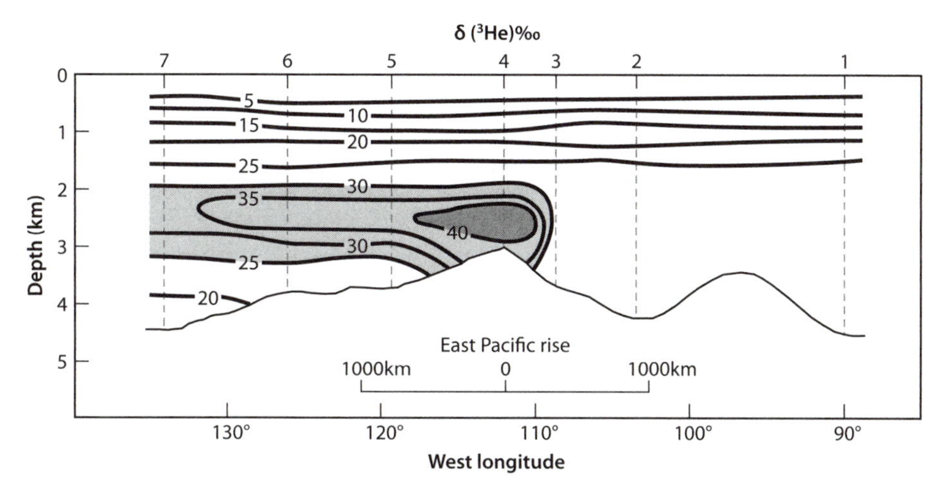

Fig. 12-5: Map of the distribution of ^{3}He in waters of the Pacific Ocean. The extensive plume with significant ^{3}He concentrations shows the extent of hydrothermal fluid distribution from the East Pacific Rise spanning the Pacific basin. (Figure modified from J. Lupton and H. Craig, *Science,* 1981 v. 214, no. 4516, 13–18)

that rise above the ocean ridge until they reach a level of neutral buoyancy in the water column, similar to the behavior of plumes coming from utility smokestacks. These plumes spread far and wide in the deep sea. This became evident from study of the distribution of the isotope ^{3}He. ^{3}He is present in gases released from ocean ridge basalts and incorporated in the hydrothermal plumes but is steadily lost from seawater as it diffuses into the atmosphere and eventually out to space. Because ^{3}He is created only during nucleosynthesis, and not as a result of radioactive decay, and all old helium has escaped from the top of the atmosphere, there is no ^{3}He naturally in the oceans. That allows ^{3}He to be a reliable tracer of the amount and extent of hydrothermal fluids in the oceans. The trace of the plumes can be mapped by looking at the distribution of ^{3}He dissolved in the water. Figure 12-5 shows that the ^{3}He plume from the East Pacific Rise extends far across the Pacific Ocean.

The most surprising aspect of the deep-sea vent systems was the discovery of lush and diverse ecosystems of animals surrounding the hydrothermal vents. Abundant life was thought to require sunlight, and most of the deep sea was very limited in biological productivity, relying on scavenging of organic materials falling form the surface. Some early photographs showed pictures of what looked like clams and mussels present more than 2 km beneath the surface. Clams and mussels are ordinarily found in the tidal environment! While many of the animals bore superficial resemblance to those we are familiar with, most were unique new species. Some of the animals were dissimilar from any species found on land. Most colorful were the tall white tubeworms with bright red tops that could rapidly extend and retract from their white stalks (Fig. 12-6). The density of life in the immediate vicinity of the vents is staggering. Living in an environment considered to be the most barren part of Earth, isolated from the life-giving sun, the deep-sea vent communities can have densities of life that are locally as rich as tropical rainforests, though of course far smaller in extent.

The base of the food chain for these communities is sulfur-oxidizing bacteria. The sulfide in the hydrothermal fluids is out of equilibrium with oxidized seawater, and this coupled with the high temperatures provides the conditions for the sulfur-oxidizing bacteria to flourish. These bacteria form the base of the food web of the vent ecosystems, and they have developed symbiotic relationships with many of the animals that frequent seafloor near the vents. Some of the bacteria have been found to thrive at temperatures higher than 100°C, something that was previously considered impossible. The ecosystems are robust and well adapted to an environment that on land would be toxic and uninhabitable. Organisms on land look to the sun as the ultimate source of their nourishment and well-being. If intelligent life at vents evolved, the sun would be a minor and distant inference, and their gods of nurturing and destruction would be the active volcanism that controls all aspects of their life cycles. The simple fact of spreading plates creates a linked system from mantle to microbe, where life is sustained by the vertical movement of energy and mass from the mantle to the exterior.

The revolutionary aspect of discovery of life at hydrothermal vents is that the base of the food chain is supported by volcanic energy from the planetary interior, rather than energy from the sun. That this was possible greatly expanded views of the potential habitats for life on other planets. Even distant from a star, or in the complete absence of starlight, life is possible.

Fig. 12-6: Picture of a vent community around a hydrothermal vent in the same region of the East Pacific Rise as Fig. 12-2. Some of the animals bear superficial resemblance to animals found in tidal zones. Others, such as the long thin "tubeworms" have a unique appearance. All have a metabolism very different from surface organisms, depending on symbiotic relationships with sulfur-oxidizing bacteria supported by the vent fluids. See also color plate 24. (Photograph courtesy of Cindy Lee Van Dover)

Ocean Ridges and Habitability

While ocean ridges are clearly central to plate tectonics, and support their own unique biosphere, are they also central to habitability? Yes, chemical processes during formation, transport and recycling of the ocean crust are integral aspects of a habitable Earth. Four aspects of their formation and recirculation are particularly important, none of them apparent from consideration of the tectonic aspects of plate movements alone:

> (1) Geochemical processes at ridges sustain the chemical composition of the oceans;
> (2) ridges store and transport water and other elements to the subduction zone that permits volcanism and continental growth to occur there;
> (3) ridges play an important role in the water and carbon cycles that provide long-term climate stability for Earth (discussed in Chapter 9);
> (4) ridges may have played an important role in the origin of life on Earth, with implications for the viability of life elsewhere in the galaxy (discussed in Chapter 13).

THE PUZZLE OF SEAWATER COMPOSITION

From the pre-plate tectonic perspective, the ocean mass balance appears to be rather straightforward: the ocean receives inputs from river water and wind-driven dust and creates outputs from evaporation of salts in isolated seas and the deposit of sediments to the seafloor. The evaporated water that rises to form rain is very pure. It first collects some elements by reacting with gases and dust in the atmosphere. Then, when it rains and runs over the rocks of the continents, it weathers the rocks and dissolves some of their mineral constituents, so that upon its return to the ocean it has a much higher mineral content. Table 12-1 compares the compositions of seawater, rainwater, river water, and hydrothermal fluids. While river water is far less mineral rich than seawater, it contains far higher quantities of various elements than does pure rainwater. And seawater has high concentrations of many elements, but far less than a saturated solution such as the Great Salt Lake.

Table 12-1
Compositions of Earth's water (concentrations in parts per million)

Element	Rain	Rivers*	Seawater	Hydro-thermal Fluid**	Hydro-thermal/ River	Hydro-thermal Flux/ River Flux
Ca	0.65	13.3	412	1200	90	0.0675
Mg	0.14	3.1	1290	0	0	0
Na	0.56	5.3	10770	—	—	—
K	0.11	1.5	380	975	650	0.4875
Sulfate	2.2	8.9	2688	28	3.15	0.00
Cl	0.57	6	19000	—	—	—
Si	0.3	4.5	2	504	112	0.08
Fe	0	0.03	0.002	168	5600	4.20
Mn	0	0.007	0.0002	41	5857	4.39
Li	0	0.002	0.18	5	2500	1.88
H_2S	0	0	0	255	Infinite	Infinite
Mg/Na	0.25	0.58	0.12	0.00	—	—
Ca/K	5.91	8.87	1.08	1.23	—	—
Si/K	2.73	3.00	0.01	0.52	—	—
Flux to ocean		$4 \cdot 10^{16}$ Kg/yr		$3 \cdot 10^{13}$Kg/yr	—	—

**R. Chester,* Marine Geochemistry *(Oxford: Blackwell Science, 2000); and H. Elderfield and A. Schultz,* Annu Rev. Earth Planet Sci. *24 (1996):191–224.*

***Elderfield and Schultz (1996); hydrothermal fluids have a considerable range in composition. Na and Cl in the hydrothermal fluid are similar to seawater.*

In this simple version of events, pure water is removed from the oceans by evaporation, and more mineral-rich river water is added to the oceans in an equal amount. Therefore the water cycle causes more and more chemicals to be added to the ocean, and the mineral content of seawater should steadily increase with time. The ratios of elements would be the same in river water and seawater. It would be a bit like having a tub full of water that you distill to take a shower while standing in the tub. The distilled water you shower with is always clean. At first the tub water is clean, but the dirt and soap would steadily accumulate in it as you took shower after shower, and ultimately the tub water would be so rich in soap and salt that it would become saturated.

The ocean, however, is not overly mineral rich and is not saturated with salt and other minerals. Mineral-saturated water, such as the Great Salt Lake in Utah, has much higher element concentrations than seawater. In fact, there is so little Na in seawater that the entire budget of Na would be added by rivers in only 47 million years. (This was actually one of the early methods that some geologists used to calculate Earth's age.) Somehow seawater is maintained at steady state far below saturation. Furthermore, ratios of many elements in seawater are vastly different from the river water that supplies the oceans (see Table 12-1). This requires active *sinks* that are removing elements as fast as they are added, creating a kinetic balance, or steady-state disequilibrium, that has been maintained over limited ranges over Earth's history. The steady-state composition of seawater below saturation requires a balance between sources and sinks.

The water cycle also does not fractionate radiogenic isotopic ratios. If oceans received all their inputs from the continents, then the radiogenic elements in seawater should have the same radiogenic isotopic ratios as the continental crust that is weathered by the rivers. The most abundant of such elements in seawater is our old friend $^{87}Sr/^{86}Sr$. The average $^{87}Sr/^{86}Sr$ ratio of continental crust is >0.712, while the seawater ratio is much lower, near 0.709. The Sr isotope evidence shows that continents cannot be the only *source* of material added to the oceans. Seawater requires some other process contributing to *both* sources and sinks!

One sink is life in the ocean. Organisms with siliceous shells remove Si and those with carbonate shells remove Ca, leading to low Si/K and Ca/K in the remaining water. But then why does seawater have low Mg/K as well? And life does not fractionate radiogenic isotope ratios and cannot explain the Sr isotope data.

The mystery process that balances seawater composition is the ocean ridge hydrothermal circulation. The disequilibrium encountered by seawater as it passes through hot rock causes some new minerals to form, some elements to be removed from seawater, and others to be added. The modified solution, which is the hydrothermal vent fluid, then encounters the ocean. Some elements are added to the ocean, some immediately precipitate in the chimneys, and some create reactive particles that scavenge other elements from the water column and remove them to the sediments. The water circulation through the crust removes a substantial proportion of the riverine input of Na, and it quantitatively removes the Mg from the seawater that circulates (note the zero concentration of Mg in the hydrothermal fluid), contributing to the low Mg/Na of seawater (~0.1) compared to river water (~0.6). And the $^{87}Sr/^{86}Sr$ composition of the ocean crust and of vent fluids is about 0.703. Mixing of Sr in this fluid with continentally derived Sr (0.712) leads to the intermediate Sr isotope composition of seawater (0.709).

While each individual hydrothermal vent is small, the total hydrothermal system including the thousands of vents along the ocean ridge is large. High temperature flow completely processes the volume of the ocean in tens of millions of years. Lower temperature flow off axis is far more extensive, and processes the ocean volume in a few hundred thousand years. Both of these fluxes are small, however, compared to river water, which supplies one ocean volume every 30,000–40,000 years. Since rivers are a much greater volume, how can the hydrothermal flux be so important? The answer lies in the very high concentrations of some elements in high-temperature vent fluids, as is apparent in Table 12-1. Some elements have concentrations a thousand or more times higher than river water and for many elements hydrothermal fluxes are as large or larger than the global river flux.

Two elements that are very concentrated in vent fluids are Fe and Mn, and yet these elements have essentially zero concentration in seawater. How is this possible? As we shall discuss at length in Chapter 17, Fe and Mn are quite soluble in water when they have a 2+ valence, and they are readily dissolved in hot, acidic, reduced hydrothermal fluids. When these elements encounter alkaline and oxidized seawater, they are oxidized to their 3+ form, and immediately precipitate in the hydrothermal plumes above the vent chimneys. These precipitated particles have very reactive surfaces and scavenge many other metals from the water before falling to the seafloor as sediment. The hydrothermal fluids are a large source that then produces a large sink. The sink includes other metals in the oceans, because the plumes from these chimneys spread out over large expanses of the ocean, and the particles encounter large volumes of seawater. The plumes have an ocean processing time of

4,000–8,000 years and are an important means of interaction between hydrothermal systems and the larger ocean.

It is one of the remarkable aspects of plate tectonics that geochemical consequences of volcanism include a pivotal partnership in the maintenance of seawater composition that is a central aspect of Earth's habitability.

ELEMENT TRANSPORT TO THE SUBDUCTION ZONE

The mantle rising to melt beneath the ocean ridge has very low water content, and the lavas making up the crust at ocean ridges (mid-ocean ridge basalts, referred to as MORB) are made up of minerals like plagioclase, olivine, and pyroxene that do not contain any water at all. Any other volatiles that were in the mantle were also removed from the mantle during melting, and from the magmas by degassing as they solidify. The new plate when it forms is essentially anhydrous and volatile free.

Hydrothermal interactions with seawater change the composition of the plate. Interaction with seawater at both high and low temperatures "alters" the rocks, transforming the original dry minerals to hydrous minerals such as amphibole and layer silicates. These minerals, akin to mica, are not damp to the touch but contain water as an essential part of their mineral structure. For example, in the amphibole mineral formula—$Ca_2(Mg,Fe)_5Si_8O_{22}(OH)_2$—and chlorite mineral formula— $(Mg,Al,Fe)_{12}(Si,Al)_8O_{20}(OH)_{16}$—the OH group reflects the structural water that is part of the solid mineral. Formation of these minerals changes the rock from shiny black basalt to a metamorphic rock, called *greenschist* or *amphibolite*, depending on the mineralogy, that contains several percent water locked in its minerals (Table 12-2). As the crust ages, the interactions with water continues, albeit at lower temperatures, and there is some hydration of the mantle beneath the crust, where olivine and orthopyroxene are modified to serpentine $Mg_3(Si_2O_5)(OH)_4$.

As the crust traverses the ocean basin on its way to the subduction zone, a rain of sediments gradually accumulates leading to typically 500–1000 m of sediment that form a part of the subducting plate. The sediments come from several sources—from weathering of continents both from rivers and from wind, from accumulated particles originally from hydrothermal vents that settle through the water column, and from accumulation of dead organisms. Most of the minerals in these sedi-

Table 12-2
Compositions of mantle, altered mantle, oceanic crust, altered oceanic crust, and global subducting sediment

Wt%	Primitive mantle[a]	Altered serpentinite[b]	Ocean Crust[c]	Altered ocean crust (side 801)[d]	GLOSS[e]
SiO_2	45.00	40.14	49.71	49.23	58.57
TiO_2	0.20	0.01	2.02	1.7	0.62
Al_2O_3	4.45	0.79	13.43	12.05	11.91
FeO	8.05	7.46	12.92	12.33	5.21
MnO	0.14	0.12	0.19	0.226	0.32
MgO	37.80	40.83	6.83	6.22	2.48
CaO	3.55	0.97	11.41	13.03	5.95
Na_2O	0.36	0.09	2.56	2.3	2.43
K_2O	0.03	0.00	0.14	0.62	2.04
P_2O_5	0.02	0.01	0.17	0.168	0.19
CO_2	<0.1	{ 8.61	~ 0.02	{ 6.31	3.01
H_2O	<0.01	—	0.20	—	7.29
ppm	—	—	—	—	—
Rb	0.6	14.56	1.46	13.7	57.2
U	.022	1.51	0.02	0.39	1.68

[a] *Chemical compositions of average primitive mantle (W. McDonough and S. Sun (*Chemical Geology *120 (1995) 223–253).*

[b] *Average altered serpentinite (harzburgite aver. (OM94) K. Hanghoj et al.,* J. Petrol. *51 (2010) 201–227).*

[c] *Average ocean crust (Gale, Langmuir, and Dalton, in press).*

[d] *Altered ocean crust (SUPER, K. Kelley and T. Plank,* Geochem. Geophys. Geosys. *4(6) (2003) 8910).*

[e] *GLOSS (Global Subducting Sediment composition (T. Plank and C. Langmuir,* Chemical Geology *145 (1998) 325–394).*

Note the very high volatile contents of altered materials and sediment.

ments have even higher proportions of water than the altered ocean crust. CO_2 is also added to the plate locked in solid form in carbonate minerals, both through deposition of carbonate sediments and precipitation of carbonates in veins in the ocean crust and mantle. An average composition of subducting sediment, GLOSS (global subducting sediment), is included in Table 12-2.

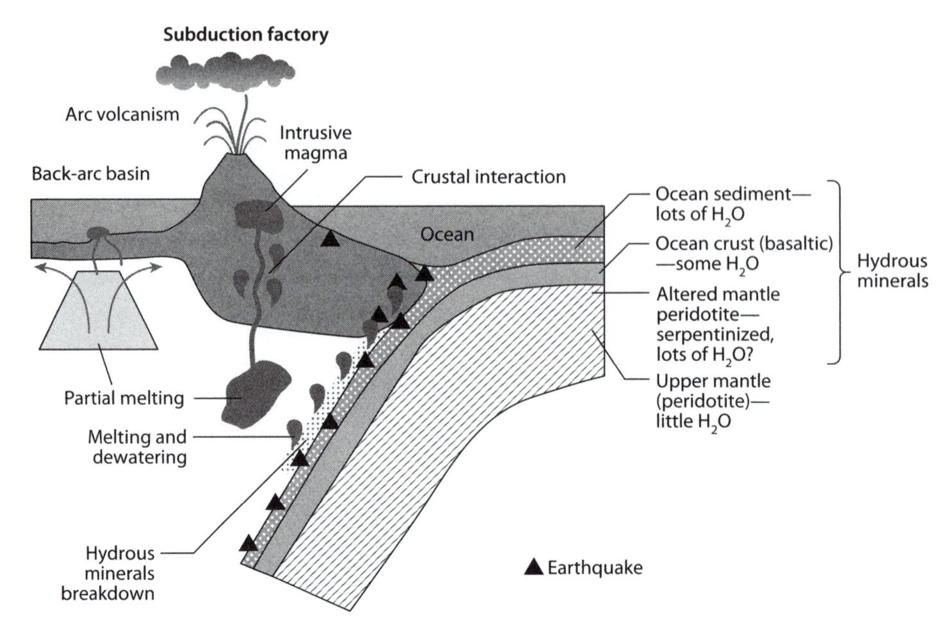

Fig. 12-7: Illustration of the change in plate composition leading to the volatile-rich package of materials that enters the subduction zone.

Volatiles are not the only materials added to the plate through alteration and sediment accumulation. Many elements, such as U, Rb, Ba, K, and B, are taken up by the hydrous minerals (see Table 12-2). Elements such as Pb, Cu, and Zn are accumulated in the sulfides formed by deposition from hydrothermal fluids. The reactions with oxidized seawater also convert the oxidation state of the crust so that much of the Fe^{2+} is converted to Fe^{3+}.

This entire package of material—altered mantle, altered basalt, and sediments—then moves down into the mantle at subduction zones (see Fig. 12-7). The descending plate is very different from the water- and CO_2-poor magma that came out at the spreading center tens of millions of years before, because of all the interactions with the surface reservoirs. Interactions with seawater modified the crustal composition, adding volatiles and other elements. Continental erosion, biogenic accumulation, and deposition of particles from hydrothermal vents created a rich and diverse sedimentary package above the crust. Movement of the plates transports this diverse assemblage to the subduction zone, creating flux from the surface to the mantle. As we shall see in the following section, this flux is what permits volcanism at convergent margins, creates a heterogeneous mantle, and leads to the formation of the continental crust.

Geochemical Processing at Convergent Margins

The kinematic description of plate tectonics associated the downward moving plate with the seismically defined Benioff zones. Standing above the Benioff zones with striking regularity are linear chains of conical volcanoes, such as the Pacific "Ring of Fire." When they are built on the seafloor, their great heights barely rise above sea level. When built on the high plains of central Mexico or the Altiplano of the Andes, where the thick crust results in a base level of 2,000–3,000 m, the volcanoes rise to 5,500 m or more, forming some of the highest peaks in the world. The existence of most of these volcanoes was well known hundreds of years before the advent of plate tectonics. Plate tectonics showed their ubiquitous relationship to subduction zones. More careful examination of the detailed locations of earthquakes revealed a remarkable regularity with respect to the down-going plate—when the positions of the volcanoes were compared to the seismicity below, most were found to occur about 110 km above the Benioff zone. Subduction and volcanism are systematically related (Fig. 12-8).

CAUSE OF MELTING AND VOLCANISM AT CONVERGENT MARGINS

There is a predicament when trying to understand why subduction leads to volcanism. At ocean ridges, depressurization of hot material causes melting. At subduction zones, in contrast, the cold plate is descending and produces a predominantly downward flow of the mantle in the mantle wedge, as was illustrated in Figure 11-9. Downwelling of cold material is the opposite of the hot upwelling at ocean ridges and makes it more and more difficult for the mantle to melt. At subduction zones, why isn't melting suppressed?

The key to melting at convergent margins lies with water. Several lines of evidence show that convergent margin magmas, unlike those at ocean ridges, are water rich.

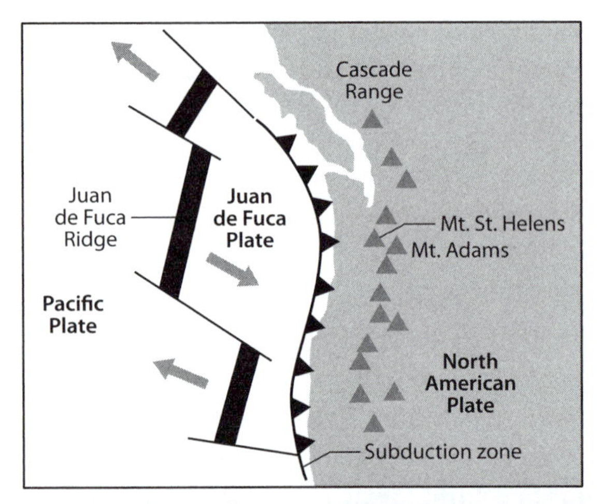

Fig. 12-8: Top: Map of the Juan de Fuca Plate and Cascades volcanoes showing their linear alignment and relationship to the down-going plate. *Bottom:* Aerial photograph of the Aleutian volcanic front. Mt. Kanaga is in the foreground. Moffett in the middle distance and Great Sitkin further away. (Photo courtesy of Chris Nye, Alaska Division of Geological and Geophysical Surveys and Alaska Volcano Observatory)

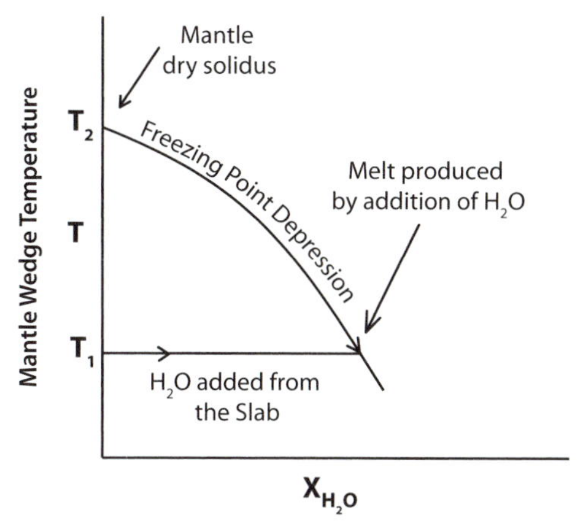

Fig. 12-9: Temperature vs. water concentration diagram illustrating how water affects melting temperature for mantle peridotite and subducted basalt and sediment. T_2 is the melting point of dry peridotite. The freezing point depression effect depends on how much water can be dissolved in the magma, which is pressure sensitive. Low-pressure magmas cannot contain significant water, and therefore water has no effect on melting. At higher pressures, magmas can dissolve 20% or more of water, leading to a very strong depression of melting temperatures, as illustrated at T_1, a wet melting temperature.

- Analyses of small inclusions of magma trapped in crystals that preserve initial volatile contents contain 5% or more of water (as well as high CO_2).
- The volcanoes are often explosive, and dissolved water converting to gas is a major cause of such explosive eruptions.
- Differentiation under hydrous conditions leads to higher silica magmas (andesites, rhyolite, granites), which would account for the preponderance of such magmas at convergent margin volcanoes.

High water contents are the key to how melting can occur at convergent margins (Fig. 12-9). Melting experiments of mantle peridotite show that water is an extremely effective agent of freezing point depression (see Chapter 7). The extent of freezing point depression is proportional to the amount of water that can be dissolved in the liquid magma. If too much water is added, it simply creates a fluid or gas and does not further lower the melting temperature. Because water vapor is so compressible, the maximum water content (the "solubility" of water in magma) increases greatly with pressure. At one atmosphere, almost no water can stay dissolved in magma; but at pressures of 10–30 kb in the mantle wedge (30–90 km below the surface) as much as 20% water can enter the magma, lowering the melting temperature by hundreds of degrees. The vast differences in water content cause melting at divergent and convergent margins to occur by different mechanisms. Whereas melting at ridges is by decompression of hot mantle, melting at convergent margins is caused by *flux melting* from lowered melting temperatures due to the freezing point depression caused by the addition of water.

For flux melting to occur, water must somehow be transported to the mantle beneath convergent margin volcanoes. The obvious source is the volatile-laden crust of the down-going plate. The water-rich minerals that formed near the surface are not stable at high pressures and temperatures. They undergo mineral transformations that lead to new structures and minerals with lower water content that release the excess water as a fluid. This overall process is called *metamorphism*. Metamorphism is solid-state transformation of rocks in response to changes in temperature and pressure. With increasing pressure and temperature, metamorphic reactions usually involve the progressive dehydration and decarbonation of rocks and the release of H_2O and CO_2 to the surroundings. During subduction, hydrous phases such as amphibole and chlorite that formed at the ocean ridge are transformed by a series of reactions to the anhydrous assemblage of pyroxene and garnet, making the dense rock eclogite, which helps to drag the slab down into the mantle (see Fig. 11-11). Carbonate minerals also become unstable, and high pressures and temperatures cause carbonates and silicates to react and release the CO_2 as a gas. For example,

$$CaCO_3 + MgSiO_3 + SiO_2 = CaMgSi_2O_6 + CO_2$$

As we learned in Chapter 9, this decarbonation reaction is one of the vital pathways for preservation of Earth's equable climate on long timescales.

The melting temperature of the basalts and sediments of the ocean crust in the presence of water is far lower than the melting temperature of mantle peridotite (~800°C rather than the ~1,500°C solidus of anhydrous mantle peridotite at a depth of 100 km). Because of these low melting temperatures in the presence of water, portions of the slab may also melt during subduction. The slab temperature becomes hottest where it is in contact with the mantle wedge. Since sediments are the uppermost slab layer, the sediments are most likely to melt. Geochemical evidence suggests such melting commonly occurs in the sediment. Melting may also occur in the ocean crust, depending on the detailed thermal environment of subduction. For example, very slow subduction rates give the slab much more time to heat up as it descends, making slab melting more likely. In the ancient Earth, when the mantle temperature in the wedge was considerably higher, slab melting was inevitable. Melts of the slab would have high concentrations of H_2O and CO_2 and would be another mechanism of volatile transport.

The fluids and melts from the slab that form at the relatively low temperatures at the top of the slab are much less dense than the mantle and will rise into the mantle wedge. The mantle wedge at this depth has an inverted temperature gradient—temperatures are much higher in the "core" of the wedge than they are in the older slab (see Fig. 11-9). Although the mantle immediately adjacent to the cold slab is too cold to melt, once the water rises far enough it enters the region of hotter mantle, where the added water lowers the melting temperature sufficiently to cause mantle melting to occur (Fig. 12-9). Whereas human beings usually associate melting with increased temperatures, and the mantle beneath ridges melts by decreasing pressures, melts at convergent margins are formed by a third mechanism—flux melting, where melting temperature is lowered by addition of another chemical component. We make use of the same principle to make roads safe for driving in the winter. Adding salt to our roads lowers the melting temperature of the ice even when the temperature is below the freezing point of pure water. At convergent margins, the migration of CO_2 and H_2O from the slab lowers the melting temperature of the mantle to produce the volatile-rich magmas that rise to the surface to form the explosive arc volcanoes.

The high concentrations of water also explain some of the major differences between the eruptive behavior of divergent and convergent margins. Ocean ridge basalts do not erupt explosively, and most flows are small and move slowly across the seafloor. In contrast, continental volcanoes are renowned for their explosive behavior. The tops of many of these volcanoes consist of large craters that are the remnants of the explosive removal of the volcano summits. Such eruptions became well known when Mount St. Helens erupted in May 1980 (Fig. 12-10). The

Fig. 12-10: Mount St. Helens before it erupted in May 1980 (*top*) and after it erupted (*bottom*). (Courtesy of U.S. Geological Survey)

solubility of water in magmas explains much of this contrast in behavior. The high water content of convergent margin magmas can all remain dissolved in the magma as long as the pressure is high. But as the pressure drops, the solubility decreases and some of the water escapes as a vapor. This vapor creates bubbles in the magma, which can lead to an extreme buildup of pressure, and the volcano filled with magma is a bit like an out-of-control pressure cooker or an overgassed bottle of champagne. If an earthquake makes a crack in the surface of the volcano, or the pressure builds up enough that the rocks near the surface of the volcano cannot support the stress any longer, the volcano explodes catastrophically. It is owing to their high water content, ultimately derived from the subducted slab, that convergent margin volcanoes are explosive while ocean ridge volcanoes are not. This differential solubility of volatiles is also what leads to the flux of gas from convergent margin volcanoes that is essential to the composition of the atmosphere and climate stability.

ELEMENT TRANSPORT TO THE CONTINENTAL CRUST

Water is not the only element transferred from the down-going slab. Many elements are strongly enriched in sediments and altered ocean crust relative to the mantle. Some elements are soluble in hot water-rich solutions at high pressure, and these elements should be efficiently extracted from the slab. Other elements are carried by slab melts and are effectively transferred. And sediments are enriched in many elements by a factor of one hundred or more compared to the mantle, so any sediment contribution leads to high abundances of certain elements.

Is it possible to prove that elements subducted at the trench are recycled through the mantle to come out at convergent margin volcanoes? Fortunately, there is a magic bullet in the geochemical arsenal that provides such proof. The interaction of cosmic rays with Earth's atmosphere produces a number of radioactive isotopes, the most familiar one being ^{14}C. Another cosmogenic radionuclide with a much longer half-life of 1.6 million years is ^{10}Be. The ^{10}Be formed in the atmosphere falls to the surface leading to small but measurable amounts in the sediments that fall to the seafloor to create the sediment layer of the ocean crust. The topmost oceanic sediments have the most ^{10}Be, and the concentration

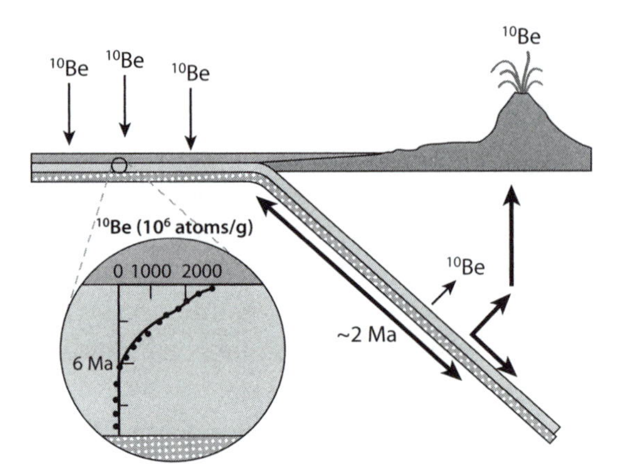

Fig. 12-11: ^{10}Be content of deep-sea sediments. Because of the 1.6 Ma half-life of ^{10}Be, it gradually decays away in older sediments as younger ones accumulate on top of them. (Figure courtesy of Terry Plank)

gradually decreases with depth as the ^{10}Be decays away. In sediments that are older than 10–15 Ma all of the cosmogenic nuclide has decayed away to its daughter product ^{10}B. This decay occurs no matter where the ^{10}Be resides. In the column of sediments being subducted, the ^{10}Be exists only in the youngest sediments of the top few meters (Fig. 12-11). There is no ^{10}Be in the mantle, and none in ocean ridge basalts. Nonetheless, many recently erupted convergent margin magmas contain some ^{10}Be! The presence of ^{10}Be in freshly erupted arc volcanics requires that Be from the uppermost sediments is carried down the trench, off the slab, into the wedge, and out the volcano in a few million years. This is conclusive evidence that elements from subducted sediments contribute to arc magmas. An evaluation of many other elements shows good correlations between element ratios in the sediments of the down-going slab and elements in the volcanic rocks of the overlying arc, confirming that subducted sediments are recycled to contribute to convergent margin volcanism.

More subtle evidence that we will not delve into here shows that elements from ocean crust are also recycled. The flux of elements from the various layers of the slab then makes the chemical compositions of arc volcanics vastly different from those of ocean ridge basalts. Elements that are enriched in sediments or are easily transported in metamorphic fluids or partial melts of the slab are greatly increased in abundance in the sources of convergent margin magmas. Whereas the volcanics at ocean ridges have very low abundances of magmaphile elements, those at convergent margins have large enrichments (10–100 times of many magmaphile elements, e.g., K, Rb, Cs, Ba, Th, U, and Pb). Over Earth's history, the fluxes from the slab have lead to progressive concentration of magmaphile and fluid-loving elements toward the surface, including Na, K, and P, three elements that are essential for living organisms.

High water contents also lead to higher SiO_2 contents of magmas, and to the precipitation of minerals that cause magmas to evolve toward higher SiO_2 as they cool. Basalts at ocean ridges all have relatively low SiO_2, near 50%, while convergent margin magmas commonly have 55% SiO_2 or more. The higher silica magmas also have low density. Convergent margin magmatism then leads to light crust, filled with magmaphile elements, that is not able to be subducted and becomes a permanent resident of the surface, riding buoyantly on top of plates. Continental crust formation can then be viewed conceptually as arising from the solid Earth geochemical cycle: first, basaltic crust forms at an ocean ridge; then, it is modified and hydrated by interaction with the surface; when subducted, it serves as the agent for delivery of the necessary elements to form the continental crust, where most of advanced life on Earth now resides.

Water clearly is the central element in this process. Recycling of water to the mantle is necessary to form continental crust. And at the same time, for the longevity of the oceans, the water that is subducted must be returned to the surface. Were the water in altered ocean crust and sediment simply returned to the mantle, the oceans would gradually empty over just a few hundred million years as water was returned to the interior. Since water is essential for life, there would be no planetary habitability.

Final Consequences of Plate Recirculation

After the plate has been processed at convergent margins, it continues its journey into the mantle. Also, not all of the material transferred to the mantle wedge makes it out to subduction zone volcanoes, so some of the material exiting the slab remains in the mantle. The end products

Fig. 12-12: Left: Illustration of the "marble cake" concept where the mantle is a folded mixture of diverse layers. *Right*: An outcrop of mantle rock that is layered with diverse veins. Some of these veins may reflect recycled crust. (Photography courtesy of Peter Kelemen)

of subduction are over time gradually mixed back in to the mantle and play a pivotal role in mantle evolution.

One of the most important consequences of this recycling stems from the role that H_2O plays in mantle viscosity. While most of the water is returned to the surface, a small fraction of recycled water is able to be included in nominally anhydrous mantle minerals—those that have no "OH" in their formula but nonetheless can include tens of parts per million of water in their mineral structure. Mantle minerals thus become "moistened" by their contact with fluids from the slab, and the slightly hydrated minerals are much weaker than perfectly dry minerals. The small amount of water causes the viscosity of the mantle to decrease by 1–2 orders of magnitude. The change in viscosity has a significant effect on the Rayleigh number, and hence the vigor of mantle convection. Some scientists think that Venus has no plate tectonics because its mantle is dry and therefore stiff. The presence of an ocean on Earth, coupled with the plate tectonic geochemical cycle, may be what permits steady, active convection and exchange between Earth's interior and the surface.

The process of plate recirculation also influences the major element composition of the mantle and how it is distributed. The recycled plate differs from what melted at ocean ridges by having separated crust from residual mantle and being processed at convergent margins. The 6-km-thick crust in particular has mineralogy, density, and physical properties different from mantle peridotite. While convection is "vigorous" from a planetary point of view, it is still a rather slow stirring and folding that is taking place in the solid state. The recycled material is not efficiently stirred and homogenized into the mantle but instead will be slowly deformed and thinned, in a fashion described by Claude Allègre and Don Turcotte as a "marble cake." This process leads to extended veins of diverse thicknesses distributed through the mantle (Fig. 12-12).

There is also the possibility that some subducted plates, or portions of them, are sufficiently dense that they fall through the mantle and accumulate at the core/mantle boundary. If this boundary is also the location where mantle plumes form, the crust would preferentially contribute to the plumes. These plumes then rise to the surface, generating magmas and releasing volatiles. Some of the outpourings are so intense that they significantly impact climate and life (see Chapter 17). Plumes relate subducting crust and the heat output from the core to surface reservoirs.

Since the recycled crust also has a low melting temperature relative to the surrounding mantle, it also melts more easily during mantle convection, so that low-degree melts of the subducted plate, highly enriched in magmaphile elements, are a possible consequence of plate recirculation. All these diverse processes lead to *mantle heterogeneity* on various scales—variations in mantle composition that are created and preserved through plate recycling. Such heterogeneity then leads to a diverse set of basalt compositions appearing at the ocean ridge and at ocean islands hundreds of millions to billions of years later. The complement of continental crust formation and maintenance of seawater composition is the formation of the heterogeneous mantle through plate recirculation.

Summary

The plate tectonic geochemical cycle relates mantle, crust, ocean and atmosphere. Magma emplacement at ocean ridges produces a large temperature contrast between crust and the deep ocean that drives vast hydrothermal systems at the ridge axis. These systems provide important sources and sinks for seawater, greatly influencing the chemical composition of the ocean. At the same time, the crustal mineralogy is transformed to contain volatile-rich minerals such as chlorite and carbonate. During passage of the plate to the subduction zone, low-temperature hydrothermal circulation and sediment deposition continue to influence the chemical composition. As the plate descends into the mantle at the convergent margin, the volatile-rich minerals undergo metamorphism and release their H_2O and CO_2 as well as many other magmaphile elements, such as Na, P, K, and Pb. The water depresses the melting temperature of the slab, possibly causing melting, particularly of the upper sedimentary layer. Fluids and hydrous melts are light and percolate upward into the hot mantle wedge. There they lower the melting temperature of the mantle and cause the formation of water-rich magmas, which ascend to the surface to form arc volcanoes. The high water contents make these volcanoes very explosive, as the water degasses during eruption.

Geochemical data show the importance of the subducted materials from the oceanic plate for the creation of convergent margins. The mystery of convergent margin volcanism—how do volcanoes form in a cold, downwelling environment—can be understood as a simple consequence of volcanism at mid-ocean ridges, with the attendant hydrothermal interactions adding volatiles and other elements to the crust which are then transported by plate tectonics to subduction zones. The water exploding out of convergent margin volcanoes originated through the hydrothermal systems circulating two miles beneath the surface at ocean ridges.

High water also contributes to lower-density magma compositions higher in SiO_2, and the low-density crust is stabilized at the surface and is not subject to subduction. Since convergent margins are the locations where continents are being formed, the origin and continued existence of the continental landmasses depends on the functioning and recirculation of the oceanic plates hidden beneath the sea. The distinctive compositions of continental rocks owe much to the transport of elements in fluids off the slab, and the recycling of sediments through erosion, deposition in the deep sea, and subduction into the mantle. Continents then become a natural outgrowth of oceanic volcanism and plate tectonics taking place beneath an ocean. As we saw in Chapter 9, the ocean itself is able to persist because of the climate feedbacks involving interactions of the atmosphere with the sun coupled with the CO_2 recycling controlled by subduction. The persistence of plate tectonics and the convective vigor of Earth's mantle may also be maintained by the plate tectonic geochemical cycle, as recycled water lowers mantle viscosity by 1 to 2 orders of magnitude. This also contributes to the viability of mantle plumes rising through the entire mantle to the surface to contribute to surface reservoirs. Atmosphere, ocean, ocean crust, mantle, core, continents, and plate tectonics form a linked system that sustains the conditions of our habitable planet.

Supplementary Readings

Special Issue on InterRidge 2007. *Oceanography* 20, no. 1.

J. D. Morris and J. G. Ryan. 2003. "Subduction Zone Processes and Implications for Changing Composition of the Upper and Lower Mantle." In H. W. Carlson, ed., *The Mantle and Core,* vol. 2 of *Treatise on Geochemistry.* Oxford: Elsevier Science. Pp. 451–70.

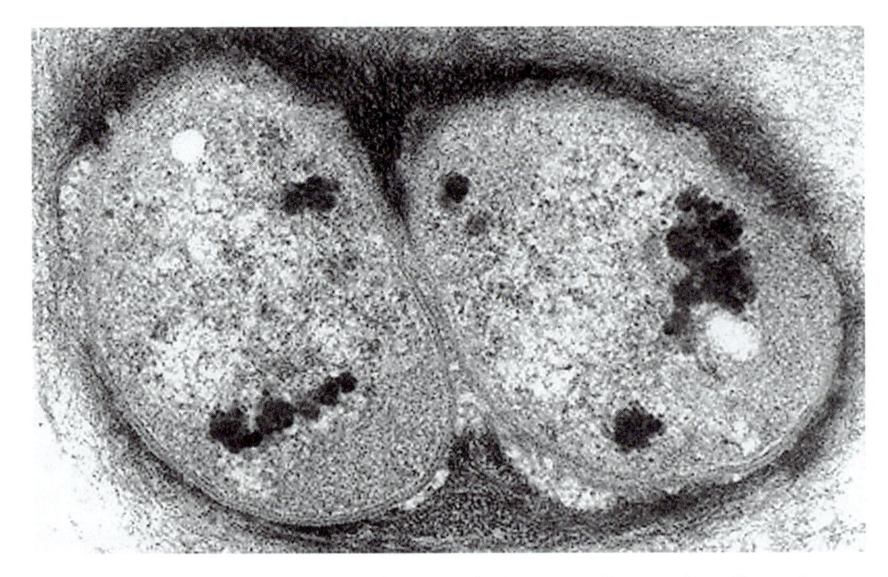

Fig. 13-0: TEM image of a complete division of a bacteria cell. Note the cell membrane delimiting their cytoplasm from the extracellular matrix. Magnification 20,000×. (Image reprinted by permission from the author, M. Halit Umar, and the copyright-holder, © Microscopy UK or their contributors, Copyright 2000)

CHAPTER 13

Colonizing the Surface

The Origin of Life as a Planetary Process

How life began is the least understood aspect of the development of our habitable planet. There is evidence for primitive life between 3.5 and 3.0 Ga, but a good fossil record exists only for the most recent 10% of Earth's history, since the beginning of the Cambrian period, 543 million years ago. Earth's earliest history, when life likely began, has no direct record at all. Understanding life's origin then requires careful detective work to infer a process hidden in the deepest planetary past.

The most important evidence comes from life itself. Life depends on the alpha-particle nuclide carbon, which can build three-dimensional molecules of immense variety and size and has many oxidation states that facilitate the transfer of electrons that are essential for life's energetic processes. The other critical elements that make up more than 98% of living matter are H, O, N, and S. Phosphorous plays a critical role but is low in actual concentration. All of these elements are made in abundance by nucleosynthesis. With a few notable exceptions such as hydrogen and the heavy elements that make up rocks, the chemical composition of life is very similar to the sun, showing the strong influence of the laws of nucelosynthesis on life's development.

A single origin for life is evident from the commonalities among all living organisms. All life is cellular, and the earliest evidence for life is unicellular organisms that have similarities to the most primitive organisms living today. All cells today have the same sets of molecules as essential building blocks—carbohydrates, lipids, amino acids, and nucleic acids. The amino acids have a particular chirality—i.e., they are "left-handed." All cells also have the same chemical machines, the most central of which are the pathways from DNA to RNA to proteins that govern cellular operations, the role of DNA in inheritance, and the processes that store and release energy through adenosine triphosphate (ATP).

The unity of life and the gradual changes in life through time point to a first common ancestor, the primitive cell from which all subsequent life evolved. The origin of life can then be viewed as a series of steps leading to this first cell. These steps involve (1) formation of the molecular building blocks in the correct state of matter; (2) construction of complex molecules from simpler components; (3) development of an outer membrane to contain the cell contents; (4) a chiral selection process; and (5) self-replicating chemical cycles. There is clear evidence for the first three of these, and examples of emerging possibilities for the remaining steps.

Life is often viewed as going "against nature" because life involves increasing order and decreasing entropy, which appear to violate thermodynamics. Life also has many "chicken and egg" paradoxes. The increasing order is possible because life is a nested system, transforming energy from sun and Earth. Life facilitates this energy transformation and leads to faster production of entropy in the larger system than would happen absent life. The "chicken-egg" relationship is inevitable when seen as a progressive evolutionary sequence of chemical cycles that become dependent on one another. Such processes have the great advantage of being self-sustaining.

The origin and evolution of life is a solar system process, deriving energy from Earth and sun and fully dependent on planetary cycles. The origin of life cannot be solved absent understanding of the planetary conditions that made it possible. If life is viewed as an efficient and natural planetary process, then it is likely to occur widely throughout the universe.

Introduction

Earth today is fully inhabited. On the microscopic scale, millions of species, most of them unidentified, occupy every ecological niche, even apparently hostile ones such as saline brines in oil fields, toxic waste dumps, or cracks deep in Earth's crust. Life is so successful that each milliliter of seawater contains more than 10 million micro-organisms, and every square centimeter of our skin is home to a zoo of millions of minute cells, thriving off our waste products. Where did all this life come from, and how did it start? Is life a planetary accident or part of normal planetary functioning? Is life a passive passenger on Earth's surface or an integral part of the planetary system? Have the inhabitants influenced and modified the planet's habitability? The next chapters address these issues that are central to our understanding of Earth as an inhabited planet.

Life and the Universe

Life is a chemical phenomenon based on molecules that transfer material and energy in complex cycles within and among organisms and in exchange with the environment. Like Earth itself, life is a system and shares the characteristics of natural systems outlined in Chapter 1. Life is distinguished from other natural systems, however, by being capable of and undergoing Darwinian evolution. Life is also is based on a fundamentally different chemical structure than the rock and metal that make up the solid planet. Both life and rock, however, have the commonality that they depend upon the chemical behavior of a single element that occupies a position in the middle of the Periodic Table, with a 4+ valence that makes bonds in three dimensions and can build three-dimensional building blocks. For rocks, the 4+ element is Si, and the fundamental building block is the silica tetrahedron discussed in Chapter 4. For life, the 4+ element is carbon, and the structures are the organic molecules on which all life is based. As alpha particle nuclides, both C and Si are very abundant in the universe. What about other elements that are also centrally located in the periodic table? Beneath C and Si in the periodic table is the element germanium (Ge), which is also a 4+ element and can also form a complex class of three-dimensional molecules, called the *germanates*. Germanium, however, has a mass of 72, higher than ^{56}Fe, and very little Ge is made in stars. Even though it is a refractory element, the terrestrial abundance is only 1 part per million—250,000 times less than Si. The stellar processes of nucleosynthesis relegate Ge to trivial rather than planetary importance.

Carbon has five major advantages compared to silicon that permit more complex three-dimensional structures, more diverse chemical reactions, and easier chemical transport:

(1) Under normal planetary conditions carbon can bond to itself as well as to other elements—carbon-carbon bonds are at the backbone of many organic molecules.
(2) Organic molecules can bend and fold, creating large, complex three-dimensional structures like proteins and DNA that are central to life processes, whereas silicate minerals are relatively rigid and inflexible.
(3) Carbon forms various common molecules that can be solid (e.g., bone, limestone, and wood), liquid (e.g., alcohol, gasoline, and acetone), and gas (e.g., carbon dioxide and methane [natural gas]) at the same temperature and pressure. This allows transport and exchange of carbon among solid, liquid, and gaseous reservoirs.
(4) Carbon can form molecules, some of which are soluble in water (e.g., sugar, alcohol) and some insoluble (e.g., wood and oil), permitting the coexistence of and exchange between solid and liquid.
(5) Carbon can have multiple valence states (e.g., 4+ in CO_2 , neutral in C, and 4– in CH_4) that enable electron transfer reactions permitting energy flow and storage.

The greater flexibility of carbon bonds leads to a far greater assemblage of molecules than is possible with silicates, and millions of different organic molecules are known (Fig. 13-1), compared to thousands of silicate minerals. In contrast to silicates, organic molecules are almost infinite in their capacity for variation and modification. Their ability to be present as different states of matter and in one form to be stable in the presence of fluid and in another to be transported by the fluid permits chemical cycles. The electron transfer processes allows collection of energy from the environment (e.g., by eating or by photosynthesis) and its storage and transfer, permitting energy systems within an organism and in ecosystems. While the solid earth transfers mass and energy on geological timescales of thousands to millions of years, and relies mostly on large changes in temperature and pressure to implement changes in state and transfer of energy, organic compounds can perform similar functions over the diversity of timescales that are pertinent to life—shorter than microseconds in terms of energy transfer within a cell and

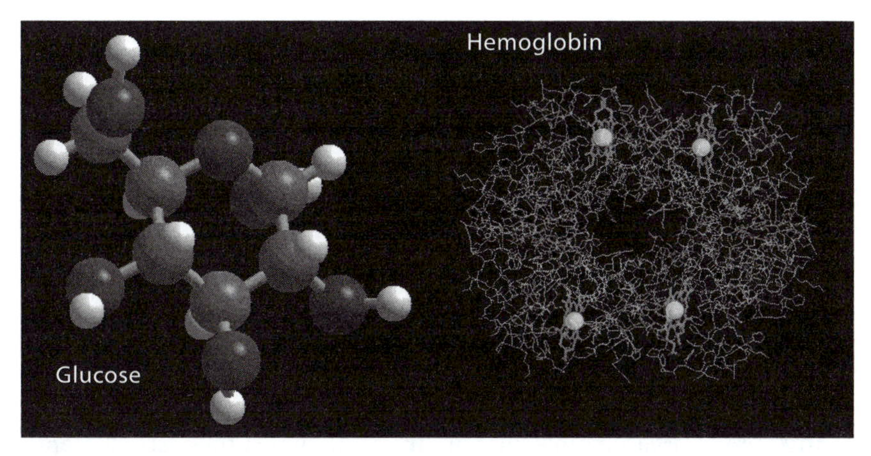

Fig. 13-1: Ball-and-stick models of a simple and a complex organic molecule. On the left is glucose $C_6H_{12}O_6$, consisting of twenty-four atoms with a molecular weight of 180. The gray balls that have four bonds are the carbon atoms. On the right is hemoglobin $(C_{738}H_{1166}N_{812}O_{203}S_2Fe)_4$, with a molecular weight of about 67,000. Each tiny dot is one of the atoms. The large gray balls are the four Fe atoms in the hemoglobin. Hemoglobin is a protein made up of 574 amino acids of twenty different types. (University of Arizona)

as long as years to decades for the storage of food and transfer of matter and energy through ecosystems.

One of the largest questions about life in the universe and early life on Earth is whether other forms of life very different from our own (e.g., not based on carbon) might be possible. While one cannot definitively rule out such a possibility, from the perspectives just given the chance of a life system based on an element other than carbon seems remote indeed. Silicate-based life in particular would have huge disadvantages compared to carbon-based life systems. Instead, it appears from our limited sample that there is a straightforward consequence of the abundances of the elements produced by nucleosynthesis and the nature of the properties of elements in the Periodic Table. Silicates and metals with a central role for Si and Fe form the three-dimensional structure of rocky planets. Organic molecules with a central role for C form the structure of life. Both of them are produced in abundance in the universe and depend on the unique characteristics and three-dimensional capabilities inherited from the fundamental atomic structure revealed in the Periodic Table.

In Chapter 6 we were able to understand the abundances of the different elements in the entire Earth by comparing them to nongaseous elements in the sun and in chondrites, and we noted the overall correspondence between the chondritic meteorites and the solid Earth, adjusted for the relative volatility of the different elements. This led to an understanding of the dominance of Fe, Mg, Si, and O, which make up more than 90% of the planet.

As the solid planetary system is dominated by only four elements, organic life also consists largely of a small number of elements. H, O, and C make up 98% of the human body in terms of numbers of atoms, and 93% by weight. The next three most abundant elements are N, S, and P, and 99% of organic molecules are made up of just these six elements. It is clear that life has a fundamentally different chemical composition than rock and metal, and we are not simply made up of representative fragments of Earth.

If we view the chemical composition of life from a cosmic perspective, however, the dominance of H, C, O, N, and S is not surprising. For the solid Earth we could understand the dominance of the elements Fe, Mg, Si, and O by considering the amounts made during nucleosynthesis and the loss of volatile elements during planetary formation (see Table 5-5). We can similarly understand the chemical composition of life, recognizing that life relies on volatile elements and largely excludes the refractory lithophile elements that go predominantly into rocks and the siderophile elements that go into the core. Let's revisit the first twenty-eight elements of the Periodic Table from this perspective (refer to Table 5-5). Hydrogen is the most abundant element in the universe, and also very important for life. Helium and the other noble gases are not chemically reactive and therefore do not participate significantly in low-temperature chemical systems such as life. Li, Be, and B are only produced in tiny quantities during the Big Bang. The next two most abundant elements produced during nucleosynthesis are the alpha-particle nuclides ^{12}C and ^{16}O, and ^{14}N is the even nuclide that occurs between them on the Periodic Table. All of these are produced in abundance by nucleosynthesis and are central to life. F and Na are odd, and Ne is a noble gas. Mg, Al, and Si are made in abundance, but are refractory lithophiles that go into rocks. Phosphorous appears at first to be more of a puzzle. It is next to Si and below N on the Periodic Table, and substantial amounts are formed during nucleosynthesis, but much less than the alpha-particle nuclides and nitrogen that are otherwise dominant. This paradox is partially resolved when we note that while phosphorous is a common constituent of vital organic molecules, it is a minor constituent in terms of number of atoms. For example, the molecule of adenosine triphosphate that lies at the heart of energy transfer in life contains three P atoms and forty-four C, H, and N atoms. The role of phosphorous is important; its overall abundance is small. Sulfur, the next most abundant element in life, is also an alpha-particle nuclide. While central to life, it is actually low in relative abundance, which may be understood by its combination with Fe and incorporation into the core. All of the elements heavier than S are either odd, noble gases, or lithophile and/or siderophile.

Figure 13-2 shows graphically that for all elements but noble gases and the rock forming elements Fe, Mg, Si, and Al, there is an approximate correspondence between solar abundances and human abundances. The chemical composition of life makes some sense from a cosmic perspec-

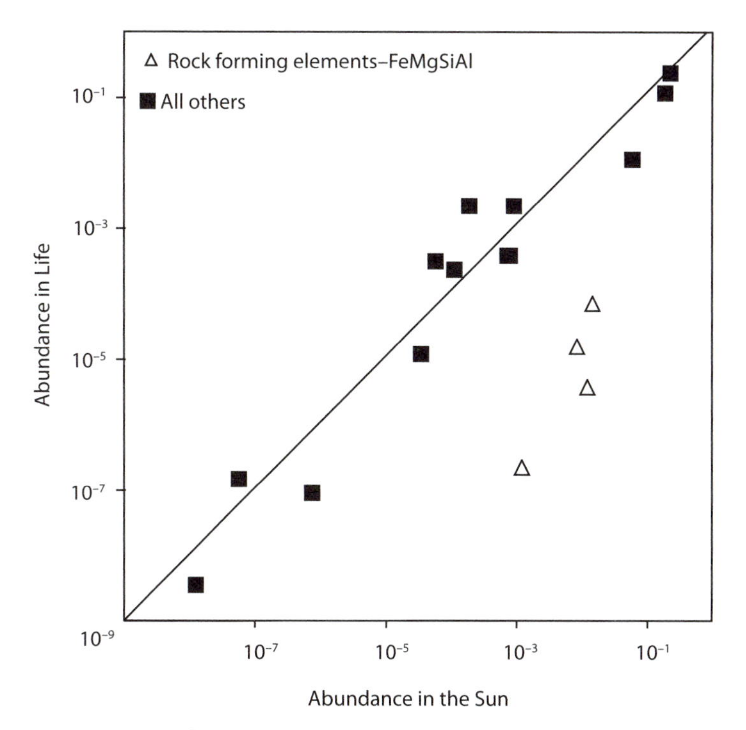

Fig. 13-2: Comparison of relative abundances of Li, Be, B, C, N, O, F, Na, Ca, Al, Si, P, S, and Fe in living tissue and in the sun. Values are normalized to H/1000. Note that with exception of the strongly lithophile Si, Al, Mg, and siderophile Fe, there is broad similarity between life and sun in relative abundances.

tive. From a chemical perspective, life is the complement of the solid Earth. The solid Earth is representative of the sun and solar system minus most of the volatile elements; life is representative of the sun and solar system minus the rock- and metal-forming elements that make up the solid Earth.

Of course, just as minor elements such as H_2O and CO_2 play an important role in planetary systems, minor elements such as Fe, Ca, and Zn play a very important role in living systems. Bones and skeletons require Ca. Hemoglobin has Fe as its central and all-important molecule. About one-half of all enzymes have a metal atom as an important constituent. Life is fully planetary in its chemical composition.

The Unity of Life

For most of us, the impression of the life that surrounds us is one of great diversity. Mold growing on old food, giant sequoias, oysters, cobras, cockroaches, and human beings seem very dissimilar from one another. At the same time, we see great commonalities among different types of life—mammals share many characteristics, as do many flowering plants.

While we tend to see the differences among living organisms and can marvel at life's diversity, examination of life at the microscopic and molecular levels provides a very different perspective, showing that all of life shares essential characteristics. It is this fact that allows the question of the origin of life to be reduced to the origin of the simplest single-celled organism that has the essential characteristics shared by all of life today. What are those characteristics?

LIFE IS CELLULAR

All of life is made up of cells with similar attributes. Whether an organism is a single-celled bacteria or complex assemblage of tens of trillions of cells of some 210 different types that make up a person, life is cellular.[1] Figure 13-3 illustrates this simple fact, where a unicelled fungus is com-

[1] Viruses are a notable exception to this observation, but they do not live independently of cellular life on which they depend for replication.

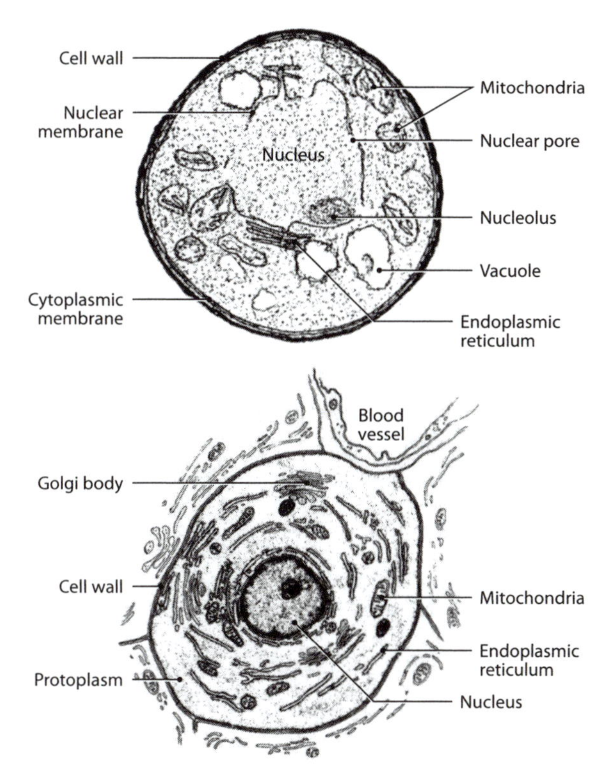

Fig. 13-3: Comparison of two eukaryotic cells. The top sketch is the fungus cell; the bottom cell is a human cell. Notice that broad similarity in appearance and structure, with an external cell membrane, a nucleus, organelles, and protoplasm.

pared to a cell from the human body. Examine any organism under a microscope, and all of them are made up of cells with an exterior membrane to provide a boundary with the external world, and across which selective transport takes place, and an interior where similar molecules and similar geochemical reactions and cycles carry out metabolism and replication. Animal and plant cells have important differences—plant cells also have an external cell wall and cellulose is a most important molecule—but the commonalities are much greater than the differences.

ALL LIFE USES THE SAME GROUPS OF MOLECULES

The second half of the twentieth century saw the rise of biochemistry, where life could be investigated on the molecular level. Close examination of life revealed still more stunning similarities among all organisms. On the atomic level, this similarity is expressed in terms of the small number of elements that make up all of life. And in turn these elements combine to form a small number of building blocks, such as H_2O, CH_3, NH_3, CO_2, PO_4, and so on that combine to form the enormous diversity of larger organic molecules.

These larger molecules, while showing huge variety in detail, nonetheless can be classified into four groups of macromolecules, common to all cells, that fulfill the basic functions of the cellular machine. These four classes are carbohydrates, amino acids, lipids, and nucleic acids.

Carbohydrates are the fuel source for cellular operations. Carbohydrates are hydrated carbon atoms—carbon atoms that combine with whole numbers of H_2O molecules. In terms of the chemical reaction that summarizes the creation of organic carbon through *oxygenic photosynthesis*,

$$CO_2 + H_2O + \text{energy} = CH_2O + O_2 \qquad (1)$$

Note that the valence state of the carbon changes from +4 in CO_2 to +0 in CH_2O, as the valence state of two oxygens changes from –2 to 0. This results from a transfer of electrons to the carbon atoms. Such a transfer of electrons is at the heart of most organic reactions. Carbon is special in this respect because it can lose or gain up to four electrons and have a completed electron shell.

A simple carbohydrate such as the sugar glucose (Fig. 13-1) has a simple formula, $C_6H_{12}O_6$, and structure. Fructose has the same formula with the atoms arranged in a different structure. Combining fructose and glucose together makes sucrose. There are also very large and complex carbohydrates, such as the starches or cellulose, which have formulas consisting of a hundred atoms or more.

Oxidation of carbohydrates releases energy that can be used by the cell through reactions that can be simplified as, for example:

$$C_6H_{12}O_6 + 6O_2 = 6\,CO_2 + 6\,H_2O + \text{energy} \qquad (2)$$

The organic carbon is converted to CO_2 as electron transfer goes in the opposite direction from reaction (1), and energy is produced. When we burn wood in our fireplaces, we are facilitating this electron transfer and producing heat in the process. Our bodies "burn" carbohydrates in a more controlled fashion to produce the energy needed for our cellular metabolism.

Lipids have much less oxygen than carbohydrates, and the carbon is in a more reduced form. There is a higher potential energy content, because of the larger electron transfer that occurs. Lipids are a very efficient way to store high energy per molecule, and our bodies convert carbohydrates to fat in order to store the extra food energy in compact form. Lipids are the fats found in animals and the oils found in plants, and they also have other functions, such as creating the basic structure of cell membranes.

Amino acids are the twenty-two molecules that are the building blocks of proteins. Amino acids have a particular chemical structure consisting of a central carbon whose four bonds are connected to an "amino" group (NH_2), a carboxyl group (COOH), a hydrogen atom, and a side chain that is called an *R group* (Fig. 13-4). The first three are the same in all amino acids. The identity of the molecule that makes up the side chain is what distinguishes one amino acid from another. Their chemical formula can be written as $H_2NCH\mathbf{R}COOH$. The R group molecules may be hydrophobic—i.e., they do not want to coexist with water—or hydrophilic—they do want to be next to water molecules (this latter class of amino acids is called the *polar* class). A third type of amino acid, called the *charged* class, has an R group that contains a positive or negative

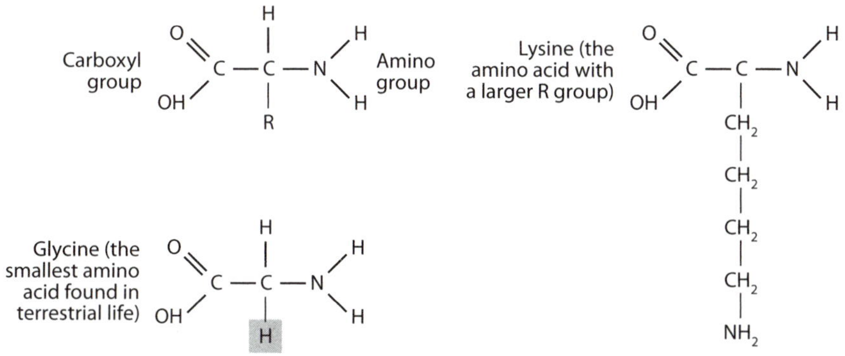

Fig. 13-4: This diagram shows the general structure of amino acids. The amino group and acid group are common to all, the identity of the *R-group* changes from one amino acid to another, as illustrated with the simple R-group for glycine and the larger R-group for lysine.

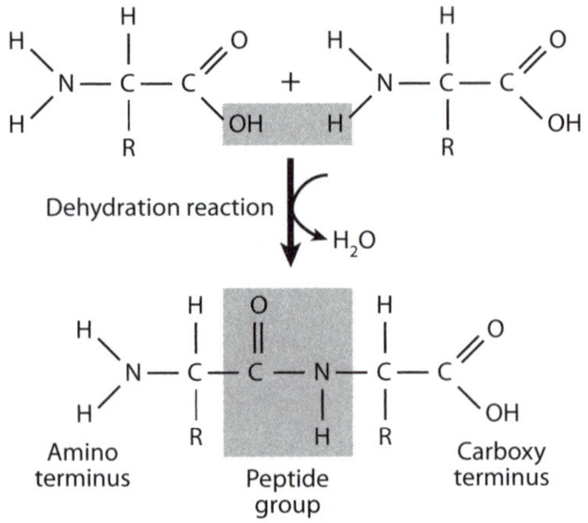

Fig. 13-5: Illustration of the peptide bond reaction that permits amino acids to join together to form proteins. Note that the reaction involves dehydration, the removal of a water molecule as the amino and acid groups of the two different amino acids become bonded together. Proteins commonly consist of hundreds of amino acids joined together through peptide bonds.

charge. Within these classes, the amino acids can also be of different sizes and shapes. The largest (and rarest) amino acid found in terrestrial life, for example, tryptophan, has a side chain consisting of eighteen atoms. The smallest, glycine, H_2NCH_2COOH, has only one additional hydrogen atom as its "R group." Many more amino acids can be made in the laboratory than exist as the protein building molecules of terrestrial life. Amino acids are also commonly found in carbonaceous chondrites, showing an important organic component to molecule creation in interstellar space.

An important aspect of amino acids is that the carboxyl and amino groups can join together to form a *peptide bond* (Fig. 13-5). This universal capacity of amino acids to bond together is what makes the construction of huge protein molecules possible. The proteins form more than 10,000 different molecules that include the basic structure of the organism as well as enzymes, hormones, etc. They are involved in oxygen transport, muscle contractions, and countless other metabolic activities. If proteins are thought of as "words," the twenty-two amino acids found in living organisms are the protein alphabet, and their combination is able to create the immense diversity of proteins found in living organisms. Very complex molecules can be built by amino acids. For example,

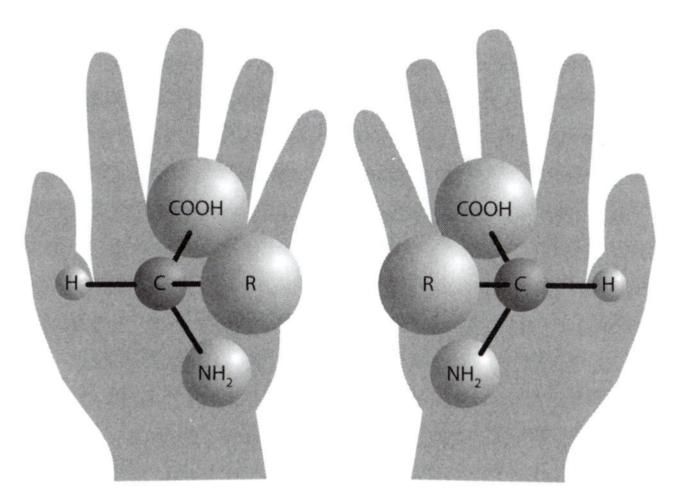

Fig. 13-6: Illustration of the chirality of amino acid molecules. Both left- and right-handed forms are made in most natural processes, but life uses only left-handed amino acids. This requires an early selection process that was able to distinguish between the two. Notice how the "shape" of a protein where these two forms were joined together with a peptide bond would be very different depending on the chirality. (Image courtesy of NASA)

hemoglobin is a protein with a chemical formula of $C_{2952}H_{4664}N_{812}O_{832}S_8Fe_4$, consisting of more than five hundred amino acids surrounding four Fe atoms (Fig. 13-1).

Each type of amino acid can also occur in left-handed and right-handed forms—i.e., they can be mirror images of each other (Fig. 13-6). This *chirality* is very important for how the amino acids can fit together. For example, it is very difficult for human beings to shake hands in the conventional way if one person uses a left hand and the other a right hand. And it is not possible to stack left and right hands on top of one another so that the shapes match. Molecules with different chirality can also have very different effects on the body. Thalidomide, for example, in left-handed form, was an effective antidepressant that was used for pregnant women in the 1950s. The right-handed form, produced in small amounts during the manufacturing, produced birth defects in babies.

Many amino acids can be made in the laboratory—some seventy are known, and all of them occur in right-handed and left-handed forms. The remarkable feature of life on Earth is that all organisms make use of only twenty-two left-handed amino acids.

Nucleic acids carry out information, communication, and memory functions within the cell. Like the amino acids, nucleic acids also have commonality of structure, with a sugar backbone, a phosphate group, and five basic building blocks consisting of two classes of molecules: the *purines*, adenine and guanine, and *pyrimidines*, thymine, cytosine, and uracil. Deoxyribonucleic acid, DNA, makes use of the adenine, guanine, cytosine, and thymine. Ribonucleic acid, RNA, substitutes uracil for thymine. The important characteristic of the nucleic acids is that they make complementary chains. This permits the molecules to replicate and communicate. RNA can match up with DNA to carry the information necessary to make proteins to other parts of the cell. DNA can split and replicate to pass almost identical information and instruction from one generation to the next.

The four classes of molecules can be looked at from an organizational standpoint, where each class of molecules provides a different set of functions: energy source, energy storage, structure, and instruction and communications. Carbohydrates are the immediate energy source, and oxidation of the carbohydrate back to CO_2 plus H_2O provides the basic fuel for cellular operations. Lipids allow excess energy to be stored efficiently for possible future use. Fats build up when an excess of carbohydrate accumulates in the organism, and then can be released during times of food shortage. Lipids have other important functions (such as cholesterol in the blood) and are also important constituents of cell membranes. Amino acids combine to form the remarkably diverse proteins and make up the physical structures of life. They also act as important enzymes, the catalysts that enable efficient cellular functioning. Nucleic acids provide the instruction kit for cellular operations and the means of communication both within the cell and from one generation to the next. All of life—all plants, animals, and single-celled organisms—use the same molecules and same basic structural organization. From these perspectives, life is a unity.

ALL LIFE USES THE SAME CHEMICAL MACHINERY

In addition to the commonalities of basic cellular appearance and the same restricted groups of molecules present in cells, a limited number of chemical machines are fundamental to the operation of all cells.

Perhaps the most fundamental machine is the relationship among nucleic acids and proteins, whereby the instruction kit contained in DNA is put into operation in the cell. The DNA carries code that specifies which amino acid will be added to a protein. The RNA reads the code and then carries it to the protein where the appropriate amino acid can be placed. DNA to RNA to protein is the "central dogma" of cellular operations. Each one of the amino acids is coded within DNA by a series of three distinct bases, called a *codon*. Since there are four bases, the total number of codon instructions is 4^3, or 64, which code for the twenty-two amino acids, as well as "start" and "stop" commands, which are essential because a given strand of DNA may code for many proteins, so the RNA needs to know when the job is complete. Since the number of possible commands is larger than the number of amino acids, there is some redundancy, with different codons able to specify the same amino acid. This mechanism of protein synthesis, information storage, and transfer of genetic information from one generation to the next operates in all cells.

Each cell also has a fundamental energy driver, which is an electrical charge set up across the cell membrane. This electrical potential acts as a microbattery that causes electron flow, which is necessary for the basic chemical reactions of cellular function.

The chemical cycles that mediate energy conversion in cells are also remarkably similar. The currency of cellular energy, discussed at length in Chapter 15, is the conversion between adenosine diphosphate (ADP) and adenosine triphosphate (ATP) involving the addition or removal of a phosphate molecule. In most cells, a fundamental mechanism for this conversion is the *citric acid cycle*, a complex series of chemical steps that converts between ADP and ATP and can run in both directions, depending on whether energy is being used or created.

These shared characteristics show the great commonality of all of life on Earth. All of life uses the same chemical building blocks, down to the detail of having a limited number of amino acids with the same chirality, and consists of cells that operate using the same fundamental machinery for protein building, transfer of information from one generation to the next, and energy production, storage, and use. The discoveries of the late twentieth century have revealed the remarkable unity of life from the cellular to the atomic realm.

Earliest Life

Our view of the history of life on Earth is informed by detailed study of living organisms, and by fossils—the remains of once-living organisms preserved in sedimentary rocks (Fig. 13-7). The fossil record reveals a remarkable diversity of organisms, most of which have no living examples today. What is largely unappreciated to the nonspecialist is that the visible fossil record begins only 543 million years ago, at the boundary between the Precambrian and Cambrian periods. In fact, it is the appearance of mineralized skeletons that produce visible, macroscopic fossils that defines this boundary. While 543 million years is a long time by human standards, it is only 12% of Earth's history. If we visited Earth a billion years ago—less than 25% of Earth's history—the planet would be unrecognizable to us: no grasses, trees, or shrubs, no plants; no mammals, no fish, worms, or insects. There would be nothing for us to eat and very little to see aside from barren landscape. From this perspective, time travel has its disadvantages.

There is abundant evidence, however, for the presence of earlier life that did not have the hard parts that could be preserved as fossils. While plants and animals were absent, the most abundant form of life was thriving and omnipresent. That life is the millions of species of unicellular organisms, the building blocks for the more complex multicellular organisms that exploded onto the scene in the Cambrian.

Unicellular life can be divided into two major groups, called *prokaryotic* and *eukaryotic* (fig. 13-8). While both groups share the common characteristics of life discussed above, the two types of cells are very different from one another. Prokaryotic cells are usually small, less than 1 micron (one-thousandth of a millimeter) in diameter. They have a minute quantity of DNA, have no cell nucleus, and can divide and double population in twenty minutes. The inner structure of these cells is not differentiated. They are essentially a membrane sack with the basic ingredients necessary for cellular metabolism and reproduction. In these primitive organisms, many of the important aspects of photosynthesis or respiration take place in the cell membrane.

Prokaryotic life continues in unbelievable abundance on, around, and within us. While our eukaryotic cells make up the parts of our anatomy

Fig. 13-7: Trilobite fossil from the Cambrian period from Schlotheim, Czech Bohemia. Prior to the Cambrian, before 543 million years, no fossils with hard body parts have been found. (Photograph courtesy of Museum of Comparative Zoology, Harvard University)

that we recognize, there are ten times as many prokaryotic as eukaryotic cells within a human being. Prokaryotes surround us and inhabit us by the billions, too small for us to see. Each square centimeter of our skin is home to a million of these organisms. Our armpits have ten times more. More of them reside on the surface of each human body than there are people on Earth. Each cubic centimeter of seawater contains 10 million

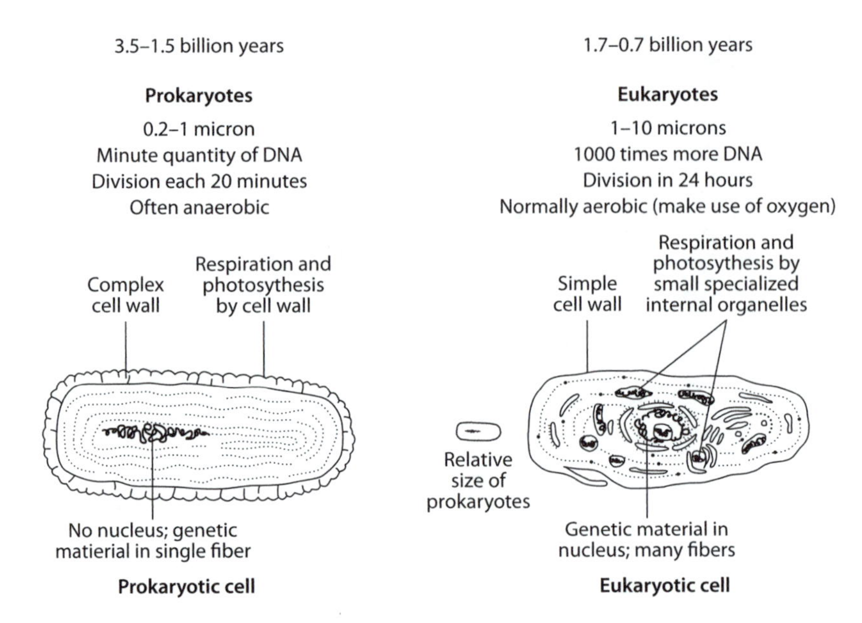

Fig. 13-8: Schematic illustrations of prokaryotic and eukaryotic cells. While the images are the same size, the actual sizes of the cells are very different. The small symbol to the left of the eukaryotic cell gives an indication of relative size. Prokaryotic cells are usually small, generally less than 1 micron in length, while eukaryotic cells are commonly 10 microns. This makes a difference of a factor of a thousand in terms of volume, and provides credence to the idea that eukaryotes evolved from incorporation of or symbiotic relationships among prokaryotes.

of them. Each cubic centimeter of soil is a thriving metropolis of 100 million. In their diversity, population, and flexibility they beat out plants and animals hands down. They are an invisible world that is fundamental to most geochemical cycles and are the invisible backbone that makes life sustainable. Their effects are everywhere—from the health of soil, to the photosynthetic capacity of the oceans, to the proper operation of our digestive system, to mold on old food and many diseases.

Eukaryotic cells are complex factories compared to their prokaryotic cousins. They are far larger (1–10 microns in diameter, so a thousand times larger in volume) and have a complex inner structure, with DNA contained in a cell nucleus and the interior populated by a series of molecular machines called *organelles* that undertake functions such as respiration (in mitochondria) and photosynthesis (in chloroplasts). In contrast, prokaryote cell interiors are much less differentiated. Eukaryotes have a thousand times more DNA than prokaryotes, and replicate in about twenty-four hours.

The organelles that carry out important functions within eukaryotic cells have their own DNA and a strong kinship with some prokaryotic cells. Lynn Margulis pioneered the idea, now largely accepted, that organelles in eukaryotic cells developed by evolving symbioses among diverse prokaryotic cells, which eventually fully merged into discrete individuals that preserved the essential functions of their ancestors. Eukaryotes may be evolved products of earlier prokaryotic communities. That said, there are still major debates on how the original eukaryotes arose.

The multi-cellular life that appeared about 600 million years ago is made up of groups of eukaryotic cells, which themselves had become specialized to fulfill specific functions. In our bodies, for example, kidney cells, liver cells, nerve cells, blood cells, muscle cells, and so on are all eukaryotic cells that have become adapted to their own specialized and coordinated function. The overall development of life can then be viewed in simplistic terms as an early stage of primitive prokaryotic cells, which combine and are transformed to form the larger and more complex eukaryotes, which in turn combine and transform to form the multicellular, macroscopic life that emerges in the Cambrian and now forms the visible life we see all around us.

WHEN DID LIFE BEGIN?

These overall trends in the history of life need a timescale, including the time of appearance of the first organism in the geological record. While individual microorganisms are too small to see without a good microscope, large communities of microorganisms do make visible communities. Particularly important microbiological communities for the geological record are those that create rocks, called *stromatolites*. Stromatolites are growth structures preserved in carbonate sediments and are most often created by photosynthetic microbiological communities that live in shallow seas. The metabolism of some bacteria causes precipitation of calcium carbonate between the cells. As the carbonate precipitates, the living bacteria propagate upward toward the sun and then precipitate another thin layer of carbonate. Over thousands of years, this creates a characteristic rock structure. Sometimes under special circumstances these structures even preserve the cellular remains of the microorganisms that caused them to be deposited, though in most cases the progressive cementation by carbonate tends to destroy these remains.

Stromatolites may be present in some of the oldest rocks, become common through billions of years of early Earth's history, and are preserved in rare environments on Earth today. Therefore it is possible to examine present-day examples and try to relate their sedimentary structures to those that can be observed in the distant past. Figure 13-9a shows modern examples of stromatolites from the classic locality of Shark Bay, Australia, and sedimentary structures similar to stromatolites that are preserved today and in some of the earliest rocks of the geological record from 3.5 billion years ago (Fig. 13-9b). The stromatolite evidence is considered by some to be evidence for thriving bacterial life at 3.5 Ga, about the same age as the oldest rocks.

Further evidence for earliest life comes from the stable isotopes of carbon. Life significantly prefers the light ^{12}C isotope over the heavier ^{13}C isotope, by 2.5%. Using the stable isotope nomenclature discussed in Chapter 9, the 2.5% preference means that carbon compounds made by life have $\delta^{13}C$ (a normalized measure of the $^{13}C/^{12}C$ ratio) about 25 per mil lighter (more negative) than inorganic compounds such as $CaCO_3$. Some carbon compounds separated from ancient rocks with an age of 3.5 Ga have just this signature of "light" carbon.

Another line of evidence comes from *biomarkers*, complex organic molecules that do not break down easily and would be the result only of life. Roger Summons and colleagues found evidence for biomarkers made during photosynthesis in rocks 2.7 billion years old. On the basis of this evidence, there was confidence that Earth was inhabited by 3. 5 Ga. It even appeared that photosynthesis had evolved as early as 2.7 Ga, substantially before the rise of oxygen in the atmosphere near 2.4 Ga (discussed at length in Chapters 15 and 16).

However, all of this evidence for very ancient life and its capabilities has become subject to doubt. Structures similar to ancient stromatolites have been shown by John Grotzinger to be able to be formed by inorganic processes. Evidence from carbon compounds also has the problem that the rocks in question have existed in the crust for billions of years of Earth's history, and during all of this period, life has been present. All

Fig. 13-9: Top: Modern stromatolites from Shark Bay Australia (photograph courtesy of Paul Hoffman and Francis Macdonald, Harvard University). *Bottom:* Right panel shows an example of stromatolite structure found in the 3.45 Ga Warawoona formation from Australia. A variety of evidence suggests these stromatolites were formed in association with microbial mats. Scale bar: 15 cm (photograph courtesy of Andrew Knoll, Harvard University, based on Allwood et al., *Proc. Natl. Acad. Sci.* 106 [2009], no. 24: 9548–55).

the water circulating through the cracks and porosity of Earth's crust contains microorganisms, and it would be exceedingly difficult to keep a rock isolated from these effects over billions of years. Furthermore, petroleum compounds, made from living matter and therefore containing "light" carbon, are formed at depth in the crust and migrate here and there, providing further sources of contamination. For these reasons, evidence from carbon compounds cannot be regarded as definitive. Indeed, careful recent work has shown that the biomarker evidence in the 2.7 Ga rocks is from young rather than ancient compounds, taking away the main line of evidence for a very ancient date for the beginning of photosynthesis.

Definitive evidence for earliest life then needs a combination of evidence and reasoning—reliable visual evidence, carbon isotopes, how pristine the rocks are, what geological environment they come from, etc. Textural evidence that lacks carbon isotopes, microfossils and biomarkers comes from 3.45 Ga stromatolites (Fig. 13-9b). The most definitive evidence (as of 2010) comes from 3.2 billion-year-old rocks (Fig. 13-10a), where structures such as cell membranes are still preserved from ancient microorganisms. Many geoscientists also consider that visual and carbon isotope evidence from 3.5 Ga rocks indicates life existed at that time. Photosynthesizing *cyanobacteria* are found in rocks as old as 2.0 Ga, and must have occurred earlier than that to account for the first rise of O_2 in the atmosphere. If the 3.45 Ga stromatolites included photosynthetic bacteria, photosynthesis may have had a much earlier start. Beautiful visual evidence for eukaryotes occurs in rocks at 1.5 Ga (Fig. 13-10b), and there is substantial evidence for them at 2.0 Ga.

The progression from prokaryote to eukaryote to multicelled organisms is crudely correlative with the maximum size of organism. Jonathan Payne and others have looked at the maximum size of organisms from diverse lines of evidence to produce a plot of maximum size through time, which shows well the progression of life over time (Fig. 13-11). All of this evidence indicates that the earliest life we can recognize was similar to the prokaryotic life-forms whose ancestors still thrive today. From the perspective of the geobiological detective, this is great good fortune, because study of prokaryotic cells and the communities they form today gives clues of what to look for in the rock record that would indicate when primitive ancestral prokaryotes first appeared. The question of the

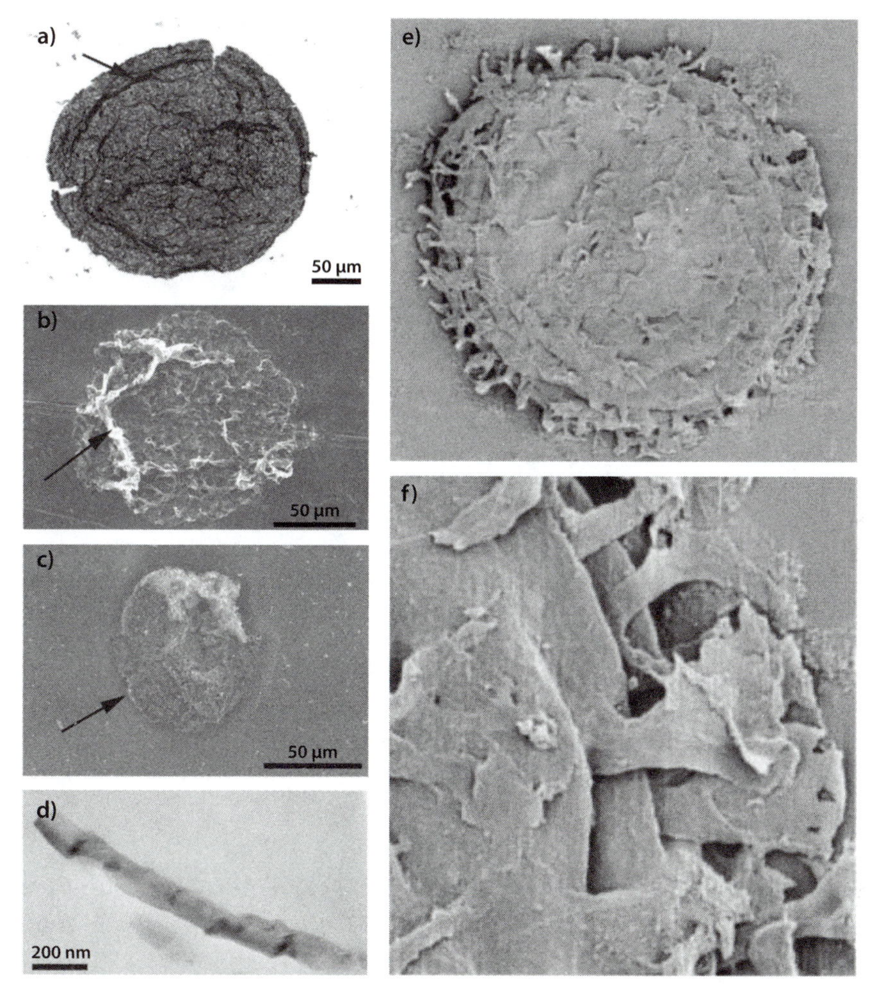

Fig. 13-10: Images of early micro-fossils. Images were produced with a transmitted light microscope, a backscattered environmental SEM, and a TEM. The images on the left, (a)–(d), are features interpreted as indicating cellular life in rocks as old as 3.2 Ga (Javaux et al., *Nature* 463[2010]:18). Images on the right are of the eukaryotic Shuiyousphaeridium macroreticulatum from the Ruyang Group, northern China, showing definitive evidence for eukayotic cells at about 1.5 Ga. Image size of (e) is about 300 microns across, and for (f) showing a feature of the cell, about 40 microns across (Javaux et al., *Geobiology* 2 [2004], no. 3: 121–32).

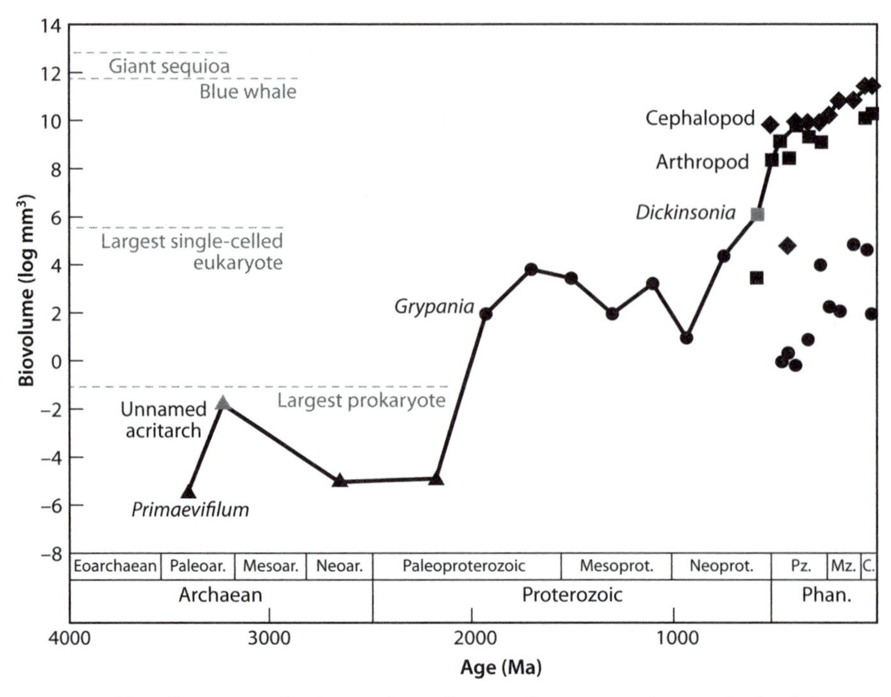

Fig. 13-11: Plot of maximum body size through time of organisms from rocks (and the present day.) Triangles are prokaryotes, circles are eukaryotes, squares are animals, and diamonds are vascular plants. (Modified from Payne, et al., *Photosynth. Res.* (2010) DOI 10.1007/s11120-010-9593-1)

origin of life can then be reduced to the origin of the simplest life we know—a cell that has the necessary characteristics to be the common ancestor from whom all subsequent life has evolved.

Life's Origin

The background given in the previous section provides a framework for an understanding of life's origin. Life evolved in early Earth's history, most likely before 3.5 Ga where direct evidence is lacking. All of life today has a startling unity of chemical composition that reflects the solar system and a great specificity of process that shows the relationship among all living organisms. Evolved life can be understood as a progressive development from the simplest prokaryotic organism, which we can examine today and evidence for which we see in the rock record. The origin of life then becomes the question of how the early planetary environment gave rise to the most primitive common ancestor of all of life in a fairly short period of time early in planetary history. What are the steps that would be necessary for this simplest life form to develop?

Fifty years ago this question was not readily accessible to scientific investigation, and while there are still great challenges, that is no longer the case. The increasing understanding of the molecular machinery of cells makes the chemical description of life's operations ever more precise, and the developing understanding of early planetary environments leads to specific laboratory experiments where the potential development of this machinery can be investigated. The origin of life then stops being a vague puzzlement—how did this mystery of life begin?—and can be broken down into a precise series of more specific questions. Given the increasing knowledge of the conditions of early planets, how can the various components of life arrive naturally in the absence of living organisms? How can the basic organic building blocks form? How can these molecules join together into the larger polymers that are needed? How can the basic structure of a cell membrane be formed? How can steady state chemical cycles come into being?

This change from imponderable big questions to focused, specific questions is characteristic of the progression of science. Throughout this book we have encountered questions that not so long ago were imponderable mysteries, now revealed to be subject to quantitative understanding. How the universe began—now revealed from evidence of the Big Bang. Where the elements came from—revealed in the understanding of stellar interiors. How old Earth is, and how it has stayed hot enough to be geologically active for so long—revealed by the discoveries of radioactivity and convection. How characteristics such as family resemblance are passed from one generation to another—revealed in the structure of DNA. Why continents fit like a puzzle across the Atlantic—revealed in the precise operation of plate tectonics. "How did life begin?" is another of these great questions, and one for which a satisfying understanding is not yet determined. What we will see in the remainder of this chapter, however, is that from the ability to pose increasingly precise questions, the architecture of an understanding of life's origin is rapidly developing. By "architecture" we mean the overall structure of a solution, with a much more satisfactory understanding likely to emerge in the coming decades of this century. This framework comes from two directions:

(1) a more thorough understanding of how life works through the continuing revolution in chemical biology;
(2) a more thorough investigation of present and past planetary environments that might provide the raw materials and flow of energy that would make early life possible.

Steps in the Path to Life

Our understanding of the history of life shows us that the most primitive living organism we know, and that found most deeply in the fossil record, is the prokaryotic cell, which contains the fundamental chemical machinery that is the basis of all complex life. The fossil record reveals the progressive diversification and growing complexity of life, the theory of evolution provides the framework or understanding for this development (see Chapter 14), and the understanding of DNA provides the detailed chemical mechanism by which it takes place. Fossils, evolution, and chemical biology all conjoin to relate the most primitive cell to the present diversity.

The situation is not quite so simple, however, because there is no known organism that can be placed at the base of the Tree of Life, from which all other organisms can be descended. All of present life is evolved, and the earliest record of life's evolution remains hidden. Prokaryotes today are not the representatives of early life, but rather their very distant descendants that have undergone billions of years of evolution. Nonetheless, life points in the direction of a common ancestor, whose characteristics we can infer. This unknown organism can be referred to as the *universal common ancestor* (UCA). The UCA would have the principal characteristics shared by all of life:

All cells are made up of the same limited set of elemental building blocks—H_2O, C, N, P, and so on that are in the appropriate state of matter.

All cells have a cell membrane that isolates the organism from its surroundings and across which chemical exchange can take place.

All cells operate with the same set of organic chemicals to carry out the basic mechanisms of life—carbohydrates, amino acids, nucleic acids, and lipids.

All cells work with amino acids that are left handed, and nucleic acids that are right handed, i.e. they have a definite "chirality" that is not random.

All cells have a cellular organization that manages chemical cycles within the cell, permitting a steady-state existence in the face of environmental change.

All cells have the means to replicate and pass information from one generation to the next.

We now turn to an exploration of how each of these steps may have developed.

ELEMENTAL AND SIMPLE MOLECULAR BUILDING BLOCKS

Hydrogen, carbon, oxygen, and nitrogen make up >98% of the mass of cells. As noted previously, the basic building blocks for life are among the most abundant elements produced by nucleosynthesis and are not in short supply throughout the galaxy. A greater challenge than the presence of particular elements is that the elements need to be in the correct chemical form and state of matter. Most scientists currently believe, for example, that carbon needs initially to be in a reduced chemical state, with some C combined with H rather than O, because it is much easier to build organic molecules from carbon in reduced (CH_4) form than in oxidized (CO_2) form.

An even more important requirement is liquid water. All living cells are composed dominantly of water. By weight, cells are about 70% water. Water has unusual chemical properties that give it an important role for life as well as for climate stability. It is a polar substance, which means that many substances can dissolve in it. We will see below that the polar property is also essential for forming the first cellular containers. It has high heat of fusion, heat of vaporization, and heat capacity, which allow it to remain in a liquid state over wide changes in conditions. And its solid form is lighter than the liquid form, which enhances convection and vertical circulation of bodies of water. These characteristics make water an indispensable medium for life. Molecules dissolve in it and are transported by it, and it provides a persistent and stable environment in which the various reactions necessary for life to begin could take place. Water is clearly in great abundance on Earth's surface today. Early Earth's history, filled with meteorite bombardments and even the massive bombardment that is believed to have led to the formation of the moon, would have caused surface temperatures that would boil water, and therefore liquid water is not likely in earliest Earth's history. As we saw in Chapter 9, however, there is evidence from the oxygen-isotope compositions of zircons for a liquid ocean as early as 4.4 Ga. Liquid water appears not to be an impediment to an origin of life far earlier than the first evidence for it in the geological record.

MAKING THE ESSENTIAL BIOCHEMICAL INGREDIENTS

With the necessary elements and molecules in place, the next step is to form the organic molecules that are the basis of living processes. The word *organic* is applied to these molecules because it was initially believed that they could be constructed only by living organisms and not by purely physical mechanisms. We now know that there are many physical processes that create organic molecules—there are so many, in fact, that it is difficult to know which of them may have been most important. In Chapter 3 we provided the evidence for organic molecules in the galactic interstellar medium and remarked that carbonaceous chondrites that arrive on Earth today contain organic molecules, including amino acids. Studies of comets have also revealed organic molecules. One possibility, then, is that Earth does not need to make the organic precursors to life, because they were delivered from the very reduced environment of the solar nebula during planetary formation. We cannot be sure to what extent they would not be destroyed by impact, nor can we be assured that they were present in sufficient quantities. As it turns out, Earth itself also has diverse means to make simple organic molecules.

One of the most important experiments in this direction was carried out by Stanley Miller in 1952, when he was a student in the laboratory of Nobel prize–winning chemist Harold Urey. Miller designed an apparatus (Fig. 13-12) in which electrical discharges were applied to a mixture of water vapor, CH_4, H_2, and NH_3 that underwent cycles of evaporation

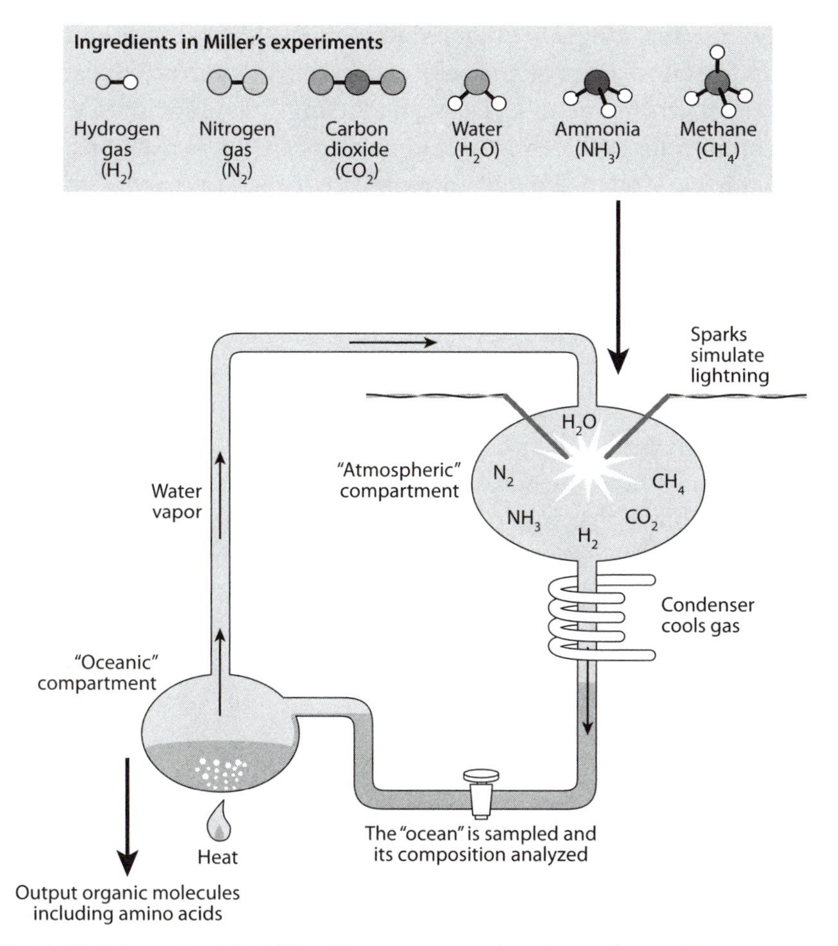

Fig. 13-12: Schematic of the Miller-Urey apparatus where it was shown that atmospheric processes were capable of producing a large variety of organic molecules, including amino acids.

and precipitation. This experiment produced a wealth of amino acids and other organic molecules. Subsequent experiments under varying conditions have been able to produce all the necessary amino acids, sugars, the bases essential for nucleotides and adenosine triphosphate (ATP) through reactions taking place in conditions appropriate to those which might have existed in certain environments on the early Earth.

Another environment that is promising is deep-sea hydrothermal vents. The vents have favorable attributes for chemical reactions—ubiquitous seawater, warm temperatures and large temperature gradients, gaseous compounds including H_2, other reduced molecules, disequilibrium conditions, metals that are often useful catalysts to organic reactions, abundant and diverse mineral surfaces, and mixing of fluids of diverse chemical compositions, all of which lead to many chemical reactions. These conditions are often difficult to replicate in the laboratory, since the reactions are often not at equilibrium and occur at high temperatures, high pressures, and with large temperature gradients. There has been substantial effort, however, to calculate from thermodynamics the diverse organic chemicals that could be produced under a variety of conditions. These calculations have shown that diverse synthesis of organic molecules can be generated by fluid mixing of hot, moderately reduced submarine hydrothermal solutions with seawater.

The various experiments, observations, and thermodynamic calculations have shown that amino acids, lipids, carbohydrates, and the building blocks for nucleic acids can be formed in a variety of potential planetary environments. Abiotic synthesis of many of the essential molecular ingredients of life has been demonstrated, proving the feasibility to form the fundamental building blocks of the complex organic molecules needed for life.

BUILDING COMPLEX MOLECULES

The molecules involved in life today are mostly more complex than the simpler organic building blocks generated in experiments. Some of the smaller building blocks need to join together to make more complex monomers. The nucleotides that make up RNA and DNA, for example, require joining of the nucleobases with the sugar backbone and phosphate groups. The monomers then need to come together in long chains called *polymers*, groups of the simpler molecules joined together like links on a chain. Amino acids combine to form the remarkable diversity of proteins and enzymes, following very specific rules. Nucleotides need to join in long chains to make up the nucleic acids of RNA and DNA. Simple sugars combine to form complex carbohydrates, and simple fats combine to make the large group of lipids, which in turn combine to make membranes. The essential next step to life is to form the more complex monomers and to join them together in polymers.

A plethora of possible processes to form polymers have been proposed and are under active investigation, but the problems are not straightforward ones. Forming nucleotides from the base, phosphate, and sugar components is not a problem for which there is yet a clear solution. Polymers do not form automatically after the monomers exist. There needs to be a high concentration of the monomers, so they need to be concentrated in some manner. Many of the reactions involve the loss of water, and this is difficult when dissolved in seawater. Furthermore, the polymers of amino acids have a left-handed chirality, and those that make DNA and RNA are right handed. Not only do polymers need to form; there needs to be a selective process that distinguishes right from left. None of these challenges are definitively resolved.

Most of the processes that generate the organic building blocks of more complex molecules create them in dilute concentrations, and further chemical reactions among these molecules require that they become concentrated. For example, if an amino acid is made in the atmosphere by a Miller-Urey process and then is carried into the ocean by rainwater, its concentration in the ocean is minuscule. Furthermore, many molecules have short lifetimes, because they are progressively modified by other chemical reactions or by heat or cold. Therefore, it is necessary not only to make the necessary molecular building blocks but also to concentrate them, and the available time is limited before they break down.

Since water is the common solvent needed for all life-forming processes, freezing and evaporation of water are two possible mechanisms of concentration. Amino acids can become strongly concentrated by evaporating the water from a dilute solution of them. The bond that combines amino acids, called the *peptide bond*, is produced by dehydration, and therefore the same process that concentrates the amino acids could also facilitate their combination. For example, tide pools, which may have been more abundant on the early Earth because of the much larger tides that would have resulted from the moon being much closer to Earth than it is today, could undergo repeated replenishment and evaporation. These concentration processes also lower the water content of the system, making dehydration reactions more possible. For example, letting amino acid–laden water evaporate on hot rock can lead to the formation of the peptide bond and the formation of amino acid polymers.

Concentrating the organic ingredients is not alone sufficient to create the characteristic chemistry of living organisms, because life as we know it today is highly selective in "handedness" of the molecules. All of the amino acids but one (glycine) can occur in both right-handed and left-handed forms, and the natural planetary processes that are able to make amino acids produce approximately equal amounts of the left-handed and right-handed varieties. But only the left-handed form appears in living organisms. Chirality is essential because an important property of proteins is their physical shape, and this depends on how they angle and bend as the many amino acids that make them up are combined together. The chiral uniformity of amino acids is central to the operation of life as we know it.

Chirality is also important for RNA and DNA, because the sugar ribose can occur in both left-handed and right-handed forms. Only the right-handed form is found in nature. This selectivity is what permits the formation of the right-handed double helix, always turned in the same direction, and symmetrically opposite to a left-handed helix. Selective chirality is an essential aspect of terrestrial organisms.

The origin of selective chirality is not yet clearly understood. Some experiments have shown that incorporation of left-handed components can terminate the growth of a right-handed helix. In this case, only chirally uniform helices would grow, leading to their ultimate success. For amino acids, the origin of the chirality is even less clear. In living cells, the chirality is controlled by the enzymes that are involved in protein synthesis—they are chiral and therefore preserve the uniform handedness of protein synthesis. The initiation of such chiral selectivity thus seems to require a chiral template that would selectively bond only with one of the two forms of amino acids, even though both were present in the initial chemical soup.

One speculation is that mineral surfaces may pose a possible solution to both the concentration and chirality questions, and also aid in the formation of much larger molecules from the simpler organic building blocks. Experiments have shown that minerals can form single layers of molecules on many mineral surfaces. Clays are of particular interest because their layered structure and fine grain size provide many regular surfaces with properties that are useful for the arrangement of molecular layers on the surfaces of the minerals. These surfaces provide a mechanism for concentration of molecules from water, for interactions between the surface-bound molecules and other molecules in the water, and for formation of polymers as one monomer after another is absorbed onto mineral sites that have the same configuration. Clays are also of fine-enough grain size that they are often suspended in the water, enabling greater scope for interaction and transport. An intriguing possibility is that mineral surfaces may also contribute to monochirality. Some mineral surfaces are chiral, and layers absorbed onto them could also have a single chirality. Therefore, a mineral surface could be a site of molecular concentration, select for particular molecules, provide an environment for the formation of polymers, and accept only molecules with the same chirality.

One could even speculate that there might have been left- and right-handed forms of early life. They would not be able to interact effectively with each other—organic reactions taking place in one form would likely be inert or lethal to the other. Having both survive would not be stable, and therefore inevitably one died out and the other persisted.

A CELLULAR CONTAINER

All cells are contained within a cell membrane that isolates the chemical contents from the external environment and allows selective transport of matter across the membrane to maintain stability, import nutrients, and export waste. Creation of a suitable container is essential for life as we know it.

The characteristics of the membrane container have much to do with the particular chemical properties of water. Water has two small, positively charged hydrogen atoms on one side of the molecule and a large, negatively charged oxygen on the other end. This makes water a polar molecule with positively and negatively charged ends, and these ends align like little magnets to create an ordered array in the liquid. The polar properties of water have a large influence on what other molecules can be dissolved in it—polar molecules generally dissolve easily in water, while nonpolar molecules, like fats and oils, do not. There are a class of molecules that have one end that is polar, and hence hydrophilic, and

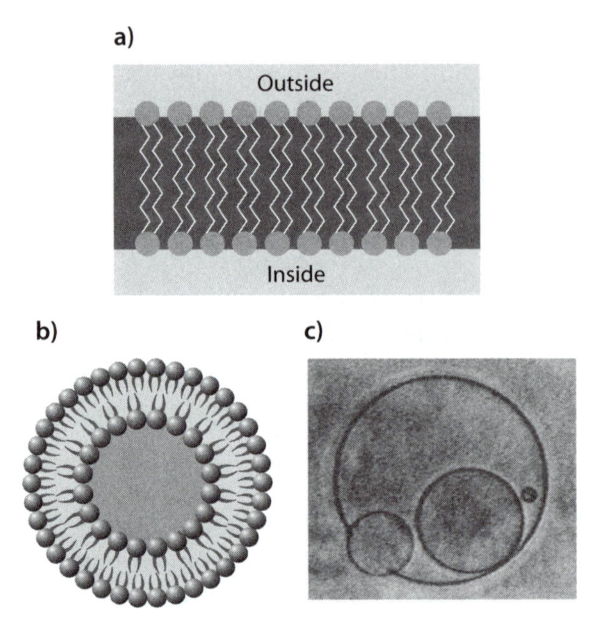

Fig. 13-13: Illustration of how early cell containers may have formed. Fatty acid molecules with hydrophobic and hydrophilic ends form bilayers (a) so that only the hydrophilic ends are in contact with water. These then bend into spheres, called *liposomes* (b) that completely isolate the hydrophobic ends of the layer. Modern cell membranes have this sort of structure. Image (c) shows the experimental formation of such structures, and how they can combine and incorporate other liposomes. They are also able to split and divide.

another end that is nonpolar, and hydrophobic. The hydrophilic end likes to be dissolved in water, while the hydrophobic end wants to avoid the water. Detergents have such properties, which explains their tendency to form bubbles.

Cell membranes are made up of fatty acids. Once formed, the fatty acids are very stable and are not easily destroyed, and their longevity gives them ready availability for prebiotic processes. These acids have hydrophilic and hydrophobic ends. When placed in water, the fatty acids prefer to have their hydrophilic ends in contact with the water and their hydrophobic ends isolated. This leads to the formation of *bilayers*, with hydrophilic ends on the outside and hydrophobic ends on the inside (Fig. 13-13). An even more stable configuration is to wrap the bilayers into a sphere, with the inner and outer surfaces consisting of the hydrophilic ends of the molecules, which completely isolates the hydrophobic ends from water. These little spherules, called *liposomes*, are very similar to essential features of the cell membrane, though modern membranes have evolved much intricate cellular machinery to facilitate transport between the cell interior and the external environment.

These various considerations suggest a simple process for the formation of early fatty acid containers that could incorporate other organic components in the initial steps leading to cellular precursors.

THE MISSING LINKS

At this point we reach the largest gap in our understanding of life's origin. Thus far we have discussed the plenitude of the necessary elements for life, the diverse environments for making the simple organic building blocks, the potential for concentration of monomers and the formation of polymers, possible mechanisms for selection for one chiral form, and the arising of a cellular membrane. All of these are essential steps for life—but life is a self-sustaining and self-replicating process that is well beyond any of these steps. None of the steps discussed thus far lead us to the universal common ancestor from which all life evolved. There remains a large gap between the view looking backward in time from the present toward the universal ancestor that no longer exists and forward in time through early planetary processes that can lead to the essential building blocks of life. The unresolved gap contains the essential steps where all the necessary ingredients work together to form self-replicating chemical systems (Fig. 13-14).

A number of aspects make this gap in our understanding exceedingly difficult to fill. The first is that we are looking at a vast expanse of time, possibly as much as a billion years, with no good historical record. A lot can happen in a billion years—for example, in about half that time life has evolved from single celled organisms to the complex ecosystems that we see around us today. The recognition of the vast expanse of time that is available is difficult to encompass. Within this time period, hosts of processes could have taken place, and events with minuscule statistical probability may become inevitable. If, for example, there were a one in a million chance of some event, but there were millions of opportunities,

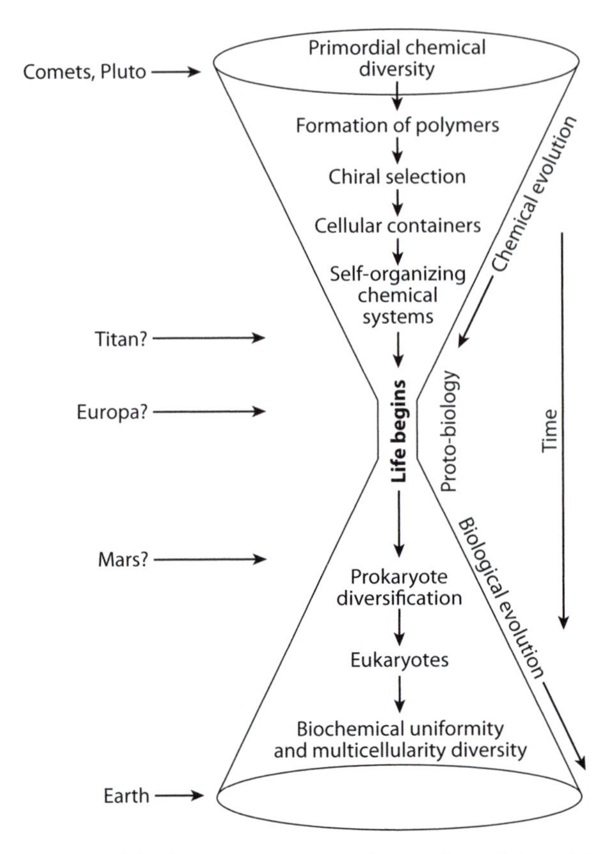

Fig. 13-14: Illustration of the bottleneck in our understanding of the origin of life. (Modified from Jonathan I. Lunine, *Earth: Evolution of a Habitable World* (Cambridge: Cambridge University Press, 1999))

the event moves from unlikely to inevitable. It would be as if one could purchase a lottery ticket every week for a million weeks—eventually one would become a winner. The long timescales create a difficult problem for laboratory experiments that must take place on the timescale of weeks to months.

Life's development must also have involved many sequential steps, taking place in environments for which we have few constraints. Laboratory experiments normally try to explore and constrain single environments—experiments with clay surfaces, or with hot water interacting with sulphides, or reactions in atmospheres subject to ultraviolet radiation—and emerging life probably involved exchanges and reactions depending on the diversity of environments offered by the Earth and the cycles among them. Chemicals made in one environment would be transported and react with others formed in an entirely different environment. And there could have been profound catastrophic events that were pivotal, such as impacting comets and meteorites, massive volcanism, and abrupt climate change. We know that such events have profoundly influenced life in the last billion years, and their frequency would have been much greater in early Earth's history.

It is evident that modern life as we see it is too complicated and interconnected to relate it simply to the basic ingredients discussed above. There must be a long series of intermediate steps that connect the fundamental building blocks of life to a fully functioning and self-replicating system. One of the intricacies of all of modern life is the relationship among DNA, RNA, and protein. DNA carries the cellular memory and instructions; RNA reads the instructions and enables the building of the proteins and enzymes. The proteins and enzymes are needed in turn to enable the communication between the DNA and RNA. Therefore, we have a classic "chicken and egg" problem—DNA is needed to code for protein formation, and proteins are needed to enable DNA to give the instructions. How could such a system evolve?

It seems almost inevitable that certain steps that were essential to the life-forming process disappeared as life moved on to the next step. We can consider this in an overly simplified and schematic outline. Imagine there are different environments where a series of chemical reactions take place ending up with the production of molecules A and B. These environments are specific to a certain period of Earth's history and no longer exist. Then A and B react to form molecule C, C converts back to A and B, and they become part of a stable cycle involving production and release of energy, sustained by sunlight or a hydrothermal vent, for example. From present observation, A and B are needed to make C, and C is needed to make A and B. This chemical cycle has, again, a "chicken and egg" character. The cycle survives because it is a cycle, not a one-way reaction that goes to completion. This cycle can then interact with other cycles to create far more complex relationships, which will also have precursors that disappear from view. Now imagine the scientist who comes along at the end trying to figure out how it all began, with no direct knowledge of the environments that gave rise to the precursors to the current system.

"Chicken and egg" paradoxes arise from a temporal series of reactions and relationships that become progressively linked and evolve with time.

The DNA-RNA-protein cycle is a complex cycle involving innumerable steps with highly specialized enzymes to make it efficient. It surely resulted from a complex history of thousands of small changes. This recognition has led to the idea of a simpler form of replication and protein formation that might have preceded the fully developed system that survives today in living organisms.

One idea is that the DNA-RNA-protein connection was preceded by an "RNA world" where DNA did not yet play a part. RNA has the advantages that it carries information in its nucleotides, like DNA (only one of the four nucleotides of the DNA and RNA differ from one another); it can serve as the facilitator of protein construction; and it is more amenable to prebiotic synthesis. What gave additional support to the idea of an RNA world was the discovery that one form of RNA could also serve as an enzyme, called a ribozyme. Therefore, RNA alone has the potential to fulfill the major necessary functions for a primitive cell—memory, replication, protein synthesis, and enzymatic activity. This led to the idea of the "RNA world," which may have preceded a more evolved world in which DNA took over the function of inherited memory from one generation to the next. DNA is more stable than RNA and is a better storage device for cellular memory. In the long run, a DNA-based system will have an evolutionary advantage, while an RNA-world may be a necessary but no longer existing precursor.

Some General Considerations on the Origin of Life

The various steps that we have discussed show the progress and existing problems with developing a verifiable understanding of life's origin. Important new developments in this field appear frequently in the scientific literature. One of the difficulties is the diversity of environments on Earth. If life is a planetary process, then it is not a test tube process. For example, some have proposed that ancient hydrothermal vents are favorably locations for the origin of life. Some of the most primitive bacteria are thermophilic (liking heat), consistent with a vent location. And hydrothermal vents are a source of concentrated energy and chemical gradient, shielded in large part from the ultraviolet radiation and destructive impacts that would characterize near-surface reservoirs. Such environments are very difficult to replicate in a laboratory—most modern vent organisms cannot be cultivated in the lab. The natural scientific method is to strictly control variables and experimental parameters, where it may be fluctuating and diverse conditions that are important for aspects of life's origin. It is perhaps even more likely that it is the interactions among diverse planetary environments that are necessary. With this diversity and the hundreds of millions of years of time available, the human laboratory challenges are daunting.

There is a larger question, however, which is whether there is a fundamental tendency toward life on planetary systems or whether life on Earth is a rare and highly unusual occurrence, requiring a whole series of statistically unlikely steps. For this larger question, two aspects of life can seem particularly perplexing. One is the argument that life violates the laws of thermodynamics. In thermodynamics you always lose—there is an inevitable tendency toward increasing disorder, or *entropy*, and there is an inevitable energy loss in every process, so that you never get as much energy out as you put in. Life appears to violate these principles because life is the appearance of order from disorder, and with evolution one could argue that the extent of order has increased with time. And life also creates energy—plants take raw ingredients from the air and soil and through photosynthesis turn them into more energetic compounds whose consumption provides the basis for animal life. Separation of carbon from oxygen creates an energy potential that can be used up the food chain or burned and used to power a modern civilization. So how can we understand the creation of order and energy within a universe that is bound by the laws of thermodynamics?

The question of energy is solved by the fact that order and higher energy compounds are created in nested systems, where the smaller system is making use of the energy from the larger system. There is enormous energy flow coming into the atmosphere from the sun, and coming from Earth's interior. Life on Earth's surface lives off these external energy sources. Within the solar system as a whole, energy is flowing downhill, and most of the energy is lost. Plants make use of solar energy, but the efficiency is not complete, and less energy is produced by photosynthesis than is received from the incoming photons. Therefore, life makes use with inevitable inefficiency of the dissipating energy of the universe. Life is possible only because it is a small part of a larger system. We owe our existence entirely to ongoing energy flow from the Universe,

The question of arising order is perhaps more intriguing, because it seems to violate the law of increasing entropy. Life makes sense from a thermodynamic point of view, however, if the scale is again adjusted so that there is maximum entropy production in the larger system, i.e., that the rate of change in entropy is maximized. Processes are more successful if they are able to generate entropy with greater efficiency. This principle can be easily appreciated from simple physical systems. If a pan has two holes in it, one larger than the other, most of the water flows out of the larger hole. If there are two water wheels, one with less friction than the other, the lubricated wheel will spin faster, process more water, and generate more energy per unit time. Whether liquids convect or not depends on which process more efficiently dissipates the available energy. Rocks take the steepest path if they fall down a slope. Processes that make use of the available energy with most efficiency garner that energy, and "win" relative to less efficient processes.

Life can also be viewed in this context. For example, a piece of wood is out of equilibrium with the atmosphere—the organic compounds in the wood are thermodynamically unstable in the presence of oxygen. Dry wood in a sterile environment, however, decays very slowly. We build long-lived houses from it. If the wood is put in the ground with moisture and bacteria, the bugs use the available energy and cause the wood to rot much more rapidly. Termites are even more efficient entropy producers than bacteria. Similarly, bacteria are essential for efficient operation of a compost heap. Worms in compost heaps are even better. Our human bodies are very efficient in this respect. We ingest plant and animal matter and in a few hours convert it to lower energy forms, while decay of the same matter sitting on the kitchen counter or even in a compost heap takes weeks or months.

What about plants? If you imagine green-colored ground with the same reflection characteristics as a leaf, and no photosynthesis, and you leave it in the sun for a brief period of time, it will warm up, and then the heat will gradually dissipate if it is taken out of the sunlight. A leaf undergoing photosynthesis stays cool because it is immediately converting the sunlight to chemical energy. The leaf causes greater entropy production compared to inert matter.

Any process that can take a potential energy and make the reaction run more efficiently to completion is favored, i.e., it succeeds relative to less efficient processes. Viewed from this perspective, life is a process that maximizes processing of energy. Even evolution might be viewed from this context, where evolutionary change is a series of steps of progressively greater rates of entropy production. Far from struggling against entropy production, life maximizes it! And this explains its great success. Life occupies every ecological niche, makes use of a host of energy sources, and appears in abundance out of nowhere once water, energy and nutrients become available. If we view life as an inevitable outcome of maximizing energy use, then this characteristic of life seems natural. Our own success as a species makes sense in this context—by use of tools and fuels we are able to harness the available energy from our environment far more efficiently than any other organism.

If one views life as a process that leads to more efficient dissipation of energy, then the origin of life no longer seems a statistical improbability but rather a natural outcome of the energetics of the universe. From the perspective of entropy, rather than defying the fundamental thermodynamic law of increasing entropy, life ruthlessly obeys it. The creation of order entails creation of greater disorder in the larger system, so the combination of energy flow and order creation maximizes entropy production. The series of steps leading to life's origin would then be driven by their success at maximizing energy use and the rate of production of entropy. Enzymes, the efficient drivers of the chemical reactions of life, succeed because they allow the available energy to be processed more efficiently. Symbioses occur because the interactions allow each organism to process energy more efficiently. And systems that are able to grow and replicate and make use of cycles to replenish the necessary raw materials will be more successful than systems that use up the energy and material without cycles and reproducibility. From this perspective, while many of the detailed mechanisms of life's formation remain to be elucidated, the driving force for planetary life is a fundamental feature of the Universe.

Summary

The chemical composition of life roughly corresponds with the solar composition, and the basic chemical ingredients for life are present in abundance. Central to life is the element carbon and the water molecule. Carbon has unique attributes among all the elements of the periodic table, making it ideally suited for life. Water also has exceptional chemical properties that give it a central role. Both water and carbon molecules were present in earliest Earth's history, providing suitable environments for life's origin.

All of life is remarkably coherent in terms of its fundamental molecules, molecular machinery, and cellular structure suggesting a common origin. Earliest life, likely beginning before 3.5 Ga, was prokaryotic cells, and evolution of life since that time can logically be related to them. The task for discovery of the origin of life is to find a candidate for a universal common ancestor that could give rise to the simplest prokaryote, from which all subsequent life could have evolved over billions of years. A possible framework of steps leading to life's origin can be laid out, and many of the steps can be approximated in laboratory settings. Because there must also be many unknown steps that depended on the details of the planetary environment prior to the oldest rock record, a true biochemical replication of an origin of life remains problematic. The most difficult step is the passage from polymers contained in a "cellular" container to a self-replicating organism that can evolve by natural selection.

By the creation of order and increased efficiency of energy dissipation, life enhances the creation of entropy in the larger system. In this sense, life is a natural response to fundamental thermodynamic laws. Given a suitable planetary environment, life can then be viewed with some justification as a natural consequence of planetary evolution.

Supplementary Readings

Andrew A. Knoll. 2004. *Life on a Young Planet: The First Three Billion Years of Evolution on Earth*. Princeton, NJ: Princeton University Press.

Jonathan I. Lunine. 2005. *Astrobiology: A Multidisciplinary Approach*. Boston: Pearson Addison-Wesley.

J. William Schopf, ed. 2002. Life's Origin: The Beginnings of Biological Evolution. Berkeley: University of California Press.

Alonzo Ricardi and Jack W. Szostak. 2009.The origin of life on Earth. *Scientific American* 301(3):54–61.

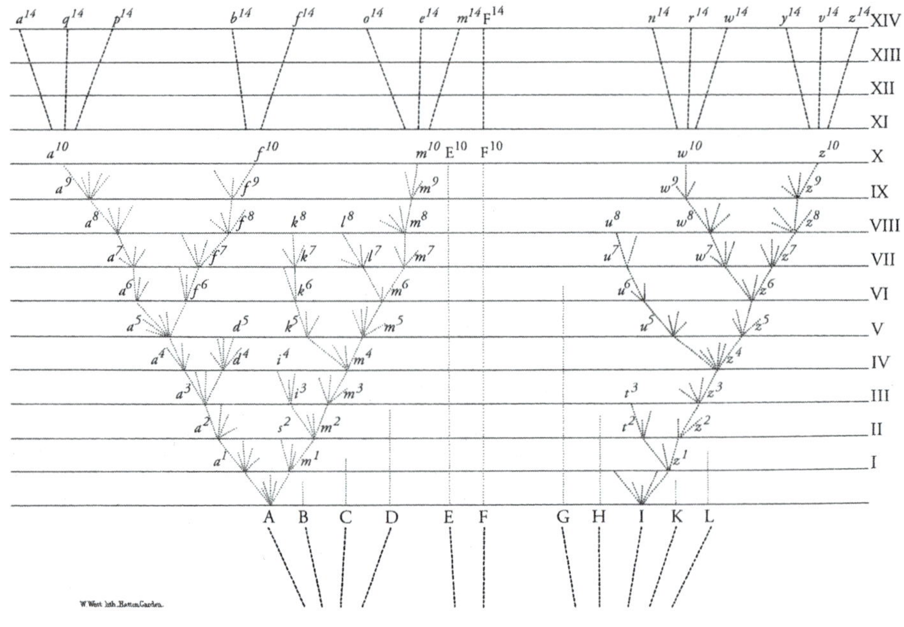

Fig. 14-0: Darwin first introduced the concept of "the Tree of Life" with this figure in *The Origin of Species*, published in 1859. He wrote: "From the first growth of the tree, many a limb and branch has decayed and dropped off; and these lost branches of various sizes may represent those whole orders, families, and genera which have now no living representatives, and which are known to us only from having been found in a fossil state.... As buds give rise by growth to fresh buds, and these, if vigorous, branch out and overtop on all sides many a feebler branch, so by generation I believe it has been with the great Tree of Life, which fills with its dead and broken branches the crust of the earth, and covers the surface with its ever branching and beautiful ramifications."

CHAPTER 14

Dealing with the Competition

The Roles of Evolution and Extinction in Creating the Diversity of Life

All of life has remarkable unity in terms of its cellular structure, metabolism, and chemical pathways. As we look around us, however, we can only marvel at life's diversity. Early naturalists categorized the diversity of life by a classification system, ranging from kingdom at the broadest scale to species for unique organisms. Originally plants and animals were the two kingdoms, but the deepening revelation of the microbial world led to five or six kingdoms within three "domains" of life—bacteria, archaea, and eukarya. Both plants and animals are made up of the complex eukaryotic cells, and are in one domain. The single-celled prokaryotes have much of the genetic diversity of the planet and are so different from one another that they make up two domains, bacteria and archaea. How could such diversity arise from a common origin?

The links between past and present are preserved as the fossils of the geological record. The progressive changes in life through Earth's history were revealed through the careful study of layered sedimentary rocks, revealing a biological history where generations of organisms went extinct and new species appeared. Current organisms can be traced back to common ancestors in the geological record. The geological timescale is divided largely on the basis of sudden changes in the life assemblage. For this reason there are far more divisions after the first fossils with hard body parts appeared 543 million years ago, separating the Phanerozoic from the Precambrian. The Phanerozoic is divided into three eras, bounded by mass extinction events. Charles Darwin investigated current life and the fossil record. He observed vast differences in life between different continents and smaller differences between neighboring islands. He proposed the theory of evolution—that life would progressively diversify through time by the process of natural selection, where competition and environmental change would stress populations and lead to more successful reproduction of individuals with certain characteristics. Small changes over long periods of time could lead to the diversity of life. Darwin's intuitive theory was vindicated by the discovery of DNA that provided a mechanism for inherited characteristics and for steady small changes in DNA sequences. Current DNA sequences also provide a living record of ancient change. Organisms with similar DNA have a recent common ancestor, while those with very different DNA diverged from each other in the ancient past.

Evolution of macroscopic life is too slow to be seen on human timescales, because new species appear by very gradual modification of existing ones. For bacteria that can undergo multiple generations in a day, evolution is fast enough that it can be observed in the laboratory. Concrete experience of evolution can be seen for macroscopic life by understanding that evolution requires both species origination and species extinction. The geological record shows that barring mass extinctions, the background extinction rate is about 0.00001% per year, which leads to 99% species reduction in 40–50 million years. During geological time, however, the number of species has greatly increased, showing that species generation has outpaced species extinction. While species origination is gradual, species extinction can be abrupt. Human domination of the planet has vastly increased the extinction rate, by a factor of 10,000, and consequently the extinction half of evolution is very evident to us. If current human-induced extinction rates continue, much of the current diversity of macroscopic life will be gone in a few hundred years.

Introduction

In the previous chapter we emphasized the unity of life when viewed on the cellular level and considered how the formation of a first protocell, the universal common ancestor to all of life today, might have come into being. These cells would have been so small that they would be visible only under a microscope, and life might not have been obvious to the untrained and unaided eye examining Earth's surface.

What a contrast to life today! Life is everywhere, and its diversity can be awe inspiring: giant sequoias, oysters, cobras, honey bees, human beings, and mold growing on old food—everywhere we look, even in deserts and glaciated regions, life can be found. Clearly, habitability has increased markedly over Earth's history. Our habitable planet is not just one where life was able to begin, but where life has flourished to its current omnipresent state. The building of a habitable planet includes the initial conditions that have been discussed thus far, and the evolution of the planet from the first primitive life more than three billion years ago to the complex and diverse ecosystems that we can observe today. Our task then is to describe this change and to explore the mechanisms by which it has come about.

How can we adequately characterize life's diversity? At the same time as we see diversity, we also see great commonalities among different types of life. Mammals share many characteristics, as do many flowering plants. We intuitively perceive a scale of similarity, with some organisms very similar to one another, others slightly similar, and others vastly different. Such obvious observations have long led people to try to classify life systematically. Naturalists in the eighteenth century went to great lengths to try to describe, categorize, and make sense of the diversity of life that they could see (they were only vaguely aware of the vast microbial world). Carolus Linnaeus, a Swedish naturalist, was one of the most influential and created a hierarchical classification of plants and animals whose essential form has survived to the present day. The term *Homo sapiens*, for example ("man wise"), is an example of his *binomial nomenclature* for the naming of different species.

Linnaeus linked species by their common characteristics into groups, in ways that we can readily understand. Human beings, chimpanzees, and gorillas share many characteristics, as do wolves, dogs, and coyotes or bobcats, leopards, and tigers. A dog is more similar to a coyote than to a gorilla or a leopard. And all of these have common characteristics not shared by fish and lizards, let alone ants, flowering plants, and seaweed. Very careful consideration of these issues allows a classification of macroscopic life, where some characteristics (like plant and animal) are very broad and common to huge groups and others (such as having a single hoof and walking on four legs) are shared by few, down to the species level where the collection of characteristics is unique. Early naturalists had a passion for a careful classification of all of Earth's life that they could see. They were willing to undertake long arduous voyages, stopping at ports plagued with incurable diseases, in order to explore, collect, systematize, and try to understand life's diversity, and to preserve the specimens in newly established museums. The diversity of life and its organization into distinct groups based on shared characteristics was a great and exciting process of discovery of the life of the planet. These early systematists organized life into a hierarchical structure in order of decreasing generality and numbers of organisms—kingdom, phylum, class, order, family, genus, species (Fig. 14-1). Since they were little aware of the microbial world, the two kingdoms were plants and animals.

The development of the microscope permitted the discovery and investigation of microbes, and the discovery of the cellular structure of all of life. In the mid-nineteenth century a new kingdom, protista, was suggested for the microbial world. Once the prokaryotes and eukaryotes were clearly differentiated, it became clear that both plants and animals had a kinship in that they were both made up of eukaryotic cells. Ultimately, in the late twentieth century DNA analysis revealed that prokaryotes have fundamentally different genetic lineages, and they were divided into *bacteria* and *archaea*, also called *eubacteria* and *archaebacteria*. One is left then with three domains of life (eukaryotes, bacteria, and archaea), which can be divided into five or six kingdoms—two that are prokaryotic (eubacteria and archaebacteria), two that are the unicellular or simple multicellular eukaryotes (protists and fungi), and two for the diversified eukaryotic organisms (plants and animal) (Fig. 14-1).

DNA provides a rigorous and quantitative measure of diversity. A small but rapidly growing number of organisms have had their DNA sequenced, and the results can be compared to the visual and qualitative classification carried out by the early naturalists. The macroscopic changes we can see correspond to changes in the detailed molecular sequences of the DNA molecules that direct the appearance and metabolism of all of life. Human beings individually have 99.9% of their DNA in common, and the human species has 96% DNA in common with chimpanzees,

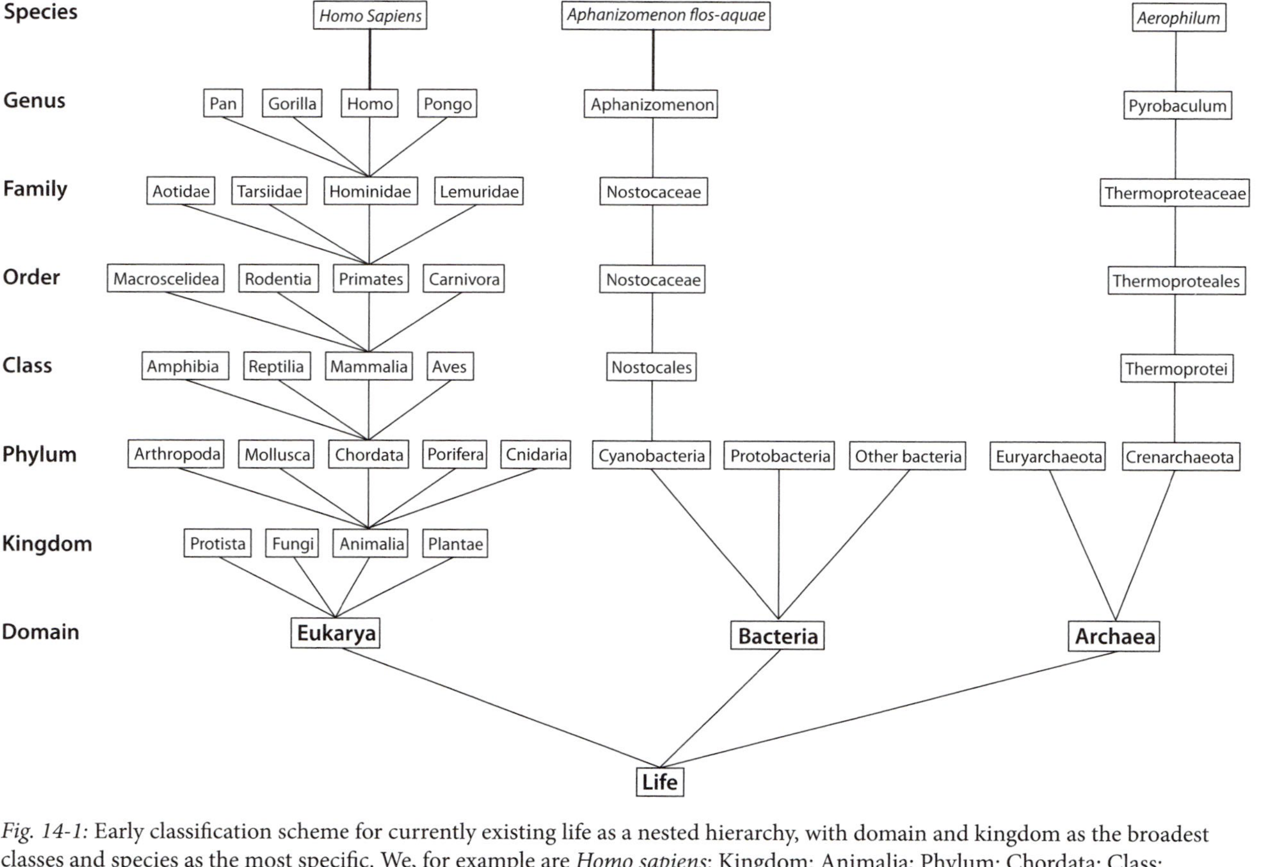

Fig. 14-1: Early classification scheme for currently existing life as a nested hierarchy, with domain and kingdom as the broadest classes and species as the most specific. We, for example are *Homo sapiens*: Kingdom: Animalia; Phylum: Chordata; Class: Mammalia; Order: Primates; Family: Hominidae; Genus: Homo; Species: Sapiens.

90% with mice, and 80% with the distant mammal, the platypus. Mammals and plants have only 22% of their genes in common. The correspondence between visual differences and DNA variation validates the comparison of living organisms with extinct organisms, for which visible features are preserved as fossils but DNA is no longer present. The discovery of the microbial world, where visual differences are few, adds a new dimension to this picture, which we will discuss further later in this chapter.

History of Life and Earth Revealed through the Rock Record

During the same time period that early biologists were identifying and classifying the diversity of modern life, geologists were working out the stratigraphic record, where they would try to order different rock layers from oldest to youngest. This also became practical and important to predict where coal seams might be found to fuel the burgeoning industrial revolution. Since radioactive dating did not exist in the eighteenth and nineteenth centuries, it was not possible to assign exact ages to particular rock formations. Instead, geologists had to rely on the relative ages of sedimentary rocks. The simple principles of relative age were deduced by the Danish geologist Nicolas Steno in the seventeenth century. He noted that sediments were generally formed by particles falling from or precipitating from water, and that individual sediment layers could be traced for long distances. Furthermore, the layer on top conformed in its shape to the layer beneath, showing that it was deposited after the deeper layer was already in place. The law of superposition codifies this principle: for a series of flat-lying strata, the ages become progressively more recent upward in the stratigraphic section. Of course, in detail there are exceptions to this principle. Sediments can be deposited on slopes, and can be thrust on top of one another by faults, but the principle is fundamentally sound. Undeformed sequences abound in the geological record (Fig. 14-2). The law of superposition can also be used to unravel complex sequences of deformed and faulted rocks.

Fig. 14-2: Two examples of horizontal strata, laid down one on top of the other following the *law of superposition*—in undeformed sequences, the younger rock is found on top of the older one. *Top panel:* sedimentary rocks from the Grand Canyon; *bottom panel:* basaltic lava flow from the Columbia River basalts in Oregon. (Courtesy of U.S. Geological Survey)

This procedure worked well for limited areas where the layers could be continually traced, but how could rocks from different places be compared to one another? The rocks themselves were rarely distinctive—one shale, sandstone, or limestone can look very much like another. Fossils provided the Rosetta Stone (Fig. 14-3). The thousands of animal and plant species identified in fossils could be carefully catalogued into fossil assemblages, the groups of animals that lived together in a particular time period. Fossil assemblages were found to vary in the same stratigraphic order on every continent. Even though the exact sequence of rock types might differ from one region to another, the changes in fossil assemblage were always regular. While no one location contained a complete stratigraphic record, the fossil assemblages permitted the partial columns to be cross-linked into a single whole.

The oldest sedimentary rocks did not contain fossils, and this nonfossiliferous time was initially all lumped together as Precambrian (Fig. 14-4). Well-preserved fossils with hard body parts appeared suddenly in the stratigraphic record, and these first layers were called the Cambrian period, for which a characteristic fossil is the trilobite (seen in Fig. 13-7, Chapter 13). Trilobites vary in detail as one ascends through the stratigraphic column, and then they disappear, never to be found in younger rocks. Much later, the first fishes appeared. Trees were even younger. In this way, a global stratigraphic column, ordered in relative time series with the oldest rocks at the bottom and the youngest at the top, was able to be constructed (Fig. 14-4).

In the absence of precise dates, either in our personal lives or in ancient history, we refer to time periods with names, such as "teenager," the Bronze Age, or the Middle Ages, brackets of time defined by beginning and ending events. Geologists used the same convention for their relative timescale, applying names to periods with characteristic fossil assemblages and creating a boundary where the fossil assemblage showed a distinct change, usually associated with the sudden disappearance of abundant species. The names were applied to a hierarchical system, eons, eras, periods, epochs, and—the finest time division—stages. (Only specialists have a detailed knowledge of the epochs and stages.) The appearance of animals with hard body parts was a defining event in Earth's history, and it was used to divide the Phanerozoic eon from the great expanse of earlier time initially called the Precambrian. Within the Pha-

Fig. 14-3: Fossil assemblages differ in major ways for different periods of Earth history. On top is an artistic rendition of an Ordovician ecosystem (500–425 Ma) where invertebrates dominate, and there were no vertebrates or plants. (Figure © C. Langmuir) On the bottom is an assemblage from the Permian, with advanced plants and dominant reptiles. (Image © by Karen Carr (www.karencarr.com)) Rocks from these two periods collect completely distinct fossil assemblages that allow rocks on different continents to be place in stratigraphic order.

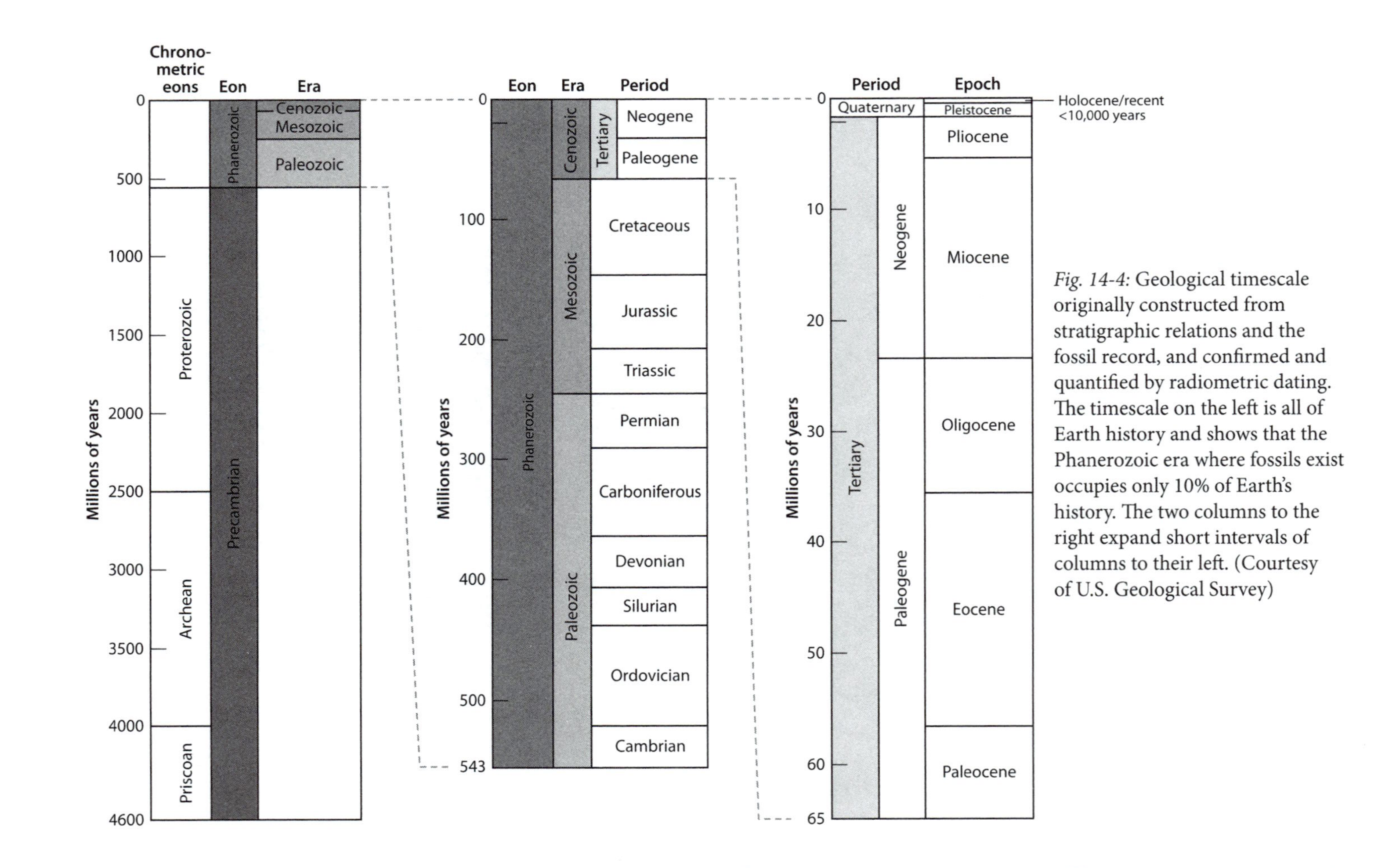

Fig. 14-4: Geological timescale originally constructed from stratigraphic relations and the fossil record, and confirmed and quantified by radiometric dating. The timescale on the left is all of Earth history and shows that the Phanerozoic era where fossils exist occupies only 10% of Earth's history. The two columns to the right expand short intervals of columns to their left. (Courtesy of U.S. Geological Survey)

nerozoic, there were occasional disappearances of large numbers of species, now referred to as *mass extinctions*. The largest extinctions were used to define the eras. The extinction between the Permian and Triassic led to the demise of some 80% of existing genera and separates the Paleozoic and Mesozoic eras. The disappearance of the dinosaurs (and about half of all other species) at the end of the Cretaceous is defined as the boundary between the Mesozoic and Cenozoic eras. The large numbers of extinctions, deforestation, and environmental change associated with the rise of human civilization probably merits another era boundary, as the whole surface of Earth is transformed by our activities.

For early geologists this record of temporal change was relative. Radiometric dating, as discussed in Chapter 4, allowed the hypothesis of a biologically constructed stratigraphic timescale to be tested. While sediments were made up of diverse particles weathered from rocks of different ages, igneous rocks found in the same intervals provided discrete dates. These dates verified the validity of the stratigraphic column and also assigned exact times to the geological periods (Fig. 14-4). The Phanerozoic/Precambrian boundary was found to occur at 542 Ma—only 12% of Earth's history!

Radiometric dating was gradually able to reveal more about the Precambrian, where the fossil record did not permit accurate cross-calibration of sections from different regions. The quantitative timescale has allowed the Precambrian to be divided into intervals that correspond with the eon-era-period framework. The expanse of time between the formation of Earth and the oldest known rocks is called the Hadean. (The younger bound of the Hadean is subject to change as older rocks are found.) The oldest rocks, currently the Acasta Gneiss in Canada dated at 4.0 Ga, define the beginning of the Archean eon, which extends to 2500 Ma. The next youngest eon is called the Proterozoic and spans 2500 Ma to the start of the Cambrian at 542 Ma. Because the Archean and Proterozoic encompass such vastness of time, they are each divided into three somewhat arbitrarily defined eras, and the Proterozoic also has well-defined periods. The youngest rocks of the Proterozoic eon are the Neoproterozoic era, from 1000 Ma to 542 Ma, a time of great importance. During this time there were multiple "snowball Earth" episodes in the aptly named Cryogenian period, and the early development of multicellular life in the youngest period of the Neoproterozoic, the Ediacaran, which immediately underlies the Cambrian period of the Phanerozoic eon. While the names initially seem somewhat arcane, they become like old friends with frequent usage. This overall naming of Earth time provides us with the vocabulary we will use to discuss events in Earth's history.

Relating Fossils to Present-Day Life: The Theory of Evolution

Early nineteenth-century scientists developed two macroscopic avenues to explore the diversity of life—existing organisms, showing diversity today, and the fossil record, showing diversity through time. How could these be related? How does life in the fossil record compare to the newly discovered systematics of living organisms, and has life changed through time? The remarkable discovery was that most fossils had no living counterparts. There are no living trilobites or dinosaurs, and even large mammals present in the most recent fossil record, such as mastodons and saber-toothed tigers, are extinct today. Then the task became to systematize the changes in the fossil record, using techniques similar to those which had been used for living organisms. Since most classification of organisms is based on morphological differences (e.g., how many legs does the organism have, does a beetle have five or six segments to its antennae, and so on), and fossils also reveal morphology, it is possible to relate current life to the fossil record. This comparison reveals that much of the current diversity of characteristics of similar species can be related to a common ancestor in the geological record. Working backwards, this common ancestor can be related morphologically to common ancestors of other species, finding a mutual common ancestor even deeper in geological time. These relationships give rise to the concept of the Tree of Life (Fig. 14-5), where the current diversification of species are the growing ends of the individual twigs of the tree and the ultimate source of life would be at the base of the trunk. Note that the Tree of Life shows changes through time, not to be confused with the treelike classification of modern life (Fig. 14-1), which deals only with currently living species.

The great leap in understanding the history and diversity of life was made in the mid-nineteenth century. At this time there was convincing

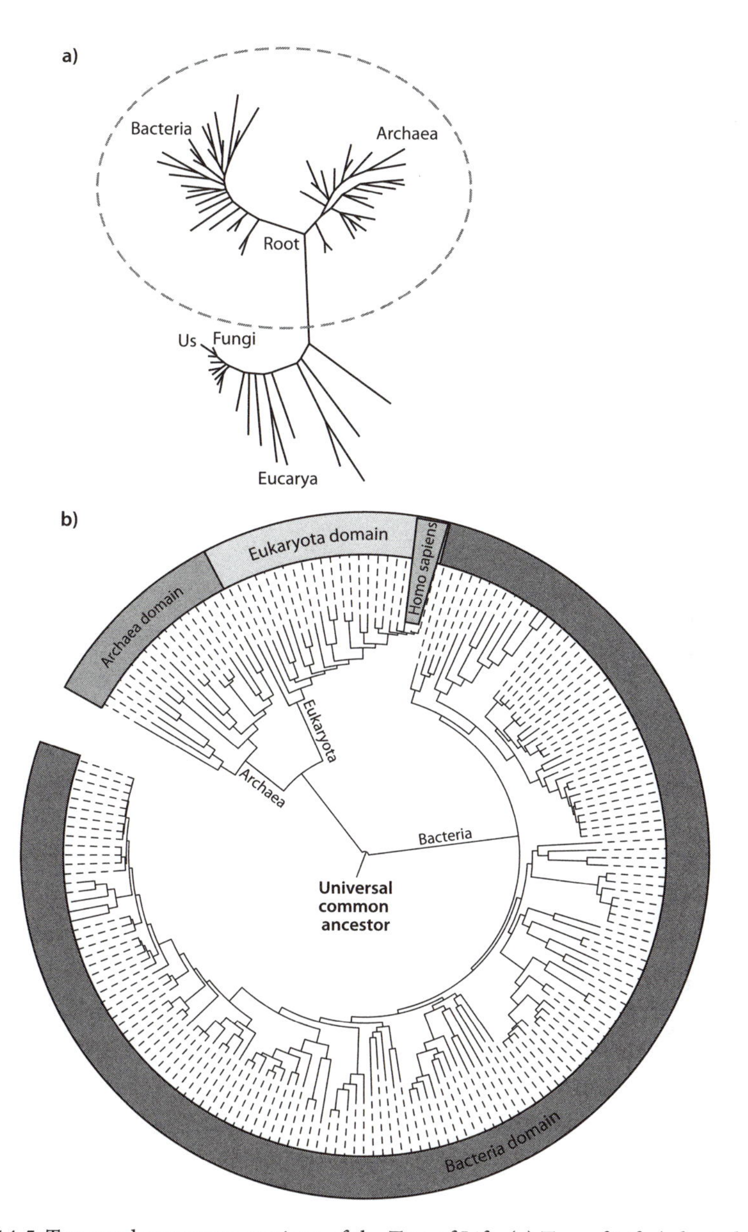

Fig. 14-5: Two modern representations of the Tree of Life. (a) Tree of Life (adapted from Pace, *Science* 276 [1997]:734–40); (b) Alternative Tree of Life based on genome sequencing. The center is the root of the Tree of Life and corresponds to the universal common ancestor of life. The different shadow areas represent the three domains of life: dark gray represents bacteria, gray represents archaea, and light gray represents eukaryota (protista, fungi, animalia, and plantae). Note the presence of *Homo sapiens* (humans) second from the rightmost edge of the light gray segment, and that most of the genetic diversity of life rests with the single-celled organisms (modified after iTOL: Interactive Tree Of Life; http://itol.embl.de/itol.cgi).

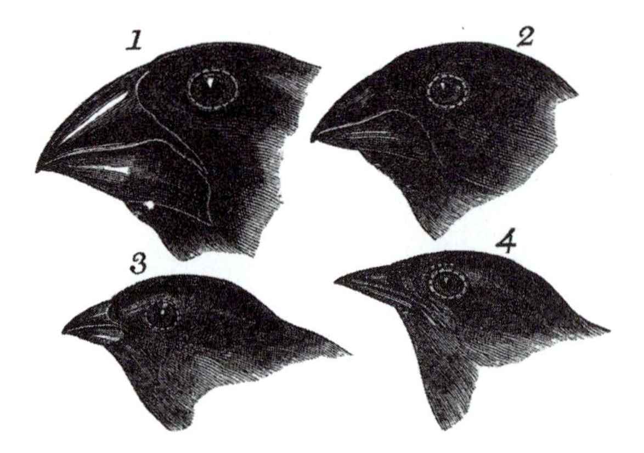

1. Geospiza magnirostris. *2. Geospiza fortis.*
3. Geospiza parvula. *4. Certhidea olivasea.*

Fig. 14-6: Darwin's original illustration of the slight differences in finches found on different islands of the Galapagos archipelago off the coast of Ecuador. Darwin noted that the finches had only slight differences from one another and were also similar to related species in South America, even though the climate was very different. He wrote: "There is nothing in the conditions of life, in the geological nature of the islands.... in fact there is a considerable dissimilarity in all these respects. On the other hand, there is a considerable degree of resemblance in the volcanic nature of the soil, in climate, height, and size of the islands, between the Galapagos and Cape de Verde Archipelagos: but what an entire and absolute difference in their inhabitants! The inhabitants of the Cape de Verde Islands are related to those of Africa, like those of the Galapagos to America."

evidence of a vast expanse of geological time and a developing understanding of the details of life's diversity as a whole. Charles Darwin in his travels saw that physical separation of continents like Australia led to large differences in living species, and in the Galapagos Islands he observed that small changes emerged for populations from neighboring islands that were barely isolated (Fig. 14-6). The accumulating body of information and recognition of the vast expanse of time suggested by the geological record gave rise to the grand synthesis of Darwin's theory of evolution. Darwin proposed that life would progressively diversify through time by the process of natural selection, where competition and environmental change would stress populations and lead to more successful reproduction of individuals with certain characteristics. Small changes over long periods of time could progressively lead to the diversity of life. This could be seen by differences that corresponded with the physical separation of living organisms, and also through the progressive change observed in the fossil record. In this way both the diversity today and the change through time could be related to a common process.

Darwin's ideas were controversial in the eighteenth century and remain difficult to understand and accept for many nonbiologists today. One of the difficulties has been that evolution operates over millions to billions of years, and our lifetimes are more than ten thousand times shorter than that, so we do not see new species emerging in clear ways within our human experience. Instead, life seems relatively unchanging—humans have always been here, oak trees are oak trees, dogs are dogs. The theory of evolution when it was developed was based on differences in life in different locations, on historical inference and reasoning, rather than direct observation of change and experiment. Perhaps we can accept the likelihood of small changes. But how can these lead to the large diversity of living organisms? Could a whale and a tree really have a common ancestor?

The same genre of issue was faced with geological time, where Earth seems relatively unchanging within human experience, and it is difficult to fathom the consequences of hundreds of millions of years of change. A similar disconnect made it difficult to comprehend plate tectonics. We are now forced to accept as fact the geological timescale, based on radioactive dating. And plate tectonics is a proven fact from the accurate measurements of plate movements today that correspond precisely with the inference from magnetic anomalies. Is there similar conclusive evidence for evolution?

The DNA Revolution

A stunning contribution to the understanding and development of the ideas of evolution was the discovery of DNA as the genetic material that controls the specificity of species and provides the mechanisms of inheritance. Since the turn of the century, as the genomes of various species become precisely described, we understand in ever greater detail the origins of differences within and among species. Mutations are no longer the hypothetical qualitative concept they were in Darwin's era;

they reflect precise changes in the base sequence of DNA strands. Mutations can come about in several ways. Replication efficiency is not 100%, and sections of DNA can be snipped, pasted, or duplicated, providing inevitable mechanisms of progressive biological change. Some changes will lead to fatality, some to a competitive advantage. The advantageous changes survive. DNA provides the mechanism for evolution and the inevitable transfer of changed characteristics from one generation to the next (Fig. 14-7).

There is also one domain where evolution can be observed experimentally on laboratory timescales, and that is in the microbial world. The number of generations and DNA replications that occur along with the selective pressure of environmental change is the control on the rate of evolution. For human beings with a generational change of roughly 30 years, a thousand generations is 30,000 years, well beyond our historical record or even perspective. To see evolution in action requires organisms with lifetimes short enough that thousands of generations can occur on human timescales, and environmental change can be manipulated. The simplest bacteria can divide every hour or so, leading to the possibility of twenty or more generations in a day. A single bacterium, call him Adam, could in a week go through one hundred generations, giving rise to a population of billions or more, depending on the ratio of reproduction rate to the lifetime of an individual bacterium and the availability of nutrients. A crisis such as lack of nutrients, change of physical conditions, or introduction of a toxic disease could lead to precipitous decline in population with only a few survivors with the special mutations that allowed them to survive, from which again a genetically distinct population could arise. If this species of bacteria were writing their history, it would be filled with grand events of population rise and decline occurring over hundreds of generations—an apocalyptic history in a test tube on a timescale less than a human summer vacation. In some laboratories these kinds of experiments have been carried out over a decade or more, leading to substantial genetic diversity.

Through these detailed experiments and observations, evolution has been observed in the laboratory. Biologists have been able to start with a single cell and watch it evolve through tens of thousands of generations, developing diverse capabilities and behaviors in response to environmental change, and then mapping these changes in the DNA of the

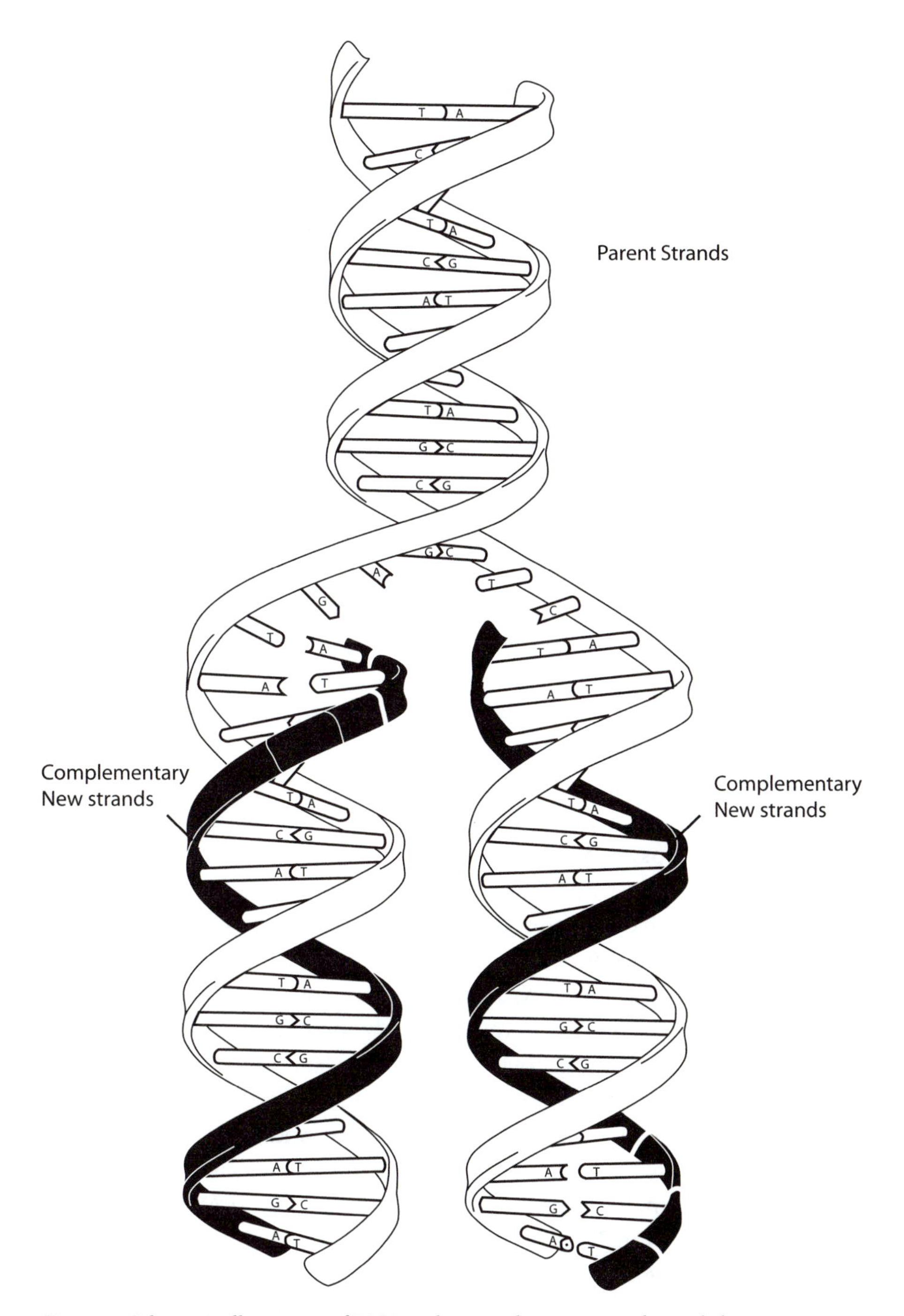

Fig. 14-7: Schematic illustration of DNA replication that permits inherited characteristics and mutations to be passed on from one generation to the next. (Illustration by artist Darryl Leja, courtesy of National Human Genome Research Institute)

evolved organism. Gene transfer of various types and the development of dependent symbiotic behavior have also been able to be observed experimentally. Thus, while changes in large organisms with long lifetimes are poorly accessible to us by direct experience, use of microorganisms with very short lifetimes reveals the reality of evolution by genetic change.

On the macroscopic scale, changes in complex species on the scale of centuries are evident to us upon reflection. The diversity and specialization of dogs and domestic animals and plants, and the common use of breeding to develop particular characteristics, are evident. Some of these changes approach species differentiation—breeding between St. Bernards and Chihuahuas is virtually impossible physically and would likely lead to the death of the mother if she were the Chihuahua. While less familiar to most city dwellers, the changes in crops through human intervention are also great. And the appearance of new diseases, as well as the development of new strains that are resistant to antibiotics, is evolution of great importance to the future health of human beings.

Knowing that progressive changes in DNA sequences are inevitable, and are observed in laboratory experiments with living organisms, evolution can be understood as an inexorable statistical process. If two populations of a species are isolated on two different islands, or isolated into two different ecosystems by environmental parameters, each one will gradually change in different directions. As time passes, these changes become progressively greater, until the point where the genetic similarity is too disparate for interbreeding. The extent of the genetic difference will increase with time, providing a kind of clock that dates the time since the two new species were genetically identical. This process, repeated over and over again over millions of years, leads to groups that had a recent common ancestor, and hence have similar DNA, and groups that only have a very ancient common ancestor, and greater differences in their DNA. It explains why species on nearby islands are subtly different, and why animals in Australia, such as the kangaroo, are very different from those found in North America. This process explains beautifully the fact from the fossil record that organisms in Africa and South America were the same at the time the two continents were joined together and then progressively changed to the distinct flora and fauna that are present on these continents today.

The quantitative character of species identification through DNA allows the Tree of Life to be constructed using quantitative methods rather than inferences from physical characteristics. The degree of similarity of the DNA quantifies the differences among species and allows estimates of the time since there was a common ancestor. In this way the present genetic diversity also contains information about past history. How many genetic changes would be necessary to make the genome of organism A identical to that of organism B? If genetic changes happen at a more or less regular rate, that makes possible a genetic clock providing an estimate of the time when two different branches on the tree of life divided (Fig. 14-5).

This concept shows that no two living species are descended from each other. A common misconception about evolution is illustrated by the statement, for example, that "man is descended from the apes." We often interpret this statement to mean that humans were descended from gorillas or chimpanzees, which is incorrect. Instead, humans and chimpanzees are cousins descended from a common ancestor. The two species now have about 96% of their DNA in common, fully quantified thanks to the sequencing of the human and chimpanzee genomes. Using the mutation clock, the differences require about 6 million years. The identity of the common ancestor can then be related to specific species in the fossil record, linking current diversity, established fossil history, and quantitative DNA measurements. The places in the DNA where we differ from chimps will eventually reveal details of the mutation paths that have led to the diversification of the two species from their common ancestor.

DNA quantification has also permitted the detailed investigation of the microbial world, where visual differences are not so evident as for macroscopic, multicellular life. While human beings have more than 3 billion base pairs, a typical prokaryote contains on the order of 1 million base pairs. As more and more bacterial species were sequenced, the remarkable discovery was that these unicellular organisms have such variable genes that they contain most of the genetic diversity of the planet. Carl Woese pioneered the new view of life's diversity based on DNA, and the eukarya that represent all the life we can see are only a small fraction of the total genetic diversity that has developed over Earth's history (see Fig. 14-5b).

Examining the bacterial DNA relationships in detail has also revealed another mechanism for evolution besides gradual mutations. Bacteria can sometimes transfer a gene from one organism to another. This type of change in DNA is referred to as *horizontal gene transfer* because it occurs laterally between different branches of the tree of life, rather than linearly along a single branch of the tree. Viruses are another mechanism of gene transfer other than mutation, since some viruses inject their DNA into a cell where it is incorporated in the host cell DNA. Horizontal gene transfer is abundantly observed in microbial life, to such an extent that many biologists think the strict linear tree-of-life concept no longer is relevant to the microbial domain and that the deep roots of the tree have many exchanges and interconnections, allowing one branch to exchange genetically with another. Microbial life is a vast storehouse of genetic information that can be transferred among organisms as they adapt to competitive pressures and diverse environments.

EVOLUTION OF LANGUAGE

The evolution of language is a useful parallel to consider on a shorter timescale the effects of small changes and interactions leading to diversity over time. While we are scarcely aware of language changes in our lifetime, the contrasts between modern English and Shakespearian English are clear to anyone who reads a Shakespeare play, and these changes have occurred in less than four hundred years. The differences between modern English and Chaucer, and Chaucer and Beowulf, are even greater, occurring over a thousand years. The differentiation of European languages from Latin has occurred in only two thousand years. The common Indo-European root that has led to languages as diverse as Hindi, Farsi, Russian, and Gaelic was about 5000 BC. Some of the changes occur through slow progressive change in a single isolated language. Some occur when there is an invasion or development of communication between two different language groups, and a new language that makes use of ingredients from both groups evolves. Longer times and greater isolation lead to greater differences. Greater communication leads to more sharing and commonality. The global language exchange that is now taking place is leading to the extinction of many minor languages. Languages also contain a history of where they came from, when they diverged from ancestral tongues, and how they were combined with other languages. And language is passed quite faithfully, but not quite perfectly, from one generation to the next. Language diversity demonstrates clearly how small changes and combinations that occur progressively through time lead to large differences, such that two populations no longer are able to communicate with one another.

The analogies with biological evolution are evident—small changes over time lead to separate languages where communication is no longer possible. Change is fostered by time and separation. New languages do not suddenly appear but emerge from small changes over time. Current languages have ancestral roots that can be traced to their common parent, and a "tree of language" can be constructed with many similarities to the tree of life. In mathematical terms, language differentiation has many similarities to biological evolution. DNA is the "language" of the cell. While biological changes happen more slowly than spoken language, there is much in common in the principles and outcome.

The magic of DNA from the perspective of the theory of evolution is that it provides a mechanism for inheritance, a simple and quantitative basis for understanding of gradual evolutionary change, and the possibility of rigorously quantifying diversity. It also contains history—a record of how the current diversity has evolved through time. DNA thus provides a very precise and quantitative mechanism for evolution. Scientists in the lab today can manipulate DNA strands to change the characteristics of organisms and produce new evolutionary mechanisms—human-mediated gene transfer.

The Extinction Half of Evolution

Another perspective on evolution is to realize that there must always be two complementary processes occurring—the extinction of old species as well as the gradual development of new ones. The total number of species on the planet will depend on the balance between those going extinct and those coming into being. The overall trend of numbers of

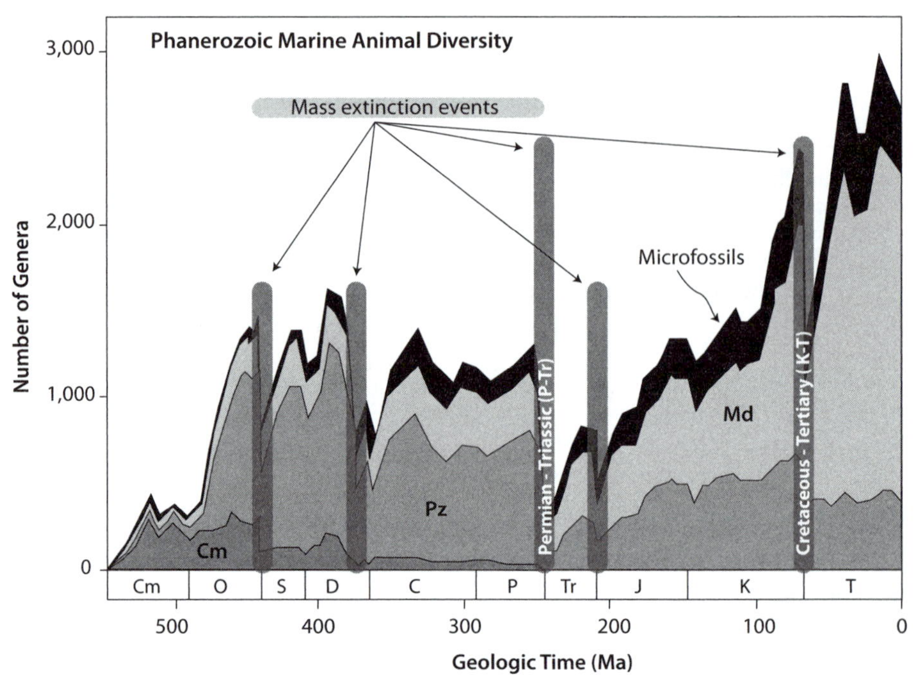

Fig. 14-8: Quantification of the number of distinct genera through geological time. Note that the number of genera has increased substantially over the Phanerozoic, but this overall increase has been punctuated by periodic abrupt extinctions, called *mass extinctions*. The five largest mass extinctions are indicted by the gray bars. The distinctively Cambrian fauna went permanently extinct at the Permo-Triassic boundary that separates the Phanerozoic and Mesozoic eras. (Modified after Sepkoski, *Bulletins of American Paleontology* 363 (2002). See also strata.geology.wisc.edu/jack)

species through time that emerges from the fossil record is shown in Figure 14-8. The geological record shows that there is a background extinction rate, punctuated by short periods of mass extinctions. During the largest of the mass extinctions, at the boundary that defines the transition from the Paleozoic to the Mesozoic eras, some 70–90% of species went extinct in only a few million years. After this precipitous decline, diversity and the number of species increased during the Mesozoic, only to undergo another sudden decline at the boundary between the last period in the Mesozoic, the Cretaceous, and the first period of the Cenozoic era, the Tertiary. Since that mass extinction, the number of species again climbed substantially until recently, when the latest mass extinction precipitated by humans began to influence the planet.

Studies of the fossil record suggest that the background extinction rate is about 0.00001% per year. That number can be viewed analogously to a decay constant for radioactive decay—the rate per year at which species disappear is analogous to the proportion of atoms that will decay in a given year. In this terminology, the decay constant for species extinction is 10^{-7}, which corresponds to a mean life of species of 10 million years. Using this number as a "decay constant" leads to a half-life for all species of 6.9 million years. With this extinction rate, 99% of all species go extinct in 43 million years.

The data shown in Figure 14-8, however, show that the number of species has not decreased with time. No complex organisms were even present long before the Cambrian, and there is an overall increase in numbers of species through the geological record. If species only went extinct, diversity and numbers of species would inevitably decline, so species creation has to have been occurring at a faster pace than species extinction. Using the background extinction rate, this suggests almost complete species turnover in the time periods of 43 million years where 99% of extant species go extinct—and this is less than 1% of Earth's history. Biological change has a very dynamic pace on Earth timescales. Such change is fortunate because it allows life to adapt to inevitable environmental change, provided that change is not too abrupt.

On our human timescale, however, emergence of new species appears to be a very slow process. For every million species, only sixteen new species would emerge each century. And this emergence is not like some new species popping up and appearing out of nowhere—it would be a gradual evolutionary, mutation-by-mutation change where two species that used to be approximately the same have now become just different enough to be identified as separate. Emergence of new species is thus subtle, gradual, and almost invisible to us on human timescales.

The extinction half of evolution, however, is easier to see, because extinctions can be abrupt. Extinction on human timescales is obvious to all of us, as the rapid changes in habitat imposed by the exponential increase in human population and associated civilization lead to large numbers of extinctions. Biologists who are specialists in this area estimate current extinction rates of about 0.1% per year for macroscopic life, or a decay constant of 10^{-3}, ten thousand times the background level. Human beings have thus accelerated extinction rates by ten thousand times, allowing this aspect of the evolutionary process to become obvious to us. If emergence of new species had been similarly accelerated, then more than 20% of Earth species would be new in the past two centuries, and this aspect of evolution would also be obvious. Extinctions are subject to rapid rate changes, while species emergence is limited by the inexorably slow molecular process of DNA mutations. This accounts for the observation that species emergence is not obvious to us, while the extinction half of evolution is so much in evidence.

The evidence for evolution is thus firmly based in the geological record of life, understood in molecular detail through DNA, obviously necessary from observations of extinction, and subject to laboratory verification. Like the Big Bang and plate tectonics, it ranks a 10 on our theory scale and is one of the firm foundations of scientific understanding of the world we inhabit.

Summary

From the perspectives of current diversity of life, the fossil record, the theory of evolution, and the modern evidence from DNA studies, we can understand the present complexity and diversity of life as a progressive evolutionary process whose molecular basis can be understood and verified. Going backward through time we find common ancestors for all current species, which themselves have common ancestors that occurred early on Earth. The unity of all of life points to a common ancestry. The details of the evolutionary process show the mechanisms by which diversity has emerged within this unity.

It is difficult to encompass the intuitive genius of Darwin's concept. He intuited that characteristics were inherited, that they could undergo small changes progressively through time, and that the selection of favorable (i.e., successful in terms of surviving and reproducing) changes would lead to gradual changes over time, as revealed in the fossil record and both the diversity and commonality of life as we know it. These ideas had no known mechanism when Darwin proposed them. DNA provided the precise description of the mechanism of transfer of inherited characteristics from one generation to the next, and the means of mutation and gradual change through small mutations in the specific DNA sequences. DNA thus provided a link between rigorous, detailed modern biochemistry and the overall understanding of life and its evolution proposed by Darwin. This synthesis is one of the great moments of the history of science, where two independent approaches converged to provide an understanding that united biochemistry, biology, paleontology, and the history of Earth.

The two halves of the evolutionary process are the emergence of new species and the extinction of existing ones. Both must occur for ancient species to no longer exist, and the total number of species on Earth to have increased during the Phanerozoic. Emergence is so gradual as to be almost imperceptible for macroscopic life on human timescales, though it is readily apparent in the microbial realm. Extinctions are abrupt, and the massive environmental change caused by human domination of all ecosystems has increased the extinction rate by a factor of 10,000, making the extinction half of the evolutionary process all too evident.

Supplementary Readings

Lynn Margulis and Michael F. Dolan. 2002. *Early Life; Evolution on the Pre-Cambrian Earth.* Sudbury, MA: Jones & Bartlett Learning.

Andrew H. Knoll. 2003. *Life on a Young Planet: The First Three Billion Years of Evolution on Earth.* Princeton, NJ: Princeton University Press.

Richard Dawkins. 2004. *The Ancestor's Tale: A Pilgrimage to the Dawn of Evolution.* Boston: Houghton Mifflin Harcourt.

Charles Darwin and E. O. Wilson. 2005. *Darwin's Four Great Books (Voyage of the Beagle, The Origin of Species, The Descent of Man, The Expression of Emotions in Man and Animals).* New York: W. W. Norton & Co.

Fig. 15-0: Forest fire, which is an uncontrolled release of the energy stored in Earth's planetary fuel cell, as reduced organic carbon molecules react with oxygen. This image is from the Biscuit fire, which was Oregon's largest forest fire of the last century, consuming almost 500,000 acres. (Photo © Lou Angelo Digital on Flickr, with permission)

CHAPTER 15

Energizing the Surface

Coevolution of Life and Planet to Create a Planetary Fuel Cell

Biological evolution has been closely coupled to planetary evolution throughout Earth's history. The origin of life required reducing conditions and the earliest Earth provided them, with a surface devoid of free oxygen. Modern life requires oxidizing conditions and free O_2 *for its metabolism. Between the ancient and modern Earth, during the Archean and Proterozoic, a planetary transformation progressively oxidized the atmosphere, ocean, and crust.*

This transformation was the result of life. To make its constituent organic molecules, life requires hydrogen and "reducing power," electrons that can be added to the oxidized carbon in CO_2*, where C has a 4+ valence, to make the reduced carbon of organic matter,* CH_2O*, where carbon has a neutral valence. This change requires a source of hydrogen and a source of electrons to reduce the carbon. Early life was limited by sources of hydrogen and reducing power, but at some point in the Archean life developed photosynthesis to take advantage of the ubiquitous water molecule as a source of both hydrogen and electrons. Photosynthesis converts the energy of stellar nuclear fusion to an electron flow that stores energy in the chemical bonds of organic matter.*

The complement of the reduced compounds of organic matter is the production of oxidizing power. Every molecule of CO_2 *converted to reduced carbon and stored in the Earth creates an equivalent molecule of the highly reactive* O_2*. The oxidizing power released by life reacted with other planetary materials* *and gradually oxidized the ocean, soils, and atmosphere. The* O_2 *waste product of photosynthesis was initially toxic to organisms, but the formation of oxidized and reduced reservoirs also created a large potential energy source, and life evolved to take advantage of it. The development of aerobic respiration provided eighteen times as much energy per glucose molecule, empowering new biological potential. After a long process of progressive planetary oxidation,* O_2 *was able to reach high enough concentrations in the atmosphere for complex multicellular life to evolve, and for an ozone shield that protects from the harmful effects of ionizing radiation, permitting abundant life on land.*

The progressive oxidation of the surface has transformed Earth from a homogeneous oxidation state in the interior and exterior to one where the planet has become a kind of giant fuel cell, with a reduced interior and largely oxidized exterior, whose combination produces energy. In this sense, life has energized the planet, using solar energy through photosynthesis to separate electrons and create reduced and oxidized reservoirs whose reaction powers life and planet. The development of chemical mechanisms that transitioned from a reduced early Earth, where no oxygen was produced, to a modern Earth, where high levels of oxygen are essential for multicellular life, was a linked biological and planetary evolution. This process is the gradual transformation of planet and life, a story that closely links Earth, life, and the sun.

Introduction

Life at the beginning was a planetary process. In Chapter 13 we found that life's origin was not an isolated biological event but depended on the existence of an ocean, a stable climate, an appropriate atmosphere, volcanism, and mineral surfaces—all planetary phenomena. Today, terrestrial life and Earth are also inextricably interdependent. Life depends on water, soil, air, and climate, which are planetary, and those domains are also influenced by life. The oxygen in the air and ocean that makes animal life possible is biologically produced, as is the organic matter that makes soils fertile for plants. Climate stability depends on the carbon cycle, which links life and climate to volcanism and the rock cycle, and life also enhances weathering through breaking down minerals. Biological and geological processes are linked through the cycles of most elements through the various earth reservoirs. Life at the beginning was a planetary process; life today is a planetary process.

There are nonetheless vast differences between the early Earth and Earth today for both life and planet. The early Earth was a barren landscape populated by unicelled organisms even simpler than the most primitive prokaryotic cells, and O_2 was toxic for these early organisms. The atmosphere had a very different chemical composition, much higher in CO_2 and lacking in oxygen. Today multicellular organisms dominate every nook and cranny of the planetary surface, and the atmosphere has 21% oxygen, sustained by plants and essential for modern animals.

There is thus a long journey from early barely colonized, anoxygenic Earth to modern fully inhabited, oxidized Earth—*the story of planetary evolution*. To understand the gradual development of the habitable planet we experience today—exquisitely attuned with the survival of the species that abound on the surface—we need to unearth the planetary history that transformed ancient past to familiar present, which is our aim in the next three chapters. A central aspect of this history is how life and the planetary surface coevolved to generate reduced and oxidized reservoirs whose interactions provide the energy for modern life.

Life as an Electrical Current

Life's metabolism produces organic molecules and processes energy. The energy for life can be viewed as a kind of slow electric current involving electron transport. Carbon is an essential medium for this transfer of electrons. With valence states between +4 and –4, carbon contains the maximum potential of any element in the periodic table for electron transfer. The source of carbon for most life is CO_2 from volcanic outgassing, where carbon has a 4+ valence. Organic molecules in contrast are made up of carbon in more reduced states and have carbon-hydrogen bonds. The generic formula for organic matter is CH_2O, where carbon has been reduced from a 4+ valence in CO_2 to a neutral valence—e.g., a common product of organic synthesis is glucose, $C_6H_{12}O_6$ or $6(CH_2O)$. Carbon can become even more reduced, as in the compound CH_4, where carbon has a 4– valence and has eight more electrons than carbon in CO_2. To reduce the carbon and form organic molecules life requires a source of electrons and of hydrogen. Schematically the overall production of organic matter on Earth can then be summarized as:

$$CO_2 + \text{electron donor} + \text{hydrogen} \rightarrow CH_2O + \text{oxidized by-product} \quad (15\text{-}1)$$

The electron donor element, called the *reductant*, becomes oxidized by transferring electrons to carbon.

Reaction (1) is the formation of organic molecules, and such formation requires energy, which must come from the sun, Earth, or some disequilibrium that has energetic potential. Reaction (1) can also run in reverse, releasing energy. Plants use energy from the sun to run Reaction (1) in the forward direction; animals then consume the CH_2O to make energy for metabolism by running Reaction (1) in reverse. Both involve electron transport—the slow electrical current. All of life involves such transport.

Chemical reactions that involve electron transport are called *oxidation/reduction reactions*, and to maintain charge balance every molecule that is reduced by electron addition must be balanced by another molecule that is oxidized by electron removal. This means that life, forming reduced molecules, requires partners that can be oxidized. This is the chemical coupling between life and planet, as most of the elements that can become oxidized are molecules such as Fe and S (as well as O^{2-}) that we associate with rocks and the solid Earth rather than life itself. Life and Earth process energy in an energy partnership.

The separation of reduced and oxidized compounds then creates an energy potential that is released when the compounds come into contact. The amount of energy produced depends on the number of electrons that are transported. Maximum electron transport occurs when highly reduced molecules encounter highly oxidized ones. An example of such release, Reaction (1) running in reverse, is when we heat our homes with natural gas. Natural gas is methane (CH_4), where carbon is in its most reduced –4 valence. When we burn it, it reacts with the highly oxidized O_2 molecule:

$$CH_4 + 2O_2 \rightarrow 2H_2O + CO_2 \quad (15\text{-}2)$$

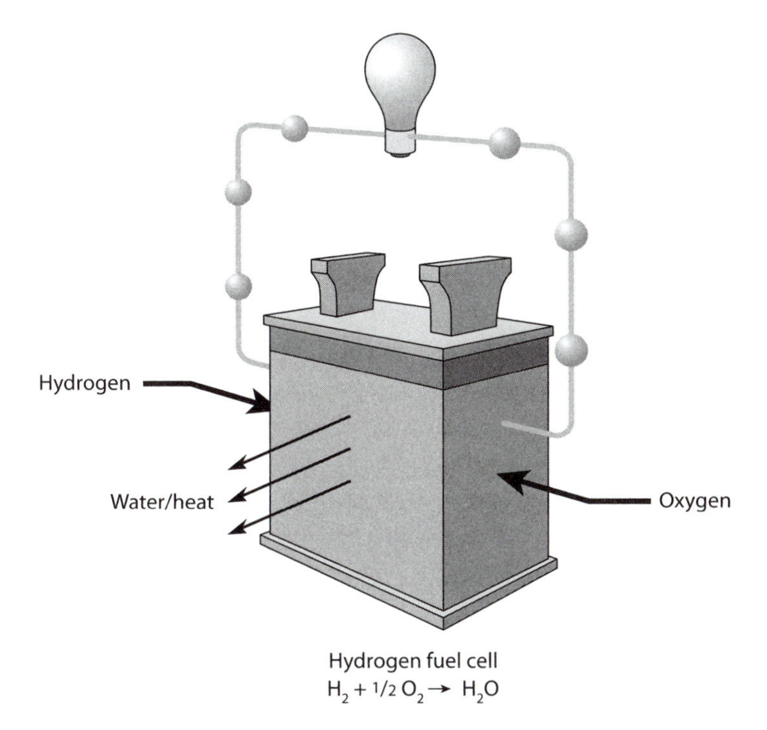

Fig. 15-1: Schematic illustration of a fuel cell, where the oxidation/reduction reaction of H_2 plus ½ O_2 to make water is used to generate an electrical current.

Each oxygen accepts two electrons to change from neutral valence to 2–, so eight electrons are transferred from the carbon, releasing large amounts of energy. Heating homes involves flames and release of heat. Another way the energy potential can be released is directly into an electrical current, as takes place in a fuel cell (Fig. 15-1). The fuel cell analogy can apply to all reactions between reduced and oxidized compounds that produce controlled electron flow. All of modern animal life depends on such controlled reactions between highly reduced molecules (food) and O_2.

A Reduced Early Earth

The starting point for Earth's history is based on the ages of the meteorites and moon near 4.55 billion years ago. Apart from the tiny zircons discussed in Chapter 9, there is no terrestrial rock record for what happened between that time and tiny continental terrains that contain the oldest reliably dated rocks—the Acasta Gneiss in Canada, with an age of 4.03 billion years. The Hadean is an apt term for this portion of Earth's history, since Earth would have been a kind of fiery hell from our current perspective, with much more abundant volcanism and frequent meteorite impacts. For this time period much of the scientific interpretation is inevitably guesswork, because there is little direct evidence of life and Earth's surface conditions apart from what can be gleaned from zircons and planetary science.

We are slightly better off for the Archean era, covering the time span from 4 Ga to 2.5 Ga. "Slightly" is the appropriate word, because the volume of Archean rocks preserved today is only a few percent of current continental crust. The few rocks that have survived have been substantially modified during their long history in the crust, leaving a record that is difficult to interpret. To put the early Earth problem in perspective, the Hadean and Archean are when the first continents formed, plate tectonics may have begun to operate, and life first appeared. This two-billion-year time span is four times longer than the entire record of animal life in the Phanerozoic. With no preserved rocks for the first 550 million years of Earth's history, and few metamorphosed rocks for the next billion years, the challenge to understand earliest Earth's history is daunting. Nonetheless, the evidence that does exist is sufficient to show that the early Earth was a very different place from Earth today.

In particular, all the evidence suggests the early Earth was lacking in free O_2. The present atmosphere composition with 21% oxygen is a striking disequilibrium state, because O_2 is such a highly reactive molecule. Left to itself, oxygen cannot persist at equilibrium as a separate gaseous molecule as long as there are reduced compounds with which it can react. Oxygen reacts with metals, carbon, sulfur, and other atoms to form oxides. Some reactions, such as the weathering of rocks, are relatively slow by human standards, others are rapid enough to create fire or even violent explosions. Only the continuous production of oxygen by plants allows the oxic present conditions of Earth's atmosphere to persist. Absent such production, the oxygen in the atmosphere would react away, ridding Earth's surface of organic matter in a few hundred years, and then gradually disappearing over a few hundred thousand years as the remaining O_2 reacted with rocks and reduced gases coming from Earth's interior. How can we tell if the early Earth might also have had free O_2?

There are no samples of ancient atmospheres remaining, but there is an abundance of other evidence showing that free O_2 was lacking from Earth's origin through much of the Archean. Because O_2 cannot be measured directly, the evidence comes from other elements with multiple oxidation states that would have reacted with O_2 if it had been present.

Many atoms can have multiple valence states and combine with varying amounts of oxygen. As more and more oxygen is consumed, the valence state of the metal progressively increases. For Fe and O, two of the major planet-forming elements, some of the possible molecules are Fe, FeO, Fe_3O_4, and Fe_2O_3. Notice that in this series the oxygen proportion of the molecule increases as the valence state of Fe increases from neutral to 2+ to 3+. For this reason, we call the higher valence states the oxidized forms of the atom. Rusting of iron and the red color of many soils are visible examples of iron becoming oxidized by reaction with oxygen.

Sulfur is another abundant element with multiple oxidation states under terrestrial conditions. Sulfide minerals such as troilite (FeS), the form of sulfide found in meteorites, has S with a −2 valence. Pyrite (FeS_2), has S with a −1 valence. These minerals can be oxidized to sulfates (e.g., $FeSO_4$ or $CaSO_4$) where S has a +6 valence. Many other elements have multiple oxidation states and provide additional clues for the geological detective (Fig. 15-2). Elements with different oxidation states form different minerals, and the presence or absence of these minerals indicates the oxidation state of the surface at the time the minerals formed. When free O_2 is available, only fully oxidized minerals are stable. The mineralogy of rocks thus reveals the oxidation state of Earth's reservoirs.

The first rocks of interest are the meteorites that combined to form the early Earth. Chondritic meteorites contain Fe metal, silicate minerals with FeO, and the most reduced form of sulfur, FeS. There is no excess oxygen available, so the Fe and S are in reduced states. The volatile-bearing meteorites, carbonaceous chondrites, contain iron metal and reduced carbon compounds, materials that cannot exist in the presence of O_2. We saw in Chapter 7 that early planetary differentiation involved reaction between Fe metal and silicates. Free oxygen is also lacking in

Element	Common Oxidation states	Main Species and minerals more reduced →				→ more oxidized
Fe	0, 2+, 3+	Fe Iron; FeO Wustite		FeS_2 Pyrite	Fe_3O_4 Magnetite	Fe_2O_3 Hematite
S	−2, −1, 0, 2+, 4+, 6+	H_2S Hydrogen sulfide (rotten eggs)	FeS_2 Pyrite	S sulfur	SO_2 Sulfide dioxide	SO_4^{-2} Sulfate
C	-4, 0, 2+, 4+	CH_4 Methane	CH_2O Carbohydrate	$C_6H_{12}O_6$	CO Carbon monoxide	CO_2 Carbon dioxide
H	0, 1+	H_2 Hydrogen gas		H_2O Water		
U	4+, 6+	UO_2 Uraninite				UO_3 Uranyl oxide
Mo	4+, 6+	MoS_2 Molybdenite				MoO_3 Molybdenum oxide
O_2	−2, 0			FeO, SiO_2, etc.. oxides		O_2 oxygen

Fig. 15-2: Illustration of the various oxidation states that important minerals can have during Earth processes. Reduced forms are on the left, and more oxidized forms on the right. The early Earth had all elements but carbon in their reduced forms. During Earth's history, life has taken oxidized carbon from CO_2 to make reduced organic carbon, and this electron flow has been balanced by oxidation of all other species. Because different oxidation states have different solubilities in water and form different minerals, preserved minerals in ancient rocks record the oxidation state at the time they formed.

the solar nebula, where there is always an excess of oxygen-hungry elements, such as H, C, and Fe, to combine with oxygen to form oxides. This evidence shows that materials that formed Earth were highly reduced. Furthermore, the moon, which formed from similar materials near the same time, is reduced even today. There is no ferric iron on the moon, and lunar basalts appear to have been in equilibrium with Fe metal. Since the moon is a "planetary fossil" that did not undergo co-evolution with

Fig. 15-3: Ancient river gravel containing grains of uraninite, indicated by arrows, where U is in the +4 valence state. The gravel was subsequently buried and turned into a hard rock that was recently unearthed by erosion. Uraninite can persist and be included in river gravels only under reducing atmospheric conditions, showing the Archean atmosphere had no oxygen. (Courtesy of Harvard Museum of Natural History, Dick Holland Collection)

life, it provides further evidence for a reduced state for the early Earth before life began its transformative process.

We can move forward in time by looking at some of the oldest Earth rocks. To infer past conditions of the atmosphere, it is key that the minerals formed at the surface and were not buried deep in the crust, where even today oxygen does not penetrate efficiently. Sedimentary rocks form at the surface, and river sediments in particular are inevitably in contact with the atmosphere. Some ancient river sediments, a type of gravel called a *placer deposit* (Fig. 15-3), survive from the Archean. The mineralogy of these sediments can be used to test for the presence of oxygen in the atmosphere.

The two critical minerals in the Archean placer deposits are reduced sulfides and a uranium-bearing mineral called uraninite (UO_2). Uranium can have multiple oxidation states, of which the two found in minerals are U^{4+} and U^{6+}. U^{6+} is relatively soluble in water and does not form stable minerals on Earth in the presence of oxygen. Uranium in its reduced 4+ form, however, is insoluble in water and precipitates to form uraninite. When uraninite is exposed to the surface today, the U^{4+} is oxidized to U^{6+}, the uraninite mineral rapidly disappears, and U^{6+} is carried away by the water. Uraninite persists only in reducing environments lacking in oxygen, and does not exist in modern river sediments. The placer deposits formed by Archean rivers, however, formed ancient rocks that were buried and shielded from later atmospheres. When these rocks are freshly brought to the surface, they contain uraninite, showing that when these gravels were formed, no oxygen was available at the surface to oxidize the uraninite. Similar arguments apply to pyrite. While pyrite is found in modern environments, it occurs only in reduced rocks, such as sediments rich in organic matter. Pyrite does not survive for long in contact with the atmosphere, and no river gravels today contain the mineral. The presence of sedimentary pyrite in ancient rocks exposed to the atmosphere also indicates reducing conditions with no atmospheric oxygen.

Another source of indirect evidence about the oxidation state of the early Earth comes from consideration of the origin of life. Life is made up of organic molecules that have reduced forms of carbon, and organic molecules react with oxygen and are converted to their oxidized forms where organic molecules are not possible. For life to get started, organic molecules must exist stably in the environment, and this is not possible with an oxidizing atmosphere, so reducing conditions are probably essential for the origin of life. Once the first cell formed, it would also survive by making use of molecular building blocks in its environment. The organic molecules required by early organisms would be destroyed by oxygen. Both life's origin and its early persistence seem to require *anaerobic*—oxygen-free—conditions.

This evidence and reasoning about a reduced early Earth are also consistent with our understanding of the source of oxygen in the atmosphere today. Oxygen-producing photosynthetic bacteria were absent in earliest Earth's history, there were no plants, and there was a large mass of reduced compounds such as FeO, FeS_2, and probably CH_4 and H_2 ready to attach to any trace amounts of oxygen that might have been produced—e.g., by reactions in the upper atmosphere.

All these lines of evidence—meteorites, early planetary differentiation with metal present, the oxidation state of the moon, the mineralogy of ancient sediments, the origin of life, and where O_2 comes from today—lead to a convincing conclusion of a reduced early Earth. The early atmosphere was so reducing that O_2 contents are estimated to be less than 10^{-10} of the present atmospheric level (PAL) of O_2.

In contrast to the early Earth, the habitability of Earth today and the existence of multicellular organisms depend on the presence of the high levels of O_2 in the atmosphere. Without O_2, little of the life currently on the surface would survive, and there would be no ozone shield protecting surface life from high-energy ultraviolet radiation coming from the sun. Clearly the oxygen content of the atmosphere has changed through Earth's history, and this change is closely associated with life. Why did life start to produce O_2? As we will see in the next section, O_2 production is a manifestation of the gradual development of biological technology to harvest and make use of the energy coming from the sun. This development took place through a series of "energy revolutions" that enabled life to have progressively increasing access to energy over Earth's history.

The First Three Energy Revolutions

As we saw from Reaction (1) above, formation of organic molecules requires input of energy and creates an oxidized atom. Organic molecules need to be formed not only to make living matter but also as a source of food that can be "burned" to release energy for metabolic processes. The energy currency in the cell is adenosine triphosphate (ATP), which is used to power cellular reactions. As ATP, the molecule is charged with energy; conversion of ATP to ADP (adenosine diphosphate) provides energy for the cell, and then conversion back to ATP recharges the molecule with energetic potential. The breakdown of carbohydrates such as glucose provides electron transport and chemical pathways that allow production of ATP from ADP. Organisms "burn" the sugar, using the special ATP-ADP mechanism to capture the energy so it can be used for metabolic processes.

Ultimately, formation of ATP requires an external energy source. Some organisms make use of external energy sources such as sunlight or chemicals that are out of equilibrium with their surroundings to produce ATP as well as glucose, which is later burned for cellular operations. Organisms that produce their own organic molecules by making use of an external energy source are called *autotrophs*. Autotrophs make use of available energy to create their own food. Other organisms, called *heterotrophs*, absorb (e.g., eat) organic molecules made by other organisms and make use of them for the energy for cellular processes. Plants are autotrophs because they make their organic molecules by harvesting sunlight. Animals are heterotrophs—we eat for essential molecules and for our energy source. There are also autotrophic and heterotrophic bacteria.

It is not clear whether earliest life was autotrophic or heterotrophic. One idea is that earliest life may have been heterotrophic, eating the organic chemicals that were made abiotically in the early Earth. Such life would be limited by the availability of nonbiogenic "food" provided by the planetary environment. In this case there would be distinct advantages for organisms that were able to make their own food using energy directly, rather than relying on their surroundings to provide them with food. Alternatively, the earliest organisms may have developed means to convert sunlight or other external energy to make their own food. In either case, the development of autotrophy, allowing life to make its own food through access to planetary and solar energy, was such a major step in the evolution of life that we can call it the *first energy revolution.*

Chemoautotrophs made use of chemical energy, e.g., in hydrothermal vents. Methanogens, for example, derive energy to make organic matter from the reaction

$$CO_2 + 4H_2 \rightarrow CH_4 + 2H_2O \qquad (15\text{-}3)$$

oxidizing hydrogen from neutral to +1, and releasing methane in the process.

Photoautotrophs were able to harvest light energy from the sun to gain electrons from reductant molecules such as H_2 and H_2S, which also served as a source of hydrogen. These organisms made glucose from reactions such as:

$$12H_2 + 6CO_2 + \text{solar energy} \rightarrow C_6H_{12}O_6 + 6H_2O \qquad (15\text{-}4)$$

$$6CO_2 + 6H_2S + \text{solar energy} \rightarrow C_6H_{12}O_6 + 6S \qquad (15\text{-}5)$$

In the first reaction, the H_2 is the source of electrons and hydrogen as it is oxidized from neutral to +1. In the second, H_2S is the source of hydrogen, and electrons are released by the oxidation of sulfur from −2 to neutral valence. Notice that no oxygen is produced in these reactions, and the reactions require the reduced molecules H_2 and H_2S and can take place only in reducing environments. The photosynthetic processes for these anaerobic bacteria are of two types—photosystem 1 and photosystem 2 (PS1 and PS2)—with the numbers simply indicating the order in which they were discovered.

The glucose that is formed can then be processed to produce ATP for other cellular reactions, for example, by fermentation:

$$C_6H_{12}O_6 \rightarrow 2CH_3CH_2OH + 2CO_2 + 2ATP \qquad (15\text{-}6)$$

Glucose → 2 alcohol (ethanol) + 2 carbon dioxide + 2 ATP

Another method is glycolysis, which has a more complex reaction that also produces two ATP for each glucose molecule. Both of these metabolic pathways are anaerobic methods of ATP production. The waste products are not fully oxidized, and cells that use these mechanisms need to have means to dispose of the waste. Aerobic metabolism, discussed below, is able to further process these "waste products," and through the citric acid cycle—or the Krebs cycle—makes use of the energy available in their chemical bonds to generate much more ATP.

These mechanisms permitted bacteria to make their own food and process it to gain ATP. When they died or leaked molecules into the environment, they provided resources for heterotrophic bacteria, leading to the development of early ecosystems and the first simple food webs. These mechanisms still exist in cellular metabolisms today. Prokaryotic cells that thrive in anaerobic environments use photosynthetic mechanisms that are most closely related to early life.

The disadvantages of PS1 and PS2 used in Reactions (15-4) and (15-5) are that they rely on availability of H_2S and H_2 as a source of electrons and hydrogen, and these molecules are always at low concentrations in seawater. This limits both the amount of organic matter that can be formed and the extent of the biosphere. Earliest life then likely required a reducing environment to gain a planetary foothold, but it was limited by the availability of the reducing power and the H-containing molecules that it needed to form organic matter. Given these limitations, it is unlikely that early life was very abundant.

Within the liquid environment, however, was an almost infinite reservoir of hydrogen and electrons in the H_2O molecules. One of the great evolutionary innovations occurred when organisms developed the chemical machinery to break down the H_2O molecule as a combined source of reducing power and hydrogen. This innovation required the combination and modification of PS1 and PS2 acting together, leading to the oxygenic photosynthesis that is at the base of most food chains even to the present day:

$$6CO_2 + 12H_2O + \text{solar energy} \rightarrow C_6H_{12}O_6 + 6O_2 + 6H_2O \tag{15-7}$$

The electrons to reduce the C come from converting the O^{2-} in the H_2O molecule to neutrally charged O_2, so Reaction (15-7) requires breaking the strong H-O bonds of water. PS2 was modified to be able to break the H-O bonds, and then PS1 was able to complete the process. Oxygen is the reductant, supplying electrons to reduce the carbon as oxygen changes its valence from −2 to neutral, and each atom of organic carbon produced leads to a by-product of one molecule of O_2. Cyanobacteria are the diverse descendants of early organisms that developed this capability.

The development of oxygenic photosynthesis was a *second energy revolution* that allowed much greater access to the energy of the sun, enabling it to be captured and stored in the form of reduced organic molecules. The new mechanism of oxygenic photosynthesis was no longer limited by a source of hydrogen and electrons—there was an almost infinite source of hydrogen and reducing power once the very strong bonds holding the water molecule together could be broken, and there was a vast energy supply in the form of sunlight. Of course, while oxygenic photosynthesis eliminated the problem of hydrogen availability, there were still limits to growth, owing to the limited supply of other critical nutrients for life such as phosphorous and nitrogen. Furthermore, early prokaryotic organisms would have had to rely on anaerobic me-

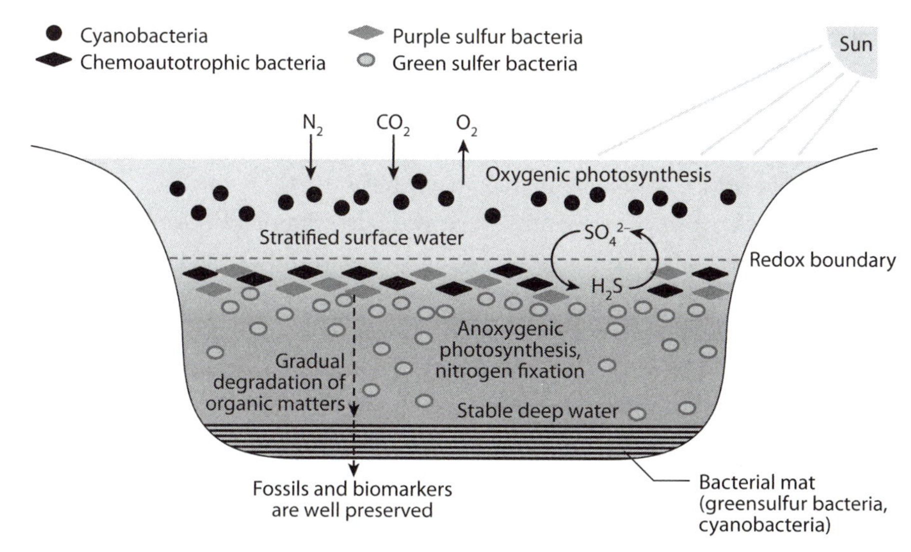

Fig. 15-4: Illustration of conditions in the Black Sea, one of the few places in the ocean where oxygen is absent just below the surface. Oxygenic photosynthesis by aerobic cyanobacteria takes place in the top layer. Oxidation of sinking organic matter uses up all the oxygen in the water below, preventing cyanobacteria from surviving. In their place, anaerobic bacteria (purple and green sulfur bacteria) make use of the variable oxidation states of sulfur to generate their own thriving ecosystem. The anaerobic conditions of the early Earth would have been fully occupied by similar anaerobic bacteria, which today are relegated to minor niches of the biosphere that have avoided oxidation.

tabolism of glucose to generate their energy. For every molecule of glucose they were able to obtain, they could still get only two ATP units of energy for cellular processes.

All of these forms of photosynthesis survive today. For example, in the Black Sea cyanobacteria occupy the shallow water where they carry out oxygenic photosynthesis. As they die and fall through the water column, the oxygen they produced is consumed in the breakdown of the organic matter, and the deeper water is anaerobic. There, purple and green bacteria that make up colorful layers in anaerobic environments are carrying out photosynthesis using PS1 and PS2 separately (Fig. 15-4).

As can be seen from Reactions (4) and (5) above, anaerobic operation of PS1 and PS2 created benign waste products like H_2O. Oxygenic photosynthesis, however, created a difficult pollution problem. While today we think of O_2 as a beneficent aspect of photosynthesis, this was not the case for early organisms. O_2 breaks down organic molecules and would have been a devastating poison for early life. Even for modern life oxygen is a potent poison, as reflected in our popular consumption of "antioxidants" to prevent deterioration of our cells. Absent the evolution of molecular protection against this poison, the waste problem would have limited the spread of oxygenic photosynthesis.

Within this crisis, however, there was also latent opportunity that contained the potential for a *third energy revolution*. Oxygen is a highly reactive molecule that spontaneously reacts with all kinds of metals and reduced organic matter, and such reactions release abundant energy. Where energy is available, some evolutionary adaptation comes along to take advantage of it. This adaptation was a major boon for life's energy, because while anaerobic breakdown of glucose by Reaction (15-5) produces only two molecules of ATP, aerobic (using oxygen) breakdown of glucose is able to produce thirty-six molecules of ATP!

$$C_6H_{12}O_6 + 6O_2 \rightarrow 6CO_2 + 6H_2O + 36ATP \qquad (15\text{-}8)$$

$$\text{Glucose} \rightarrow 6 \text{ carbon dioxide} + 6 \text{ water} + 36 \text{ ATP}$$

This ability to respire oxygen in a controlled way required major evolutionary adaptation and was the third energy revolution for Earth's life, because it increased by a factor of 18 the energy that could be gained from the glucose molecule (see Fig. 15-5). Every breath you are taking as you read this chapter delivers the oxygen and releases the CO_2 from Reaction (8), and every thought you are having is making use of the ATP released in the process.

The potential energy of Reaction (15-8) could not be fully exploited, however. To take advantage of aerobic respiration requires that there be plenty of O_2 available. And because O_2 is so highly reactive, it is gobbled up by reduced species, which were in abundance in the early Earth in the form of CH_4, H_2S, FeO, and FeS_2. Raising the oxygen level of the atmosphere to allow life to fully make use of oxygen for energy production would require producing enough oxygen to overcome the reducing power of the planetary surface. The planetary exterior needed to become oxidized before the full potential of the third energy revolution would be able to be realized.

So we see that evolution of life is closely coupled to evolution of the oxidation state of Earth's surface. The atmosphere and ocean start with

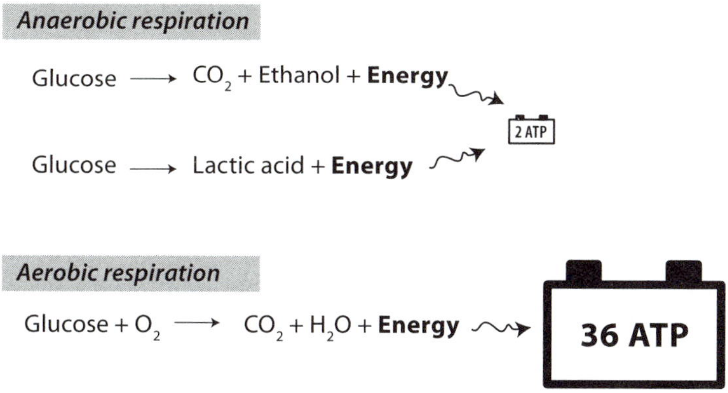

Anaerobic respiration

Glucose ⟶ CO_2 + Ethanol + **Energy**

2 ATP

Glucose ⟶ Lactic acid + **Energy**

Aerobic respiration

Glucose + O_2 ⟶ CO_2 + H_2O + **Energy**

36 ATP

Fig. 15-5: Illustration of the different energy production that can be obtained by anaerobic and aerobic metabolism of glucose to form the energy currency of life, adenosine triphosphate (ATP). Anaerobic respiration of glucose produces only two ATP per glucose molecule, indicated by the small battery. Aerobic respiration fully metabolizes the glucose to produce 36 ATP, indicated by the large battery. The greater energy production is an energy revolution for life.

essentially no O_2, and the abundance of life is kept at low levels by the limited supply of reductants that enable conversion of CO_2 to organic carbon. Then life evolves oxygenic photosynthesis to break the water molecule and have an essentially infinite supply of hydrogen and of reductants in the O^{2-} of the water molecule. The O_2 produced by the growth in organic matter is at first a pollutant and poison, but life evolves to make use of it as a much more efficient energy source. Once sufficient O_2 is available, life even comes to depend on high levels of O_2 in the environment. Ultimately, breathing and inner circulation systems evolve to transport oxygen efficiently to multicelled organisms, allowing oxygen-hungry organs like the brain to evolve. Evolution thus leads to and is influenced by a progressively oxidized surface. The entire process leads to an increasing ability to harness energy from the sun and send it through the biosphere via electron flow.

The Planetary Fuel Cell

A fuel cell is a device that converts the chemical potential of oxidized and reduced molecules into an electrical current. A fuel cell is like a battery, except that it relies on replenishment of reduced and oxidized molecules to maintain its energy potential. The fuel cell becomes charged by the separation of positive and negative electrical potential enabling electron flow.

From a planetary perspective, over billions of years of earth evolution, the electrical current generated by life has been used to "charge" a planetary fuel cell, using solar energy to convert CO_2 and H_2O to reduced carbon and oxidized species. The solid Earth also plays a role. The interior reservoir of the Earth is so vast that its overall oxidation state is little affected by organic processes at the surface. Earth's interior then provides a sustained reduced reservoir for the fuel cell. Life adds to the reduced reservoir by burial of organic carbon, and every atom of carbon sequestered by burial releases a molecule of O_2 to oxidize the surface. Surface oxidation is a long process, because the abundant reservoirs of reduced sulfur and iron provide a large sink that absorbs the oxygen. Only when this reservoir has become sufficiently saturated with oxygen is an oxygenic atmosphere able to be developed. All of this occurs while life consists only of single-cell organisms, a process taking 2 billion years or more.

Once sufficient free O_2 is available at the surface, multicellular life is able to evolve by making use of the planetary fuel cell to generate high energy during aerobic respiration, when reduced organic molecules are reacted with O_2 to produce energy. For example, at hydrothermal vents, reduced species from the mantle encounter oxidized seawater, supplying the energy potential for the microbes that lie at the base of deep-sea food chains. Life evolved to take advantage of the energy flow, leading at hydrothermal vents to the spectacular abundance of life in the absence of sunlight. Energy can be released in the absence of sunlight by the flow between the two "poles" of Earth's fuel cell, reduced reservoirs of organic carbon or Earth's interior encountering the oxidized pole of the planetary surface (Fig. 15-6).

Nonbiological processes are also influenced by contact between the reduced interior and the oxidized surface. When new reduced rock is exposed to the surface, it reacts and becomes oxidized, enhancing weathering and contributing to geochemical cycles. When oxidized ocean crust is subducted into the interior, it adds ferric (Fe^{3+}) iron and water to the

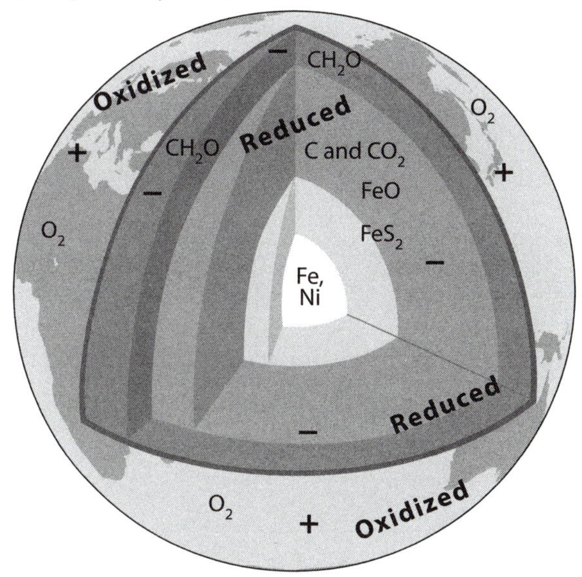

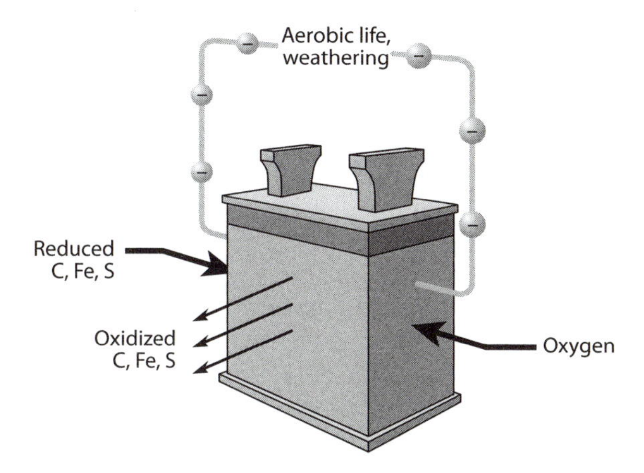

Fig. 15-6: Top panel: This illustrates the state of the modern Earth with oxidized and reduced reservoirs that when combined release energy. The reduced reservoirs are organic carbon and Earth's interior. The oxidized reservoirs are the oxidized surface rocks and the O_2 in the atmosphere. *Bottom panel:* The fuel cell concept of Figure 15-1 with the Earth's chemical reservoirs as the reduced and oxidized species that power Earth processes.

mantle wedge above the slab, oxidizing the local region of the mantle and influencing the compositions of magmas, and their cooling path, to produce high SiO^2 magmas and oxidized gases characteristic of the continental crust.)

The fuel cell can also be released in catastrophic and uncontrolled energy loss to heat. This occurs, for example, when the organic carbon buried in Earth's crust is burned by contact with the atmosphere. As we shall see in Chapter 17, extreme climate change and important punctuation points in biological evolution often result from catastrophic connections between Earth's reduced and oxidized reservoirs. And all of modern civilization depends on our access to and exploitation of Earth's fuel cell through extraction and burning of fossil fuels.

Summary

Earth's history began with both interior and exterior at a similar reduced oxidation state. This reduced state lasted more than 1 billion years and was an essential condition for the origin of life, during which reduced molecules must be stable in the surface environment. Early life discovered how to make its own food by making use of chemical energy from Earth and light energy from the sun. These autotrophs were nonetheless limited by sources of reductants. The discovery of oxygenic photosynthesis, sometime during the Archean period, vastly increased life's access to solar energy by obtaining hydrogen and electrons from the ubiquitous water molecule. Life would then have been limited only by other critical reagents such as N and P. The O_2 by-product of oxygenic photosynthesis was initially a poisonous pollutant, but it also contained the possibility of much more cellular energy through aerobic respiration of glucose, increasing energy available for life by a factor of 18. The buildup of oxygen in the atmosphere would have been limited, however, by all the reduced species resident at the surface and continually added from Earth's reduced interior. Only after life had produced enough oxygen to saturate the surface reservoirs could the O_2 content of the atmosphere rise to modern levels, permitting the development of multi-cellular life in the late Proterozoic and Phanerozoic. The progressive oxidation of the surface by production of organic matter through Earth's history has created vast oxidized and reduced reservoirs that serve as the poles of a planetary fuel cell. Reactions between these reservoirs provide the essential energy for modern life and planetary processes.

Fig. 16-0: An outcrop of of banded iron formation (BIF) in Australia. These great accumulations of iron, the source of most iron ore today, were deposited over a rather brief period of Earth's history as Earth gradually became more oxidized and converted the soluble reduced iron in seawater to the insoluble oxidized iron, which precipitated to make these spectacular rocks. See also color plate 25. (Courtesy of Simon Poulton and the Nordic Center for Earth Evoluton)

CHAPTER 16

Exterior Modifications

The Record of Oxidation of the Planetary Surface

As we saw in the previous chapter, planetary evolution involved the progressive oxidation of surface reservoirs and the eventual rise of O_2 in the atmosphere. The ancient reduced Earth surface contained H_2, Fe as Fe^{2+} (FeO), S as sulphide (S^-), and an atmosphere with no O_2. Today H_2 is minimal, Fe at the surface is Fe^{3+}, S is S^{6+}, and the atmosphere has 21% O_2. These vast changes occurred by the gradual transformation of planet and life, the exterior renovation of the planet, a story that closely links Earth, life, and the sun. When and how did this happen during Earth's history?

As we learned in the previous chapter, every molecule of CO_2 converted to reduced carbon and stored in Earth creates an equivalent molecule of the highly reactive O_2. The oxidizing power released by the stored molecules of life reacted with other planetary materials and gradually oxidized the ocean, soils, and atmosphere, converting Fe and S to their oxidized forms. Oxidized Fe and S contain almost all the O_2 produced by burial of organic carbon over Earth's history.

The rock record provides the history of this process, in large part because the solubility of iron and sulphur depends on their oxidation state. Under reducing conditions, iron is soluble and mobile in water and sulphur is immobile. For the early Earth, this led to an Fe-rich, S-poor ocean. Under oxidizing conditions, iron is insoluble and sulphur soluble. Today this leads to high S and vanishingly small Fe in the modern ocean, where seawater has an S/Fe ratio millions of times higher than the continental crust. The transition between these two states of the planetary surface caused the precipitation of massive Fe deposits known as banded iron formations (BIFs). The demise of BIFs near 2.0 Ga marks the initial rise of O_2 *in the atmosphere. From 2.0 to 0.8 Ga,* O_2 *was likely a percent or two of present levels. Then in the Neoproterozoic, a second rise in* O_2 *appears to have occurred, leading to the possibility of the Phanerozoic explosion of macroscopic life. Since ~600 Ma,* O_2 *has been within 50% of modern levels. Both the 2.0 Ga and ~0.8 Ga oxidation events mysteriously also occur near times of global glaciations.*

While this process is the surface renovation of Earth, the growth of oxygen and its steady-state disequilibrium is intimately connected to the operation of the whole earth system. The source of oxidizing power by burial of organic carbon requires steady addition of CO_2 *to the surface by volcanism. Addition of reduced materials to the surface from the mantle provides a steady sink for the oxygen. Subduction of carbon and other oxidized and reduced materials has had important long-term effects on the oxygen balance. Today the excess oxygen produced by life is compensated by subduction of oxidized iron produced by alteration of the ocean crust in the deep sea, largely at ocean ridges. While the broad outlines of the rise of* O_2 *are clear, the complete story of the rise of oxygen is like a partially completed jigsaw puzzle where many of the pieces are well defined but do not yet come together into a unified and complete picture.*

Introduction

The development of chemical mechanisms that transitioned from a reduced early Earth with no free oxygen, to an Earth where oxygen was a pollutant, to an Earth where O_2 could be a major atmospheric constituent, to a modern Earth where high levels of oxygen are essential for animal life, was the exterior renovation of the planet. It is closely tied to the development of evolutionary mechanisms in living organisms, and also to the geochemical cycles that control the abundances of elements in Earth's various reservoirs. In the previous chapter we saw how life was the process that oxidized Earth's surface, changing it from its reduced beginnings to the highly oxygenated environment that has been produced and is maintained by an inhabited planet. In this chapter our task is to try to discover the major events and mechanisms involving the inorganic Earth by which this exterior renovation took place. Were the changes steady or punctuated? When did they occur? What were the mechanisms and how were they recorded on the planetary surface?

Earth and Oxygen

To understand the history of O_2, we need to delve a bit more deeply into Earth's oxygen balance, because we can't measure O_2 in past atmospheres directly. Instead, this aspect of Earth's history is a bit like a play where the protagonist has a major role but is never seen on stage. Fortunately, while O_2 is the missing protagonist of this story, there are other starring characters that do make regular appearances.

These characters are the alpha-particle nuclides that occupy multiple oxidation states—C, Fe, and S—as well as the most abundant element in the universe, H. Of course, many other elements, such as manganese (Mn), arsenic (As), and molybdenum (Mo) have variable oxidation states and we will turn to these for additional evidence, but they are not of great importance to the overall oxygen budget (see Fig. 15-1).

Reactions among these elements reflect and control the oxidation state of Earth's layers. The photosynthetic reaction—Equation (1) of Chapter 15—liberates oxygen. As we all know, organic matter is broken down by bacteria, is eaten by heterotrophs, and burns. These processes are simply the photosynthetic reaction running in reverse (Reaction [16-1]) as free O_2 is consumed and recycled to CO_2 and H_2O. The oxygen that is not recycled in this way is available to react with other reduced species, as illustrated in Reactions (16-2)–(16-4).

$$C_6H_{12}O_6 + 6O_2 \rightarrow 6CO_2 + 6H_2O \tag{16-1}$$

$$2FeO + 1/2O_2 \rightarrow Fe_2O_3 \tag{16-2}$$

$$FeS_2 + 5/2O_2 \rightarrow FeO + 2SO_2 \tag{16-3}$$

$$2H_2S + 2\,CaO + 4O_2 \rightarrow 2CaSO_4 + 2H_2O \tag{16-4}$$

Note that all of these reactions are oxidation/reduction reactions where the valence states of Fe, S, and O are changed. And depending on the oxidation state, different solid materials are formed—reduced carbon is found in organic matter (e.g., coal). Reduced iron is in many silicates, while oxidized iron is found in hematite (Fe_2O_3) and magnetite (Fe_3O_4). Reduced sulfur is in pyrite (FeS_2) and oxidized sulfur in gypsum ($CaSO_4$). The materials found in rocks record the history of oxygen, even though oxygen itself is not measured.

Earth's interior is relatively reduced. The core is reduced Fe metal, and the mantle contains about 93% of its Fe as FeO and most of its S as iron sulfide. This reduced oxidation state is inherited by magmas and gases expelled to the surface, so the operation of the plate tectonic geochemical cycle provides a steady diet of reduced species that would react with O_2. Reduced species are also provided by exposure of the crystalline rocks of the continents. "Weathering" is the oxidation of these rocks as they are broken down by reactions with O_2 and H_2O. Without life's continual re-supply of O_2, all free O_2 would be consumed, and Earth's surface would revert to the same reduced state as the interior.

Because of this steady supply of reduced molecules from Earth's interior, the presence of O_2 in the atmosphere is a disequilibrium state, a "steady-state disequilibrium" reflecting the checks and balances of positive and negative feedbacks in the context of a very active geochemical cycle.

Today if O_2 were to increase, the oxygen-consuming reactions would take place more quickly. If O_2 were to reach something like 27% of the atmosphere instead of 21%, then fires would be rampant, putting Reaction (16-1) into overdrive, consuming organic carbon and reducing the mass of the biosphere and hence limiting O_2 production. Or if oxygen were to drop, organic matter would be oxidized more slowly and the deep oceans would become reduced, causing greater burial of organic carbon and less oxidation of the rocks of the seafloor. Both would reduce the sinks for O_2, and O_2 would rise.

These processes apply to the present-day Earth, where, despite rampant O_2 production by the biosphere through photosynthesis, atmospheric O_2 stays constant at 21% because sources are exactly balanced by sinks (consumption of O_2 by aerobic respiration, sulphide mineral oxidation, oxidation of iron, and oxidation of reduced volcanic gases). Feedbacks operate to preserve this level. Earlier in Earth's history we know that O_2 was nowhere near these levels. In fact this balance could theoretically occur at many other levels of O_2 in the atmosphere, all depending on the relative rates of O_2 production by photosynthesis and consumption by reactions such as (16-1)–(16-4).

Over the long term, for O_2 to rise, the balance has to be perturbed so that O_2 production is greater than consumption. Over Earth's history, Reaction (16-1) has moved from right to left and Reactions (16-2) to (16-4) have on balance moved from left to right. Oxygenic photosynthesis reduced carbon to create organic matter and the highly reactive O_2. But it does not matter how much O_2 plants make if all of the organic matter is destroyed by oxidation. The O_2 available to oxidize the surface environment requires the storage of an atom of organic carbon somewhere in the Earth system. Net O_2 production is the *excess* of organic matter production over destruction, and this organic matter has to end up somewhere, unoxidized. The complementary other half of the story is that the net O_2 produced ends up somewhere, either as O_2 in the atmosphere or in the oxidized species of Fe and S. Today Fe in surface reservoirs is largely Fe^{3+} and S is sulfate (SO_4^{2-}) with a valence of +6. Therefore, it is evident that the rise in atmospheric oxygen is only one aspect of the overall oxidation of Earth's crust and ocean (Fig. 16-1). Indeed, oxygen could not persist in the atmosphere if the other surface reservoirs were not already oxidized, because in that case the oxygen sinks would be greater than they are currently. Over Earth's history, progressive oxidation of the surface diminished the available sinks for O_2, permitting it to rise in the atmosphere.

It is an imperfect analogy, but we can imagine Earth as if it were a person with a large amount of debt (reduced molecules). Earth is very fortunate because thanks to the sun, money keeps coming in (O_2 generated by photosynthesis). As long as all of the income is spent (on reoxidation of organic molecules) the debt cannot be paid off. The saved portion (from sequestered organic matter) is first used to pay off the debt (oxidize the Fe and S), and then once the debt is paid off, the saved portion can go into building capital (raising the O_2 of the atmosphere). When sufficient capital is built up, entirely new enterprises (multicellular life and advanced ecosystems) become possible.

So, if we have described the system accurately, we have a simple mass balance—the O_2 production by organic matter burial should equal the

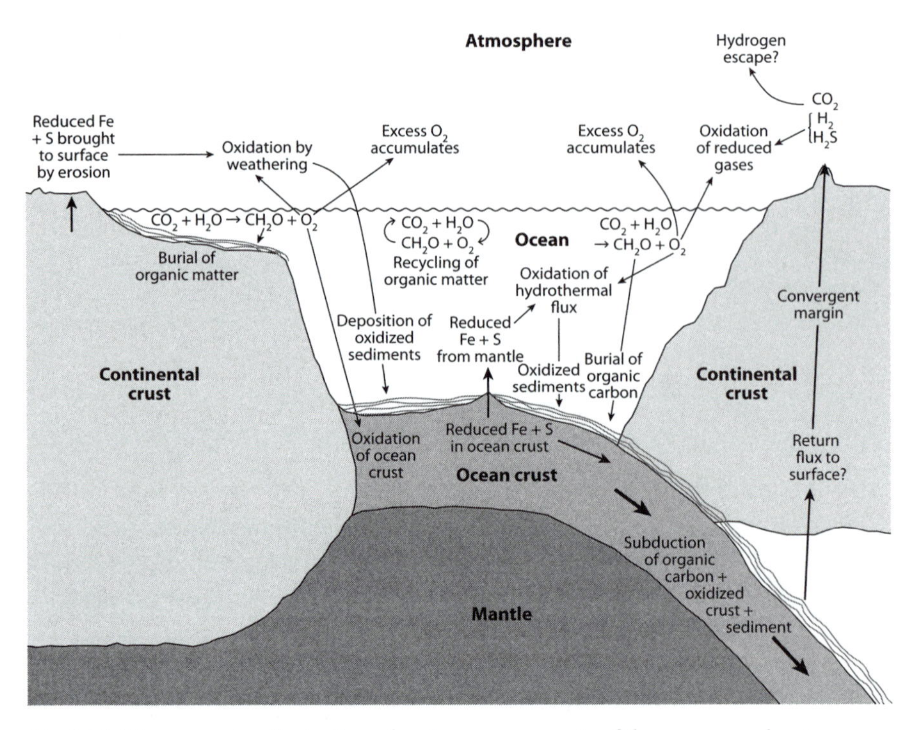

Fig. 16-1: Diagrammatic illustration of various components of the oxygen cycle. Oxygen is produced by photosynthesis. Most of the organic matter that is produced is then re-oxidized. Burial of organic matter leads to excess oxygen that is consumed by various reactions. Reduced Fe and S coming from hydrothermal vents and exposure of the continental crust are oxidized, as are the reduced gases coming out of volcanoes. Subduction recycles various molecules to Earth's interior. The oxidation state of sediments contains the record of these reactions through Earth history.

O_2 incorporated in oxidized species. This mass balance also provides us with two avenues to explore the history of oxygen, one through study of organic matter through time (the carbon cycle), which is the story of oxygen generation, and the other through study of the oxidized products (largely the Fe and S cycles), which is the story of oxygen consumption.

Carbon: The Record of Oxygen Production

Reduced carbon resides in black shales, soils, coal, oil, and natural gas, resulting from billions of years of photosynthesis. This storage is what permits the high oxidation state of the planetary surface and the exis-

Table 16-1
Reservoirs of carbon on Earth (in units of 10^{18} moles)

Reservoir	Reduced (organic) C	Oxidized C	Total
Atmosphere	—	0.07	0.07
Biosphere	0.13	—	0.13
Hydrosphere		3.3	3.3
Deep-sea sediments	60	1300	1360
Continental margin sediments	>370	>1000	>1370
Sedimentary rocks	750	3500	4250
Other rocks	100	200–400(?)	300–500
Total surface reservoirs	~1250	~6100	~7350
Mantle	—	—	27000

Reduced carbon estimates are from Des Marais, *Rev. Mineral. Geochem. 43 (2005):555–78.*

tence of animal life that depends on it. If we were to somehow recover and burn all this organic matter, we would reverse 3 billion years of planetary evolution and return Earth to its reduced state where only primitive life is possible. The current distribution of organic and inorganic carbon in Earth reservoirs is shown in Table 16-1. Notice that the biosphere, living or recently living organic matter, is only a very small portion of total organic carbon, that total organic carbon is about 17% of total crustal carbon, and that the mantle is by far the largest carbon reservoir.

Tracking sources of O_2 through time directly would require knowledge of how the mass of organic carbon has changed through time. This is an intractable problem because organic carbon resides in sediments, and these are rapidly recycled through Earth's history. So scientists have had to turn to an indirect method, where the rate of formation of organic carbon is inferred from measurements of carbon isotopes in carbonates formed from seawater and in ancient organic matter.

The reasoning behind this approach makes use of the contrast in carbon isotope ratios between organic carbon and carbonate carbon. The starting point is that we can measure the *average* carbon isotope ratio of surface carbon reservoirs, and it turns out that this value is the same as the input to the system, which is carbon coming from the mantle. The mantle-derived carbon then becomes distributed by geological and biological processes to (inorganic) carbonate and organic carbon. As we learned in Chapter 13, biological processes that make the organic carbon preferentially incorporate ^{12}C, the lighter of the two carbon isotopes, and the $^{13}C/^{12}C$ ratio of organic carbon is about 2.5% lower than the ratio in inorganic carbon. The variations are reported as per mil differences relative to an arbitrary standard (per mil is parts per thousand, as compared to percent, which is parts per hundred). These differences are conventionally assigned the symbol $\delta^{13}C$ ("del C 13"). For example, a value of $\delta^{13}C$ of –25 means the $^{13}C/^{12}C$ ratio is 25 per mil lower than the standard value.

Mantle carbon has $\delta^{13}C$ of ~–5, which must always be the value of the total carbon in the system. At the same time, there will also always be the offset between organic and inorganic carbon, which is about 30 per mil. If all the carbon were in the form of inorganic carbon, then the inorganic carbon would have to have the mean value of –5. The first organic molecule to form would be offset 30 per mil lighter to –35. Or if all the carbon were only organic carbon, then the organic carbon value would be –5 and the first inorganic carbon molecule would be offset to +25. The average is always –5 and the offset is always 30, but the carbon isotope values of organic and inorganic carbon depend on the relative amounts of each (Fig. 16-2). Mathematically, the mass balance can be written as:

$$\delta^{13}C_{or}\, M_{or} + \delta^{13}C_{ic}\, M_{ic} = \delta^{13}C_T\, M_T \tag{16-5}$$

where M_T is the total mass of carbon, M_{or} is the mass of organic carbon and M_{ic} is the mass of inorganic carbong. Then $M_T = M_{or} + M_{ic}$. It is then possible to define a parameter, conventionally called *f*, which is the fraction of organic carbon to total carbon $f = (M_{or}/M_T)$. Substituting terms, we obtain:

$$f = (\delta^{13}C_{ic} - \delta^{13}C_T) / (\delta^{13}C_{ic} - \delta^{13}C_{or}) \tag{16-6}$$

By measuring the $\delta^{13}C$ values of organic and inorganic carbon in rocks of different ages, *f* can be estimated, permitting estimates of the proportion of organic to inorganic carbon through time (Fig. 16-2).

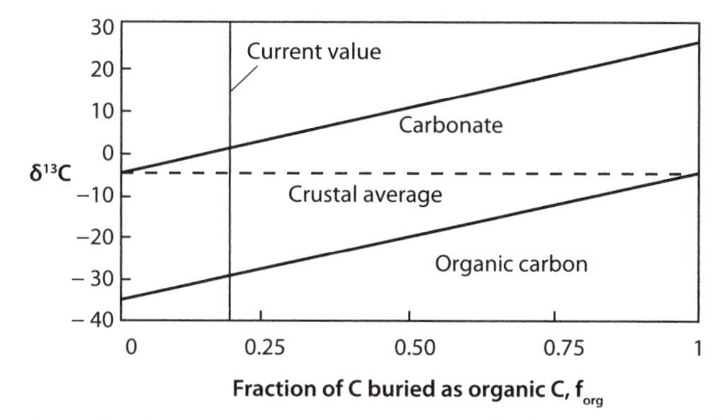

Fig. 16-2: Relationship between isotopic composition of carbonate carbon ($\delta^{13}C_{ic}$) and organic carbon ($\delta^{13}C_{org}$) as a function of the fraction of carbon buried as organic matter. The offset between the two is always 30 per mil. The total carbon must always average −5 per mil. The vertical line represents the current value for the global carbon cycle, and this value has remained remarkably constant over Earth's history. (Modified from Des Marais, *Reviews in Minerology and Geochemistry* v. 43, 555–78 (2001))

Carbon: Evidence from the Rock Record

The principle behind a carbon isotope record is straightforward—to measure carbonate and organic matter of various ages that reflect seawater composition. In practice, there are many difficulties. Samples distributed over the various periods of Earth's history with preserved carbonates and preserved organic matter are necessary. While carbonates are fairly abundant in the geological record and sometimes well preserved, organic matter is rarer and more easily modified by metamorphic processes. There are also the difficulties that the two types of material are often in different rocks from different locations, so the record requires correlation of ages across different continents. These aspects make it difficult to obtain a reliable record and to know whether measurements at particular time intervals in the Precambrian are global in extent or reflect local conditions. With such questions about the data, interpretations have "wiggle room."

Despite the difficulties in detail, the long-term history of carbon isotopes tells a compelling story. The data for both organic and inorganic carbon are shown in Figure 16-3. The horizontal line provides a reference for the mantle value of $\delta^{13}C$. If no organic carbon were produced,

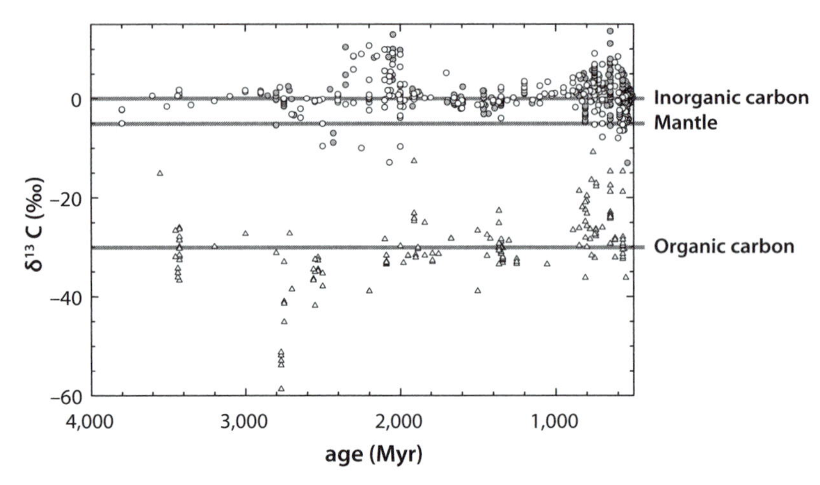

Fig. 16-3: Data for carbon isotopes from sediments over geological time. Open circles are carbonate (oxidized) carbon. Triangles are organic (reduced) carbon. Note the two periods of large variation in the carbon isotope composition of inorganic carbon near 2200 Ma and 800 to 600 Ma. (Adapted from Hayes and Waldbauer, *Phil. Trans. R. Soc. B.* 361 (2006):931–50)

we would expect carbonate carbon to have the same value as the mantle. As is apparent from the figure, throughout Earth's history carbonate carbon has had higher δ^{13}C than the mantle, and the offset on average varies within a small range. Since mantle carbon is −5, and carbonate carbon is about 0, this implies from Figure 16-2 that the carbonate carbon has made up about 80% of the active carbon reservoir, and inorganic carbon about 20%. The δ^{13}C of organic carbon should then be −30. This value inferred from carbon isotopes is essentially the same as total carbon balance seen in Table 16-1.

The contrast between carbonate carbon and organic carbon is not a simple constant, however, and while modern organic carbon varies around −30 per mil, the sparse measurements of ancient organic carbon are lower. Simplifying and smoothing out the data by averaging leads to the variations in fraction of organic carbon through time shown in Figure 16-4. The scatter in the data preclude a definitive conclusion, but the data suggest a possible increase in proportion of organic matter from about 15% in the Archean to 20–25% today.

The carbonate isotope data also show two periods of increased variability, including excursions to much higher values of carbonate δ^{13}C.

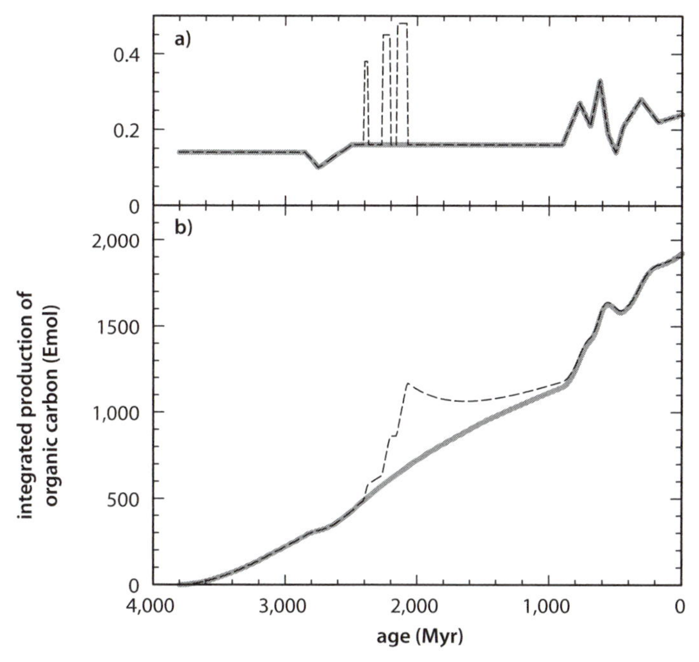

Fig. 16-4: Possible interpretation of smoothed data from figure 16-3. If CO_2 is steadily outgassed from Earth's interior, then the total amount of carbon at the surface increases with time. The constant fraction of organic carbon (*top panel*) then means that total buried organic carbon also increases with time (*lower panel*). There may have been short time periods of increased fraction of organic carbon burial (indicated by the dashed lines). Since O_2 production has a one-to-one relationship with organic carbon burial, the lower panel could also be viewed as the increase in oxidized molecules at the surface (largely oxidized Fe and S). (Modified after Hayes and Waldbauer (2006))

These occur between 2.4 and 2.0 Ga and 0.8 to 0.55 Ga. Taken at face value, these would suggest higher proportions of organic matter, and greater production of oxidizing power, during these two periods. If this were true in a simple way, however, we would expect to see in Figure 16-3 a smooth and globally uniform increase in carbonate $\delta^{13}C$ values over these time periods and a corresponding increase in the $\delta^{13}C$ of organic matter during the same time periods, and neither of these is evident. The "noise" and lack of correspondence of the organic and inorganic carbon records is problematic. Furthermore, these two time periods also show some negative $\delta^{13}C$ excursions. Many workers have interpreted these data to indicate when oxygen rose in the atmosphere, with a first-step increase in the 2.4–2.0 period and a second in the Neoproterozoic just prior to the development of multicellular organisms. What is clear is that these were times of rapid change and great variability, and that in contrast, the period from 2.0 to 1.0 Ga appears on present data to have been a remarkably long and stable period, sometimes referred to as the "boring billion."

What the geological record does show is an increasingly oxygenated surface over Earth's history. This requires steady oxygen production in excess of the sinks. If oxygen were not steadily in excess, the progressive oxidation of crust, ocean, and atmosphere could not occur. Therefore, a constant organic fraction of carbon, and the fact that the integrated organic fraction in all rocks surviving today is similar to the long-term average, requires an increase in the total mass of carbon in external reservoirs, unless there is a phantom source of oxidizing power with no surviving rock record, a possibility we will come back to at the end of the chapter. If CO_2 were steadily outgassed from Earth's interior, thereby increasing over time at Earth's surface, then a steady proportion of 20% organic matter removal would be a continuing flux to the surface, leading to a gradual increase in oxidizing power. It is not necessary to have periods of greater burial of organic matter to raise oxygen—oxygen is being produced all the time. A necessary consequence of this way of viewing the problem is that the total size of the organic matter reservoir would increase with time (as would the size of the carbonate reservoir). Determining the total carbon budget through Earth's history is a difficult problem, however, because the rates of recycling and removal through subduction are poorly known, even for modern periods.

The data in Figure 16-3 also show a great deal of scatter and no clear correlation between variations in organic and inorganic carbon. The concept is clear. The data are noisy. Therefore the carbon isotopes provide important constraints and much food for thought about the rise of O_2, but do not provide definitive answers.

Iron and Sulfur: The Record of Oxygen Consumption

The most important elements that track oxygen consumption are oxidized Fe and S. The rocks and seawater of Earth's surface today contain

Table 16-2
Where the oxygen produced by organic matter resides (in units of 10^{18} moles of O_2 equivalents)

O_2 in atmosphere	37.2
Fe^{3+}	1375
SO_4^{2-}	410
Total O_2 in oxidized compounds	1847
Total O_2 production by organic carbon from table 16.1	700–1280
Mass balance problem	500–1000 $*10^{18}$ moles of reduced carbon equivalent

Modified after Hayes and Waldbauer, *Phil. Trans. Roy. Soc.* B 361 (2006):931–50.

abundant oxidized Fe (Fe^{3+}) and S (S^{6+}), in contrast to the reduced state of the mantle that delivers these elements to the surface through volcanism. The amount of oxygen tied up in these oxidized molecules vastly exceeds the amount of oxygen in the atmosphere (see Table 16-2). Therefore, almost all of the oxygen that has been produced in the formation of organic matter does not exist as atmospheric O_2. It is largely stored in the oxidized molecules of Fe and S.

There is a special characteristic of Fe and S that makes them particularly useful for tracking the history of oxygen and that also had fundamental impact on Earth's history, and that is how their solubility varies as a function of their oxidation state.

Ferric iron is highly insoluble in water—this leads today to an oceanic concentration of Fe that is generally less than 1 part per billion, even though iron in the continental crust is about 5% (50,000,000 ppb). Ferrous iron, on the other hand, is quite soluble in water. The reduced fluids of hydrothermal vents, for example, can contain 100 ppm of Fe^{2+}. When the fluid exits the vent and encounters oxidized seawater, the Fe^{2+} is converted to Fe^{3+} and precipitates to contribute to the "black smoke" of the deep sea hydrothermal vents (recall Fig. 11-4).

Sulfur has the opposite behavior. Reduced sulfur (S^{-1}) is relatively insoluble, while oxidized sulfur (S^{6+}) forms sulfate that can have very high concentrations in water. Sulfate in seawater today, for example, leads to concentrations of about 900 ppm sulfur, even higher than the continental crustal composition of 400 ppm. Sulfide concentrations in a reduced ocean would have been only a few ppm.

This opposite behavior of Fe and S is particularly useful when we examine the geological record. A reduced Earth has mobile Fe and immobile S, leading to reduced soils with low Fe and high S and oceans with very high Fe/S. As the surface becomes oxidized, Fe becomes immobile in sediments and insoluble in seawater, while S becomes mobile from weathering and soluble in seawater. This leads to seawater composition with very low Fe/S. The differences are not small. The S/Fe ratio in seawater today is 40 billion times greater than the ratio in the continental crust, whose weathering is contributing to the ocean. In the early Earth, Fe may have been more abundant than sulfur in the reduced ocean. Oxidation state can be a powerful element separator!

These changes in oxidation state influence the distribution of Fe and S in rivers and oceans and the mineralogy of soils and sediments. Reduced iron minerals are Fe silicates, pyrite, and the Fe carbonate, siderite ($FeCO_3$). Reduced sulfur minerals are sulfides. Oxidized iron minerals are red and include silicates, oxides, and hydroxides. Oxidized sulfur minerals are sulfates, the most common of which is gypsum, $CaSO_4(H_2O)_2$. The scientific detective work is then to identify where, when, and why these minerals occur in the sediments of the geological record, and how the change in transport and deposition could be preserved in the rock record. By examining the history of Fe and S recorded in the rocks formed at the surface, we can derive constraints on the history of O_2.

Iron: Evidence from the Rock Record

The characteristics of many Fe- and S-bearing sediments have varied considerably over Earth's history. Two distinctive rock types that have very particular time distributions are banded iron formations and red beds. While both these formations are red in color, their mineralogy and composition differ strikingly, and they reveal very different stages of the evolution of oxygen at the surface.

Banded iron formations (BIFs) are distinctive, colorful rocks that derive their name from the fine laminations of the rocks, where layers of very Fe-rich sediments alternate with layers of cherts, which are almost

Fig. 16-5: Sample from a banded iron formation (BIF) showing the prominent banding of alternating Fe-rich layers with chert layers. Such rocks are the dominant form of iron ore for modern civilization. See also color plate 26. (Courtesy of Harvard Museum of Natural History, Dick Holland Collection)

pure SiO_2 (Fig. 16-5). They formed beneath shallow water and may have covered large portions of the continental shelves at the time of their maximum occurrence, at about 2,500 Ma. These unique rocks can contain up to 50% iron oxides, and are the chief source of iron ore for modern civilization.

The detailed characteristics of the BIFs change with time. The formations that occur in the oldest rocks such as the 3.8 billion-year-old Isua sediments in Greenland are generally thin (tens of meters thick) and occur interlayered with volcanic rocks. When BIFs reach their peak some billion years later near 2.5 Ga, they can be a thousand meters thick and extend over hundreds of square km. They decline in abundance in younger rocks, except for a brief reappearance at 1.8 Ga, after which they disappear from the geological record (Fig. 16-6), with one minor exception in the Neoproterozoic during the proposed "snowball Earth" episodes (see Chapter 12). The Neoproterozoic BIFs are chemically distinguished by having higher ratios of ferric to ferrous iron, with 95% of their total iron as Fe^{3+}, in contrast to the ~50% Fe^{3+} in the older BIFs.

Notable aspects of BIFs are their restriction to the Archean and early Proterozoic, and the vast concentrations of multivalent Fe that they

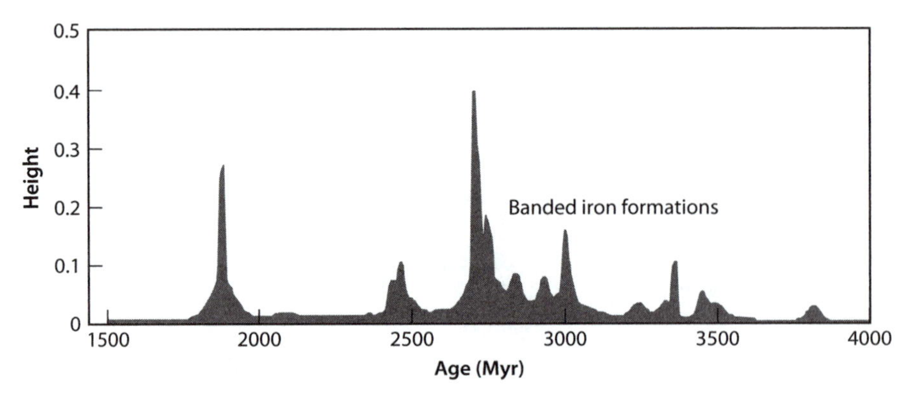

Fig. 16-6: Relative abundance of banded iron formations through time. Note the pronounced peak between 3000 and 2500 Ma, then there is a decline near 2400 Ma, and absence in rocks younger than 1800 Ma. (From Isley and Abbot, *J. Geophys. Res.* 104 (1999):15,461–77)

represent. The challenges posed by these rocks are to understand how so much Fe could be concentrated in sediments in this period of Earth's history, and never again, and what information they convey about the oxygen history of the surface.

The influence of oxidation state on solubility must be central to BIF formation. Because the BIFs are rocks that precipitate from seawater, central to any model are two conditions: a global source of reduced Fe, which leads to high seawater concentrations, and a local oxidizing environment, which causes the Fe to precipitate on the continental shelf. Within this overall framework, different hypotheses are possible.

One scenario, pioneered by Preston Cloud in the 1970s, related BIFs to the rise of oxygenic photosynthesis. Prior to the invention of oxygenic photosynthesis, the atmosphere and the entire ocean would have been reduced. Rivers would have carried reduced iron from the continents to the oceans, and reduced iron coming out of hydrothermal vents would have remained dissolved in seawater. The ocean would have been relatively rich in Fe^{2+} and contain very little S. The early fully reduced ocean would not have had the oxidizing power to convert Fe^{2+} + to Fe^{3+}, and Fe^{3+}-bearing sediments would not form in abundance. After cyanobacteria appear and produce O_2 in the shallow, sun-filled top of the ocean, the shallow ocean would have oxygen but the deep ocean would be reduced. Very little oxygen would exist in the atmosphere, because any oxygen that was present there would be reduced by reaction with volcanic gases and chemical interactions with the Fe- and S-bearing rocks of the continents.

At this stage, then, there could have been an oxygen source in the shallow ocean, particularly near continents where there would be sources of other nutrients, a reduced deep ocean, and a reduced atmosphere. When Fe-rich deep water reached shallow levels where oxygen was present, it would be susceptible to oxidation. Furthermore, the oxidation of Fe can produce energy, and bacteria likely evolved to take advantage of this reaction. Oxidation of Fe might even have made an advantageous environment for early ecosystems because it would remove the poisonous oxygen from the environment. The oxidized iron that formed would be highly insoluble and precipitate. If that occurred above the deep ocean, the Fe particles would sink to great depth, where they would be reduced and go back into solution. In shallow seas and continental shelves where the ocean depth was shallower than the depth of anoxia, the Fe could form very Fe-rich sediments. The change in oxidation state therefore would allow a kind of Fe pump to operate, where Fe is delivered to the ocean from deep hydrothermal vents and continental weathering and deposited beneath shallow, partially oxidized seas (Fig. 16-7).

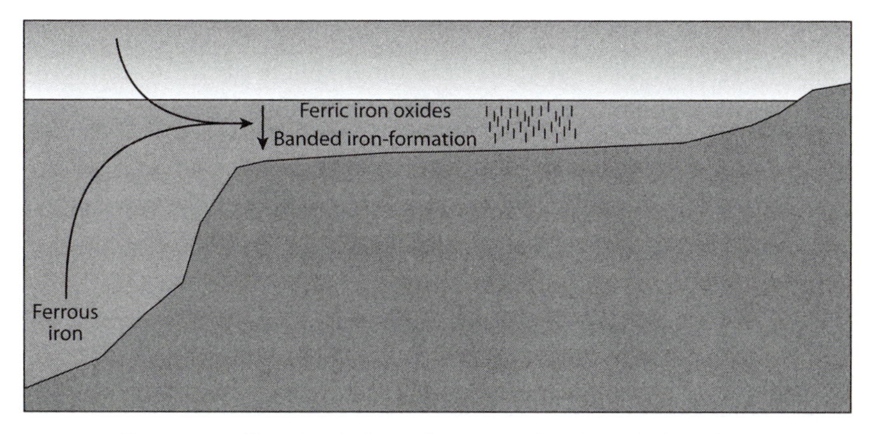

Fig. 16-7: Illustration of how banded iron formations (BIFs) might have formed. Soluble Fe^{2+} is released from hydrothermal vents and from continental weathering and remains in dissolved form in the reduced ocean. Continental shelves are sites of photosynthesis that convert Fe^{2+} to Fe^{3+}, leading to its precipitation. As long as the oceans at depth are largely reduced, then Fe can be pumped continually from the mantle to allow formation of BIFs on the continental shelves.

This chemical machine could operate only in a partially oxygenated Earth, where reduced settings mobilized Fe^{2+} and locally oxidized settings caused its precipitation as Fe^{3+}. Then BIFs would be rare and thin in the oldest Archean rocks, prior to an initial rise of oxygen, and abundant in a transitional period with small amounts of oxygen present. Today Fe-rich precipitates are rare. They form in small quantities near the ocean ridges, where hydrothermal vents continue to pump Fe^{2+} into the local environment and particles form immediately and settle to the seafloor. They cannot form in abundance today because atmosphere and the entire ocean are oxidized, the Fe content of seawater is vanishingly small, and oxidized Fe of the continental crust is not transportable. In this scenario, BIFs occur after the development of oxygenic photosynthesis and reflect a transitional stage of oxidation of Earth's surface reservoirs.

While this scenario has great appeal, it does not account easily for several aspects of BIFs. The first is the alternating layers of chert and Fe minerals. A successful model needs to account for both the Si and the Fe deposition. The second is that while BIFs do contain oxidized iron, they also contain a large amount of reduced Fe. Where does the reduced iron come from? And finally, we do not know for certain from independent evidence when oxygenic photosynthesis became important.

An alternative model, recently elaborated by Woody Fischer and Andy Knoll, forms BIFs by anoxygenic photosynthesis. Without oxygenic photosynthesis there must be another source of electrons to produce organic matter. A source that does not involve oxygen and could lead to deposition of Fe^{3+} could be:

$$4Fe^{2+} + CO_2 + 2H_2O \rightarrow CH_2O + 2Fe_2O_3 + 2H^+ \qquad (16\text{-}7)$$

In this scenario, Fe-oxidizing photosynthetic bacteria, which are known to exist today, would take advantage of the abundant Fe^{2+} in seawater to create Fe^{3+} minerals by photosynthesis, using the energy from the sun. The oxidized iron creates very reactive particles that scavenge Si from seawater. Reaction of the oxidized Fe and the reduced carbon in the organic matter in the sediment would create reduced Fe and oxidized carbon, leading to the precipitation of the reduced iron mineral siderite ($FeCO_3$). This mechanism carries both Si and Fe to the sediment, permits the reduction of some of the Fe, and leads to the formation of the siderite found in BIFs.

Whatever the detailed mechanism, the large mass of BIFs requires a high Fe content of seawater in the shallow ocean where sunlight is available, hence a much more reduced Earth than exists presently. And the BIFs also represent a major oxidation event of Earth's surface, since a great deal of Fe^{3+} was generated from Fe^{2+}. Every four moles of oxidized iron would reflect one mole of stored organic matter produced from CO_2.

When we turn to land, another piece of evidence from Fe emerges from the rare preserved ancient soils, called *paleosols*. Dick Holland studied such soils through geological time and showed that prior to 2.2 billion years soils were deficient in Fe, formed reduced Fe minerals, and did not contain oxidized iron minerals. Soils became more oxygenated in younger rocks. He used this and other evidence to propose a "Great Oxidation Event" around this time period.

The Fe evidence from these ancient sediments thus indicates a world with little free O_2. Even if oxygenic photosynthesis were underway, there were huge sinks for O_2 in the reduced Fe and S species at Earth's surface. The vast quantities of reduced species would require large amounts of formation and burial of organic matter before free oxygen was able to exist. These conditions were sufficient to keep atmospheric O_2 at such low levels that continental weathering took place under a reduced atmosphere, mobilizing Fe and immobilizing S.

Sulfur: Evidence from the Rock Record

Much more precision on the timing of the early rise of oxygen came from the tools of sulfur isotope geochemistry. We have made use several times already of stable isotope variations to reveal earth processes, but all of the variations that we have discussed thus far have been *mass dependent fractionations* that make use of slight differences in chemical behavior of two isotopes of the same element with different masses. Because this behavior is mass dependent, when there is double the difference in mass, the change in behavior is twice as great. For example, if some process changes the $^{18}O/^{16}O$ ratio by 2%, then the $^{17}O/^{16}O$ ratio changes by about 1%, because it has half the mass difference.

There are also processes that can cause *mass independent fractionation* of stable isotopes (MIF), and these can be studied in elements that have more than two isotopes. Apart from the nucleosynthetic variations produced in stars, mass independent fractionations are much less common and smaller in magnitude than mass dependent fractionations. One of the ways they can be produced is by photochemical reactions that are activated by light from the sun. For example, mass independent fractionation of oxygen occurs in the stratosphere today during formation of ozone.

Sulfur has isotopes of mass 32, 33, 34, and 36 and therefore is subject to both mass dependent and mass independent fractionation. Indeed, MIF of sulfur isotopes (SMIF) is observed during experiments on photochemical reactions and takes place to a very minor extent in today's upper atmosphere. All of Earth's major surface reservoirs today, however, show less than 0.02% mass independent fractionation. This turns out not to be the case in the ancient past. Rocks older than 2.0 Ga show larger MIF variations in sulfur, and rocks older than 2.45 Ga show variations of up to 0.4%, twenty times greater than anything seen in the last 2 billion years (Fig. 16-8).

There are two requirements for such large MIF to be preserved in ancient rocks. First, there must have been atmospheric processes that created large variations. And second, the sulfur cycle has to be in a state where multiple sulfur species with diverse isotopic composition can coexist in the atmosphere and the variations are not destroyed by mixing in the ocean. Today, for example, the ocean is a huge reservoir of sulfate, with a sulfate concentration of ~0.3% and a residence time of 9 million years. This contrasts with the ocean mixing time of thousands of years, so there is a huge mass of well-mixed sulfur in the oceans, and it takes a very large mass of distinctive sulfur to modify its isotopic composition.

Laboratory experiments and the details of SMIF suggest it is created by SO_2 and SO photolysis driven by deep ultraviolet radiation. Ozone is the present atmospheric constituent that absorbs ultraviolet radiation, and prevents both the photolysis and the coexistence of multiple sulfur species that would preserve the isotope variations. Therefore, there is no large SMIF today. The large SMIF in the Archean appears to require very low atmospheric ozone, consistent with low O_2 levels prior to 2 Ga. More detailed modelling has led to the conclusion that SMIF can be preserved only for oxygen levels 100,000 times lower than the present atmosphere.

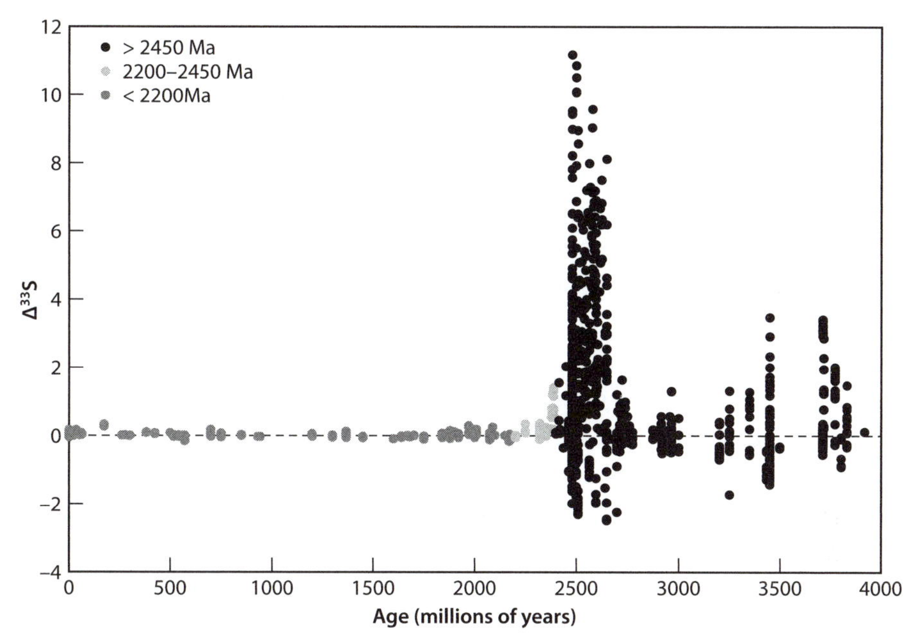

Fig. 16-8: Mass independent fractionation of sulfur (SMIF) through time. Prior to 2.4 Ga, SMIF was present on Earth, requiring anoxic conditions. The decline at 2.4 Ga, and absence of any SMIF in samples younger than 2.0 Ga, requires a rise in O_2 in the atmosphere at that time. (Courtesy of David Johnston, Harvard University)

Furthermore, the experimental results suggest that the atmospheric effects lead to SMIF variations of 6.5%, so the preservation of variations of .1–.2% requires 1–2% of sedimentary sulfur derived from the atmosphere, which would require a smaller oceanic sulfur reservoir.

How much O_2 is needed to turn off the SMIF signal? The experimental data suggest that levels of ~10^{-2} the present atmospheric level of O_2 (PAL) are sufficient to prevent the necessary UV radiation to create the SMIF variations in the atmosphere. The simplest interpretation of the SMIF data, therefore, is that oxygen levels were $<10^{-5}$ PAL prior to 2.45 Ga, gradually increased from 2.45 Ga to 2.0 Ga, and were at or above 1% PAL starting at 2.0 Ga.

Note that BIFs, SMIF, and carbon isotopes all indicate a major change occurring at about the same time near 2 billion years ago, suggesting a fundamental change at that time in the oxygen cycle. Might this change have had something to do with life itself?

A biogenic factor limiting oxygen may have been the intolerance of ancient organisms for high oxygen levels. Since oxygen breaks down organic molecules, it would have been a poison for early photosynthesizers, and too much oxygen is even a kind of poison for us today, hence, as said earlier, the popularity of "antioxidants." Anaerobic bacteria today, the likely descendants of earliest life, are killed by oxygen. The development of metabolic protection against high oxygen would be necessary for oxygenic photosynthesis to flourish. Otherwise there would be a negative feedback where high O_2 killed off O_2 production. We do not know when this evolutionary adaptation occurred, but it is conceivable that it relates to the geological rise of oxygen.

At some point bacteria did develop protection against oxygen, which would permit thriving oxygenic ecosystems. With photosynthesis providing a steady stream of oxygen to the atmosphere, we can imagine that eventually the oxygen overcame the reduced species available to consume it. At this point the atmosphere would have significant oxygen, though far less than modern values, leading to oxidizing conditions during weathering of the continents. This would impede the continental supply of Fe^{2+} by weathering and cut off one source of Fe for BIF formation. It would also lead to enhanced weathering of sulfur and a likely rise in seawater sulfate. If the oceans remained reduced at depth, the sulfate at depth would reduce to sulfide, causing precipitation of pyrite and removal of Fe from the deep sea. No longer able to be transported and unable to persist in the ocean, Fe would have dropped to low concentrations in seawater, cutting off the lifeblood of BIF formation.

The disappearance of BIFs and SMIF near 2 billion years ago appears to require only a slight rise in oxygen level, to less than 1% of the present atmospheric concentration. Of course there are important remaining details, such as the fact that SMIF declines abruptly at 2.4 billion years and stops at 2.0 billion years, and BIFs show a renewed pulse at 1.8 billion. Given the diversity of geological environments and conditions, however, the overall coherence of the two signals is quite conclusive. This change also occurs at the same time as the change in the mineralogy of soils and the great variability of carbonate $\delta^{13}C$. This time period clearly involved a major exterior modification of the surface from an anoxic to partially oxygenated world.

Evidence for High O_2 in the Phanerozoic

The other distinctive red rocks in the geological record are the *red beds*, sandstones such as the rocks that can be seen in the glorious landscapes of the American Southwest. These rocks are mostly quartz grains with a coating of fully oxidized Fe in the form of hematite. Virtually all of the iron in these rocks is Fe^{3+} and the total iron content is low, usually less than 2%. The red color does not reflect high iron content but an oxic coating on mineral surfaces. Like BIFs, these rocks begin to appear only in rocks younger than 2.0 Ga, suggesting a more oxic environment commencing at that time (Fig. 16-9). The consensus is that red beds reflect a fully oxidized surface environment that can convert all Fe^{2+} to Fe^{3+}.

This existence of an oxic Phanerozoic atmosphere is supported by other lines of evidence. We know that modern multicellular life requires high oxygen levels, and the advent of preserved fossils suggests quite high O_2 during the Cambrian and Ordovician. Large animals later in the fossil record, along with charcoal showing that organic matter burned, require high oxygen levels. Levels even higher than today during some time intervals are suggested by the existence of very large insects (e.g., dragonflies with the wingspan of eagles) and geochemical modeling. High model values are inferred from the large deposition of organic carbon (e.g., vast coal deposits) during the aptly named Carboniferous period ~300 Ma, when carbon isotopes also showed an increase supporting a higher proportion of organic carbon deposition.

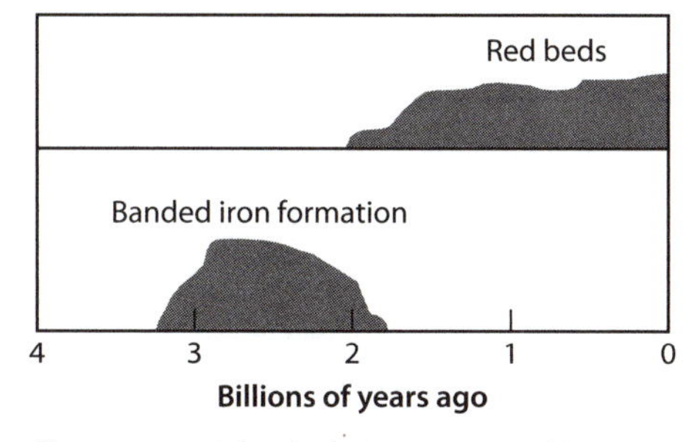

Fig. 16-9: Schematic illustration of the decline in BIFs and beginning of red beds in the geological record. Red beds are thought to occur only in an at least partially oxidized surface environment.

High Phanerozoic O_2 values can also be inferred from the compositions of *evaporites*. Evaporites form when water in closed basins evaporates, leaving its dissolved solids as precipitated sediments. We can see evaporites forming today, for example, around the Caspian Sea and the Dead Sea in the Middle East. Huge evaporite deposits formed in the Mediterranean when it dried up during the Miocene period ~5 Ma. Modern seawater composition has so much sulfate in it that modern evaporites precipitate the sulfate mineral gypsum even before halite. Phanerozoic evaporates also have an abundance of gypsum, suggesting similar oxidation conditions and a high sulfate concentration in seawater. It thus appears that Earth's surface layers have had similar O_2 levels to the present for the last several hundred million years. An Earth inhabited by plants and animals appears to have been possible only with the presence of an oxygenated atmosphere.

Oxygen *from* 2.0 Ga to 0.6 Ga

The rock record has allowed us to work forward from the Archean to see an initial rise of oxygen near 2.0 Ga, and to work backward from the present to suggest high O_2 for the last ~600 Ma. In between is the 2.0 Ga to 0.6 Ga interval of the Proterozoic, during which oxygen levels likely rose from perhaps 1% at 2.0 Ga to the 10–20% levels required for animal life at the dawn of the Phanerozoic. Did this change occur abruptly following a prolonged state at low levels, or was it a slow and gradual increase, or perhaps a series of step changes? If we look again at the carbon isotope record of Figure 16-3, it is apparent that much of this time interval appears remarkably stable, and then at the end, in the Neoproterozoic, the carbon isotopes fluctuated wildly with large positive and negative excursions, particularly during the proposed "snowball Earth" episodes (Fig.16-10). The long stability followed by great instability and a change from unicellular to multicellular biota that occurs at the same time has naturally led to the suggestion that the next big change in oxygen content of surface reservoirs took place during the Neoproterozoic.

There is emerging evidence from the chemical compositions of black shales that supports this possibility. The evidence comes from an unlikely source, the trace element molybdenum (Mo), which is another

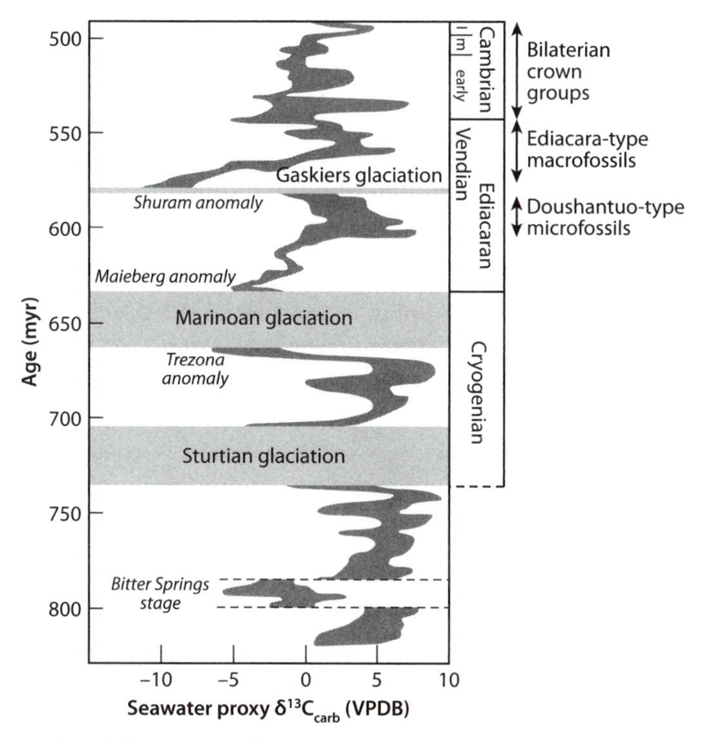

Fig. 16-10: The wildly varying carbon isotopes during the end of the Neoproterozoic. The first multicellular life (without hard body parts) occurred in the Ediacaran, just after the two major "snowball Earth" episodes. Since a snowball episode also occurred before the initial rise of oxygen near 2.3 Ga, and multicellular life likely requires high O_2, it is possible that the Proterozoic snowball episodes were also associated with a second rise in oxygen. (Modified from Halverson et al., *GSA Bulletin* 117 (2005), no. 9–10: 1181–207)

element with multiple oxidation states and changes in solubility with oxidation. Despite its low abundance in the total Earth, Mo concentrations in seawater today are higher than any other transition metal! This strange result occurs because oxidized Mo is easily weathered from continents and relatively soluble in oxidized seawater. Under more reducing conditions when the ocean would have had significant sulphide, Mo would have been efficiently removed from seawater. The record of Mo concentrations in black shales, while sparse, suggests a marked change in Mo concentration in seawater in the Neoproterozoic between 800 and 500 Ma. This result would be consistent with a fully oxygenated ocean appearing at this time, preventing Mo removal (Fig. 16-11). A

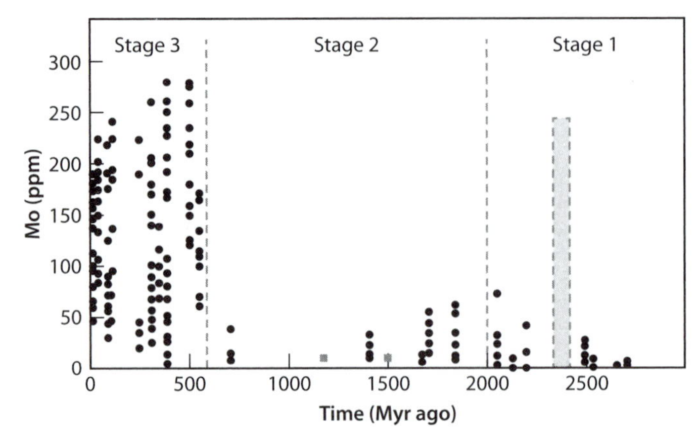

Fig, 16-11: Change of molybdenum concentrations in sediments through time. A reduced oxidation state makes Mo insoluble, hence low in concentration in seawater and oceanic sediments. A high oxidation state of Mo leads to solubility in seawater, making it one of the most abundant elements in seawater today despite its low abundance in surface rocks. The large change in its behavior near 600 Ma is consistent with a second rise in surface O_2 at that time. The shaded area shows the interval where sulfur isotopes indicate O_2 levels first rose (see Fig. 16-8). (Modified after Scott et al., *Nature* 452:456–59)

popular current point of view, therefore, is an oxidized shallow ocean and reduced deep ocean during much of the Proterozoic, and then oxidation of the deep ocean associated with a second rise of oxygen in the Neoproterozoic. The increasingly oxygenated environment then permitted the development of multicellular organisms and ultimately the Cambrian explosion of life.

These observations do not explain *why* oxygen rose to close to modern levels at this time. Various models have been proposed relating to increasing sources and decreasing sinks. Sources of oxidizing power are caused by removal of electrons from the system, normally brought about by organic carbon burial. What could precipitate excess burial of organic carbon? Once oxygenic photosynthesis became dominant and there was an essentially infinite source of electrons and hydrogen from breaking down H_2O, the limiting factor on organic matter production became supply of nutrients to the ocean. This is the case today, and very productive areas of the ocean are all areas where there is a supply of nutrients. Some tectonic factor that would increase nutrient supply could then lead to more organic matter production. One possibility is the breakup of large continents. Smaller land masses weather more efficiently because there is a greater proportion of coastline to land area. Another possibility is continental collision, since mountains weather more than flat plains. Periodic plate tectonic events could thus lead to pulses of organic carbon burial and a series of stepwise increases in O_2. Another impact on carbon burial could be sea level and the extent of inland seas and extended continental shelves, where there could be high productivity and efficient burial. All these pulses of organic carbon burial, however, should reveal themselves in the carbon isotope record, and there such signals are not clear.

There are two other ways to remove electrons from the system and cause the surface to become more oxygenated. One is the subduction of organic carbon. If the 1:5 organic/inorganic carbon ratio of subducted sediments is the same as sediments that remain at the surface, then a large amount of organic carbon could find its way to the mantle. Variations in the budget of subducted carbon would then have consequences for the oxygen budget. Variations in the ratio of subducted sulfate to sulfide might also have an effect.

Alternative models would diminish the sinks for O_2. One possibility is a locally more oxidized mantle through time at convergent margins, owing to subduction of oxidized iron. Then gases coming from volcanoes would be less reducing, diminishing the atmospheric sink for oxygen. If organic matter production were kept constant, O_2 would rise. Alternatively, metamorphic gases, which depend on the oxidation state of the continental crust, could have become more oxidizing, lessening a sink. Another diminished sink from continents would be the gradual oxidation of continental rocks so that the weathering flux contained less reduced Fe and S and had less capacity to absorb oxygen. Or the deep ocean could become more oxidizing and lessen the amount of organic carbon burial as well as sulfate reduction and pyrite burial. As sulfate concentration increases in the ocean, the flux of Fe from hydrothermal vents is also lessened. All of these are positive feedbacks—a smaller sink leads to more oxidized materials, which further diminish the sink.

These various models are not definitive because hard evidence from the rocks to support them is difficult to come by. We know the answer to the question of change in the oxidation state of atmosphere and ocean—oxygen rose. Much remains to be understood concerning the detailed mechanisms.

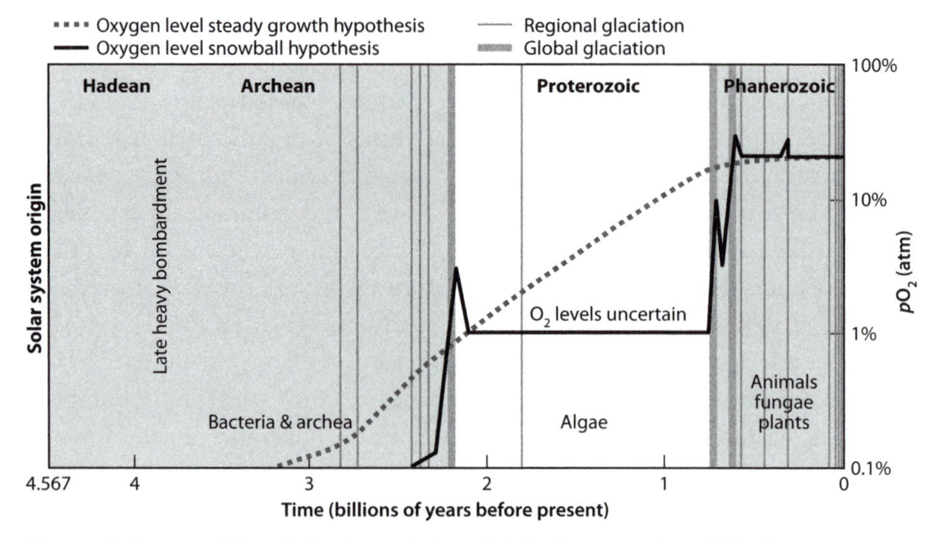

Fig. 16-12: One possible path for the evolution of O_2 in the atmosphere. This figure uses a hypothetical link between "snowball Earth" episodes (marked by vertical bands at 2.2 Ga and 0.6–0.7 Ga) and abrupt rises in O_2. It is also possible that rises were more progressive through time (the dotted gray line). Clear bounds are the absence of SMIF and decline of BIF near 2.0 Ga, and the advent of multicellular life that requires modern levels of O_2 about 0.6 Ga. (Figure modified from Paul Hofmann; http://www.snowballearth.org)

The end result of all this information is an attempt to graph the change of oxygen in the atmosphere through time (Fig. 16-12). Such a diagram has some clear fixed points, such as a zero O_2 early Earth, a significant change in atmospheric O_2 near 2.0 Ga, and a rise to modern levels at the end of the Proterozoic. To what extent the changes are gradual or abrupt, and the exact levels at the different periods of Earth's history, are less clear.

Global Oxygen Mass Balance

Whatever the details of the oxygen history, in the end there should be a one-to-one relationship between the electron addition to form organic matter and electron subtraction to create oxidized compounds, including O_2. To approach this question in the most general way, we can simply add up the oxidized and reduced reservoirs and see if the one-to-one relationship between organic carbon and O_2 is preserved.

Reduced carbon exists in myriad forms in rocks—for example, as fossil fuels—but most of it occurs in sediments, particularly in black shales that are abundant in the geological record. Total organic carbon abundance is estimated to be about 700–1,300 $\times$ 10^{18} moles (recall that one mole is 6.022 $\times$ 10^{23} molecules) . The observable oxidized reservoirs are the atmosphere, ocean, and crustal rocks. As can be gathered from Table 16-2, O_2 in the atmosphere is a trivial reservoir, and oxidized iron is the most important reservoir. Constraints on oxidized iron are quite good, because we have estimates for the Fe content and Fe^{3+}/Fe^{2+} ratio of the continental crust and sediments. We can use the oxidized Fe of the continental crust to estimate about 1,250 $\times$ 10^{18} moles of O_2 equivalents residing in Fe^{3+}. An additional contribution of 175 $\times$ 10^{18} moles equivalent comes from current oxidation of the ocean crust. As shown by comparison of tables 16-1 and 16-2, the oxidized reservoirs are then about double the reduced carbon estimates. This is the mass balance problem concerning the rise of oxygen—the oxygen sinks tell us that there has been a greater source of oxygen through Earth's history than what is preserved in the organic carbon remaining at the surface today.

These approximate calculations permit some comments on models for the rise of oxygen. The first is the great importance of oxidized iron. Consideration of the iron cycle also allows a fresh approach to the possible change in the oxidation state of the mantle, sometimes called upon to cause a diminished supply of reduced gases from volcanoes. It seems reasonable qualitatively to call upon subduction of oxidized iron to oxidize the mantle. But the upper mantle alone contains so much Fe that to raise Fe^{3+} by just 1%, less than can be resolved by measurement, would require the equivalent of 6 billion years of subduction of ocean crust as oxidized as it is today, which is unreasonable. This amount of subducted Fe^{3+} would also quintuple the shortfall of organic carbon, creating a far worse mass balance problem. And based on other evidence discussed above, the deep ocean may have been oxidized for $<$1 billion years, and prior to that time oxic alteration of the ocean crust would not occur. Oxidation of the mantle by organic matter production thus is very unlikely.

Instead, for the mantle to solve the mass balance problem, reduced atoms need to have been subducted. One possibility is ancient subduction of organic carbon. The reduced oceans of the ancient past could have led to some organic matter in deep sediments. If this material were subducted and not returned to the surface, there could be a large storehouse of electrons in the mantle. There would have to have been a mass of some 750×10^{18} moles of C subducted in excess of oxidized Fe. The amount of iron in the upper mantle is so great that this reduced carbon would have negligible impact on the iron oxidation state. It is also a small amount of carbon relative to the $\sim 30{,}000 \times 10^{18}$ moles of carbon in the mantle. Net subduction of organic carbon is thus a potential solution to the bookkeeping problem.

An alternative solution to the mass balance problem is to postulate a source of electron removal in addition to reduced carbon. The one likely candidate is loss of H_2 to outer space. Earth's early reducing atmosphere may have contained hydrogen. This creates a possible energy source for life, through the energy-producing reaction:

$$CO_2 + 4H_2 \rightarrow CH_4 + H_2O \tag{16-8}$$

Some bacteria, called *methanogens*, use this reaction to provide the energy needed to produce organic matter from CO_2 and H_2. In ancient Earth environments with H_2 available, a waste product of organic matter production would be methane, and methane is a product of anaerobic environments on Earth even today.

In today's atmosphere, methane is not a stable gas, since it reacts rapidly with the O_2 molecules that are present. Molecules of methane produced by life (e.g., cows produce an enormous amount of methane) survive in the atmosphere for an average only twelve years. In the absence of oxygen, however, methane can be stable in the atmosphere. The moon Titan, for example, has more than 1% methane in its atmosphere. In the low oxygen environment of the early Earth, methane might have been an important atmospheric constituent. Methane is also a very effective greenhouse gas and may have contributed to the early greenhouse effect that was able to keep Earth warm when the Sun was less luminous than it is today.

If methane were abundant in the atmosphere, then the ionizing radiation from the sun might have broken it up in carbon and hydrogen molecules. The H_2 formed in this process is able to escape from the upper atmosphere. Since H_2 is neutrally valent, it has two more electrons than hydrogen atoms combined in more complex molecules such as H_2O. Loss of H_2 is thus a way to have a flux of electrons from Earth's surface to space, creating a reduced reservoir of unknown size that is no longer measurable. Calculations by David Catling and colleagues suggest this flux could lead to an irreversible oxidation of the early Earth and solve the electron mass balance problem. An additional advantage of the methane formation/hydrogen loss model is that it may account for the fact that the rises in atmospheric oxygen appear to have occurred during times when Earth underwent severe glaciations. If methane were present in the atmosphere, it would keep the climate warm. A rise of oxygen would dramatically reduce the methane, causing a sudden ice age.

This model has the appeal that methanogenic bacteria are likely important parts of the early biosphere, an early reducing atmosphere would be consistent with presence of methane, and lowering methane could cause the heavy glaciations that seem to have occurred at the same times as oxygen rose rapidly. This type of model is very difficult to test, however, since the hydrogen lost to space can never be measured, and early methanogenic ecosystems likely have no exact parallels on the modern Earth.

The mass balance issues also relate to the current steady-state O_2 in the atmosphere. The carbon isotope data suggest that f has increased in the Phanerozoic, and CO_2 outgassing followed by organic matter burial should continually supply more O_2 to the surface reservoirs. Why is O_2 not then increasing? The traditional answer to this question would relate to variations in the fraction of burial of organic matter, as discussed earlier in this chapter. That fraction, however, appears to have been relatively high throughout the Phanerozoic. Furthermore, subduction of carbon in the 1:5 organic/carbonate proportions would contribute further to an increasingly oxidized surface. What is needed is a flux of oxidized material out of the system. One solution would be a modern oxidized flux to the mantle. The current outgassing flux of CO_2 is 3.4×10^{12} moles per year. If 20% of that becomes stored organic carbon, that is a flux of 0.68×10^{12} moles per year of reduced carbon. The current flux of oxidized iron in subducting ocean crust is $\sim 2.0 \times 10^{12}$ moles per year of O_2 equivalents, which more than accounts for the incremental oxidizing power produced by modern life. Plate-tectonic geochemical cycles appear to have played a very significant role in the oxygen balance in both the ancient and modern Earth.

Summary

Earth's atmosphere, soils and ocean today are oxidized environments consistent with a high level of O_2 in the atmosphere. This contrasts with the early Earth, where Earth's external reservoirs would have been in a reduced state. The journey from a reduced to oxidized surface is long, multifaceted, and complex. The broad picture of the change is apparent from the geological record, but many important details as to timing, mechanisms, and the actual levels of O_2 in the atmosphere through time remain to be fully elucidated. What is clear is that the ultimate driver of this process is life, which has made use of solar and chemical energy sources to generate a current of electrons to reduce carbon and form organic molecules. Storage or loss of these electrons has left behind the oxidized counterparts in the surface reservoirs, creating the oxic conditions that enable life to make use of aerobic respiration, which is the metabolic process that generates the most energy. Over time, through evolutionary change, life has created conditions that enhance habitability and maximize access to energy.

The exterior modification of the surface is closely related to geochemical cycles that involve Earth's interior. Because the fraction of organic/carbonate carbon stays roughly constant through Earth's history, continued generation of organic carbon and increase of oxidized molecules at the surface requires a continual flux of CO_2 from Earth's interior through volcanism. The ultimate source of surface oxidation is thus the plate-tectonic geochemical cycle. The mantle also supplied the reduced material to the surface, principally reduced Fe and S, with some H_2, which were the ultimate source of electrons to form reduced carbon. Oxygen is the intermediary of this process, donating electrons to form organic matter during oxygenic photosynthesis, and then receiving them back by oxidizing Fe and S, the long-term reservoirs of oxidized material. The steady-state O_2 in the atmosphere and ocean is the reactive reservoir involved in this transfer, and in terms of volume it is a small by-product of this overall process.

Important questions about the rise of O_2 remain unanswered. It appears that O_2 had a series of steady-state values at different levels through Earth's history, beginning at essentially zero in the Archean, rising to about 1% between 2.4 and 2.0 Ga after the rise of oxygenic photosynthesis, and then rising to close to modern values in the Neoproterozoic. The existence of long periods of steady state followed by step changes is a hypothesis that remains to be tested. Rigorous evidence of the causes of change and the detailed feedbacks that preserve different O_2 levels in each era also remain to be discovered.

Supplementary Readings

James Callen Gray Walker. 1977. *Evolution of the Atmosphere*. New York: Macmillan.

H. D. Holland. 2006. Oxygenation of the atmosphere and oceans. *Phil. Trans. R. Soc. B*. 361:903–15.

J. Hayes and J. Waldbauer. 2006. The carbon cycle and associated redox processes through time. *Phil. Trans. R. Soc. B*. 361:931–50.

D. C. Catling and M. W. Claire. 2005. How Earth's atmosphere evolved to an oxic state: A status report. *Earth Planet. Sci. Lett.* 237:1–20.

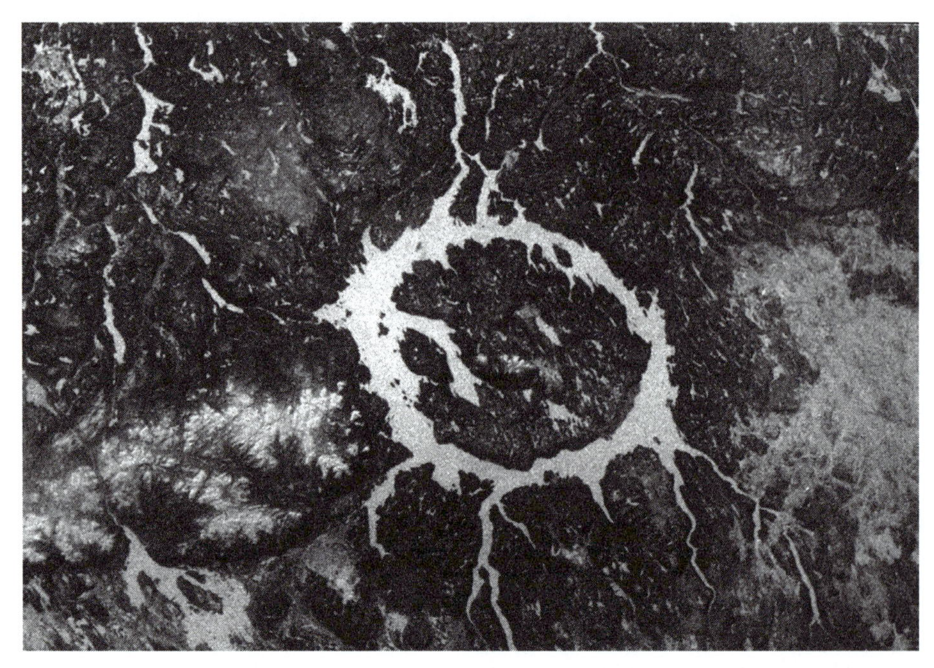

Fig. 17-0: Image of the Manicouagan Reservoir, Canada, December 1983. The reservoir, a large annular lake, marks the site of an impact crater 60 miles (100 kilometers) wide. Formed almost 212 million years ago when a large meteorite hit Earth, the crater has been worn down by many advances and retreats of glaciers and other processes of erosion. The reservoir is drained at its south end by the Manicouagan River, which flows from the reservoir and empties into the Saint Lawrence River nearly 300 miles (483 km) south. (Image and information courtesy of the Image Science & Analysis Laboratory, NASA Johnson Space Center; http://eol.jsc.nasa.gov)

CHAPTER 17

Planetary Evolution

The Importance of Catastrophes and the Question of Directionality

Oxidation of the planetary surface took place in the Archean and Proterozoic, where the limited life record does not allow high temporal resolution of events. Beginning in the Phanerozoic, the very high resolution of the fossil record reveals a variety of planetary processes that have a major effect on biological evolution. Catastrophes punctuate the Phanerozoic life record, and the causes appear to be diverse. The extinction of the dinosaurs that marks the Cretaceous/Tertiary boundary is unequivocally linked to the Chicxulub meteorite impact near the Yucatan peninsula, but it also occurs at the same time as the massive Deccan flood basalt province that marked the arrival of a mantle plume at the surface. The even larger extinction at the Permian/Triassic boundary has no known impact crater but occurred after two large flood basalt episodes, the largest of which intersected large coal beds, releasing vast amounts of gases to the atmosphere. Smaller extinction events are also closely associated with large volcanic outpourings from plumes. It may be that the largest extinctions are the result of coupled catastrophic events in a kind of "double whammy." While catastrophes lead to mass extinctions on a planetary scale, they may be necessary for evolutionary innovations to have the ecospace to succeed and an important factor in promoting evolutionary change.

Plate tectonics is another planetary process that has a major influence on life, once continents become populated. As continents join and separate, and traverse different climate environments, isolation can exist for long periods of time, permitting evolution to proceed on different pathways, creating diversity and permitting multiple experiments at innovation. Having opportunities to explore multiple pathways, evolutionary innovation may be accelerated.

While all of evolution occurred by random mutations, there are nonetheless long-term progressive changes. With time, organisms have developed increasing internal and external relationships—from prokaryotic cells to larger and more complex eukaryotic cells, to multicellular organisms, to organisms with many differentiated organs—as food chains have also expanded. Energy transformation has also increased, with the advent of photosynthesis, the change from anaerobic to aerobic metabolism, and the development of organ specialization first in organelles and later in organs. Increased relationship, organ specialization, complexity, and enhanced energy utilization may be general characteristics of more successful organisms and provide directionality to evolution. The specific pathway of genetic changes is random, but changes that lead to increased energy availability and relationship may be favored. Evolution is also enhanced by a changing planetary surface and by periodic catastrophes that permit evolutionary innovations to be expressed. These catastrophes are delivered from the solar system through meteorite impacts, from the planetary interior through volcanism, and through climate change. Global catastrophes, a changing planetary surface, climate change, directionality in terms of energy and networks are all general phenomena likely to be common in other planetary environments. It is possible that there are general principles exemplified by Earth that apply to a process of co-evolution of planet and life on other habitable planets.

Introduction

The evolution of life and planet during the Archean and most of the Proterozoic is a picture in low resolution, because the record of life itself in this period is obscure, and the absence of clearly defined and diverse fossils makes high temporal resolution very difficult. It is evident that life and planet co-evolved in the progressive oxidation of the planetary surface, but as we learned in previous chapters, the timing of the major events in the Archean and Proterozoic is only loosely constrained, and the record of unicellular life is vague and obscure. At the end of this period it is generally agreed that O_2 in the atmosphere rose to near current levels, and this step in planetary evolution permitted the formation of multicellular animals and, ultimately, the Cambrian explosion of multicellular life that marks the dawn of the Phanerozoic era. Multicellular life is not possible without the energy available from aerobic respiration—no multicellular life, for example, occurs in anaerobic environments on Earth today. Coincident with this step in planetary evolution, our ability to interpret the geological record comes into much sharper focus. The fossil record allows paleontologists to track the numbers of species, how long they last, and when the species assemblage undergoes drastic change. The fossils also initiate an era of tight temporal resolution on a global scale. The increase in resolution permits far more specific questions about co-evolution of life and planet to be considered, as major events can be pinned to specific times and causes.

Planetary Evolution during the Phanerozoic

As we learned in Chapter 14, the geological record since the Cambrian does not show the steady change that might be expected from gradual and progressive mutation of DNA. Instead, there were long periods of gradual change, with individual species existing for an average of 5 to 10 million years—only 0.1–0.2% of Earth's history—punctuated by times of more abrupt change, when mass extinctions wiped out large proportions of existing life and set the stage for new ecosystems with very different assemblages of organisms. These periods of abrupt change were used to delineate the geological timescale, giving rise to the major geological periods (Fig. 14-4).

Defining mass extinction events requires a quantitative measure of the abundance and diversity of life, and this is possible definitively only after the Cambrian when organisms developed hard parts and fossils became easily preserved. Careful work in slightly older rocks has revealed an earlier multicellular group of organisms called the Ediacaran, whose body parts are preserved in rare soft shales that never underwent significant metamorphism. These animals are sufficiently different from all later species that some paleontologists have inferred that a major extinction occurred near the Precambrian/Cambrian boundary, where the Ediacaran fauna disappeared to be replaced by early Paleozoic species. Earlier mass extinctions may also have occurred, but there is insufficient record to document them. "Snowball Earth" episodes, for example, may have led to mass extinctions.

The development of the fossil record in the Phanerozoic makes it possible to use criteria such as the overall numbers of different genera of fauna and flora to define mass extinctions. Marine organisms are the best index of the diversity of life because sediments progressively accumulate beneath the ocean and are destroyed through erosion on land. Figure 17-1 shows how the types of fauna vary along with total numbers of fauna from the careful compilation of marine genera through geological time (recall that genera are groups of related species). Examination of this figure reveals an overall growth in numbers of genera from the Cambrian to the present, but we see that this growth is not monotonic. There are periods of rapid growth, long plateaus, and sudden dips—the extinction events.

The style of life also has changed dramatically over time, starting with a typical Cambrian fauna from the early Phanerozoic that consisted of invertebrates, such as jellyfish and sponges, and the very recognizable trilobite (Fig. 17-2a). The remaining groups of invertebrate fauna all developed during an explosion of life in the Cambrian and early Ordovician. Between the Ordovician and Silurian there was a significant extinction event, reducing the number of genera by about 50%. Land plants first developed during the Silurian. Significant extinction events also separate the Devonian and Carboniferous. During the Carboniferous, land plants underwent major development, and reptiles first appeared. By the Permian most of the major fossil groups were present, with the exception of mammals. The Paleozoic era is thus marked by an explosive early growth in numbers of genera, and then a long period of relative stability where the number of genera remained relatively con-

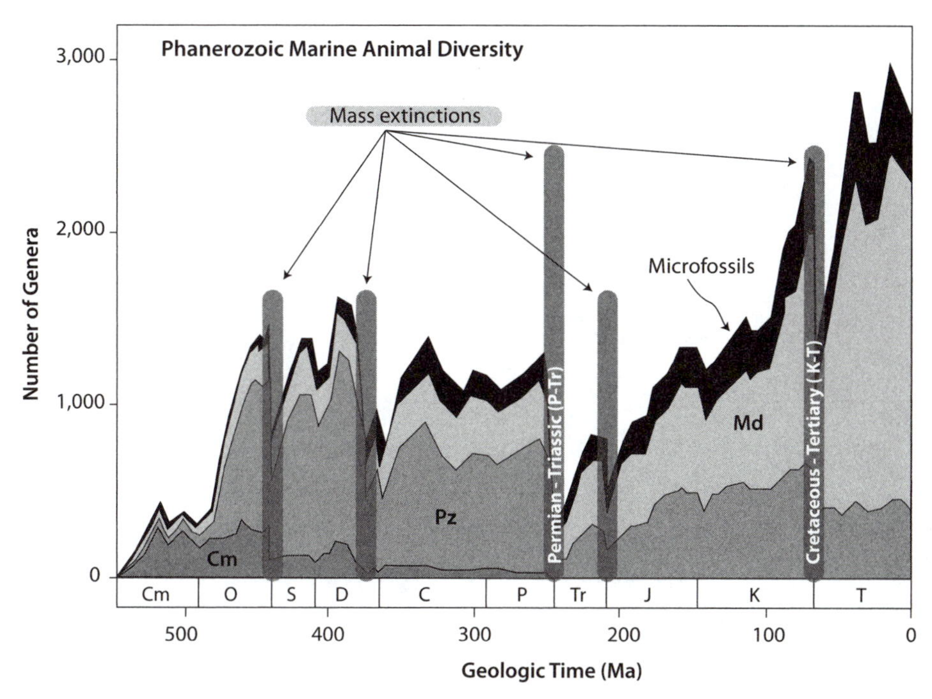

Fig. 17-1: Change in number of genera through time, showing the changes in types of life over time. Note that the Cambrian fauna, for example, peaked in the Ordovician and then went extinct at the Permo-Triassic mass extinction. *Cm* stands for Cambrian fauna, *Pz* for Paleozoic fauna, *Md* for modern fauna. (Modified after Sepkoski, *Bulletins of American Paleontology* 363 (2002)).

stant, subject to extinction events that preferentially eliminated much of the characteristically Cambrian life. Of course on the species level, extinctions and evolution continued apace through this period, but there is nonetheless a recognizable Paleozoic style to life. The difference in fossil assemblage of the Cambrian, the late Paleozoic, and modern times is apparent to the naked eye (Fig. 17-2b,c).

The late Paleozoic stability ended abruptly. At the end of the Permian there were two episodes of mass extinction, separated by about 10 million years. Combined, these two events extinguished 80% of genera on Earth. Trilobites and other families that originated in the Cambrian and had been present for 300 Ma disappeared forever. There were major extinctions on land as well, showing that the causes of this event were truly global in scale. This massive extinction is used as a time marker to

Fig. 17-2: Artist's renditions of scenes from three different periods of Earth history, all distinct from one another and from modern ecosystems. (a) Cambrian marine scene, where only invertebrate life had developed (© The Field Museum, #GEO86500-052d, with permission). (b) The Devonian period, just before the Permo-Triassic extinction, where more complex plants, fish and other vertebrates were present (© The Field Museum, #GEO86500-125d, with permission). (c) The Cretaceous, the Age of Dinosaurs, just before the Cretaceous-Tertiary extinction (© Karen Carr, with permission (www.karencarr.com)).

separate the Paleozoic and Mesozoic eras. For the first tens of millions of years of the Triassic, the first period of the Mesozoic era, life barely recovered. In the mid-Triassic mammals first appeared. At the end of the Triassic there was a smaller extinction event, followed by a second rapid growth in the numbers of families, soon exceeding what had been present even under the most fecund periods of the Paleozoic. By the

b

c

Cretaceous period, the giant reptile dinosaurs ruled the Earth at the top of food chains (Fig. 17-2c).

The Mesozoic also ended with a mass extinction, used to separate the Mesozoic from the Cenozoic eras. The end Mesozoic extinction caused the demise of the dinosaurs, as well as extinctions of other major fossil groups on both land and sea. This set the stage for another renewal of diversity, with numbers of families recovering rapidly and leading to still greater diversity of life. This new and characteristic assemblage of living organisms had food chains dominated by the newly emergent mammals, which during the Mesozoic had been small, niche creatures scratching out a living in reptile-dominated ecosystems.

From the perspective of evolution, an important aspect of mass extinctions is that they lead to subsequent more rapid evolutionary change. Andrew Bambach and others looked at both genera elimination and genera formation over the last 450 million years. The extinction events led to large negative excursions in the numbers of species. And following these events was the most rapid growth in genera diversity (apparent in Fig. 17-1). The clearing out of the existing ecospace permitted new evolutionary adaptations to flourish. When there is ecological stability, dominant organisms control the ecosystems and there is little room for genetic innovations to take hold. With the wholesale destruction of life, ecological niches open and opportunities for new forms of life appear, permitting more expression of genetic variations. Improved adaptations that were not able to be fully expressed in a dominated ecosystem have room to flourish.

CAUSES OF EXTINCTION EVENTS

It is evident that evolution on Earth was punctuated and influenced by mass extinctions, catastrophes on a planetary scale that take place in very short periods of time. How short? As shown in Figure 17-3, extinction events occur over centimeters of sedimentary thickness, so the duration must be very short indeed. For the major extinction events that separate the Paleozoic, Mesozoic, and Cenozoic eras, which have received the most attention, the time is so short that its exact duration cannot be well resolved by radiometric dating. Very careful work with zircons has progressively narrowed the timeframes revealing that the Cretaceous/Ter-

Fig. 17-3: The Cretaceous-Tertiary extinction from La Sierrita, showing the light gray claystone that marks the boundary. Notice the very brief stratigraphic interval over which the boundary occurs. (Photo courtesy of Gerta Keller, Princeton University)

tiary extinction occurred in less than 1 million years and that the Permo-Triassic extinction was the result of two events that occurred about 10 million years apart at the end of the Permian. Each of the events was very short in duration from a geological perspective. Owing to the dramatic changes in life that mass extinctions represent, there has been intensive study of their cause, and much controversy. One reason for the controversy is that it now appears that there were multiple causes of mass extinctions, rather than one single theory that is able to explain all of them.

THE CRETACEOUS/TERTIARY EXTINCTION

The most famous extinction event is the demise of the dinosaurs at the Cretaceous-Tertiary (K-T) boundary between the Mesozoic and Cenozoic eras. To try to determine the time duration of the extinction, Luis and Walter Alvarez decided to investigate the concentrations of the metal iridium (Ir) in sediments that traverse the time boundary. Iridium is highly siderophile, so most of Earth's iridium is in the core, and there are very low abundances in the continental crust. Iridium in meteorites is 10,000 times greater than the crustal concentrations. Since a little bit of iridium is continually added to marine sediments, owing to cosmic dust entering the atmosphere and sinking through the ocean, the Alvarezes thought that by measuring the iridium they could constrain the time duration that the few centimeters of sediment represent.

To their surprise, Ir was exceptionally abundant exactly at the K-T boundary, so abundant that it could not be explained by the tiny amounts of cosmic dust entering the atmosphere (Fig. 17-4). This serendipitous discovery led them to propose that a large asteroid hit the Earth at that time, bringing in huge amounts of Ir that spread in a large dust cloud around the globe and accumulated exactly at the boundary. The meteorite impact would be such a catastrophe that it might extinguish much of life on Earth in a matter of years.

This idea was not universally accepted, and an alternative hypothesis was proposed that unusually abundant volcanism at the K-T boundary was the cause of the extinction. Over less than a few hundred thousand years that span the K-T boundary, a flood basalt province appeared on Earth's surface. As we learned in Chapter 11, when a plume first arrives at the surface, the "plume head" has a massive volume that produces overwhelmingly abundant volcanism. Many flood basalt provinces represent these initial outpourings, after which a much reduced "plume tail" leads to a long ridge on the surface as the plate migrates over the plume. The first outpourings of the Deccan plume produced millions of km^3 of basalts in northern India, called the Deccan Traps, and a hot spot track that is still active today at Reunion Island (Fig. 17-5). This event was large enough to more than double global subaerial volcanism for hundreds of thousands of years, potentially leading to very serious environmental effects that could impact global life. The volcanic hypothesis, however, could not easily explain the Ir evidence.

The meteorite impact hypothesis at the K-T boundary became fact by the discovery of the Chixculub impact crater (Fig. 17-6) near the Yucatan peninsula in Mexico that is the exact age of the boundary. Drilling of sediments off the coast of Florida (see Fig. 17-4b) also showed an abrupt change in ocean species coincident with the Ir anomaly, evidence for a huge tsunami, and the telltale signs of impact discussed in Chapter 8—shocked quartz and tektites.

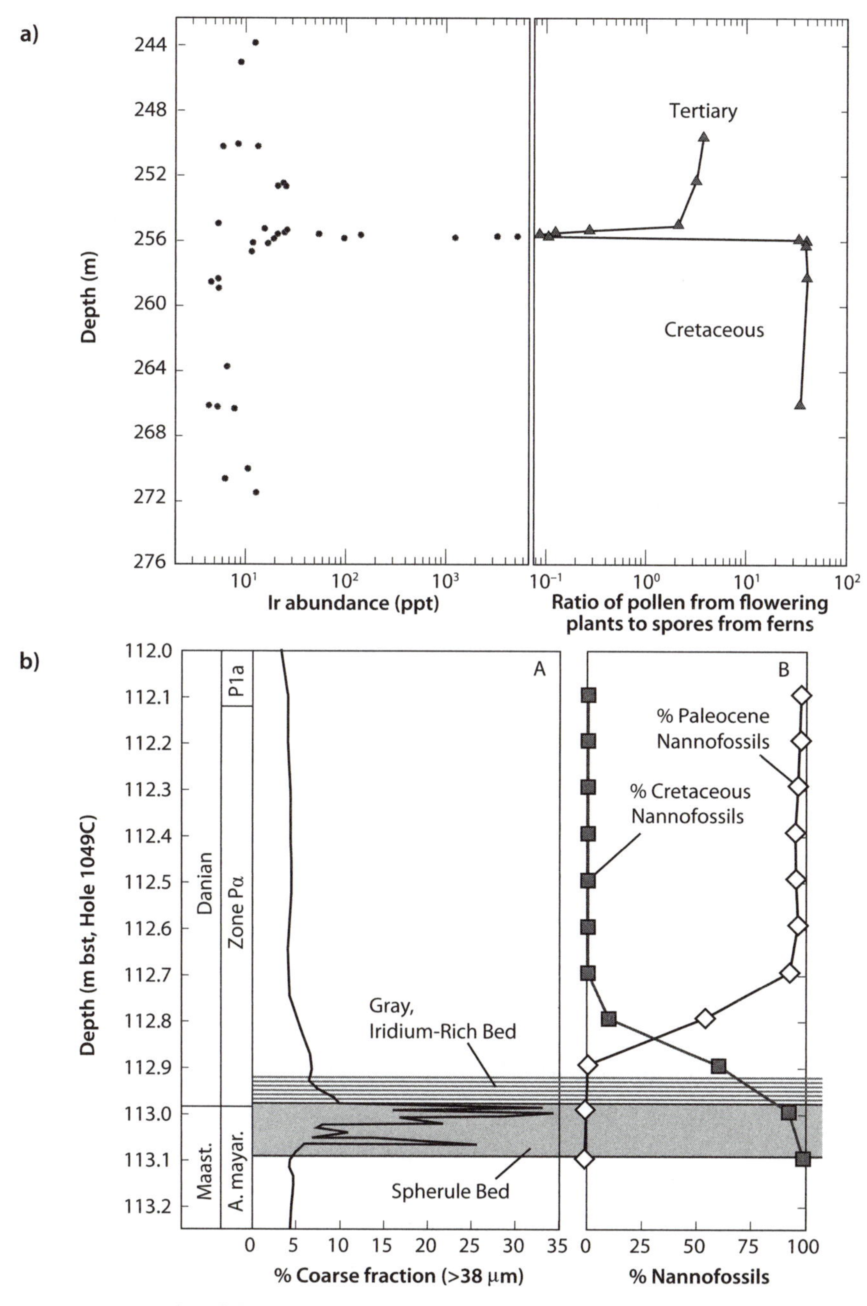

Fig. 17-4: Examples of changes at the Cretaceous Boundary. (a) An example of the iridium spike immediately at the boundary, along with the data showing the dominance of ferns after the impact (creating a negative spike in the angiosperm pollen/fern ratio) (modified from Orth et al., *Science* 214 [1981], no. 4257: 1341–43). (b) Data from an ocean drilling core off the east coast of Florida showing the boundary and the abrupt change in species that occurs there (adapted from Norris et al., *Geology* 27 [1999], no. 5: 419–22).

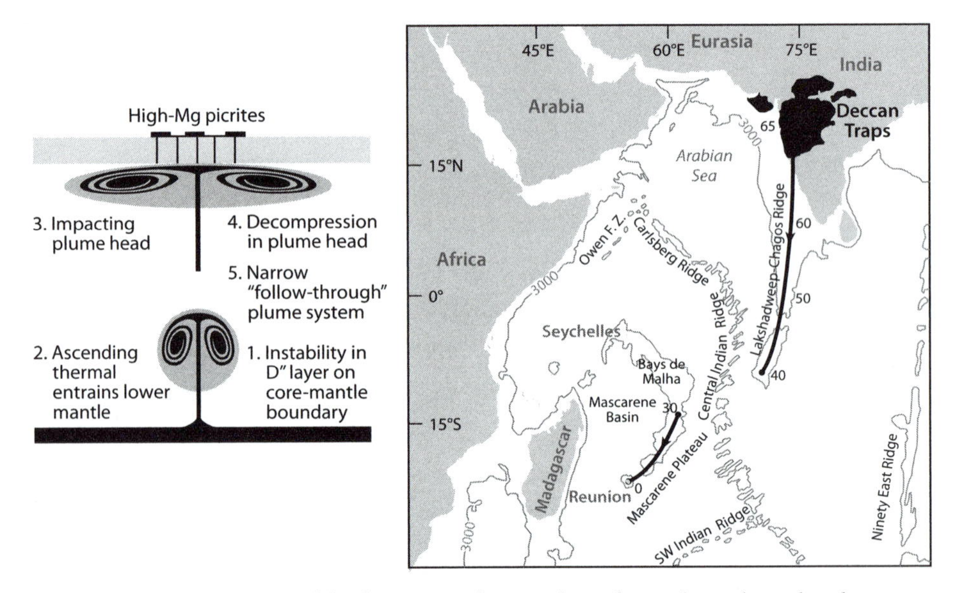

Fig. 17-5: (*Left*) Illustration of the formation of a new plume from a lower boundary layer, probably the core-mantle boundary, lifting off and heading to the surface. There is a very large "plume head" followed by a narrow plume tail. When the plume head hits the surface it spreads out and generates massive volcanism over a wide range. Following this flood basalt event, which can increase global subaerial volcanism by a factor of 2 or more during its duration, a long and narrow hot-spot track is generated as the plume tail intersects the plate moving above it. (Figure derived from Griffiths and Campbell, *Earth and Planetary Science Letters* 99, 66–78 (1990)). (*Right*) The narrow hot spot track generated by the "tail" of the Deccan plume. The current location of the plume is the island of Reunion in the southern Indian Ocean. Numbers by the track indicate the age of volcanism when that portion of the plume track was generated. The track was generated as a continuous feature and later separated by the spreading of the Central Indian Ridge. (Modified from White and McKenzie, *J. Geophys. Res.* 94, 7685–7729 (1989)).

The fact that a meteorite impact occurred, however, does not necessarily mean that it was the *sole* causal event. Other large meteorite impacts for which craters have been found, for example, are not associated with mass extinctions in the fossil record. Furthermore, some paleontologists working on the K-T boundary argued that the extinction of life on continents was not as sudden as would be predicted from an impacting event. There is also evidence of radical environmental change prior to the Chixculub impact. This has led to the idea that the mass extinction was the combined result of two major events that overlapped in time—the flood basalt eruptions stressed the global biosphere, and in the midst of this crisis the impact delivered the critical blow.

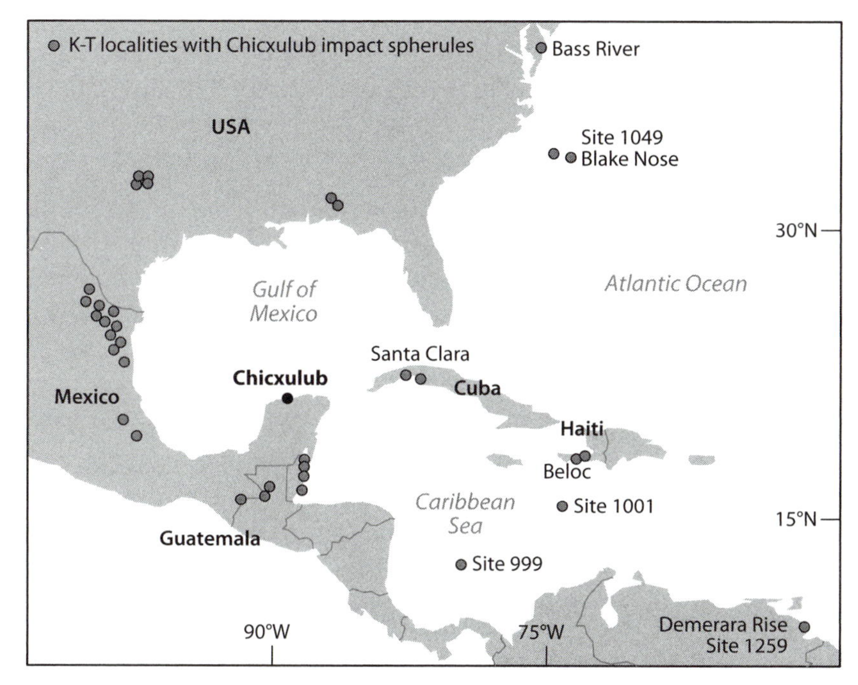

Fig. 17-6: Map showing the location of the Chixculub impact crater off the coast of Yucatán and the various sites where there is evidence for ejecta, shocked quartz, impact spherules, massive tsunamis, etc., that vary quite regularly with distance from the impact site. (Modified from Gerta Keller, Princeton University; http://geoweb.princeton.edu/people/keller/Mass_Extinction/massex.html#7)

THE PERMO-TRIASSIC EXTINCTION

The Permo-Triassic mass extinction was even more catastrophic than the K-T extinction event. Whereas some 50% of genera disappeared at the K-T boundary, 80% disappeared at the end of the Permian. The Permo-Triassic boundary is very well exposed in sediments in China, and exhaustive investigations of these sedimentary sections have revealed that the extinction was a double event, with large extinctions some 10 million years apart (Fig. 17-7). An intensive global search for impact craters has been unsuccessful. In itself this is not definitive—the oldest seafloor is much younger than the P-T boundary, so any impact crater that occurred in the ocean would been subducted long ago, and the oceans are twice as likely to receive an impact as the continents. The impact should

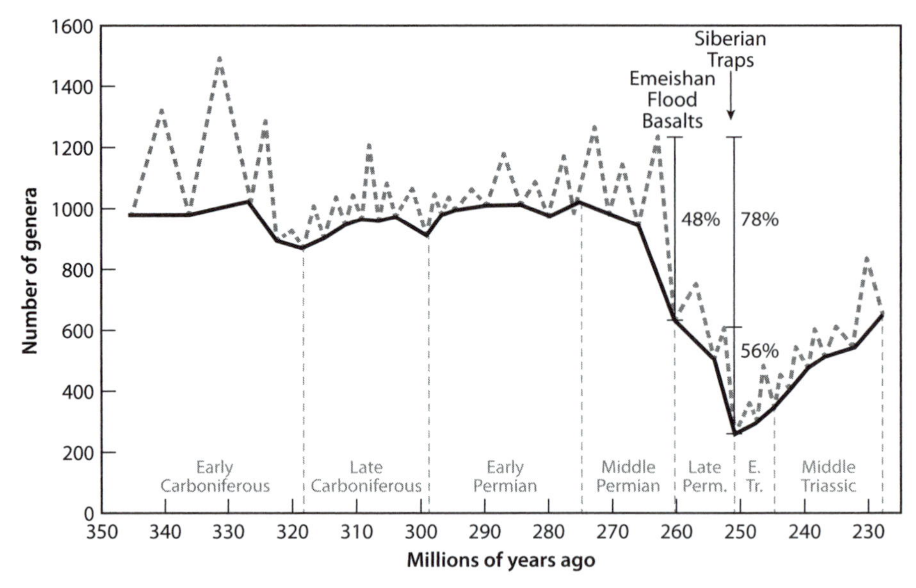

Fig. 17-7: Detailed plot of the change in numbers of genera across the Permo-Triassic boundary. Note the short timescale of the bottom boundary (in millions of years). The mass extinction is clearly seen as a double pulse, with each pulse associated with a continental flood basalt province. (Modified from Knoll et al., *Earth Planet. Sci. Lett.* 256 (2007):295–313)

have left other traces, however, such as an iridium anomaly, tektites, and shocked quartz, and these have not been found. The events that do occur at the appropriate times for the two P-T events are large volcanic outpourings akin to the Deccan Traps. The older event corresponds with the Emeishan Traps in western China. The younger event that demarks the P-T boundary is simultaneous with the largest volcanic outpouring known in the Phanerozoic geological record. The Siberian Traps (Fig. 17-8) produced more than $2 * 10^6$ km^3 of lavas in a very brief period that is indistinguishable in time.

The volcanic explanation is supported by one of the characteristics of the end Permian extinction, which is a marked change in the isotopic composition of carbon in the ocean. There is not enough carbon in a meteorite to influence the ocean's carbon budget, and the meteorite carbon also does not have the necessary isotopic composition. Instead, the carbon isotopes suggest a huge addition of organic carbon to the atmo-

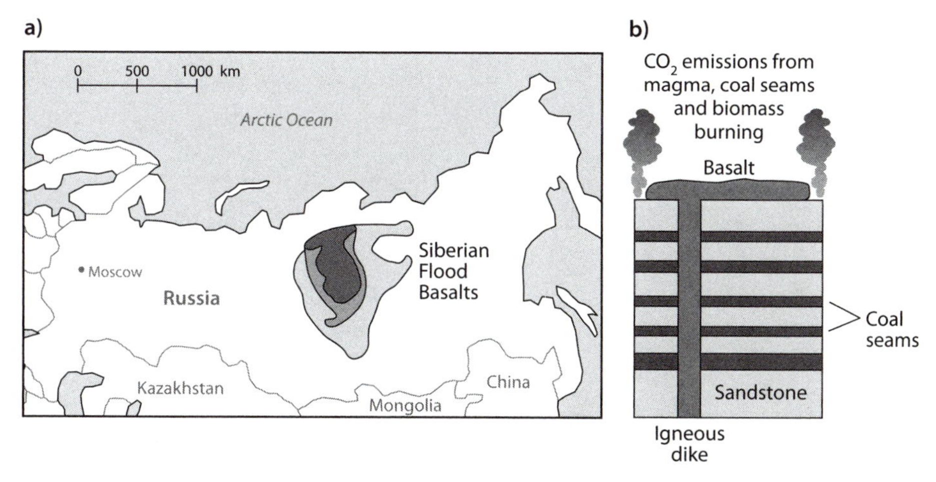

Fig. 17-8: (a) Map of the huge extent of the Siberian flood basalt province associated with the Permo-Triassic extinction. (b) Illustration of how flood basalts might interact with reduced carbon in crustal sediments to lead to huge outpourings of reduced carbon species to the atmosphere. The excess gas emissions would account both for the extinction and the large lowering of $\delta^{13}C$ at the P-T boundary.

sphere and ocean. Since organic carbon is produced by the biosphere, could volcanic outpourings somehow be associated with such an event?

In the case of the Siberian Traps, they can. The Siberian Traps were erupted into and through sediments very rich in coal, and examples elsewhere of lava flows burning through huge coal seams have been observed. Coal was formed from plant matter and has the right carbon isotopic composition. Furthermore, the burning of coal would lead to massive emissions of CO_2 that would have major climatic consequences, including a global warming episode and a change in the acidity of the ocean. The evidence from the Permo-Triassic supports the idea that massive and focused volcanic outpourings are a major driver of extinction events.

Vincent Courtillot, along with colleagues who specialize in precise radiometric dating, has argued that all the major extinction events in the Phanerozoic can be associated with massive flood basalt outpourings (Fig. 17-9a). This figure is not as convincing as it seems, however, because selection of volcanic events and extinctions can be used to force a good

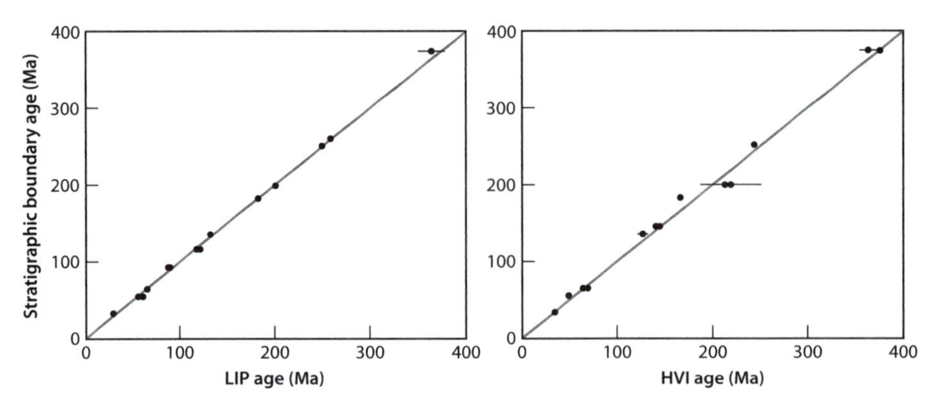

Fig. 17-9: (*Right*) Cross plots showing possible correlations between meteorite impacts (*HVI* = high velocity impacts) or continental flood basalt provinces (*LIP* = large igneous province). (*Left*) Times of massive environmental change, usually associated with mass extinctions. (Modified after Kelley, *Geol. Soc. London* 164 (2007):923–36)

correlation (Fig. 17-9b). Impacts are more difficult to date precisely because of their limited aerial extent and lack of datable materials. Simon Kelley has shown, however, that even with the dating errors the correspondence is significantly better for volcanic events than for impacts (Fig. 17-10). Therefore, with the exception of the K-T boundary, which is unambiguously associated with a specific impact, massive volcanic outpourings seem to have a more lasting environmental and biological effect. Recently Courtillot and Olsen showed an intriguing relation between the largest extinction events and the behavior of Earth's core, as apparent in the reversals of the magnetic field. The "killer plumes" that led to the largest mass extinctions occurred after long intervals where no magnetic reversals were taking place. Since large plumes are thought to originate at the core-mantle boundary, core-mantle interactions may be linked to the major evolutionary events of life at the surface.

It is probably significant that the largest extinctions, at the Permo-Triassic and Cretaceous-Tertiary boundaries, both appear to result from a double whammy, two flood basalts in close temporal proximity at the end of the Permian and a flood basalt and meteorite impact at the end of the Cretaceous. This would account for the observation that there are both meteorite impact and flood basalt events in the geological record that are not associated with mass extinctions. The biosphere appears to

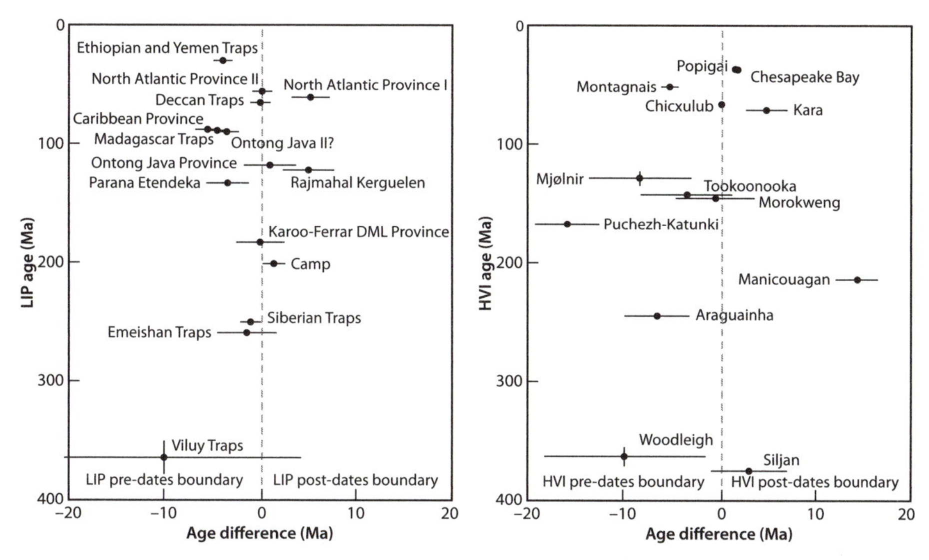

Fig. 17-10: A more detailed examination of the age offsets between dated impacts and mass extinctions. LIP stands for Large Igneous Province. Note the exact correspondence for the Chixculub impact at the K-T boundary, and the significant offsets for many other impacts, except for those with very large error bars. (Modified after Kelley, *Geol. Soc. London* 164 (2007): 923–36)

have been sufficiently robust that a "one-two punch" is necessary for the most catastrophic extinctions.

The Phanerozoic record thus shows the close linkages between biological evolution and events determined by the solar system and planetary interior. Meteorite impacts that are the remnants of planetary accretion have had a major influence on life, and core/mantle interactions that give rise to mantle plumes that are the active component of upwelling mantle convection have also greatly influenced life's history. The association of volcanism with mass extinctions also suggests a finely tuned balance of biological and solid earth processes. Volcanism brings reduced rocks and gases to Earth's surface. The vast outpourings of flood basalt provinces, doubling subaerial volcanic output for hundreds of thousands of years, may upset the normal balance of geochemical cycles and lead to massive short-term climate change. In the case of the Siberian Traps, the burning of coal would have represented an uncontrolled and catastrophic release of energy from Earth's planetary fuel cell, as millions of years of accumulation of organic carbon were released and oxidized all at once, greatly perturbing the carbon cycle.

Plate Tectonics and Evolution

Plate tectonics has also had a substantial influence on biological evolution during the Phanerozoic. The process of evolution through natural selection leads to genetic differences as a result of time and separation. Because evolution follows different trajectories when ecosystems become isolated, the longer the time of isolation the greater the differences among organisms. Separation on Earth is largely caused by tectonic processes. On the largest scale, separation of continents on different plates can lead to tens of millions of years of independent evolution. We are aware of the differences among large mammals on the different continents, which are a consequence of this continental separation. The differences are most marked for Australia, which separated from Antarctica some 50 million years ago. Australian mammals can be traced to a common ancestor with other modern mammals, but after it became separated from other continents Australia gradually developed its own characteristic fauna, including "exotic and bizarre" species such as the emu, the duck-billed platypus, and the marsupial, kangaroo. In fact, owing to its tectonic isolation and unique climatic history, from 83% to more than 90% of its mammals, reptiles, insects, and amphibians are endemic to Australia (i.e., found on no other continent). A particular feature of Australian fauna is the relative absence of placental mammals and the abundance of marsupials. The marsupials likely flourished in Australia because of their low metabolic rate, which was suited to the hot and dry climate. Human import of other mammals, such as rabbits and domesticated animals, has led to extinction of many of the marsupials and other unique Australian fauna, as well as to significant changes in Australian ecosystems.

While continental isolation has the most pronounced influence on evolution, the displacement of continents and other tectonic events must also have been important. High mountain ranges produced by plate tectonics create diverse local environments and separate vast regions of very different climates. Often today there are distinct biota on westward- and eastward-facing slopes of major mountain ranges. The plate positions and the extent of their separation also influence ocean circulation and climate, including whether or not there are ice age cycles.

Glaciation can also have a significant impact on biodiversity. An ice sheet 2 km thick eliminates macroscopic species, leading to oscillating periods of radiation during interglacials and destruction during glacials. For this reason, glaciated regions have much lower diversity of species such as plants and birds as compared to regions that were unglaciated.

If all the continents were always together in a single supercontinent with a stable climate, with no large mountain ranges created by active plate convergence, then one set of species would dominate each ecosystem, and ecosystems could become stagnant, absent external catastrophes, with little pressure (or opportunity) for evolutionary change. Displacing the continents allows greater diversity to arise. When the continents rejoin, the most successful organisms will prevail. Having a dozen opportunities on separated continents to find the most successful evolutionary innovation would lead to greater evolutionary change over long timescales than on single continents where, though time may be vast, separation is limited.

Principles of Planetary Evolution?

We often consider life as somehow a separate phenomenon from Earth as a whole, and certainly isolated from planetary processes associated with Earth's deep interior. The preceding chapters have shown some of the diverse linkages between life and planet. Can we extract principles from these interactions that might apply generally to a process of planetary evolution and would not necessarily be unique to Earth?

INCREASED RELATIONSHIP AND COMPLEXITY

We discussed in Chapter 12 the differences between prokaryotic and eukaryotic cells, with the much larger eukaryotic cells being complex chemical factories compared to their prokaryotic cousins. Inside the eukaryotic cells are organelles, such as mitochondria and chloroplasts, that carry out energy production and photosynthesis. These organelles are believed to have formed by progressive symbiosis and incorporation of bacteria, called *endosymbiosis*, where symbiotic relationships of different species become specialized and permanent. This idea is supported by the fact that such organelles contain their own genetic material that is distinct from that of the eukaryotic cell nucleus and similar to bacteria that carry out similar functions. Chloroplasts, for example, contain DNA similar to cyanobacteria (photosynthetic prokaryotes). In addition, the structure and metabolism of the organelles are similar to prokaryotic cells, and organelles replicate by binary fission, similar to bacterial cell division. Notably, these endosymbiotic organelles are no longer able to survive independently outside the eukaryotic cell, and their survival depends on other aspects of the eukaryotic cell's metabolism. This suggests a chain of events where early symbioses and incorporation lead to a partnership between initially separate organisms, and once the partnership is established, further evolution is based on interdependence.

This type of process is still evident today, as seen in the tubeworms and clams that flourish around hydrothermal vents and hydrocarbon seeps. The tubeworms are animals, but they have neither mouths nor stomachs. Instead, they have bacteria that live inside them—more than 250 billion bacteria per ounce of tubeworm tissue. The bacteria oxidize the sulfur using oxygen, and they employ this energy to reduce carbon dioxide to sugars, which feed the worm. The tubeworms have evolved to support the bacterial colonies on which they depend. For example, the red plumes of the tubeworms contain hemoglobin, which captures hydrogen sulfide as well as oxygen and transports these necessary raw materials to the bacteria living inside the worm. In this example we see a single organism whose metabolism has evolved to support an essential symbiosis. These cases demonstrate what may be a central aspect of evolutionary change, the tendency for living organisms to relate and develop symbioses that lead to more complex living machines with a more diverse set of processes and relationships.

The analogue to the development of eukaryotes from prokaryotic partnerships is mirrored in the development of multicellular from single-celled eukaryotic organisms, where individual cells become regulated and function together. Multicellular organisms in turn have gradually developed an increasing number of specialized cells. Human beings, for example, have some 75 trillion cells of 220 different types that carry out the various bodily functions, all working together to create a single functioning organism. The brain itself may be another example of such a developing network. There is little difference between the nerve cells that make up the brains of mice and human beings. The primary difference is the number of brain cells and the relationships among them. Increased relationship leads to increased capacity and ability to respond to diverse and changing environments. Relationships are also necessary to generate feedbacks, and increasing numbers of feedbacks lead to greater responsiveness and stability. And within species, it is clear that the relationship of working together often has advantages in terms of capability and survivability, such as the "superorganisms" of ant and bee colonies so well described by E. O. Wilson. Networks permit increased specialization of individual species, and complex feedbacks and networks lead to greater stability and survivability. Such trends are likely to be systemic and not restricted to Earth alone.

CHANGE IN ENERGY UTILIZATION WITH TIME

Energy is the driver of all systems, including living systems, and species that are able to harness more energy have a greater capacity to do work (e.g., chase prey or avoid predators), grow, and reproduce.

Increased access to O_2 and hence aerobically derived energy is one of the characteristics of terrestrial evolution. All animals use oxygen to some degree (and many microbes do as well). As we saw in previous chapters, life progressively developed increasing access to aerobic energy through anaerobic photosynthesis, oxygenic photosynthesis, anaerobic metabolism, and ultimately aerobic metabolism. To take advantage of aerobic metabolism requires a steady supply of oxygen. For very small organisms, this can occur by diffusion and transport through the cell wall. Larger organisms need a means of active oxygen transport and efficiency of respiration, as well as active removal of waste products. The mitochondria organelles in eukaryotic cells carry out the aerobic metabolism and serve as specialized energy factories. The development of circulation systems and breathing eventually permits far more aerobic metabolism and a greater rate of energy production in multicellular life.

Ultimately warm-blooded metabolism leads to an even higher metabolic rate. From this perspective the history of life involves a progressive increase in energy available for metabolic processes. Multicellular life cannot exist in anaerobic environments because there is insufficient energy production to support it. The triumph of mammals over reptiles would be a consequence of the higher mammalian metabolic rate and ability to generate energy independent of the external temperature environment. Ultimately this permits organs such as the brain to evolve. The brain is completely dependent on active oxygen transport. It neither makes nor stores food; its cells cannot survive on their own and require a vigorous blood supply to provide the active delivery of glucose and oxygen in order to function.

The change in energy metabolism during Earth's history has parallels with the increase in relationship and complexity discussed in the previous section. Simple processes occur at the beginning, such as prokaryotic cells and anaerobic ATP production. Subsequent processes build upon the earlier developments. While some early innovations are discarded as they become superseded, others are integrated into the more evolved forms. The primitive forms also persist in their own ecological niches. Prokaryotes are everywhere, at the same time as the processes they invented have been co-opted by subsequent eukaryotic organisms. Multicellular life, for example, provides a vast ecospace for prokaryotic organisms—the skin, mouth, and digestive systems of human beings provide environments for trillions of prokaryotes of hundreds of species. There has thus been both expansion and inclusion. This principle applies to energetic processes as well. Within aerobic organisms, anaerobic metabolic processes have been integrated. For example, anaerobic metabolism processes ATP more rapidly than aerobic metabolism, so mammals use anaerobic metabolism for short bursts of high energy.

There is another important aspect to enhanced energy production. High energy production requires an interface between reservoirs that are out of equilibrium, since it is the passage from disequilibrium to equilibrium states that generates the energy for life processes. Access of organisms to increased energy requires the parallel development of energy potential through disequilibrium, and the formation and separation of oxidized and reduced planetary reservoirs—the planetary "fuel cell" discussed at length in Chapter 15. Enhanced energy utilization requires then a coupled change in biological mechanisms and planetary environment. Evolution is a planetary process that includes life, ocean, atmosphere, and rock.

Speculations on the Possibility of Directionality to Evolution

A natural inference from this sort of discussion might be that life progressively evolves to organisms with increasing numbers of cells, an increased genome size, and more efficient energy processing, a progressionist view of evolution. This perspective is then often extended to human beings as the current apex of evolution. Very few biologists would subscribe to this view, however, because of the randomness inherent in the DNA mutation and the process of natural selection. A directional view is also difficult to justify on a quantitative biomolecular basis. Salamanders have an order of magnitude more DNA per cell than humans and other mammals. The rice genome encodes far more proteins than the human genome. The platypus cell has more chromosomes than the human cell. The elephant brain is four times the size of the human brain. On the broad scale of prokaryote to eukaryote to multicellularity to organ differentiation, there is a progression of increased relationship and complexity over the course of geological time, from which some progression might be argued, but the specifics of evolution and what makes an organism "advanced" are far richer and cannot be reduced to a simplistic directional framework.

On the other hand, there is an apparent directionality to terrestrial life that Andy Knoll and Richard Bambach have called "an increase in utilization of ecospace" (Fig. 17-11). They point out five major stages of ecospace expansion within the biological realm after life began—prokaryotic diversification, unicellular eukaryote diversification, aquatic multicellularity, invasion of the land, and intelligent life. Each step expands from the previous ecospace by increasing the dimensionality of life through evolution of organisms of increasingly complex internal relationships. At the same time, each step can expand environments for life of the previous dimensions. While there may be no direction on the

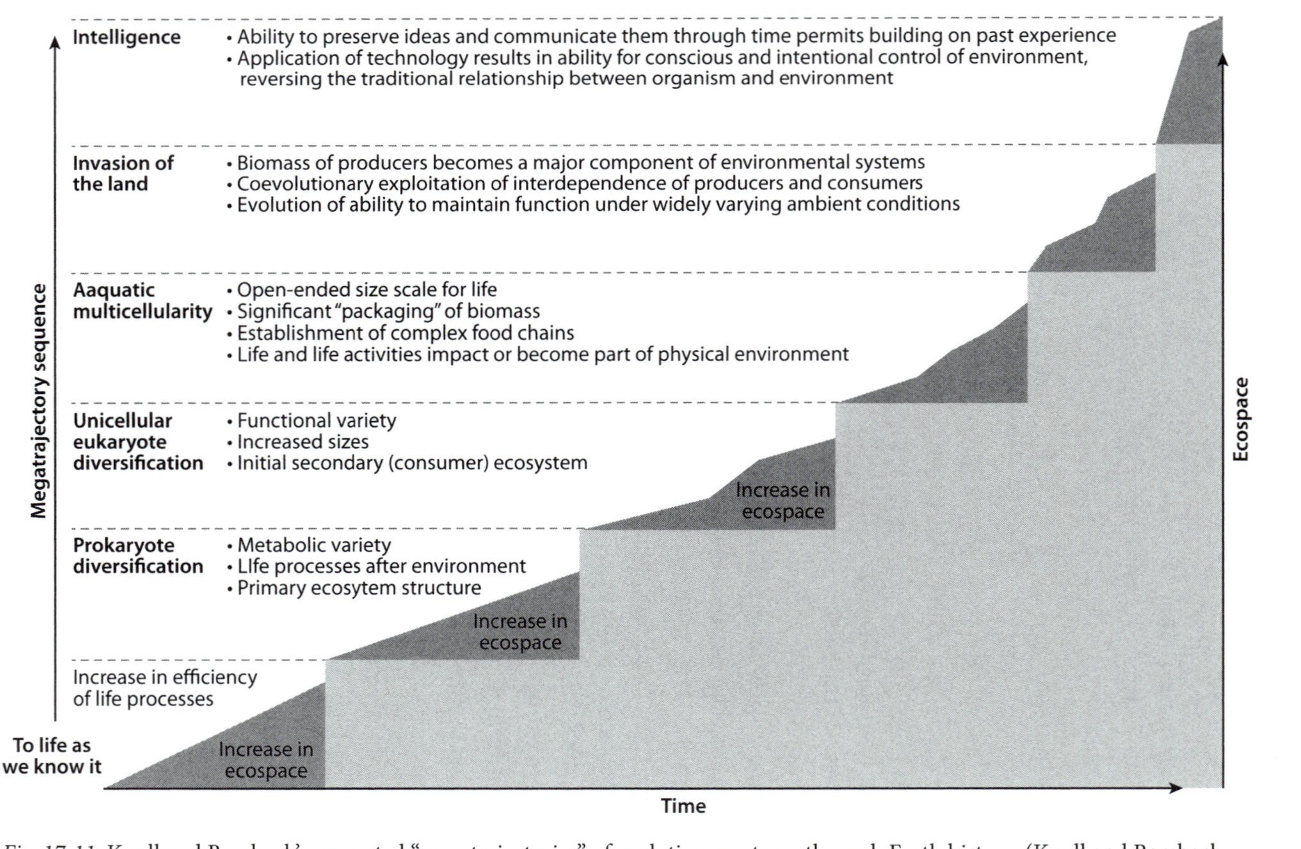

Fig. 17-11: Knoll and Bambach's suggested "megatrajectories" of evolutionary stages through Earth history. (Knoll and Bambach, *Paleobiology* 26(2000)(sp4):1–14)

reductionist scale of individual DNA mutations, the observation is that there are directional attributes to biological evolution.

Directionality can occur from random change if there are external constraints that favor particular types of changes. Imagine, for example, a large collection of dice replicating in a box, and as they replicate there are periodic random changes in the numbers on the faces. Instead of having all the die have 1–6, the change could lead to a die with 1, 1, 3, 4, 5, 6 or 1, 2, 4, 5, 6, 6. These changes would be inherited for each succeeding generation. If there were no preference for any one of the numbers, and large numbers of dice, then despite the "mutations" there would be similar amounts of all the numbers in the box at any one time. But if we put in a "natural selection" such that some of the dice are less likely to be removed from the box depending on the number on their face, the distribution will change. The changes could all be random, but if for some reason the number 4 was favored through selection—dice with a 4 facing up were less often removed—eventually the dice would be almost all 4s. That result would be inevitable, despite the random process that led to it and the fact that the specific random steps in that direction would always be different in repeated experiments and could not be predicted.

Can we imagine there might be such external constraints on evolution? One example might be the efficiency of energy processing, as discussed above. If organisms that are able to access more energy have a competitive advantage, this would inevitably lead to a directional change of increased energy usage. Changes that lead to increased system stability could also be favored, if increased stability is provided by the feedbacks that become possible with complex networks. And in the ongoing evolutionary battle between carnivores and their prey, the brain size of both has increased, suggesting there is an evolutionary advantage to larger brains (Fig. 17-12). Random genetic changes could lead to smaller or larger brains with no preference, but those changes leading to smaller brains would be selected against. Other examples might be bilateral symmetry or the development of increasing sensory systems to monitor one's surroundings. Directionality then appears from selective advantage. Viewed in this context, directionality to evolution is possible within the context of random changes on the molecular level. The fact that random

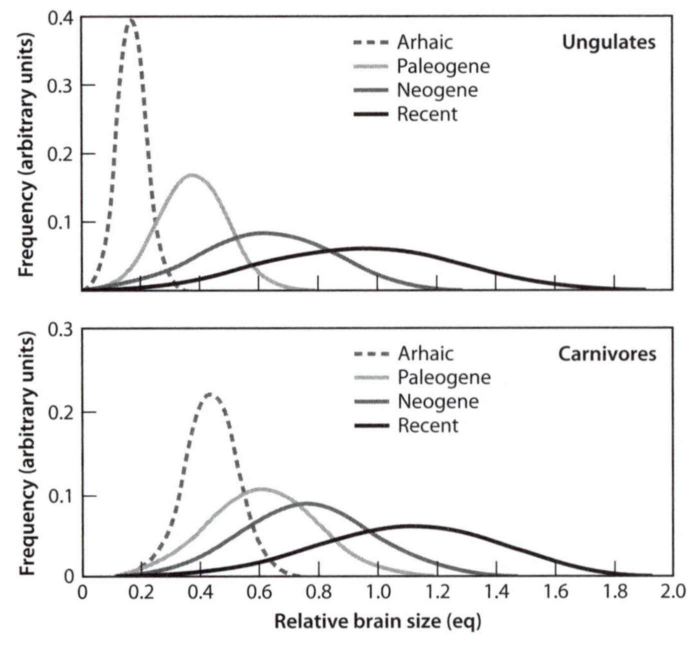

Fig. 17-12: Indications of directionality in evolution through the increase in brain size of both carnivores and their prey. (Modified from Radinsky, *The American Naturalist* 112 987 (1978): 815–83).

changes are the mechanism of evolution does not mean that larger-scale directionality could not emerge.

From such a perspective, a species capable of long-term forethought, enhanced energy use through tools and fuels, and the many advantages imparted by language would have great competitive advantages. These are all enhanced in modern civilization by global communications and transportation networks, genetic modifications for species advantage, and potential interplanetary travel. The emergence of intelligent life from the perspective of energy and networks could be a natural consequence of planetary evolution.

EVOLUTION OF HABITABILITY

This discussion builds upon the factual observation that Earth has become increasingly habitable over time. Life has created enhanced environments for life. Life has participated in major changes of the planet that have led to a far greater diversity of organisms today than existed in the past, and as multicellularity and organ specialization came into being, to greater complexity. As concrete examples of increased habitability:

> (1) no multicellular animal would be able to survive on Earth in the Precambrian, though microbes of that era might be extant in today's ecosystems;
> (2) it is likely that the mass of life on Earth today is far greater than it was during the Archean and Proterozoic;
> (3) the amount of energy flowing through ecosystems is certainly greater.

For example, prior to oxygenic photosynthesis, the capability of the biosphere for transforming energy from the sun was quite limited. Hand in hand with the evolution of the species has been an evolution of habitability. Increased habitability is of course closely coupled to changes in the planetary environment. The changes in atmospheric composition, ocean chemistry, expansion of life from ocean to land, the development of soils—all of these create enhanced environments for life. Production of organic matter provides an energy source that supports more complex food webs and—eventually—more energy-intensive organisms such as mammals. Life provides and expands the environments for life.

We can ask why such developments have taken place. Why should random DNA mutations lead to an increasingly habitable planet? On many scales one can imagine that this is a natural process of planetary evolution. If there are two competing organisms, and one of them makes an environment more hospitable for its development, or for other organisms that are beneficial companions, it will preferentially survive. And if a network of organisms creates an environment more capable of sustaining life and processing energy—i.e., more habitable—this network will have an evolutionary advantage. Organisms that make the environment less habitable will ultimately fail. One of the natural consequences of a planet with life may be a progressive increase in planetary habitability, provided the external conditions for life can persist.

Evolution of habitability also can take place only if life is able to respond to inevitable planetary change. Many aspects of planetary evolution are long and slow, such as the rise of O_2 and the development of cellular mechanisms to take advantage of it. This requires persistent life.

And in the midst of this slow evolution, there are catastrophes from the cosmos and planetary interior and inevitable change of environmental conditions. For life to continue it must be adaptable enough to survive, and even better if it can make opportunity out of crisis. Absent such capabilities, life would become extinct and the planet uninhabited. This suggests another principle, that biological evolution—the capacity for change—is a prerequisite for planetary habitability.

Summary

If we consider life as a planetary process, rather than some separate phenomenon that occurs on a planetary surface, then planetary evolution includes all aspects of the planetary system—core, mantle, crust, ocean, atmosphere, and life. Life plays a critical role in planetary evolution, by capturing and storing solar energy in planetary materials, modifying the oxidation states of near-surface reservoirs. Life is in turn greatly influenced by the physical processes of planetary and solar system evolution. Its beginning was influenced by the late heavy bombardment, and meteorite impacts have continued to have an important influence during the Phanerozoic. The convective processes of Earth's interior influenced the origin of life and climate stability, and surface expressions of active convection through plumes have likely caused many of the mass extinctions and periods of extreme climate change that punctuate the Phanerozoic record.

While biological evolution results from random changes in the microscopic DNA code, there is nonetheless progressive evolutionary change. Over the course of Earth's history, organisms have developed increasing networks and complexity, both internally and externally, and have greatly increased their energetic productivity. Directionality can emerge from random change if there are underlying principles that supply a competitive advantage. From the Phanerozoic record, it appears that the progressive changes in life are in turn facilitated by the catastrophes coming from solar system and planetary interior. These processes destroy long-term stability and permit evolutionary innovations to be expressed in a newly level playing field. Plate tectonics and climate change have also enhanced evolutionary change, by permitting separate and simultaneous evolutionary experiments and creating variable environments. Through planetary evolution, Earth has become increasingly habitable.

While Earth's history is inevitably specific as a story of one planet, principles that it embodies appear likely to apply on a universal scale. Evolution by natural selection is a general process clearly not restricted to a particular time or place. Increased stability through networks and increased access to and utilization of energy are also a kind of thermodynamic driver that would be guiding principles on any planetary body. Periodic instabilities from the solar system and planetary interior are inevitable consequences of solar system formation and planetary convection, creating the crises that destroy stasis and permit rapid evolutionary change. From this perspective, evolution toward increased habitability may be a common planetary process.

Supplementary Readings

Andrew H. Knoll and Richard K. Bambach. 2000. Directionality in the history of life: Diffusion from the left wall or repeated scaling of the right? *Paleobiology* 26(4):1–14.

Bert Hölldobler and E. O. Wilson. 2009. *The Superorganism: The Beauty, Elegance, and Strangeness of Insect Societies.* New York: W. W. Norton & Co.

Walter Alvarez. 1997. *T. Rex and the Crater of Doom.* Princeton, NJ: Princeton University Press.

Douglas H. Erwin. 2006. *Extinction: How Life on Earth Nearly Ended 250 Million Years Ago*. Princeton, NJ: Princeton University Press

Fig. 18-0: Glaciers in Alaska's Glacier Bay. Such glaciers are remnants in North America of the vast ice sheets that covered the northern half of the continent during ice ages of the past million years. (© Bart Everett with permission under license from Shutterstock image ID 5134369)

CHAPTER 18

Coping with the Weather

Causes and Consequences of Naturally Induced Climate Change

On geological timescales Earth's climate preserved liquid water at the surface, thanks to the tectonic thermostat discussed in Chapter 9. Within this stability, Earth's climate also undergoes significant change on intermediate timescales of 10^4–10^5 years that can greatly influence living conditions. These changes occur because climate is sensitive to details of planetary orbital variations. Earth spins with a tilted axis and equatorial bulge, causing it to precess like a top about every 20,000 years. Because of the gravitational tug of the major planets, Earth's orbital tilt and orbital shape change cyclically on the timescales of 40,000 and 100,000 years. Scientists now believe that the changes in the seasonal and latitudinal distribution of the sunlight reaching Earth generated by these orbital cycles (called Milankovitch cycles*) paced waxing and waning of Earth's ice caps. Extensive records of these changes over the past 800,000 years have been obtained from ice cores in Greenland and Antarctica. While the forcing functions of orbital variations would lead to sinusoidal change, the ice age cycles are not simple sinusoids. Instead, ice ages end abruptly and are entered gradually, suggesting complex feedbacks must influence them. The ice age cycles also correlate closely with atmospheric* CO_2 *and* CH_4*, probably with a slight lag compared to ice volume. While the link with Milankovitch forcing is evident, many details of the diverse manifestations of ice ages remain to be fully understood. The ocean with its large* CO_2 *reservoir and complex behavior must play an important* *role, as may induced volcanism from melting ice sheets that adds a pulse of* CO_2 *to the atmosphere during deglaciation. From 5–1 Ma, orbital forcing caused ice ages to vary on a 40,000-year (40 Ka) cycle rather than 100 Ka cycle. Prior to 5 Ma there were no northern hemisphere ice ages. Prior to 30 Ma there was no Antarctic ice sheet. In ice-free periods, orbital forcing must also have operated to induce intermediate scale climate change, though the specifics of these ancient influences on climate are currently not well described.*

The high temporal resolution of the ice-core records also reveals striking short-term climate change over 10^1*–*10^3 *years. Such changes are likely modulated by jumps in climate from one of its quasi-stable states of operation to another. The causes of such short-term change also remain to be fully understood, and they may be triggered by reorganizations of the ocean's thermohaline and atmospheric circulation. Intermediate and short-term climate change have major consequences for the habitability of continents. The exit from the last glacial advance 11,000 years ago and the following benign climate of the current interglacial have permitted the migration of human beings throughout the continents, vast population growth, and the rise of human civilization.*

Introduction

As we look at Earth today the ice sheets of Greenland and Antarctica appear to us as permanent features. The fossil record shows us that this was not always the case—alligators and lush forests once occurred in polar regions, showing that large changes in Earth's global climate have occurred during the Phanerozoic. And in the Precambrian, the proposed "snowball Earth" episodes show that climate variations can be even more extreme. These changes all occur over millions of years, extremely long even with respect to the rise of *Homo sapiens* some 150,000 years ago. On shorter timescales, is climate locked into relative stability? Might there be climate changes on timescales that would be pertinent to human history and human future? Our recognition of vast change in the positions of continents, the composition of the atmosphere, and the evolution of life were all broadened and informed by observations from Earth's geological record. As we will see in this chapter, large changes in Earth's climate that would be of great significance to modern life have also characterized the geological record. While over millions of years Earth's climate is remarkably stable in the preservation of liquid water at the surface, on short timescales the climate is fickle in ways that have major consequences for life. These changes can occur on short enough timescales to be pertinent to the history and future of human civilization.

Intermediate Term Climate Variations: Ice Ages

Evidence suggesting a recent ice age came to light through studies of Louis Agassiz in the nineteenth century. By careful examination of Alpine glaciers, he noted that striations on rocks, U-shaped valleys, and transport of large blocks of rock were characteristic of glacial behavior. Similar features occurred in central Europe and also at great distances from the modern extent of glaciers in the Alps. He noted that transport by glaciers could explain huge exotic boulders lying in open plains in northern Europe, and that distinctive heaps of debris (known as *moraines*) could be explained if they were bulldozed into place by the advancing ice. Furthermore, glaciers could be observed making distinctive landforms such as U-shaped valleys that contrasted with the V-shaped valleys of river erosion, and the mountains in Switzerland, Scotland, and Scandinavia had long and vast valleys that appeared to have been excised by glaciers. Estimates of the areal extent of these vast continental ice sheets could be constructed by mapping the extent of glacial striations on bedrock and by the moraines, showing that vast ice sheets once covered much of Europe and North America (Fig. 18-1). Estimates of the volume of these ice sheets can be made from the extent to which sea level was lowered. A record of sea-level lowering provided by fossil mangrove roots preserved along the margins of Indonesia indicate that, at its maximum, sea level was lowered about 120 m. In other words, in order to create these great ice sheets, a layer of seawater about 120 m thick across the entire ocean had to be transformed into ice. The last glaciation also had to have been recent, because the loose glacial sediments were very

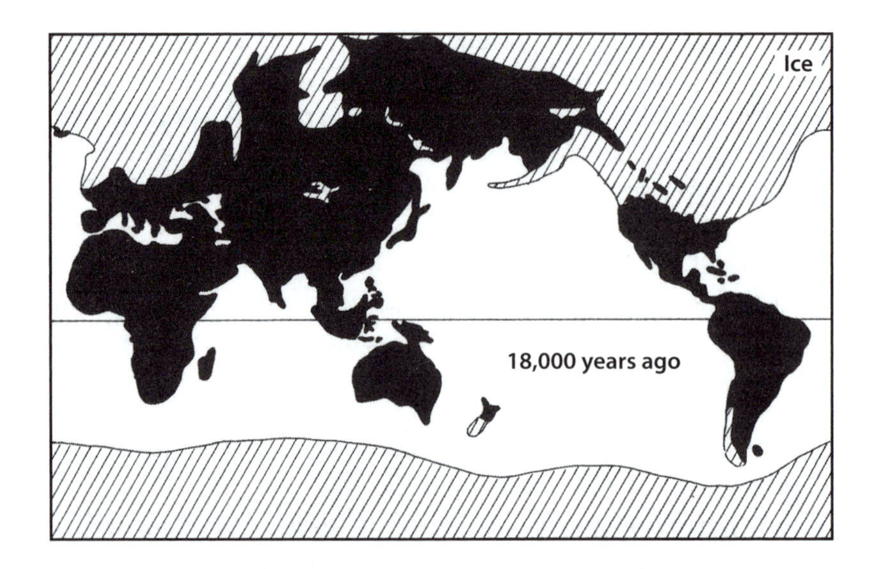

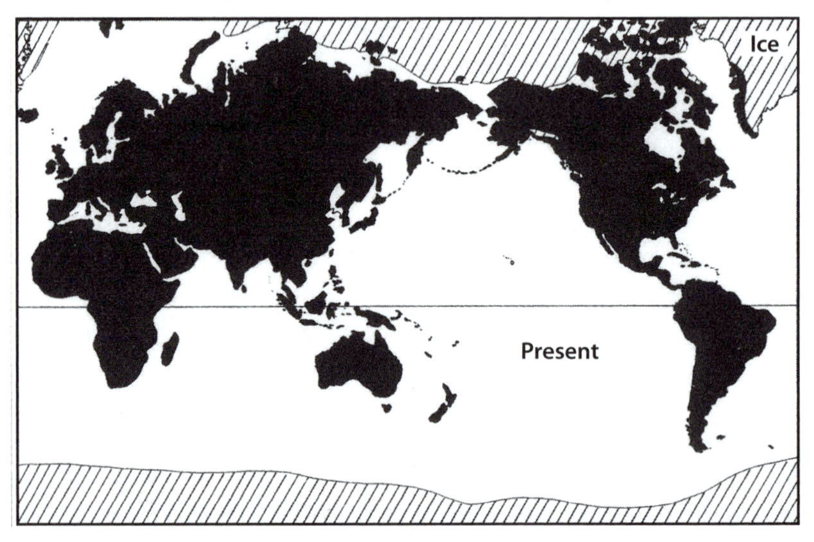

Fig. 18-1: Maps contrasting the extent of ice cover during the peak of the last period of glaciation, 20,000 years ago, with that which exists at present: Part of this ice is continental and part floats in the sea. (Map courtesy of George Kakla)

well preserved, and rivers running in the U-shaped glacial valleys had created only small incisions in the valley floors.

In the second half of the twentieth century, much greater precision on the timing and magnitude of glaciations was obtained from sediments deposited on the seafloor. The tool that revealed the glacial record with great precision was variations in oxygen isotope compositions preserved in the shells of tiny-shelled organisms, called *benthic foraminifera*, which inhabit the seafloor. As we learned previously, stable isotopes can be fractionated by low temperature processes. The process of evaporation of water near the equator and its transport and precipitation as ice near the poles causes the ice in ice caps to have an $^{18}O/^{16}O$ ratio 3.5% (35 per mil) lower than that for seawater. Like the difference in carbon isotopes between organic and inorganic carbon discussed in Chapter 16, the offset between polar ice and liquid ocean is quite constant. At the same time, the total $^{18}O/^{16}O$ ratio of all of Earth's water has a defined average value that does not change. As the ice sheet grows, and more and more water with low $^{18}O/^{16}O$ is removed from seawater, the seawater $^{18}O/^{16}O$ increases as the mean value of total H_2O is preserved (Fig. 18-2). Therefore, just as we were able to use carbon isotopes to determine the proportions of organic and inorganic carbon in Chapter 16 (see Fig. 16-2), the oxygen isotopes can be used to determine the proportions of ice and liquid water. A caveat must be mentioned here, which is that the $^{18}O/^{16}O$ of the shells is also influenced by the water temperature. Fortunately cold temperatures that would also be associated with glacial conditions also lead to higher $^{18}O/^{16}O$ in the shells, so the stable isotope record gives a clear signal of glacial vs. interglacial conditions. Work by Dan Schrag and colleagues suggests that about half the change can be associated with temperature and about half with ice volume.

Shown in fig. 18-3 is the $^{18}O/^{16}O$ record obtained from shells picked from various depths in a sediment core from the deep tropical ocean. As amazing as it may seem, these tiny organisms provide us with a record of changes in continental ice volume. Low values of $^{18}O/^{16}O$ show low ice volume and high values high ice volume. By using radioactive dating, a firm absolute timescale has been obtained for this record. What the record shows is that there have been multiple ice ages extending back about 700,000 years. The record is "sawtooth" in shape, with a long

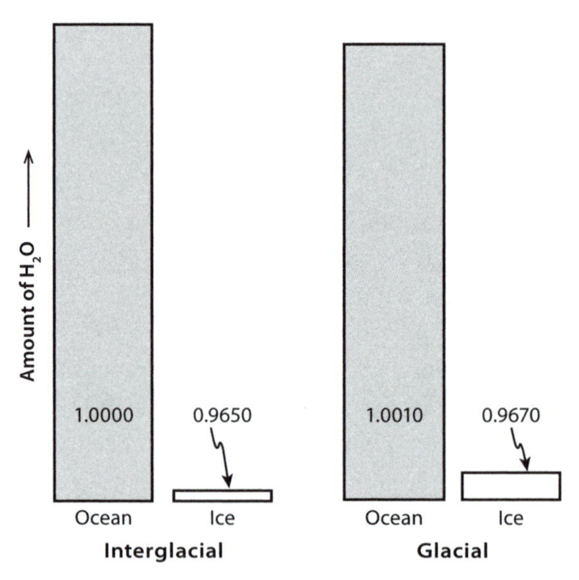

Fig. 18-2: Comparison of the amounts of H_2O in the ocean (as water) and in the continental glaciers (as ice) during times of full interglaciation (like today) and during times of full glaciation (such as 20,000 years ago): The heights of the bars indicate the amount of water in each reservoir. The numbers associated with the bars are the $^{18}O/^{16}O$ ratio for the H_2O divided by the $^{18}O/^{16}O$ ratio in today's ocean. Since the ice is about 3.5 percent depleted in ^{18}O compared to seawater, the expansion of ice sheets causes the ocean to become slightly enriched in ^{18}O. See Fig. 16-2 for an analogous relationship for the carbon isotopes and the fraction of organic and inorganic carbon.

slow slide (increasing $^{18}O/^{16}O$ in the figure) into the deepest ice age. The cooling ends with an abrupt termination to low ice volume and a brief and warm interglacial period, such as what we are now experiencing. While there is much fine structure to the record, the dominant period of change is about 100,000 years, which can be observed by noting the positions of the interglacials in the figure (times with lowest $^{18}O/^{16}O$).

Orbital Cycles

The realization by geologists that Earth's climate has undergone dramatic cycles brought with it a curiosity as to what caused these shifts. From the very first, the prime candidate has been the cyclic changes in Earth's orbit about the sun. Although when averaged over long periods

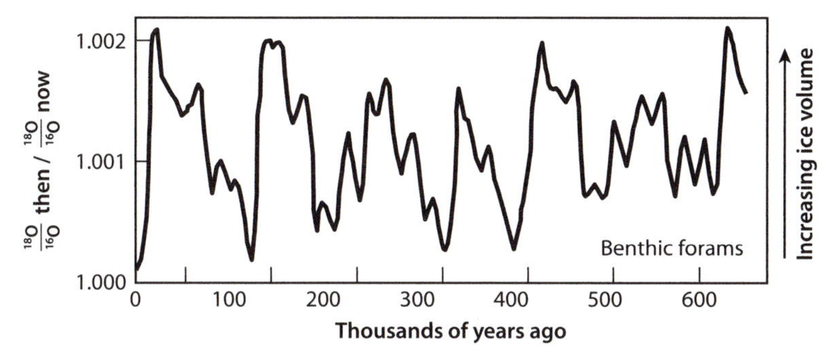

Fig. 18-3: The $^{18}O/^{16}O$ record in benthic foraminifera for a sediment core from the deep Pacific Ocean. The ^{18}O enrichment in the glacial-age shells is partly the result of growth of the ice sheets and partly the result of colder deep ocean temperatures. This record reveals that the 100,00-year cycle, which has dominated climate for the past 750,000 years, is asymmetrical. Long intervals of bumpy cooling are terminated by abrupt warmings.

of time Earth's path around the sun remains unchanged, its orbit does show some repetitive deviations from this mean. The importance of these orbital changes lies in the fact that they alter the contrast between the seasons. Changes in the seasonal distribution of the sunlight received at any given location on Earth's surface are driven by two characteristics of its orbit. The first is associated with the tilt of Earth's spin axis with respect to the plane of its orbit about the sun (see Fig. 18-4). Earth's spin axis does not stand straight up. Rather, it leans away from the vertical by an angle of about 23°. Consequently, on June 21 the Northern Hemisphere faces the sun. Six months later, on December 21, the Southern Hemisphere faces the sun. If Earth stood straight up, its equator would always face the sun and there would be no tilt-induced seasons. The larger the tilt, the greater the seasonal range in the amount of radiation received at any point on the globe.

Also shown in Figure 18-4, a second source of seasonality exists. It relates to the fact that Earth's orbit is slightly out of round. In geometric terms, the orbit is an ellipse. As you remember, circles can be drawn by tying a string to a piece of chalk. The end of the string is held at the center (focus) of the circle and the circle is drawn by swinging the chalk around. An ellipse has two foci. To draw an ellipse, both ends of the string are pinned down, one at each focus. The chalk is not attached; rather it is nested freely in the vee made by the stretched string. The ellipse is

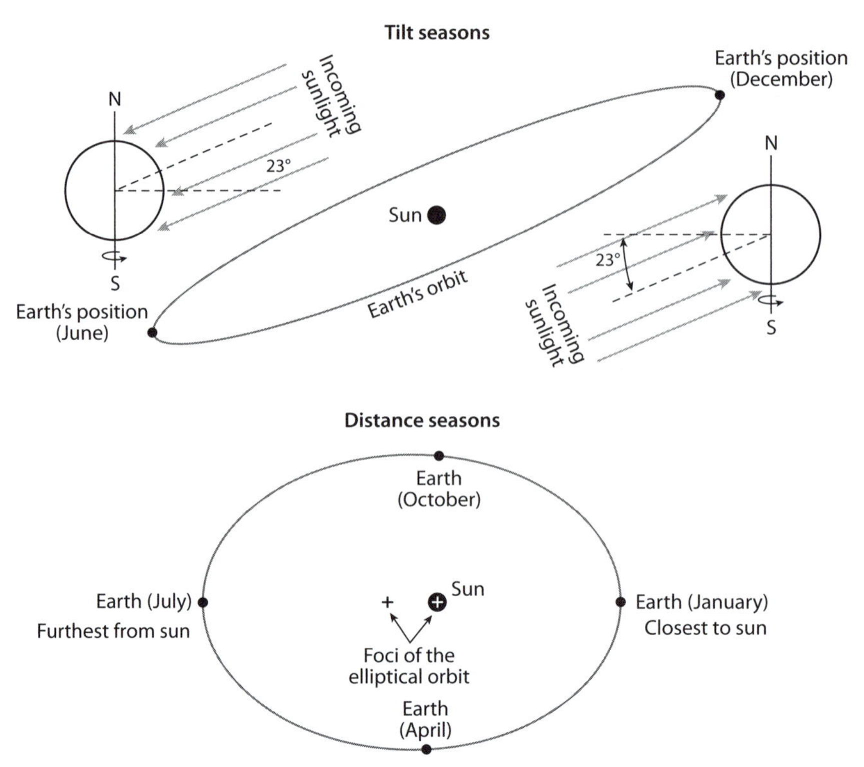

Fig. 18-4: Seasonality and its cyclic changes. This diagram shows the factors that influence the seasonal contrast in the radiation received at any point on the globe. The primary cause of seasons is the tilt of Earth's axis with respect to its orbit about the sun. *Upper panel:* As shown here, this tilt leads to higher illumination of the Northern Hemisphere during June and to higher illumination of the Southern Hemisphere during December. Our calendar is set so that June 21 and December 21 correspond to the days on which maximum radiation falls on the Northern Hemisphere and on the Southern Hemisphere, respectively. A second contribution to seasonality stems from the fact that the Earth's orbit is elliptical. Because of this, the Earth-sun distance changes during the course of the year. *Lower panel:* As shown here, currently Earth is closest to the sun early in January and farthest from the sun early in July. Thus Earth as a whole receives less sunlight during July than during January.

drawn by swinging this vee around. The farther apart the foci, the more eccentric the ellipse—i.e., the more it deviates from circularity. No planet has a perfectly circular orbit, all are elliptical. The laws of gravity require that one of the two foci of the planet's orbit corresponds in location to the sun. The consequence of the elliptical shape of Earth's orbit is that the Earth-sun distance changes over the course of the year. When

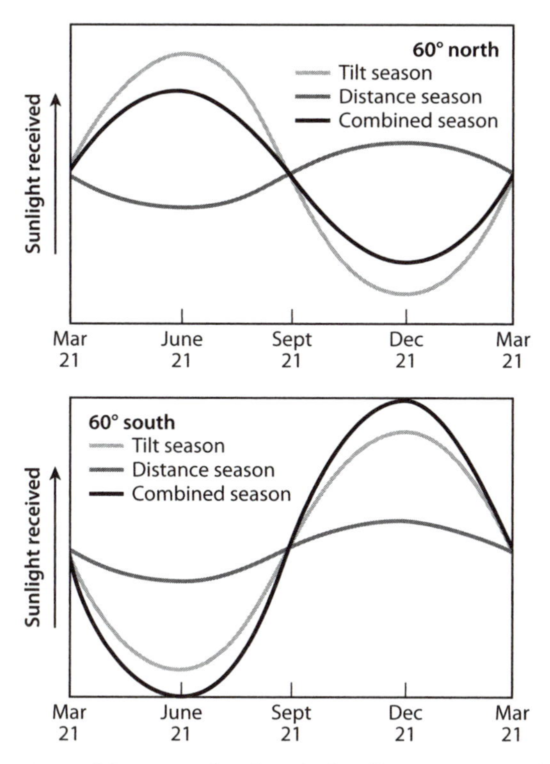

Fig. 18-5: Comparison of the seasonal cycles of solar illumination at 60°N and 60°S: In the Southern Hemisphere the tilt and distance seasons currently reinforce one another. In the Northern Hemisphere they currently oppose one another.

closer to the sun, Earth receives more radiation; when farther away, it receives less.

Shown in Figure 18-5 is the relationship between the two seasonal cycles. Currently in the Northern Hemisphere they oppose one another. The reason is that the Northern Hemisphere leans toward the sun as it traverses that portion of its orbit farthest from the sun. By contrast, the Southern Hemisphere leans toward the sun as it traverses that portion of its orbit closest to the sun. Hence, the two sources of seasonal contrast (tilt and distance) currently oppose one another in the Northern Hemisphere and reinforce one another in the Southern Hemisphere.

Our interest in all this arises because the situation changes with time. The reason for this is that Earth precesses as does a spinning top. Both precess for the same reason: their spin axes tilt. The leaning top precesses

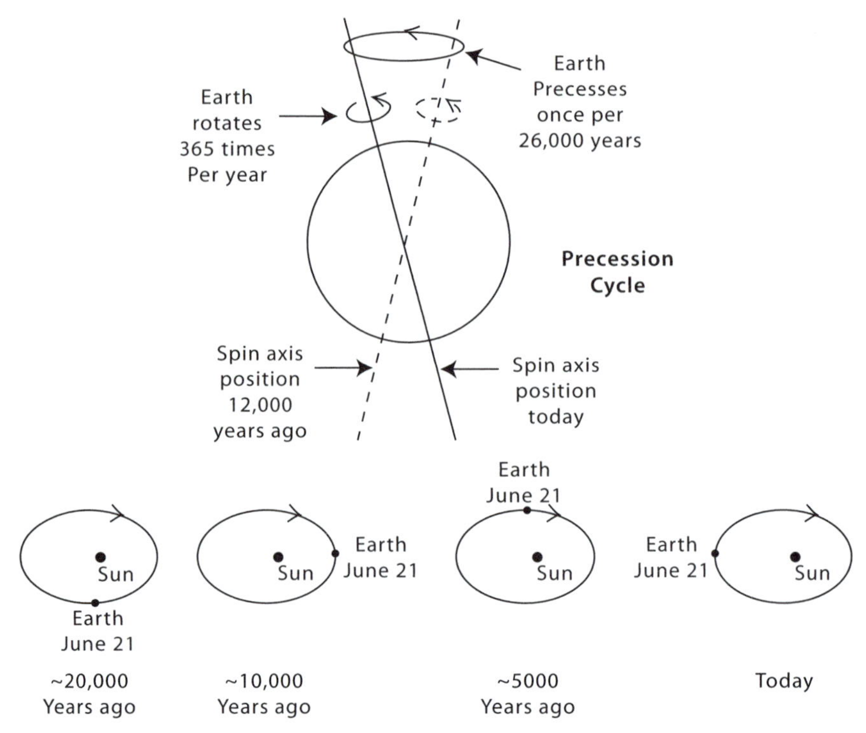

Fig. 18-6: Like a top, Earth's spin axis precesses: It takes about 26,000 years for one complete precession cycle. Precessions change the point on Earth's orbit where maximum Northern Hemisphere illumination occurs. Today it occurs at the "long" end of the ellipse, reducing full summer illumination in the Northern Hemisphere. The ellipse itself is also rotating and the net effect of the two leads to a 21,000 year cycle. One-half precession cycle ago (i.e., ~10,500 years ago) it occurred at the "short" end of the ellipse, increasing full summer illumination a bit. The precession of Earth's axis relates to the sun's pull on the equatorial bulge produced by Earth's rotation. Just as Earth's gravity seeks to tip over a spinning top, the sun's gravity seeks to remove the tilt of Earth's axis. As does a top, Earth compensates for this pull by precessing.

to keep from falling. The leaning Earth precesses to keep its equatorial bulge from becoming aligned with its orbit, i.e., pointing directly at the sun. Just as a top spins much more rapidly than it precesses, so also does Earth. It takes Earth 8,760,000 days (26,000 years) to complete one precessional cycle. At the same time the elliptical orbit rotates slowly, and the combination of the two leads to approximately 21,000 year period in the variations of solar insolation from precession.

The importance of precession to seasonality is that it causes a progressive change in Earth's location in its orbit on June 21 (Fig. 18-6).

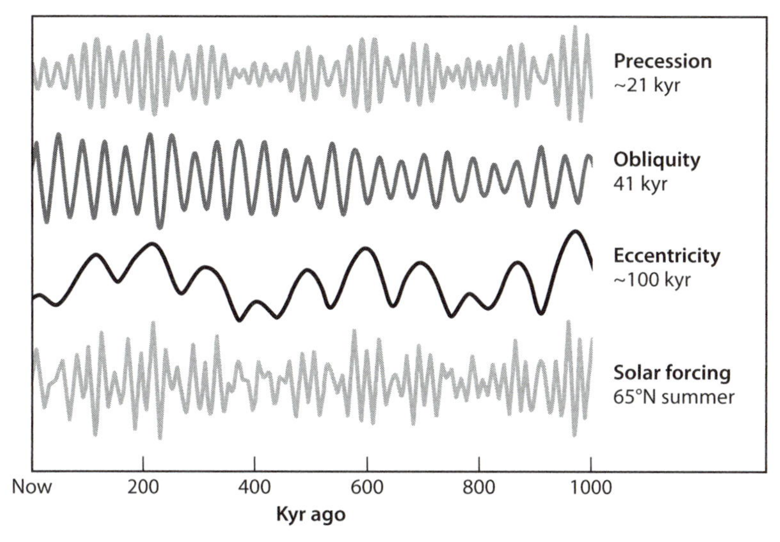

Fig. 18-7: The precession and obliquity of Earth's spin axis and the eccentricity of its orbit as a function of time, and their effect on summer insolation at 65°N. Axial tilt varies by two degrees. Eccentricity varies from 0.01 to 0.05. These records are reconstructed from calculations based on the law of gravitation and the present-day orbits and masses of the planets. (modified from en.wikipedia.org/wiki/File:Milankovitch_Variations.png)

Eleven thousand years ago the Northern Hemisphere faced the sun as Earth traversed that part of its orbit nearest to the sun. Hence, at that time the tilt and distance seasons reinforced one another in the Northern Hemisphere, and thus the precession of Earth causes the contrast between the seasons to change in a cyclic manner.

In addition to its precession, Earth's orbit is subject to two other cyclic changes. Both have to do with the gravitational pull of our sister planets, particularly Jupiter. These tugs cause the eccentricity of Earth's orbit to change with time. At times it has been even more eccentric than it is today and at times it has been less. The more eccentric the orbit, the stronger the distance seasonality. The interplanetary tugs also lead to changes in Earth's tilt. The greater the tilt, the greater the seasonal contrast; the smaller the tilt, the smaller the seasonal contrast. The time histories of both eccentricity and the tilt of the orbit can be calculated very precisely from knowledge of the masses of the planets and their orbits. The results of these reconstructions are shown in Figure 18-7. As can be seen, the tilt cycle is quite regular, the highs being spaced at

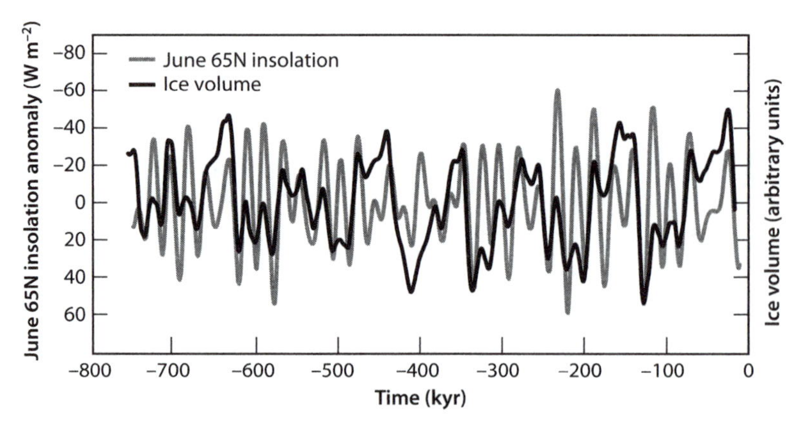

Fig. 18-8: Comparison of summer insolation with the record of ice volume, with the ice volume shifted with a lag of 6,000 years to maximize the fit between peaks and valleys in each record. The insolation axis is reversed so that low values of insolation give high values on the y-axis. While there is a general correspondence, the fit is by no means perfect. (Figure modified from G. Roe, *Geophys. Res. Lett.* 33 (2010), L24703)

40,000-year intervals. While the eccentricity changes have a more complex pattern, the highs are spaced at intervals of about 100,000 years.

Earth's precession, the changes in its tilt, and the changes in the eccentricity of its orbit combine to yield a complex history in seasonal contrast. This record is different for different latitudes. The influence of changes in tilt is most prominent at high latitudes, while the influence of distance changes is the same at all latitudes. In Figure 18-8 is shown the average daily radiation received during the month of July at 65°N (the latitude around which the ice caps of glacial time were centered) compared with changes in ice volume. At this latitude the amount of radiation received varies by about 100 watts/m^2.

The benthic ^{18}O record (i.e., ice volume) is compared with the summer sunshine record for 65°N in Figure 18-8 with a time offset of 6,000 years between the two records. While the two curves are quite different, they show some intriguing similarities. Low ice volume corresponds with the periods of high anomalies of solar insolation (e.g., near 600 Ka and 200 Ka), and the second-order "wiggles" in the record match closely in the time that they occur (but not the amplitude). John Imbrie quantified these correspondences and showed that the characteristic wavelengths of the climate record corresponded with the precession and tilt

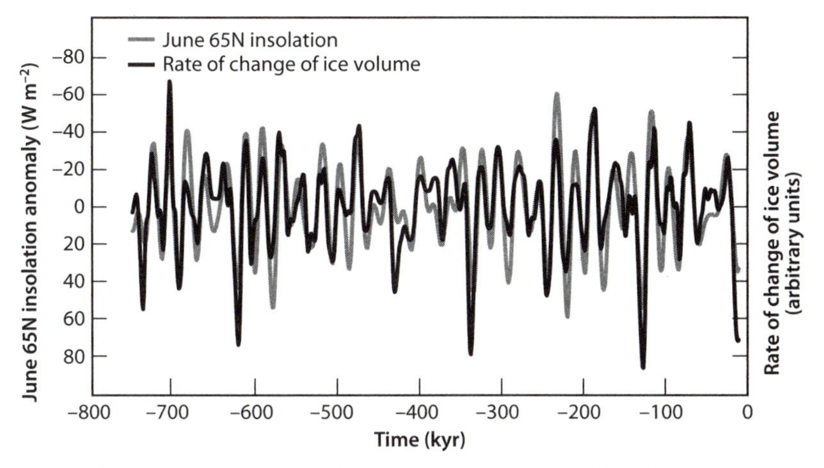

Fig. 18-9: The same insolation record as Fig. 18-8 now compared to changes in ice volume rather than absolute ice volume. High values of solar insolation (low values of the light gray line in the figure) correspond to ice sheets losing volume more rapidly. Note the close correspondence of the ~20 Ka cycles. No lag is required in the ice volume record to match the insolation record. (Figure modified from G. Roe (2010))

frequencies. This result was accepted as evidence for an orbital cause to climate variations.

More recent work adopts a slightly different approach with an even more compelling result. Because the continents are largely in the Northern Hemisphere, Northern Hemisphere insolation has the major effect on ice sheets. But ice sheets are big, and therefore one bright summer will not suddenly cause them to disappear. Instead, peaks in solar insolation should cause the most rapid change in the ice sheet volume. So, rather than the absolute size of ice sheets, it should be the temporal change in size of the ice sheet that matches variations in solar insolation. Gerard Roe of Washington University carried out such an analysis, which is shown in Figure 18-9, with no time lag between records. The correspondence is obvious—more summer sunlight corresponds with maximum changes in ice volume. These results clearly pose the Milankovitch hypothesis of orbital variations on climate: changes in solar insolation caused by orbital variations have a major influence on the temporal change in ice volume. In this form the Milankovitch theory ranks a 9 on our theory scale—orbital variations induce climate change.

While these correspondences between the summer-sunshine record and the ice-volume record have convinced most geophysicists that changes in Earth's orbital characteristics have somehow paced glacial cycles, the detailed nature of the connection remains incompletely understood, and many puzzles remain.

One puzzle is the rapid exit from glacial maxima to interglacial conditions. The orbital variations are all sinusoidal in form, so that increases and decreases in solar energy are symmetrical. The $^{18}O/^{16}O$ record, however, is not sinusoidal. Instead it shows a long slow decline to a glacial maximum, followed by an abrupt termination to an interglacial warm period, creating a "sawtooth" ice volume record rather than a sinusoidal one. This result can also be seen in the rate of change in ice volume. Inspection of Figure 18-9 shows that generally the black line of change in ice volume has lower amplitude than the gray insolation line, except at the times of the 100 Ka deglaciations, when particularly large changes in ice volume occur.

Another puzzle is the change in composition of the atmosphere that corresponds with changes in ice volume. Tiny gas bubbles trapped in the ice in Greenland and Antarctica have permitted a detailed record of change in atmospheric composition over the last 750,000 years. CO_2 and CH_4 vary coherently with ice volume, with high CO_2 and CH_4 corresponding to the warm interglacials and low values corresponding to glacials. Careful investigation of the records suggests that the atmospheric changes lag slightly in time the ice volume changes, so that ice volume decreases first, followed by atmospheric change. Why should formation and destruction of ice sheets, which are made up only of water, be associated with variations in CO_2?

These results suggest there are positive feedbacks associated with the ends of ice ages. Lowering of ice volume causes rise in greenhouse gases CO_2 and CH_4, which would then lead to further warming and loss of ice volume. A great deal of energy has gone into trying to fathom the causes of the sawtooth temperature and CO_2 records. One likely possibility is the ocean, because it contains fifty times more CO_2 than the atmosphere does. If interglacial changes somehow led to decarbonation of the ocean, then that could cause a rise in atmospheric CO_2, which would lead to further warming and cause a rapid termination. The exact mechanism of such an oceanic response, however, remains unclear.

Another possibility is that the glacial terminations are influenced by a feedback between ice sheets and volcanism. As we learned previously, magma production is sensitive to pressure changes. When ice sheets melt, they rapidly remove mass from the continents to the oceans, depressurizing the continental mantle and pressurizing the oceanic mantle. On Iceland, this led to a massive burst of magmatism that followed the removal of an ice sheet 2 km thick that covered the island during the last glaciation. Working with Icelandic colleague Dan McKenzie and coworkers has shown that this volcanic outpouring can be explained by increased melting due to the pressure changes in the mantle. The link with CO_2 would come from volcanic degassing. Because most of the CO_2 flux to the atmosphere comes from convergent margins, increased volcanism there could have a significant effect on the CO_2 budget. Data on volcanic eruptions shows that global continental volcanism increased by three to five times during the last deglaciation (Fig. 18-10). The extra volcanism would contribute more CO_2 to the atmosphere, leading to greater warming, more ice sheet melting, and more CO_2 addition. This process appears capable of accounting for about half the positive feedback that would be needed to explain the sawtooth records. It also has the appeal that it provides a single mechanism to explain rapid glacial terminations and the lag in the rise of atmospheric CO_2.

A final puzzle of Earth's response to orbital variations is the change in the period of ice ages over the last 2 million years. Figure 18-11 shows a 1.8 Ma record of glaciations based on sediment cores. Prior to 0.8 Ma, the dominant period of ice age cycles was 40,000 years, corresponding with changes in tilt. Then suddenly the dominant period changed to 100,000 years. This change is mysterious because the 100,000-year cycle corresponds to eccentricity, and eccentricity variations have very small effects on solar insolation. Revisiting Figure 18-8, for example, the 100,000-year cycle is not particularly evident in the changes in solar insolation but completely dominates the ice volume record. How can such small signals be amplified to produce such a large glacial signal, and why did the period change 1 million years ago?

On a final note, the last 2 million years have provided a clear record of orbital variations because they have been expressed as ice ages (e.g., Fig. 18-11). In other periods of Earth's history, ice sheets were not present (Fig. 18-12). Why? It seems likely that positioning of the continents at

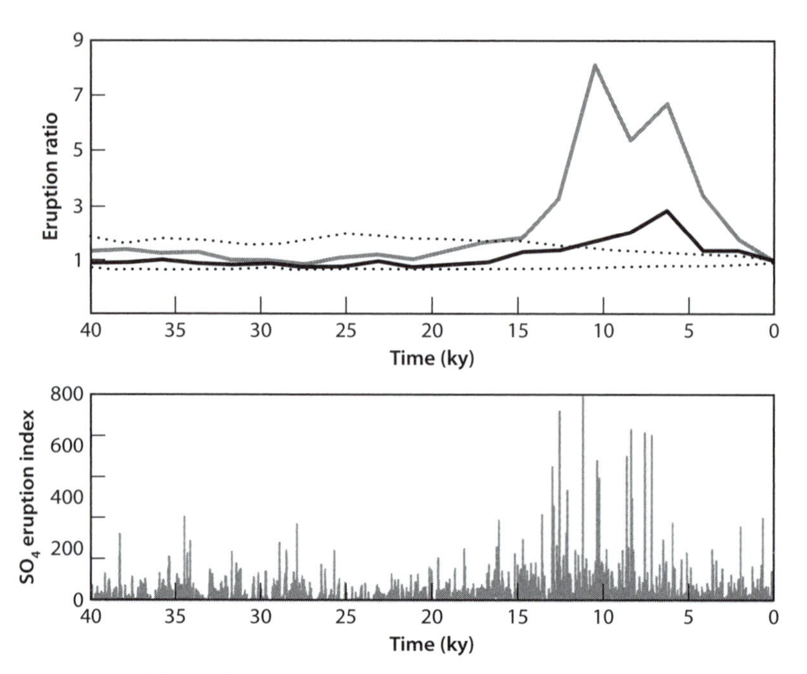

Fig. 18-10: Plot of global changes in volcano eruption frequency during the exit from the last glaciation. The last glacial maximum at 17 Ka is marked by low volcanic output. From 15 Ka to 5 Ka volcanoes erupt about four times more frequently than the glacial background, pumping CO_2 into the atmosphere and contributing to temperature rise and further ice melting. Volcanoes erupt also large amounts of SO_2, and the lower panel shows the volcanic pulse is recorded as sulfate in the Greenland ice core. (Figure modified from Huybers and Langmuir (2009) *Earth Planet. Sci. Lett.* 286 (2009):3-4 479)

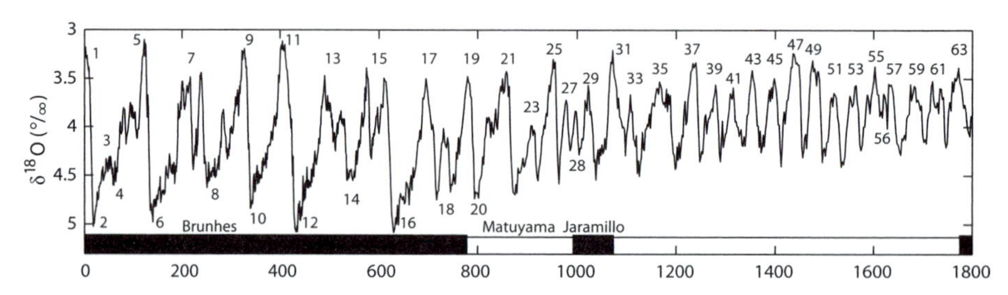

Fig. 18-11: A 1.8 Ma record of ice volume showing the change from a 40 Ka cycle prior to 1 million years ago to the 100 Ka cycle evident in more recent data. The 40 Ka cycle persists all the way back to the beginning of recent glaciations 5 million years ago. These changes remain to be fully understood. (Figure from Lisiecki and Raymo, *Paleooceanography* 20 (2005), PA1003)

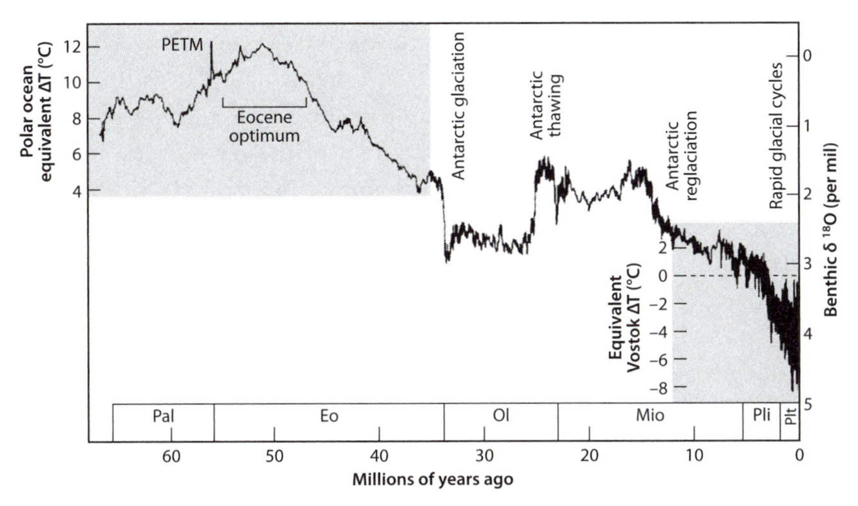

Fig. 18-12: A broader time perspective on climate change. The benthic oxygen isotope record records ice volume. Prior to 30 Ma there was no Antarctic ice sheet and no ice ages. Northern Hemisphere glaciations began about at about 5 Ma. The increased importance of orbital forcing to ice volume beginning at that time is apparent from the $\delta^{18}O$ record. (Figure from Wikipedia Commons, based on data from Zachos et al. *Science* 27 April 2001, 686–93)

high latitudes may be a crucial influence, but the overall understanding of why some periods of Earth's history have ice ages and some do not remains unclear. How orbital variations may have influenced climate in ice-free conditions is also an open question. Sediment records from Eocene and Cretaceous rocks have been found to show regular variations in composition that can be "tuned" to Milankovitch frequencies (it is very difficult to get very precise ages in these time intervals). It appears that even in the absence of ice, global climate change may have been important, but the mechanisms and specifics remain to be fully elucidated. While orbital variations are clearly important for climate changes on the 10^4–10^6 year timescale, much remains to be understood in this rapidly evolving field.

Abrupt Climate Change

An even shorter timescale of climate change is revealed in the detailed climate records preserved in ice cores from Greenland and Antarctica.

Long borings extending from the surface to bedrock provide an extremely detailed record covering the last 110,000 years (i.e., back to the middle of the last interglacial period) in Greenland and back to 750,000 years in Antarctica. The $^{18}O/^{16}O$ ratio in the ice provides a proxy for local air temperature, the calcium content of the ice a proxy for dust infall, and the methane content of air bubbles trapped in the ice a proxy for tropical wetness. In addition to the stately cycles of tens of thousands of years in duration, steep-sided millennial oscillations occur with surprising frequency. Even more surprisingly, these proxies reveal that during glacial time a series of large and abrupt changes in climate occurred (see Fig. 18-13). Intervals of extreme cold, high dust, and low methane alternated with intervals of lesser cold, lower dustiness, and higher atmospheric methane content. The transitions between these climate states were abrupt, taking place in as little as a couple of decades.

Why does this record look so different from the stable isotope record of most seafloor sediments? Two reasons can be given. First of all, in ice cores annual layers are preserved, while the deep-sea sediment record has been smoothed by the stirring action of bottom-dwelling worms, which churn these muds to depths of up to 10 cm. As the accumulation rates for the majority of these sedimentary records are in the range 2–5 cm per thousand years (as opposed to 10–20 cm per year in Greenland ice), signals associated with the millennial-duration events have been obliterated. Second, the large glacial-age ice sheets in Canada, Scandinavia, and Antarctica were too sluggish to respond significantly to millennial-duration climate changes. Hence these events are not imprinted on the benthic ^{18}O record, which is recording large changes in ice volume.

Unlike the 100,000-year climate cycle and its ~21,000- and 40,000-year modulations that are close to synchronous across the globe, the millennial record in Antarctica ice is antiphased with respect to that in Greenland ice. Across the Northern Hemisphere, however, these changes are in lockstep. Greenland's ice core reveals not only changes in local air temperature but also changes in the atmosphere's dustiness and its methane content. Measurements of the isotopic composition of strontium and neodymium serve as a fingerprint of the source of the dust and show that Greenland's dust originated in the great Asian deserts, suggesting that the frequency of intense storminess over Asia changed in harmony with Greenland's air temperature. During glacial time, the wetlands of Canada, Scandinavia, and Siberia, which currently serve as the source for about half of the atmosphere's methane, were either frozen or buried beneath mile-thick ice sheets. Because of this, during glacial time the global production of methane was smaller and confined largely to the tropics. If so, then the abrupt shifts in methane content recorded in Greenland ice suggest that tropical rainfall was lower during Greenland's millennia of intense cold than during the intervening milder intervals.

Evidence for the impacts of these millennial-duration events appears in other records. For example, deep-sea cores from the northern Atlantic reveal distinct episodes during which large numbers of icebergs were released into the sea from the ice sheets surrounding this basin. Upon melting, the rock fragments trapped in these bergs fell to the seafloor, creating ice-rafted debris. Radiocarbon dating demonstrates that these layers formed at the times of Greenland's intense millennial cold spells. Consistent with this evidence are paleotemperatures obtained for a high-accumulation-rate, deep-sea sediment core from deep ocean floor close to Bermuda. The measurements suggest that during glacial time the surface ocean temperatures jumped back and forth by 4° to 5°C in synchrony with the millennial-duration events in Greenland. The ^{18}O record for a stalagmite from China suggests that the strength of monsoonal rains varied in harmony with Greenland's temperature. The monsoons were weaker during the episodes of intense cold.

Strong evidence also comes from the most recent of these short-term fluctuations. The last glacial maximum occurred at 17,000 years, followed by a characteristic abrupt warming event. But then, from 13,000 to 11,000 years ago Earth plunged into a temporary reentry to glacial conditions called the Younger Dryas. This event is strikingly evident in the Greenland ice-core record (see Fig. 18-13). As might be expected, high mountain regions are very sensitive to cooling, leading to large down-mountain expansions of their ice caps. Radiometric dating of debris pushed out as the glaciers advanced allow the most recent epochs of mountain cold to be correlated with Greenland's most recent cold phase. So far it has been demonstrated that glaciers in the Swiss Alps, in the tropical Andes, and New Zealand's Alps underwent major readvances when Greenland ice last revealed a millennial-scale cold period.

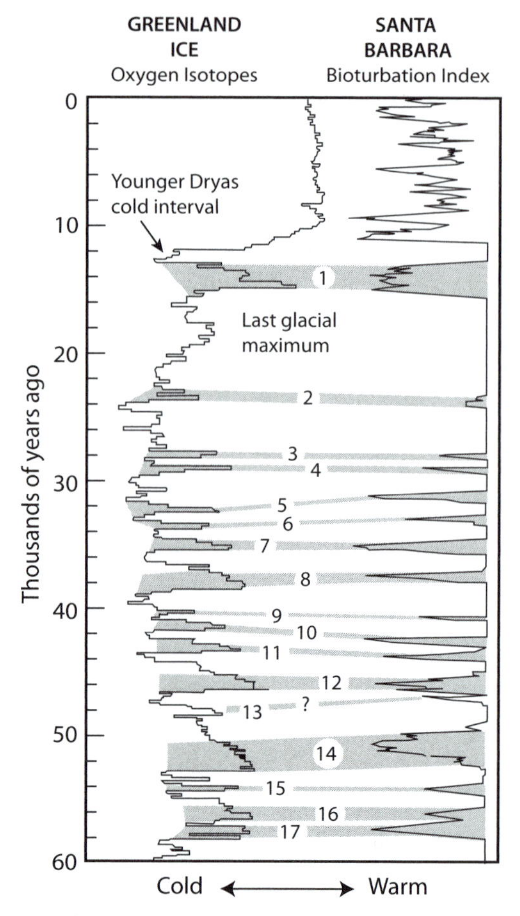

Fig. 18-13: Comparison between the air temperature record in Greenland and the O_2 content record for the bottom waters in the Santa Barbara basin: The latter is based on the extent of burrowing of the sediment by worms. At times of high O_2, the annual layering in the sediment is completely obliterated by the action of these organisms in a process called "bioturbation." At times of low O_2, the worms can't survive and the annual layering is perfectly preserved.

To this evidence from mountain glaciers can be added an impressive record from the sediment in a small basin located between Santa Barbara, California, and Santa Rosa Island (Fig. 18-13). As sediments accumulate in this basin at the rate of 1 m per 1,000 years, a record of millennial changes is beautifully preserved. At present the O_2 content of its bottom water is so low, and the rain of organic matter so high, that the water that fills the sediment pores is devoid of O_2. The anoxic conditions make the environment uninhabitable for worms, which under oxic conditions continually mix the uppermost sediment in a process called bioturbation. Under anoxic conditions, no stirring occurs and the sediments have annual laminations reflecting the seasonal variations in the mix of biological and soil debris reaching the bottom.

Interested to see how far this layered record extended back in time, Jim Kennett, a marine geologist at the University of California at Santa Barbara, initiated a program to core this basin. Annually layered sediment continued uninterrupted back to the beginning of the present interglacial (i.e., to about 11,000 years ago), but below that lies a long series of alternating zones of bioturbated sediment and annually layered sediment. Kennett quickly realized that the intervals of well-mixed sediment must represent times when the O_2 content of deep water in the Santa Barbara basin was higher than now. When he compared his record to that from the Greenland ice core, he found an incredible match (Fig. 18-13). At the times of extreme cold in the ice core, a layer of well-stirred sediment accumulated in the Santa Barbara basin. Kennett concluded that this meant that the invasion of freshly oxygenated surface waters into the intermediate depths of the northern Pacific Ocean had greatly strengthened at these times. More recently, German scientists obtained a nearly identical record in the rapidly accumulating sediment in the Arabian Sea off Pakistan.

Taken together, this paleoclimatic evidence suggests that the impacts of the millennial-duration climate episodes so obvious in the Greenland ice core were widespread. There is, however, a tantalizing exception. The record in ice cores from Antarctica reveals that during the 10,000-year-long transition from full glacial conditions to full interglacial conditions, the millennial-duration modulations of climate were antiphased with respect to those recorded in Greenland. When the Northern Hemisphere experienced the pre-Younger Dryas warm spell, the warming in Antarctica stalled. When conditions in the Northern Hemisphere were plunged into the Younger Dryas cold, Antarctica resumed its warming. Thus, while the orbital-related cycles were globally synchronous, the millennial-duration changes appear to have been antiphased. Furthermore, at least in the case of the Younger Dryas, the boundary between the two realms curiously enough appears to lie to the south of New Zealand rather than at the equator.

THE GREAT OCEAN CONVEYOR

The millennial-duration shifts in climate constitute a challenging mystery. What is it about Earth's climate system that allows it to undergo large and abrupt climate changes? Why are these changes antiphased between Antarctica and the rest of the planet? Why has no such change taken place during the last 10,000 years? What triggers these jumps? While convincing answers have yet to emerge, some clues suggest that the instigator lies in the ocean. Model simulations suggest that the ocean's so-called thermohaline circulation is capable of undergoing reorganizations. This large-scale circulation is driven by the descent to the abyss of cold and salty water from two places on the planet, the northern Atlantic in the vicinity of Iceland and the Southern Ocean along the perimeter of the Antarctic continent. Currently, the northern source floods the deep Atlantic with water that sweeps southward, passing around the tip of Africa, where it joins waters rapidly circulating around the Antarctic continent. Deep waters formed around the perimeter of Antarctica also sink into this circumpolar current, producing a mix that currently consists of roughly equal amounts of the two source waters. Portions of this mixture peel off and move northward into the deep Indian and Pacific oceans (Fig. 18-14).

These currents are important to Earth's climate because they redistribute heat. This redistribution is especially important for the land masses surrounding the northern Atlantic. Replacing the water sinking to the bottom of the northern Atlantic are warm upper ocean waters carried northward toward Iceland in the upper limb of the conveyor. As this upper limb traverses the low latitudes, it is heated by the sun. When it reaches high northern latitudes, this stored heat is released to the overlying air. During winter months, this heat takes the sting out of the cold Arctic air masses that move eastward across the Atlantic. This bonus of heat helps to sustain northern Europe's mild winters.

The magnitude of the conveyor transport by the conveyor is staggering. It equals that of a hundred Amazon Rivers and matches rainfall across the entire globe. The northward moving limb carries water with an average temperature of 12°C into the region around Iceland. The water sinking into the abyss averages only 2°C. Hence, for each cubic centimeter of water carried northward by the conveyor's upper limb, 10

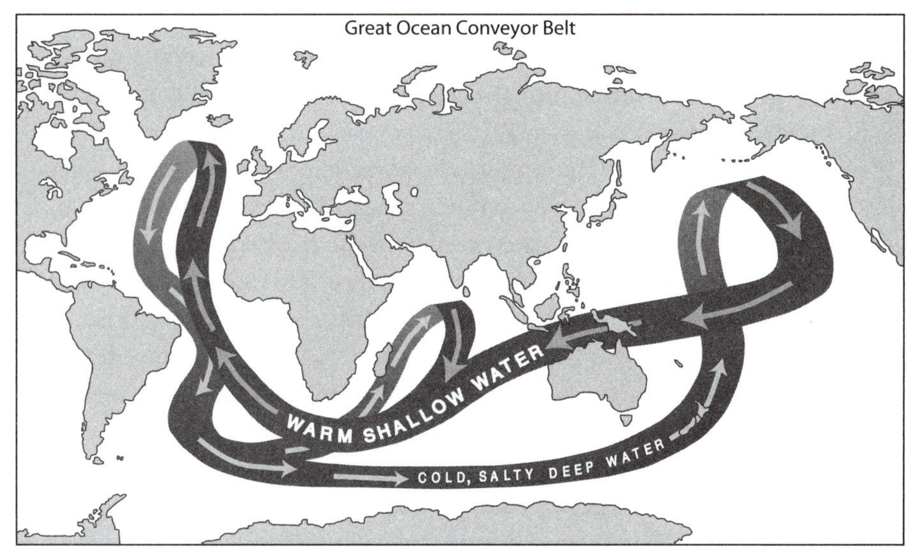

Fig. 18-14: The conveyor diagram illustrating global movements of ocean currents.

calories of heat are released to the atmosphere. This adds up to a staggering total, equaling about one-quarter of the solar heat supplied to the atmosphere over that portion of the Atlantic Ocean lying to the north of Gibraltar!

In today's ocean, a delicate balance exists between the density of deep waters generated at the opposite ends of the ocean. If this balance is disrupted, at least in the models, the system of currents reorganizes into a new pattern. Hand in hand with these reorganizations come changes in the amount of ocean heat released to the high-latitude atmosphere. While all models show this type of behavior, each is characterized by its own set of details. In some, there is a complete shutdown of the conveyor; in others, the conveyor is modified so that deep waters form farther to the south and fail to penetrate all the way to the bottom. Common to all is that deep waters from the Southern Ocean penetrate much farther up the Atlantic.

What might trigger circulation reorganizations? Although we have no firm answer to this question, one possibility is what could be termed a salt oscillator. As shown by the models, the most effective means of tampering with deep water formation is to increase the input of fresh water to one or the other of the regions where deep water forms. Such injections dilute the salt content of the surface water, thereby reducing their density. If this reduction proceeds to the point where, even during the most intense winters, waters dense enough to displace those beneath are no longer produced, then reorganization might occur. This, in fact, is why deep waters fail to form in the northern Pacific. Its surface waters carry so little salt that, even when cooled to the freezing point (−1.8°C), the water is not dense enough to penetrate into the deep sea. While evidence exists that at least one of these reorganizations (i.e., the onset of the Younger Dryas) was triggered by a sudden release to the northern Atlantic of melt water stored in a lake that formed in front of the retreating North American ice sheet, the regularity of earlier events suggests that some sort of oscillator was operative. In today's Atlantic there appears to be a balance between salt buildup as the result of water vapor export through the atmosphere from the Atlantic to the Pacific and salt export via the lower limb of the great conveyor. But it is possible that an imbalance between salt buildup and salt export could lead to an oscillation. Imagine that during conveyor "on" periods the net salt export was to exceed salt buildup. This would lead to a drawdown of the Atlantic's salinity. Eventually a point would be reached where it was too low for deep water to form. This would cause thermohaline circulation to reorganize to its "off" mode. If, in this "off" configuration, the buildup of salt due to vapor export were to outpace salt export, the salinity of the Atlantic waters would begin to rise. Eventually a point would be reached where the conveyor would snap back into action. The average duration of the millennium-duration cycles is 1,500 years. This is a reasonable time constant for a salt oscillator. The reasoning is as follows. The Atlantic exports water vapor at an average rate of about 0.25 million cubic meters a second. Were there no export of the salt left behind, the salinity would increase at a rate of 0.1 g/liter per century. Thus, in a half cycle (i.e., ~750 years), the salinity would rise 0.75 g/liter. Such a rise has a density impact equivalent to a cooling of cold polar waters by about 3°C. So while there is no way to estimate the expected frequency of these salt-induced oscillations, it makes more sense that it is on the order of one per 1,000 years than one per 100 or one per 10,000 years.

Left unanswered by this hypothesis is why the impacts on the atmosphere of these ocean reorganizations are global. In models, these reorganizations lead only to changes in the climate of the region surrounding the northern Atlantic. No significant impacts occur in the tropics and certainly not in the South Temperate Zone. Simply put, we do not understand the nature of the teleconections that propelled these messages around the Earth. It is tempting to look to the tropics and, in particular, to the water vapor carried aloft in the towering convective plumes that rise from the equatorial ocean to the base of the stratosphere. As water vapor is the atmosphere's dominant greenhouse gas, changes in its inventory could induce large global temperature changes.

For this to be the explanation, it is necessary that a link between the large-scale circulation in the ocean and convective activity in the tropical atmosphere exists. The most likely candidate is the upwelling of cold water along the equator, which currently constitutes a major component of the tropical heat budget. Perhaps the intensity of this equatorial cooling is linked to the large-scale circulation of the ocean. This is why Kennett's Santa Barbara basin record is so important. It tells us that the shallow circulation in the ocean was altered in concert with the events in Greenland. Could it be that these changes led to a modification of the supply of moisture to the tropical atmosphere?

Another possibility is that the large changes in the atmospheric inventory of dust and sea salt aerosols recorded in polar ice were the villains. When lofted into the atmosphere, soil debris reflects sunlight. Aerosols serve as nuclei for cloud droplets. The more nuclei, the smaller and more numerous the cloud droplets and, as a consequence, the greater the reflectivity of the clouds. For this to be the correct explanation, there would have to be a tie between the frequency of intense storms that carry dust and sea spray high into the atmosphere and the large-scale circulation of the ocean. One possibility is that during glacial time when the conveyor is "off" the northern Atlantic became choked with sea ice, compressing the winter temperature gradient between frigid ice-covered polar regions and the warm tropics, thereby promoting storminess.

The lesson from all this is that the climate of our planet is far from stable. Minor prods by seasonality changes and by the redistribution of fresh water have provoked large and often abrupt climate changes. We will return to this subject when we consider the poke we are currently giving to our climate system by adding large amounts of CO_2 and other greenhouse gases to the atmosphere.

Human Impacts

The termination of the last of the major 100,000-year cycle led to two extremely important events in the development of *Homo sapiens* (i.e., us). First of all, as far as we can tell, prior to this time humans had not been able to establish a foothold in the Americas. Then suddenly, between 13,000 and 12,000 years ago, there was an influx of people who very quickly spread over the entire length and breadth of the New World. The preferred explanation is that during the period of melting of the last great ice sheet, a corridor opened between the western and eastern lobes of the North American ice. Further, sea level had not risen to the point where the Bering Straits became flooded. The combination of the corridor and the land bridge allowed humans to migrate from Asia into the Americas. The rapid spread throughout the Americas may reflect the dependence of the new arrivals on large game. We suspect this because the saber-toothed tiger, giant sloth, and woolly mammoth all became extinct soon after the arrival of man. So the idea is that, as these animals were hunted to near extinction in one area, the hunters moved on to another, and soon they had decimated the entire New World. This is not the only example of man-induced extinction. A similar disappearance of large animals occurred about 50,000 years ago, when the bushmen arrived in Australia (again walking from Asia over a land bridge created by lowered sea level).

The second of these impacts was of far greater importance. It involved the transition from hunting and gathering existence for the people in the Middle East to animal husbandry and farming. This transition marks a key step in the development of our civilization. The equable climate that extended around the globe and ability to control food supply then permitted the rise of ancient civilizations in Mesopotamia and Egypt, closely following the deglaciation. It is now known that 160,000 years ago *Homo sapiens* with the same brain capacity we now enjoy lived in Ethiopia. A question that then emerges is why farming didn't begin during the penultimate glacial termination at about 125,000 years. One can only surmise that it was a lack of population pressure, or abundant food, or perhaps a lack of the sophistication of developed culture at that time. We may never know. In any case, when 100,000 years later the next great termination came along, the stage was set for the rapid evolution of what we know as civilization. Climate change has been closely coupled to the fate of human beings on the planet.

Summary

On intermediate timescales, Earth's temperature has fluctuated considerably owing to orbital variations and consequent changes in the amount of solar energy received. During certain periods of Earth's history, including the most recent one, orbital variations have led to ice age cycles. While changes in solar energy due to orbital variations would have existed throughout Earth's history, only certain periods of time, such as the most recent few million years, have been associated with widespread glaciations. Whether or not Earth is in an icehouse state may depend on positions of the plates and volcanoes that have important feedbacks on orbital climate changes. High-resolution records of the last several hundred thousand years show that there are also abrupt climate changes that can happen in as little as a decade and persist for a thousand years. The most recent such event was the Younger Dryas, which plunged Earth back into full glacial conditions as it was exiting the last ice age. Abrupt climate change cannot be caused by the tectonic thermostat or orbital variations. Instead, it is likely a consequence of changes in ocean and atmospheric circulation that can occur on very rapid timescales.

Supplementary Readings

John Imbrie and Katherine Palmer Imbrie. 1986. *Ice Ages: Solving the Mystery.* Cambridge, MA: Harvard University Press.

Gerard Roe. 2006. In defense of Milankovitch. *Geophys. Res. Lett.* 33:L24703.

Richard A. Muller and Gordon J. MacDonald. 2002. *Ice Ages and Astronomical Causes.* Reprint. New York: Springer-Verlag.

Wallace Broecker. 2010. *The Great Ocean Conveyer: Discovering the Trigger for Abrupt Climate Change.* Princeton, NJ: Princeton University Press.

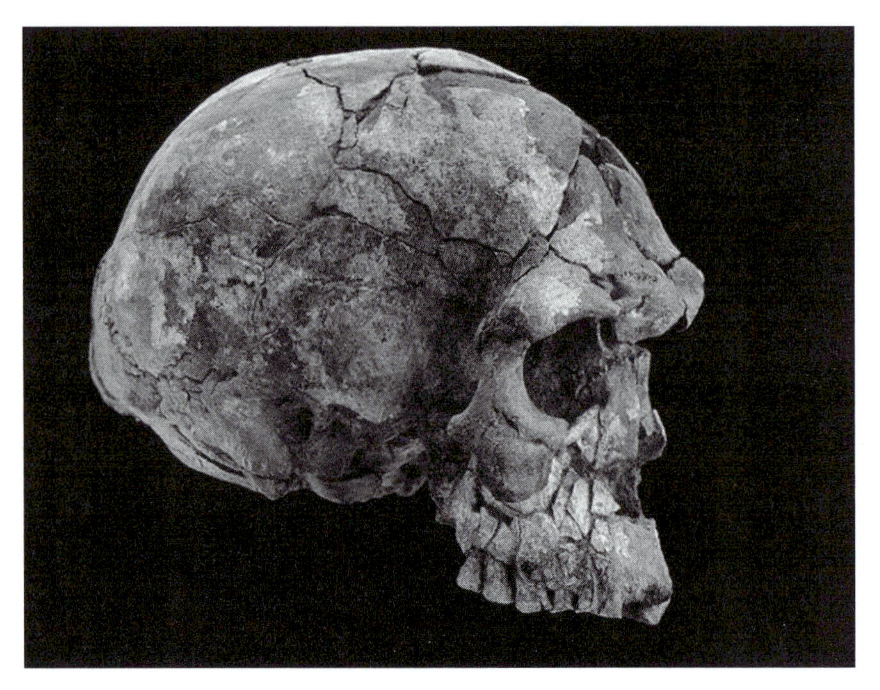

Fig. 19-0: Picture of 160,000-year-old skull of an adult male from Ethiopia, the same species as present humans.

CHAPTER 19

The Rise of *Homo Sapiens*

Access to Earth's Treasure Chest Permits a Planetary Takeover

In the absence of "intelligent" life, our planet could continue along more or less the same course it has followed in the past. The sun has enough remaining hydrogen to burn smoothly for several billions of years. The heat produced by the radioactivity of Earth's mantle will be adequate to drive the crustal plates for similar lengths of time. Evolution would change the assortment of species as they adapted to inevitable planetary changes, but barring catastrophes ecosystems would thrive in harmony with the planetary condition. The appearance of human beings has changed the character and outlook for the planet, although this was not immediately apparent. Our first tool-bearing ancestors appeared some 2 million years ago, and modern man appeared about 160,000 years ago. Throughout this time we competed on a more or less equal footing with our fellow inhabitants and lived without substantial impact on the natural environment. Environmental change at times reduced our numbers to less than 100,000 people. About 10,000 years ago, after the most recent deglaciation, all this began to change. The ability to manipulate fire permitted access to energy, large-scale modification of landscapes and ecosystems, access to metals, and far more efficient tools. The discovery of agriculture and the harnessing of other animals for our personal use gave us a great competitive advantage over plants and our fellow animals. In order to enhance our food supply, we used our ability to harness energy to reroute the flow of water, to till the soils, to eliminate weeds and pests, and to selectively breed and domesticate animals. The development of large communities where individuals developed specialized skills also greatly increased our capabilities as a species. Owing to our success, we began to modify natural systems for our own benefit, which made possible even larger concentrations of population and the great ancient civilizations of Mesopotamia, China, and Egypt.

Starting about 150 years ago, we discovered the great energy potential locked in fossil fuels. This increased energy led to industrialization that greatly accelerated this man-made modification of our planet's surface. Now each human being uses some twenty times the energy that they can obtain through food, and individuals in the most prosperous nations use a hundred times as much energy. This latest energy revolution, the human energy revolution, enabled us to access all the planetary resources that were built up by billions of years of planetary evolution—Earth's treasure chest. Some of these resources, such as most metals, are theoretically infinite in amount and readily recycled. Others, such as fossil fuels, are limited in amount and once used are gone forever. In this context, we live in the fossil fuel age, where 500 million years of Earth treasure will be consumed by a single species in a few centuries. Fossil fuels are not the only non-renewable resource. Soils that support our food supply and biodiversity that contains Earth's biological store-house of genetic potential are also being rapidly depleted.

The access to energy and the built up planetary resources has permitted a ferocious growth in population, particularly since the dawn of the industrial age, when our energy use so greatly expanded. Extinction of 97% of human beings living on Earth today would reduce us only to the number of people on Earth at the dawn of the scientific revolution some 500 years ago.

Introduction

Since the demise of the dinosaurs at the Cretaceous/Tertiary boundary, mammals have progressively evolved to dominate terrestrial ecosystems. In very recent times, one particular mammal, *Homo sapiens*, has changed the entire course of Earth's history. The appearance of intelligent life and a global civilization is planet changing both in its impact on all the various aspects of the Earth system and in the capacity of a planet in the galaxy. The various energy sources on the planet (fossil fuels, wind, sun, atom) can be harnessed to provide capabilities inaccessible to any other species and never before seen on the planet. Language permits advanced and subtle communication, among people and through generations. Industrialization allows the planet's exterior to be modified and biological evolution to be engineered. Global data systems permit "sensing" (e.g., temperature, weather, crop health, atmospheric composition, population, biodiversity, etc.) on a planetary scale. Communication of this information permits global relationships and action. Specialization permits enhanced skills and capabilities for the community that vastly exceed those of any individual. And while it is not yet within our reach, we can imagine that such a planet might also have the capacity to communicate with other planets outside our solar system as part of a galactic community, should that be technologically feasible through means we do not yet understand. A visitor who examined our planet a thousand years ago and today could only be astonished by the revolution that has taken place. What is the record of this amazing planetary transformation, and what permitted it to occur as a unique event in planetary history?

Dawn of the Human Era

The oldest human skull found so far dates back to the time of the penultimate glaciation 160,000 years ago (see frontispiece of this chapter). By "human" we mean *Homo sapiens*, the primate species with the largest brain capacity. At that time, our numbers were quite small and we likely lived only in East Africa. As recently as 70,000 years ago, genetic studies suggest the total human population may have been as little as 10,000 people. As recently as 50,000 years ago, a small number of people took advantage of the lowered sea level of glacial time and made their way across the land bridge connecting Asia with Australia. Fifteen thousand years ago we crossed the Bering land bridge and populated the Americas. At some time during the last glacial period, our main competitor, the Neanderthal, fell by the wayside.

Until the end of the last glacial period (11,000 years ago), we competed on more or less equal footing with our fellow large mammals, and human population probably did not exceed 1 million. Then, taking advantage of the dramatic warming and increase in habitable area ushered in by the present interglacial period, we transitioned from an existence based on hunting and gathering to one based on agriculture and animal husbandry. No longer forced to roam in our quest for food, we began to build ever more elaborate settlements and eventually cities, giving rise to the ancient civilizations of Mesopotamia, Egypt, and China. As prime agricultural land and reliable water sources were often geographically separated, irrigation became essential to our developing civilization. Trade flourished, permitting materials and technology to be shared over long distances. Written language and monetary currencies grew hand in hand with trade. The refinement and use of fire, providing access to high temperatures, permitted specialized activities such as metallurgy and pottery manufacture to take place in ever more complex societies. These developments also permitted increased specialization in the technical realm, in fields such as literature, music, theater, and painting, and in our capacity for war to destroy each other's existence. Humans began to specialize and work together so that the entire community had far more power and capability than any individual.

The benign climate and the rise of agriculture and civilization permitted major population growth, leading to about 200 million people at the peak of the Roman empire. Following a substantial decline after the fall of Rome, a medieval warm period led once again to a burgeoning population, followed by a steep ~25% reduction in global population caused by famine and plague. By about 1500 there were no more people than at the time of the Roman Empire. At that point population growth took off. By about 1820, the population had grown to about 1 billion, tripling in some 200 years. Another 2 billion were added by 1960 and in the last 50 years 3 billion more people were added as the population doubled again. As of the year 2012 there were 7 billion, and the estimate for the year 2050, barring catastrophes, is 10 billion.

Figure 19-1 shows population growth over the last 12,000 years, since the end of the last ice age. The vertical axis is logarithmic, so a steady rate of growth would lead to a straight line with a constant "doubling time." Between 70,000 and 10,000 BC, population appears to have grown

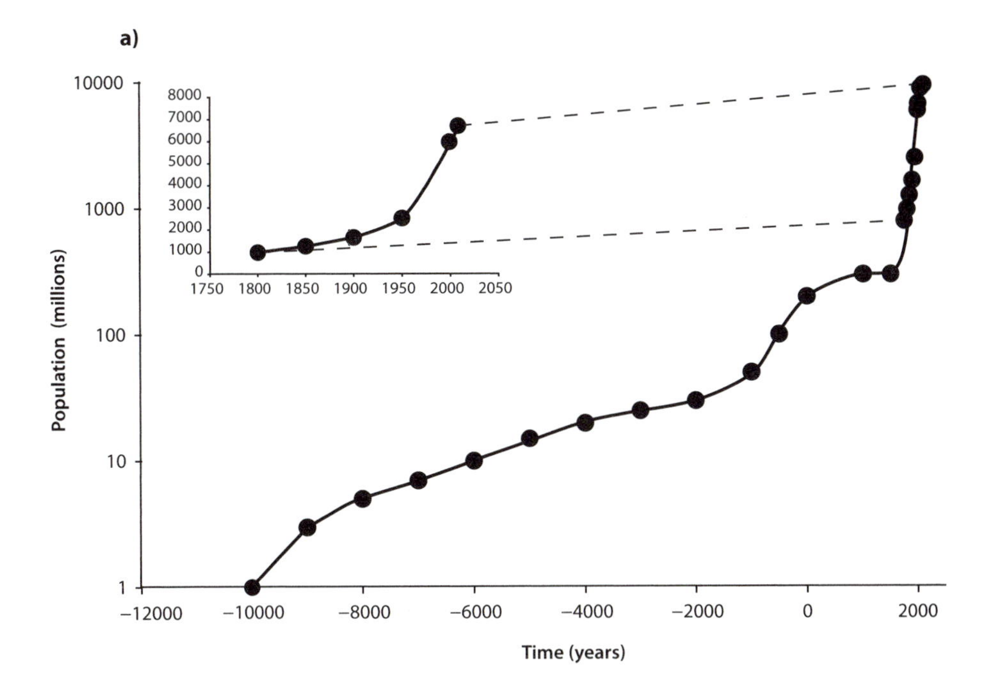

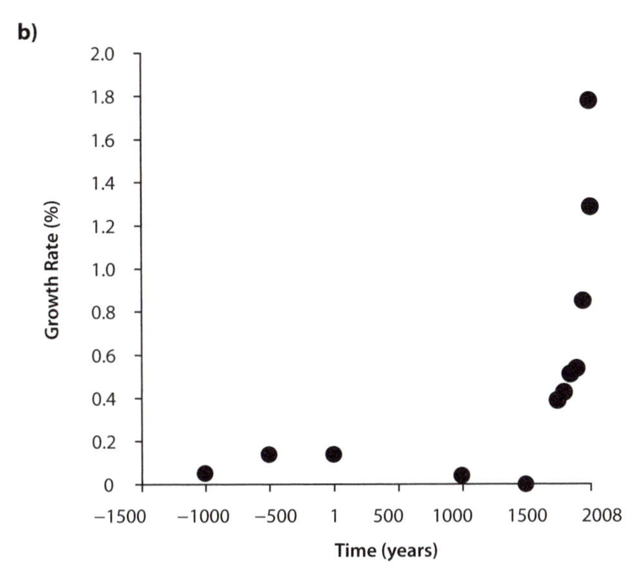

Fig. 19-1: (a) Global population growth versus time. Note that the vertical scale is logarithmic, so a straight line is exponential growth with a constant "doubling time." The expanded portion of the graph has a linear scale and shows the exponential growth between 1800 and 2010. (b) Growth rate versus time for the last 3,500 years. It is not just the total human population increase but the rate of increase that has changed with time. (Data from U.S. Census Bureau)

at about 0.007% per year on average, or a doubling time of 10,000 years. After the rise of agriculture, that rate may have increased to about 0.03% for a few thousand years, increasing to as much 0.1% during the Roman era, for a doubling time of 700 years. Then for some 1500 years population appears not to have increased, until about year AD 1600, when the rate jumped to 0.5%. After the rise of industrialization, the rate rose to more than 1% per year. It may seem like a small increase, but that changes the doubling time to only 70 years. These statistics show that it is not just the number of people that has increased, it is the *rate* at which they have increased that has marked the modern era. If global population reaches the estimated 10–12 billion people in the coming century, on the basis of population alone human beings annually will have 40 times more impact on the planet than any year prior to 1600, 4 times more than 1950, and 7,000 times more than the beginning of the modern era 12,000 years ago. Including the magnifying factor of growth in energy, the impact is even larger.

The root cause of human-induced planetary change is this population growth and its attendant demands on all aspects of the Earth system. While we see people everywhere we look, population growth is not obvious to the individual. But consider that if a terrible plague visited the Earth and wiped out 97% of the human population, there would still be as many people on Earth as there were in the year 1500. And despite a recent slowing of the rate of population growth, we are still adding more than the equivalent of an AD 1500 population every three years.

This spectacular growth would not have been possible if each of us still had to go out hunting on a daily basis for our food within walking distance of our home. Imagine, for example, if food supply was cut off to a major city for a month and the millions who lived there needed to hunt and forage to survive. Famine and riots would be inevitable.

What has made possible our spectacular growth and success is our ability to work together and access the treasure chest of Earth resources, including most importantly our ability to access energy beyond that provided by metabolizing food. Energy provides our access to all the other resources, most importantly food, but also metals, groundwater, fish, and so on. If, like all other animals, we had to rely on our muscle power alone, civilization would collapse. The rise of human civilization is the latest energy revolution for the planet—the human energy revolution.

The Human Energy Revolution

Access to energy has been a driving force for human civilization for millennia. Streams were diverted to power mills. Wind was harnessed to push ships through the sea. Wood was burned to fuel furnaces and extract metals. The whale population was decimated to provide oil for lamps. The biggest change occurred with the advent of engines fueled by the combustion of carbon. To access coal, mines were opened and canals and railroads built. With the advent of the automobile, a need for a liquid fuel led to drilling for oil.

The availability of carbon-based energy has allowed human beings to enormously supplement their energy supply. Including all energy sources, the average person globally uses 2,300 watts of power, or the equivalent of twenty-three 100-watt light bulbs burning twenty-four hours a day. The average United States citizen uses five times more, the equivalent of 11,400 watts of power continuously (including all energy sources). The food we eat provides the energy equivalent of merely 100 watts. Our bodies are so efficient that that small amount of power allows all the inner and outer activities of our bodies and makes us an organic electric blanket when we huddle together on a cold night. Access to external energy augments our average power by twenty to a hundred times compared to animal metabolism (Fig. 19-2), and much more energy is available on demand when we want it, such as when we fly, drive a car, or turn on a stove. This energy gives us the capabilities of superheroes. We can fly higher and faster than a bird, easily access building rooftops, punch through walls, excavate mountains, stop speeding bullets, send fire and destruction long distances, and communicate instantaneously with others all over the world.

From a planetary perspective, human access to external energy in the service of our species is the latest energy revolution for Earth, the *human energy revolution*. The energy revolution of aerobic metabolism discussed in Chapters 15–17 boosted energy by a factor of 18 and allowed the development of multicellular life. Our energy boost of twenty to a hundred times exceeds this even on an average basis, and of course over short times we vastly exceed it. As is obvious from examination of the world around us, this access to energy and the capability to focus our

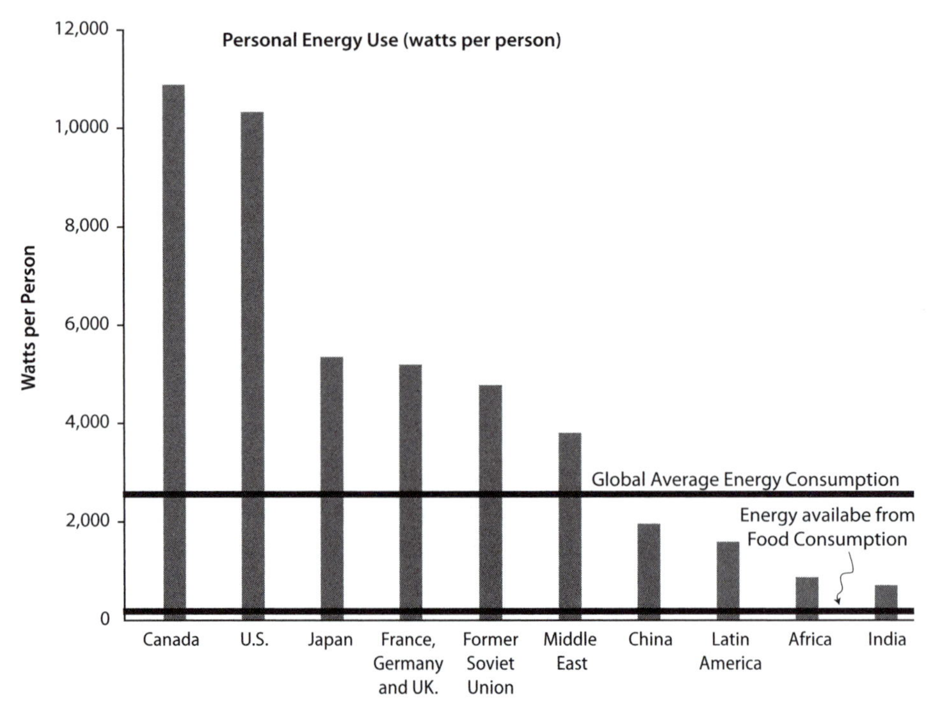

Fig. 19-2: Per-person energy use by nation. The bar at 2300 watts is the average personal energy consumption globally. The barely visible bar at 100 watts is the average energy we obtain from food. Other animals are similarly restricted. (Data from International Energy Agency, *Key World Energy Statistics*, 2009)

energy use have been able to transform the planetary system in record time.

The increased energy allowed us to extract metals from Earth's crust, pump water from aquifers and dam rivers, and entirely dominate all other animals and steadily eliminate them. While this increased access to Earth's resources has major advantages for us, it also has involved sizeable environmental and health costs. Mines destroy ecosystems and scar the landscape. Burning coal releases sulfur and mercury. Denuding land and plowing it eliminates topsoil. Industrial chemicals never produced by nature to which life is not adapted are added to air, water, and life. Groundwater use depletes water supplies. Overfishing depletes large fish from the entire ocean. The waste products of our energy production add CO_2, CH_4 and other greenhouse gases to the atmosphere and generate nuclear waste. Both the wonders and horrors of modern civilization are made possible by the human energy revolution, largely powered by the burning of fossil fuels.

A central aspect of the energy revolution is that it provides much greater access to and productivity from land, increasing our food supply. Industrial agriculture is highly energy intensive, and meat production even more so. Two pounds of meat from an industrial feedlot for a family dinner requires one gallon of oil to produce. Transporting and killing the animal, packaging and transporting the meat to market and then to the home, requires still more energy. Many of the largest cities are far from agricultural regions and could not exist without food transport using fossil fuels. Our takeover of planetary ecosystems and energy use to produce and transport food from productive to unproductive areas, as well as greatly improved health care, is what has permitted us to expand so greatly in numbers.

The human energy revolution has been possible not from a biological innovation but from our ability to directly access Earth's fuel cell, discussed in Chapter 15. We take the organic carbon stored over hundreds of millions of years of photosynthesis, and combine it with the oxygenated atmosphere developed by the billions of years of planetary evolution, reversing photosynthesis and releasing the stored energy. Were we able to access *all* the organic carbon, we would eliminate the O_2 in the atmosphere and fully discharge the fuel cell that powers advanced life. Fortunately, most of the organic carbon is in black shales where it cannot be extracted economically.

Fossil fuels have been the most important resource because they concentrate enormously the energy from the sun. This power then gives us the energy to access other resources on a grand scale, moving mountains if necessary. The human energy revolution has enabled us to open and exploit Earth's treasure chest, the vast resources accumulated gradually over billions of years of Earth's history.

Earth's Treasure Chest

The abundant resources that Earth provided are what permitted the global rise in population and spread of civilization. Rivers, food, forests, animals, and more and more as time went on, metals, fuels, groundwater, soils, and fertilizer—all were used for the economic and population growth of humans. These resources are the planetary treasure chest.

The accumulation of all these resources has resulted from Earth's billions of years of evolution. Because of the differential solubility of ions of different oxidation states, changes in the oxidation state of surface environments have been a major factor in the accumulation of high concentrations of certain metals. For example, the early rise in oxygen led to the deposition of BIFs during the period from 3.5 Ga to 1.8 Ga, as Fe was transported from soluble Fe^{2+} environments to sites of insoluble Fe^{3+} deposition (see Fig. 16-7). These rocks are the principal source of Fe ore for modern civilization, and because of them iron is so cheap that it forms the basis of modern buildings and industry. The ores of uranium are also controlled by oxidation state. Recall that the reduced form of U is insoluble and the oxidized form soluble. Ancient ores are the detrital uraninite referred to in Chapter 15 and 16 that were able to be deposited in gravels prior to the rise of O_2 in the atmosphere. At younger times, uranium was mobilized in oxidized waters, and then precipitated when these waters encountered a reduced environment, such as sediments rich in organic matter, leading to the uranium deposits in Phanerozoic rocks.

Other deposits required particular conditions of the biosphere or the thermal evolution of Earth's interior. Deposits of phosphorous, an essential ingredient for fertilizer, were largely restricted to Phanerozoic rocks when the ocean became sufficiently biologically productive to create massive deposits of phosphorite rock. Large deposits of chromium and platinum that provide most of these metals for modern civilization were formed when massive basaltic bodies of magma were intruded into the crust during the Archean, when Earth was hotter and produced more magmatism. The 3.5 billion-year-old Bushveld intrusion in South Africa, for example, contains most of the world's reserves of these two metals.

Other metal deposits are formed as a consequence of plate tectonics. Many deposits of copper, molybdenum, and tin occur in the shallow oxidized portions of hydrothermal systems that formed around granitic plutons emplaced in the crust near convergent margins. Because they occur in mountainous volcanic regions that are recycled rapidly by erosion and require oxidizing conditions, old deposits either did not form or have eroded away, and existing deposits are almost all associated with modern plate margins (Fig. 19-3).

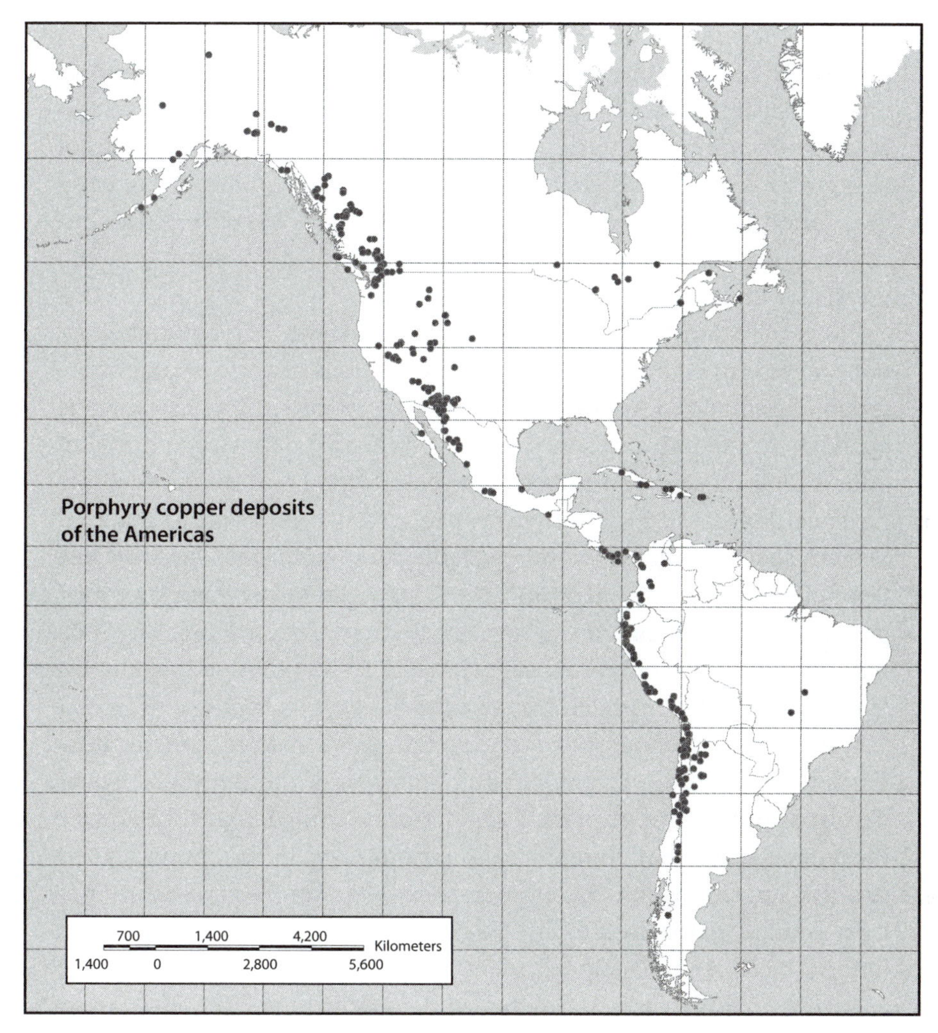

Fig. 19-3: Map of North and South America showing distribution of major ore Cu deposits. Note the preferential presence of these ore deposits near modern convergent margins, and their association with subduction volcanism. Deposits in western Canada formed prior to the subduction of the Farallon ridge when active subduction occurred in that region (see Fig. 11-7). (From Singer, Berger, and Moring (2002); http://purl.access.gpo.gov/GPO/LPS22448. Courtesy of U.S. Geological Survey)

Hydrothermal ore deposits formed at ocean spreading centers have also played a role in the rise of human civilization. Oceanic crust and its ore deposits formed at hydrothermal vents are occasionally faulted and accreted to continents, forming bodies called *ophiolites*. The ore deposits are very visible and high in grade, making them easy targets for early civilizations. Cyprus contains a large ophiolite that was rich in ore, and similar bodies throughout the Mediterranean served as a source of early metals for humans.

Fossil fuels form from burial and transformation of plant matter, so none were formed prior to the Paleozoic. Billions of years of photosynthesis were required to raise atmospheric O_2 levels sufficiently for multicellular plants to evolve. To make a commercial deposit requires still other conditions that permit the concentration of vast amounts of organic carbon. Later metamorphism causes this carbonaceous material to form coal, or to generate more volatile compounds that can flow and be trapped in reservoirs as oil and gas. Coal began to be formed in abundance in the aptly named Carboniferous period, named because of its massive concentration of coal deposits (Fig. 19-4). Oil is more common from younger Mesozoic and Cenozoic rocks.

It is amazing to note the very slow rate of formation of fossil fuels. Earth underwent a very gradual accumulation of these resources over hundreds of millions of years. The horizontal axis of Figure 19-4 gives the annual accumulation rates of the various fuels. Oil has accumulated at a few hundred cubic meters per year—less than the annual output from a single gas station today. Coal has accumulated at about 20,000 tons per year. Annual coal use today is 300,000 times that amount. We are using up this treasure at an exceedingly rapid rate.

Other resources are rapidly rejuvenated, and therefore their age is geologically young. The resources of fresh water and soil obviously depend on the longterm climate stability of the surface, but given that stability they seem to be steadily renewed, as rocks weather, organic matter decomposes, and water cycles on the short timescales of weather and seasons. Even these resources are time dependent, however, because they are influenced on longer timescales by the glacial cycles. Glaciers carve great basins that become lakes filled by the massive amounts of meltwater from receding glaciers, such as the Great Lakes and Finger

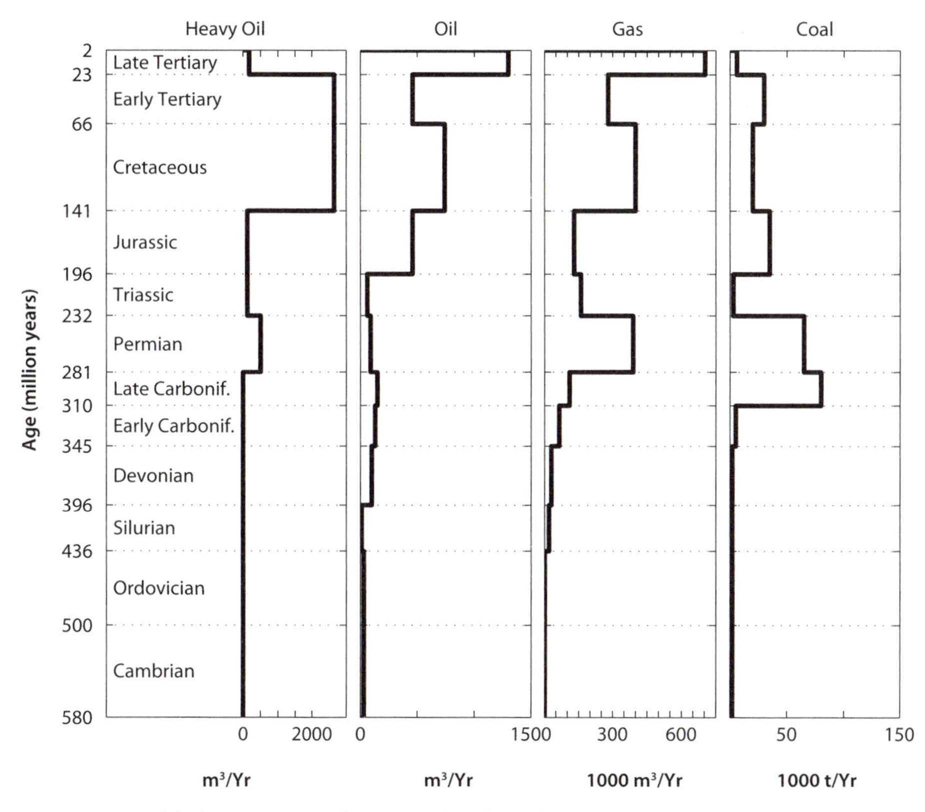

Fig. 19-4: Fossil fuel reserves as a function of geological age. Note the massive amounts of coal made in the Carboniferous period, and the young age of most oil formation. (Modified from Pimentel and Patzek, *Rev. Environ. Contam. Toxicol.* 189 (2007):25–41)

Lakes of the northern United States. Wetter climates also contributed to rich resources of groundwater in deep aquifers.

This knowledge of Earth's history shows us that a highly diverse set of processes operating over billions of years has created the resources on which human civilization depends. Some resources, such as rainwater and rivers, are rapidly recycled and depend on recent climate. Others, such as lakes and groundwater, have longer timescales associated with glacial cycles. And many resources formed under specific conditions that were restricted to short intervals of Earth's history. These diverse processes have over a very long time led to the accumulation of the planetary treasure chest of resources. No species had ever made use of mineral and fossil fuel resources prior to the appearance of human beings, and upon our arrival we found a well-stocked planet. With no knowledge of geological time and Earth's history, no sense of a planet of finite size or of geological time, ancient peoples could have no concept that planetary resources were limited in quantity and reflected a long process of planetary evolution and storage. Today we face a fully inhabited planet where understanding the different types of resources in a planetary context has become necessary.

Classes of Resources

Earth's resources have different characteristics, depending on their abundance in the earth, whether they are destroyed by human use, and the timescales of replenishment by earth processes. On this basis, we can divide resources into three different classes:

(1) Resources of vast extent and short recycling times (e.g., air and surface water).
(2) Resources that can be recycled and for which there are such vast amounts that availability is simply determined by price (e.g., most metals).
(3) Resources that are finite and not replenished on human timescales. Once used they are gone for thousands to tens of millions of years. "Thousands of years" applies to soils and groundwater. One million years of Earth's coal production or 15 million years of oil production would be necessary to provide enough fuel for three years of our current use. Biodiversity is perhaps the most precious planetary resource, for which the timescale of replenishment, known from past mass extinctions, is tens of millions of years.

RESOURCES WITH SHORT RECYCLING TIMES: AIR AND WATER

Air and water are vast in amount, essential for all of life, and rapidly recycled. The turnover time of O_2 in the atmosphere by life is 5,000 years. Variations in our inputs and outputs are small relative to the size of the reservoir, and we are in no danger of running out of oxygen. Of course, as we burn fossil fuels we use up some of the O_2 in the atmosphere, but increasing CO_2 from 350–700 ppm on a short timescale only changes O_2 by similar amounts, trivial compared to its 210,000 ppm abundance. On more local scales, atmospheric changes are rapid enough that a major cleanup of pollution in an urban environment can have positive results on the scale of years. Air pollution is a tractable problem, given political will, because human management can have rapid results, and the resource continually replenishes itself.

Water is a resource of greater complexity. With the exception of long storage times in glaciers, the recycling time of water is rapid, as evaporation from the oceans to rain to return to the oceans occurs in weeks. On land much of the precipitation evaporates (or is transpired by plants) before seeping into the soil to become groundwater or runoff. The residence time of water in rivers is less than a year.

Even though water is rapidly recycled, it is a vital and limiting resource for human civilization. As population has grown and lifestyle improved, water has been in ever-greater demand. In most of the world, a sizable part of this water is intercepted and used for irrigation of agricultural lands, for processes carried out by industry, and for household use by municipalities. Each of these activities degrades the water's quality, making it largely unusable for other activities. Evaporation from irrigated farmlands enriches the residual water in salt, industries have traditionally disposed of their spent chemicals in wastewater, and sewage is carried away via municipal sewers. While to an ever-greater extent industrial and municipal waters are purified before being released to rivers, their quality is still usually not up to drinking-water standards. This treated water is generally not "reused."

Despite nature's generous supply, in many areas of the world there are shortages. The biggest users are farmers. Of the world's grain, 40% is now grown on irrigated land. Plants transpire many hundreds of molecules of water for each molecule of CO_2 they fix by photosynthesis. A high-technology hectare of farmland produces 100 bushels of grain each year. To do so requires about 300,000 gallons of water. To the extent that is not supplied by rain, it must be supplied by irrigation. Since the 1960s there has been only a 16% increase in arable land, and virtually all of it is irrigated.

Cities developed in arid to semiarid regions are also a significant demand on water. Moving water from areas that have excess to those which have too little has become an integral part of our civilization, and many cities in some countries are built in arid or semiarid regions with insufficient water to support the population. Los Angeles, for example, is a semi-arid region. The city was able to expand in the early twentieth century by diverting water from the Owens valley, causing Owens Lake to dry up and converting fertile farmland to the north into a desert. Subsequent growth caused Los Angeles to take water even farther to the north, including the watershed of Mono Lake, causing Mono Lake to begin to dry up. This desert water body is used as a "fueling" station for a million migrating birds. Each year they fattened up by eating the tiny brine shrimp that teemed in the salty water of the lake. But after the aqueduct was opened in 1941, the lake supply of mountain runoff was largely cut off, and because of this, the lake started to evaporate away. As it did so, its salinity increased, endangering the brine shrimp, and its water level declined, opening a land bridge that threatened the island bird rookeries with plunder by predators. Los Angeles also diverts large amounts of water from the Colorado River and from the Sacramento–San Joaquin delta, where the water removal has threatened the delta smelt. Only 11% of Los Angeles water comes from local groundwater. The rest is imported. Almost two-thirds of the water use in Los Angeles is residential.

Access to surface water can also be problematic for entire nations. When Egypt was ruled by the pharaohs, life depended on the Nile. Fed by the monsoons' rains in the Ethiopian Highlands, its waters irrigated the croplands, which supplied the food for the few million inhabitants. The solutes carried by this water provided the nutrients needed for plant growth, and the mineral matter it carried was material for the manufacture of bricks. As the population grew, ever more land had to be irrigated. To accomplish this, an increasingly complex series of channels were constructed, to deliver the water to the fields and to carry the salt-loaded effluent to the Mediterranean Sea. In order to permit a second crop to be grown during the dry season, reservoirs were constructed to store the plentiful monsoon runoff.

Eventually it was deemed necessary to build the huge Aswan Dam, whose reservoir could store enough water to supply agricultural needs for several years. It also allowed the flow to the lower Nile to be regulated, providing a year-round irrigation supply. So efficiently was water handled that only a few percent was allowed to go unused into the Mediterranean. Further, initially the electricity provided by the dam's generators supplied most of Egypt's needs. All was well.

When the Aswan Dam construction began, Egypt's population was 27 million. By 2010 it had reached 80 million. Even though Egypt's crop yields are among the world's best, only about half the required food can be grown. The rest must be purchased from abroad (paid for by sales of Egyptian petroleum). Also, the nutrients that used to accompany the Nile water are now being consumed by algae in Lake Nasser behind the dam. To have sufficient nutrients for high grain yields, fertilizer has to be manufactured (the energy comes from burning Egyptian petroleum). Further, the demand from electricity now outstrips by twofold that generated at Aswan. The rest has to be generated by burning Egyptian petroleum.

No longer is Egypt able to support its existence solely from the Nile. Egypt's recent growth depends on its petroleum reserves, which are limited. Further, as is the case for all reservoirs behind dams, over the next few hundred years, Lake Nasser will fill with silt, gradually reducing its storage capacity and eventually eroding the tunnels and turbines used to generate electricity. As the amounts of silt are enormous, dredging is not an option. In this example we see how water use, food, petroleum, fertilizer, and population are all related. While water and petroleum are emerging as limiting resources, the primary driver of the problems is the population increase.

Groundwater is another matter. Groundwater is also a replenished resource, but the timescale of deep ground water replenishment is thousands of years. Much of the deep groundwater at northern latitudes was created by moister climate at midlatitudes from the last ice age and has very long replenishment times because replenishment by rain is very slow. While a time of thousands of years is short from a geological perspective, it is essentially infinite for modern water issues. Some agriculture has been maintained by unsustainable groundwater removal from aquifers. Farmers drill into the deep aquifer and "mine" the water for present use. When extraction exceeds replenishment, the water level declines and deeper wells must be drilled, until the aquifer becomes too poor in water or too saline. For aquifers near coastlines, lowering the

Fig. 19-5: Shaded area is the Ogallala Aquifer, which underlies semiarid regions of the western United States. Groundwater withdrawal from the aquifer is what has permitted agriculture to thrive in this region. Large portions of the aquifer are being depleted by extraction rates that exceed replenishment. (Courtesy of U.S. Geological Survey)

water table can also lead to penetration of saline ocean water, making the aquifer unusable. Groundwater can be used indefinitely if it is used at the rate it is replenished. This is feasible in some regions where replenishment rates are rapid. In other areas, however, groundwater formed in wetter conditions thousands of years ago and is not replaced on timescales that permit the agriculture to continue.

Groundwater depletion for irrigation to grow food is an emerging problem throughout the world. The vast Ogallala aquifer in the United States underlies parts of South Dakota, Nebraska, Wyoming, Colorado, Kansas, Oklahoma, New Mexico, and Texas (Fig. 19-5). Extracting this water and using it for irrigation in these semiarid regions has led to a high production of corn, wheat, and soybeans. While the aquifer was originally considered to be "infinite," extraction of water from it has exceeded its replenishment, leading to steady declines in the level of the water table. Groundwater depletion is also occurring in India and China. Large regional scale measurements of groundwater depletion have recently become possible through very precise measurements of changes in Earth's gravitational field from space. Figure 19-6 shows the decline

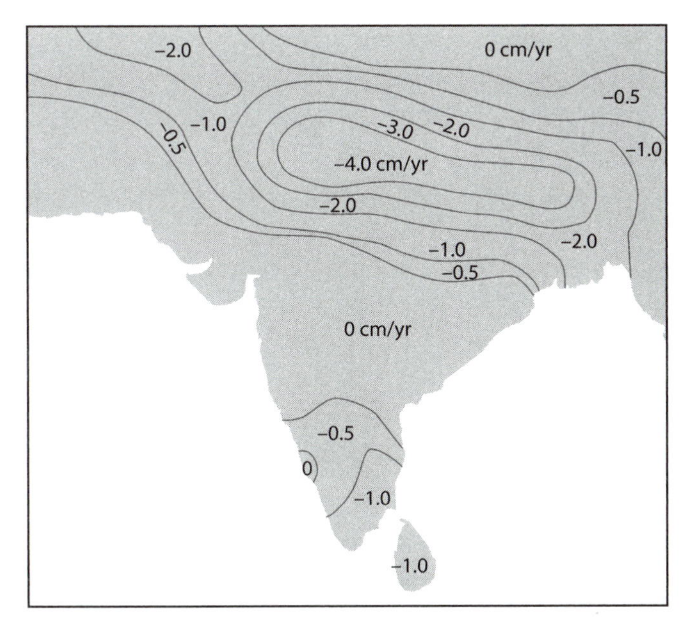

Fig. 19-6: Decline in groundwater in northern India and Bangladesh determined by satellite gravity measurements. Northern India has 600 million people and is heavily irrigated. The loss of some 55 km^3 per year is the largest groundwater loss in the world in comparable regions. (Modified from Tiwari, Wahr, and Swensen, *Geophys. Res. Lett.* 36 (2009), L18401)

in water from the world's most irrigated and populous region in northern India. Groundwater is being depleted at the rate of 50 km^3 per year, equivalent to an annual drop of several cm in the water table.

A clear example of unreplenished groundwater occurred in Saudi Arabia. Saudi Arabia is arid, and the nation has no year-round rivers or lakes. During the glacial periods, the climate was much wetter, leading to well-stocked aquifers at depth, which are not being replenished by current rainfall. In order to become self sufficient in agriculture, Saudi Arabia initiated a large agricultural development program in the 1970s, making use of their groundwater resources. They increased agricultural land by a factor of 20, requiring an equivalent increase in irrigation with groundwater. Water use and agricultural productivity peaked in the early 1990s, after which the ability to extract water began to decline (Fig. 19-7). In 2008 Saudi Arabia announced that it would end wheat production and depend on wheat imports in order to conserve water.

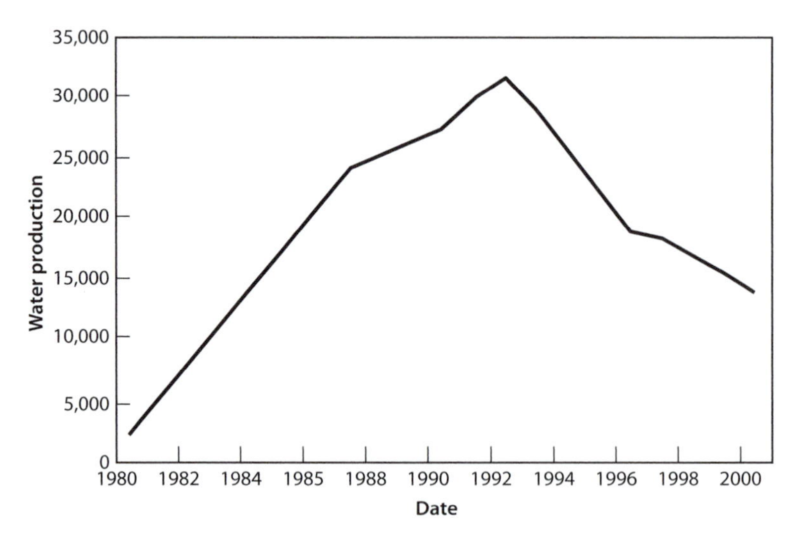

Fig. 19-7: Water production from Saudi Arabian aquifers. Saudi Arabia has no year-round surface water. Growth of agriculture in the 1970s led to unsustainable use and the permanent decline in groundwater. Owing to lack of water, the Saudis ceased wheat production in 2008. Note the similar form of this curve to the oil depletion curves in Fig. 19-10. (Created from data in Abderrahman, Water demand management in Saudi Arabia, in *Water Management in Islam,* IDRC (2001); http://www.idrc.ca/cp/ev-93954-201-1-DO_TOPIC.html)

VAST RESOURCES WITH RECYCLING POTENTIAL: METALS

Most metals are present in every rock to some extent, so the amounts theoretically available are essentially infinite. For example, igneous rocks commonly contain 5–10% FeO and tens to thousands of parts per million of manganese, copper, nickel, zinc, and so on. An "ore deposit" is defined by the grade of ore that is necessary for economic extraction. As the price rises, rocks that were not ore become ore, and additional supply becomes available because lower grades of the resource can be economically extracted. Furthermore, if the price rises enough, it becomes economic to recycle the metals, which is straightforward because most metals are not destroyed by use. Mining and recycling also compete in the marketplace, so consumption can go up even if mining does not. For example, about one-half of the copper used in the United States each year is recycled copper, and 88% of steel is recycled after use. These factors permit steady growth in production that has been sustainable

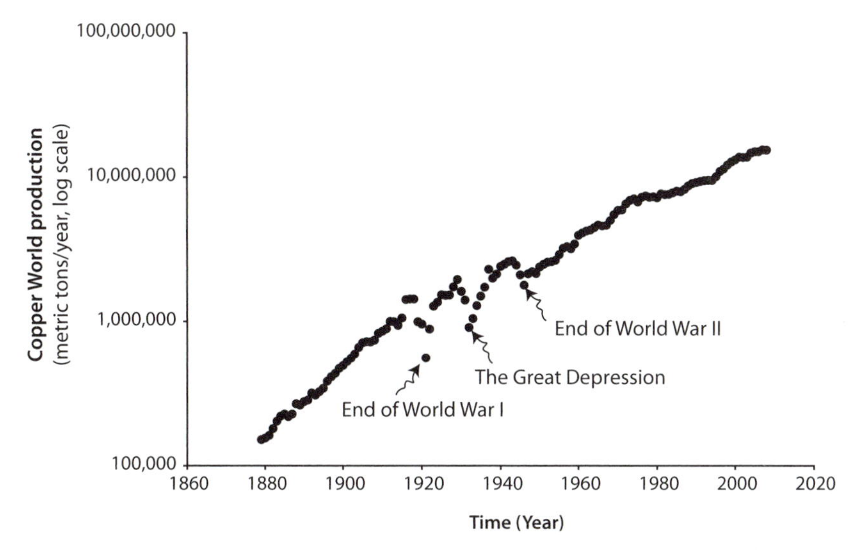

Fig. 19-8: Steady growth in world copper production, marked by declines after WWI, WWII, and the depression of the 1930s. Note the log scale shows that copper growth has been able to increase exponentially for more than one hundred years. (Data from U.S. Geological Survey)

over the past century (Fig. 19-8). While there are economic consequences of mining, they are largely limited to the local region, and modern mines are subject to environmental regulations that cause much of the environmental cost to be incorporated into the price. The infinite supply, ability to recycle, and an economic model that includes much of the environmental impact leads to use that is self-regulating.

Phosphorous is more problematic because it is an irreplaceable component of fertilizer and is used in vast amounts (more than 150 million tons in 2008) to enable the harvests that feed the world. Because fertilizer must be low in cost, only very rich deposits of phosphorous can be mined economically. These deposits were emplaced during a few limited periods of Earth's history, with one peak of deposition near the Cambrian–Precambrian boundary and an even larger peak at the end of the Mesozoic (Fig. 19-9). Growth of population and deterioration of soils increases the need for fertilizer; estimates are that in the next sixty-five years half of Earth's phosphorous reserves will have been depleted. While some phosphorous can be recycled, it is far more difficult than

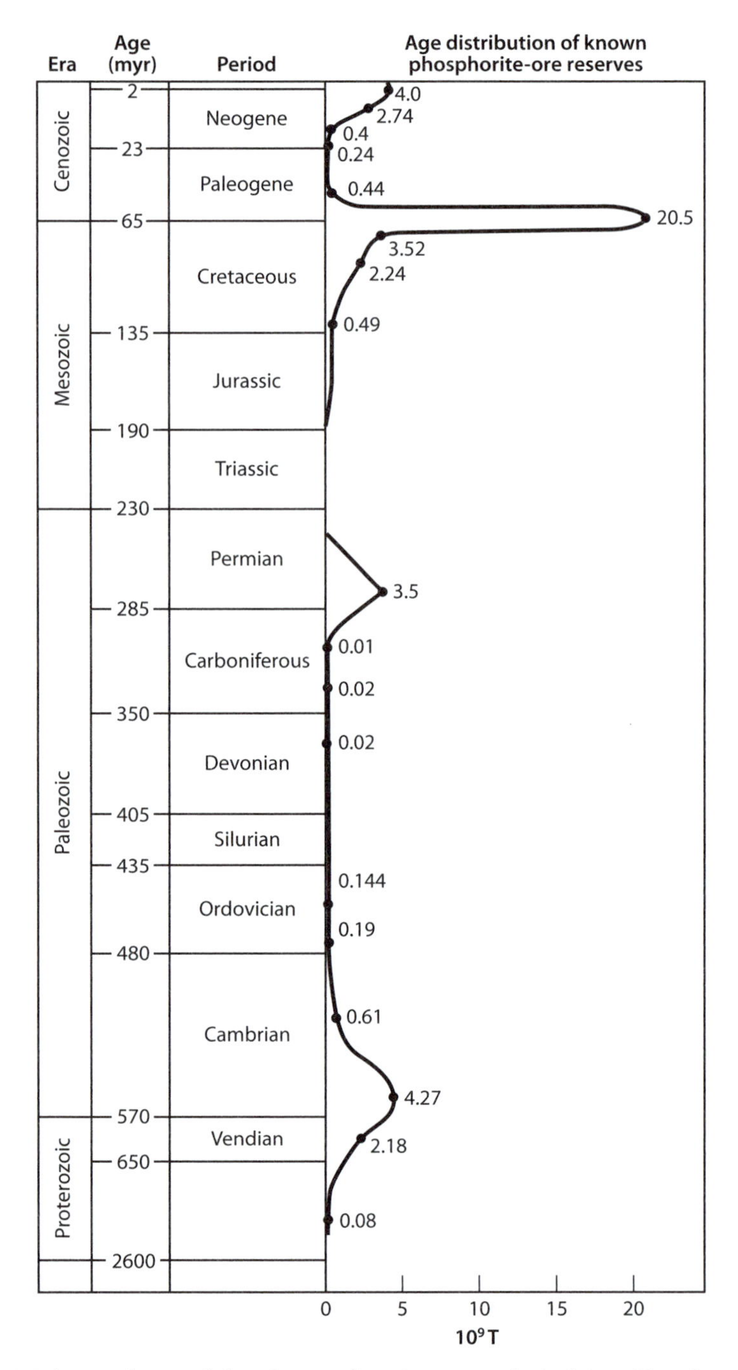

Fig. 19-9: Accumulation of phosphorous deposits over geological time. Note the discrete intervals of time where most deposits were able to accumulate. (Modified after Yanshin and Zharkov, *Intl. Geol. Rev.* (1986))

for metals, because phosphorous is water soluble and is carried by run-off into rivers, lakes, and oceans. Concentrations in these vast bodies of water are then too diluted to be recycled. Metals are concentrated by mining and use and are solid. Phosphorous is deconcentrated by mining and use and is carried away by water.

Phosphorous also has deleterious environmental consequences. Because it is a limiting nutrient for life, phosphorous addition to bodies of water causes algal growth, and when the algae die the oxidation of the organic matter uses up all the oxygen in the water, making further life impossible. This creates the eutrophication of lakes and the dead zones in parts of the ocean, e.g., in the Gulf of Mexico near the outflow from the Mississippi River. Poultry manure also contains large amounts of phosphorous, and this phosphorous has led to the gradual eutrophication of Chesapeake Bay. With current practices, most valuable phosphorous is lost and becomes environmentally destructive. There are methods to recover the phosphorous from wastewater, however, so the potential for recycling a portion of phosphorous exists.

Finite Resources with No Recycling

In contrast to our first two classes of resources, there are other resources that have a limited supply, and once used are gone "forever" in terms of human timescales. These are the *nonrenewable resources*. The most obvious of these are the fossil fuels, but soils and biodiversity also fall within this class.

FOSSIL FUELS

Metals are simply elements, and the whole Earth is made up of them. In contrast, fossil fuels are complex organic molecules largely formed by life only at the surface. Their utility resides not in the elements themselves, but in the energy stored in the molecular bonds. Once this energy is released, it is gone forever, replaced only on timescales of millions of years by further photosynthesis and storage of organic carbon. Fossil fuels are also limited by the amount of organic matter produced over Earth's history and the need to modify and concentrate it for it to be useful as fuel. Unlike metals, the amount is limited. Once used, the resource is eliminated. There is no recycling.

Fossil fuel resources also have a usage profile very different from metal resources. Metal resources for an individual mine often increase as price increases, because more and more metal becomes profitable to extract. The ore is solid rock that is recovered. In contrast, oil fields are pumped, and at some point the pumps go dry. Gas and oil are much less subject to reserve expansion than metal resources. M. King Hubbert noted that each oil field had a finite lifetime with a characteristic life cycle. When a new field is opened and the resource is abundant, extraction goes through a period of growth. Then the field tops out and goes into a steady decline. This characteristic has been verified repeatedly for individual fields, for North American oil production, for the North Sea, and so on (Fig. 19-10). Discoveries of new fields have also been decreasing, so that new supplies of oil available for their growth phase are diminishing. These facts lead to the concept of *peak oil*, where global oil production will follow the same trend as U.S. or North Sea production. Given the vast resources expended on oil discovery and the decreasing rate of discovery, oil production is likely to go into a steady decline sometime in the early twenty-first century.

For coal there is generally no fear of imminent shortage on the timescale of a human life because there are likely several hundred years' supply available. This is also the case for lower-grade or less accessible resources such as oil shales, tar sands, and the abundant gas clathrates on the continental shelves. While several hundred years is long for an election cycle, it is very short on the time frame of human civilization. Figure 19-11 shows the profile of fossil fuel usage on two planetary timescales, one associated with the Phanerozoic era where fossil fuels are gradually accumulated and the other with the 10,000-year timescale of human civilization. We live in the *fossil fuel age*—the very brief interval in Earth's history when 500 million years of resources were consumed and destroyed by a single species. We are using up these resources at a rate a million times greater than Earth produced them. Our ancestors are likely to look with wonder at our profligate waste of Earth's storehouse of organic molecules, because they can be put to so many uses—plastics, artificial joints, lubricants, etc. What were we thinking, simply burning such treasure?

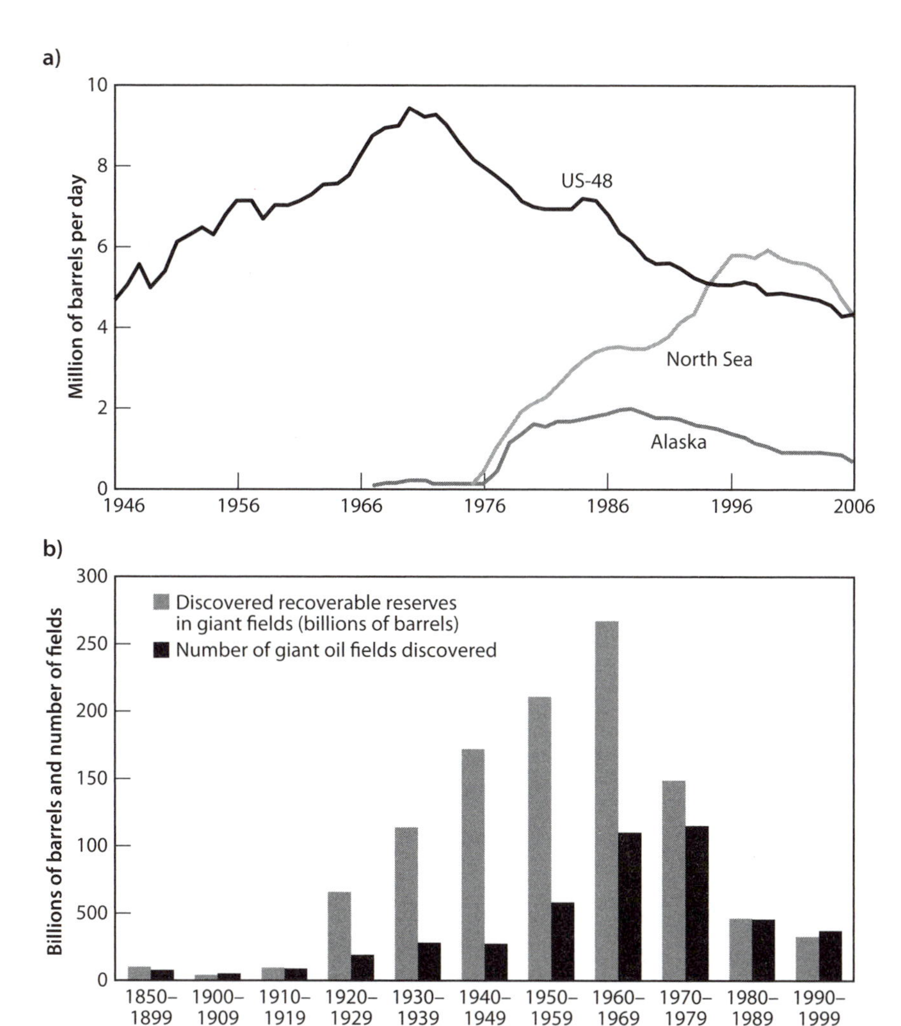

Fig. 19-10: (a) The characteristic life cycle of oil fields identified by M. King Hubbert. Initially the field undergoes rapid growth in production as it becomes fully exploited. The production then peaks and goes into a steady decline. This observation is true for individual fields such as the Alaska North Slope, for larger regions such as the North Sea, and for entire nations such as the United States. This is the concept behind "peak oil" (U.S. Energy Information Agency). (b) The bottom panel shows the discovery rate for new giant oil fields that provide the bulk of the world's oil. Discoveries have been declining steadily since 1970. Each of these fields will ultimately decline in production similar to the more mature discoveries shown in the top panel. Indeed, global oil production has flattened between 2007and 2010, one of the reasons for the oil price spikes in 2008 (American Association of Petroleum Geologists, Uppsala Hydrocarbon Depletion Study Group).

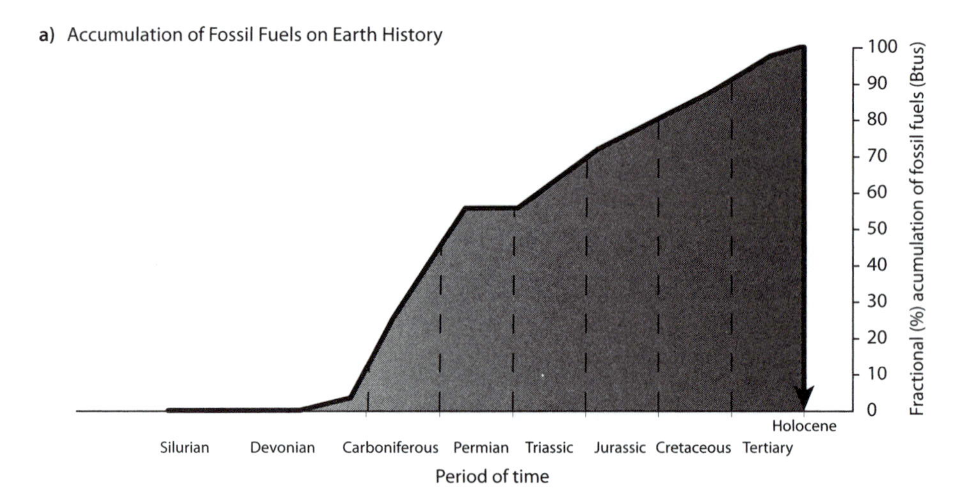

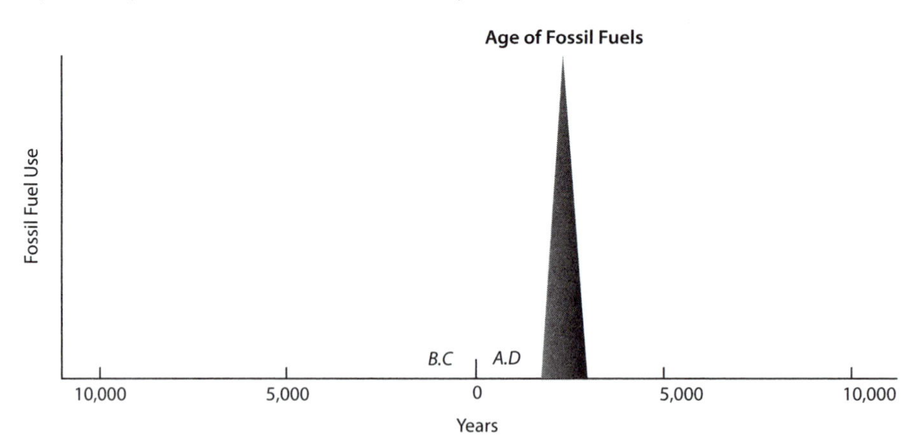

Fig. 19-11: Time scales of fossil fuel formation (a) and depletion (b). Five hundred million years of accumulation are being depleted in a few centuries in what will come to be known as the fossil fuel age. ((a) data from Pimentel and Patzek, *Rev. Environ. Contam. Toxicol.* 189 (2007):25–41)

SOILS

Soils are also a nonrenewable resource on human timescales. Soils form by the slow weathering of rock and the modification of this weathered product by life to create a markedly complex soil ecosystem. Once land is denuded of soil, it can no longer support significant life. Left to itself, Earth builds up soil resources, and humans have made use of this treasure to grow crops for food. Soils naturally do not erode significantly because there is continuous agricultural ground cover, and deep soils are not turned over to reach the surface. Agriculture removes this ground cover and exposes soil to rain and wind, and turns over deep soils, reducing their cohesion and exposing them to the surface. Soil loss in the United States is about ten times the replenishment rate, annual loss is about ten tons per acre, and half of midwestern topsoil has been lost since the 1800s. Some estimates place global topsoil loss at about 1% per year. Loss of soil leads to increased fertilization to maintain productivity, but the ammonia in fertilizer is a fossil fuel product and phosphorous is a limited resource. In some regions, it is less expensive to abandon denuded fields and clear new forest rather than fertilize. The pressure of population, limited new habitat, and modern food needs is leading to progressive deterioration of global soils. This is a problem that is well distributed across both developed and undeveloped nations. The agricultural heartlands of the United States, Europe, and China, for example, are all experiencing topsoil loss that is influencing agricultural potential.

BIODIVERSITY

The ultimate nonrenewable resource is biodiversity. All of our food comes from Earth's total genetic library. Many modern medicines and industrial processes also depend on the genes of organisms, from bacteria to mammals. Ecosystem stability depends on the diversity of life that maintains it, and ultimately the habitability of Earth depends on the viability of ecosystems. Earth's response to change and catastrophe also depends on the evolutionary potential of life, and this potential scales with the overall genetic diversity. Catastrophes of the past show us that biodiversity recovers only on timescales of tens of millions of years, and some ancient innovations may never be recovered. Destruction of biodiversity, discussed in greater length in the next chapter, is the destruction of billions of years of evolutionary potential.

Summary

Human beings appeared and became the dominant species on the planet in a remarkably short period of time. This dominance is reflected in the massive population growth from perhaps only 10,000 people about 70,000 years ago to a population soon to approach 10 billion—an increase of a million times. Human beings dominate every ecosystem, sit at the top of every food web, and claim ownership of every piece of habitable land.

This planetary domination was made possible by the human energy revolution, where one species was able to access energy far in excess of that available to any other species. This revolution in turn was possible only because humans found a planet fully stocked with easily accessible energy, produced by billions of years of planetary evolution. The access to energy also permitted exploitation of all the other planetary resources—water, metals, land with its fertile soils, and a rich, abundant, and diverse biosphere. First destroying other large carnivores that competed for food and land, human beings then proceeded to destroy and adapt habitat for their use, leading to massive extinctions of other species. All resources were treated as equivalent—freely provided by the planet and there for the taking without payment or consequence.

Resources are of different types, however. Some commodities are unlimited in potential supply, can be recycled, and have local environmental impacts that can be rectified. Others, such as phosphorous, are more limited in supply and are much more difficult to recycle. Environmental impacts are often distant from the source, leading to eutrophication of downstream rivers, lakes, and marginal seas. Fossil fuels, soils, and biodiversity are limited, irreplaceable, and once destroyed or lost are gone "forever" with respect to pertinent human timescales. Environmental impacts of fossil fuels and biodiversity are global in scale, such that one nation's behavior can impact another's halfway around the globe.

The differences in types of resources are not generally recognized by the modern marketplace. Our attitude toward use of Earth's bounty encompasses time frames of only a few years and is based on price of immediate extraction and maximizing short-term profit—not on the total amount that is available, nor the potential for recycling, nor environmental impacts. This leads to inevitable rapid extraction and use of irreplaceable treasure. In a few brief centuries, human civilization will have used up billions of years of accumulation of planetary resources. We arrived on the scene to find an eminently habitable world. Our actions in the last two centuries have made the world far more habitable for our own species, allowing us to live in comfort even in arid or frigid regions and permitting vast population growth and concentration in urban centers. At the same time, we have made Earth far less habitable for most of the millions of other species with whom we share the planet. Ultimately, it remains a pressing question whether we are also making it less habitable for our own descendants.

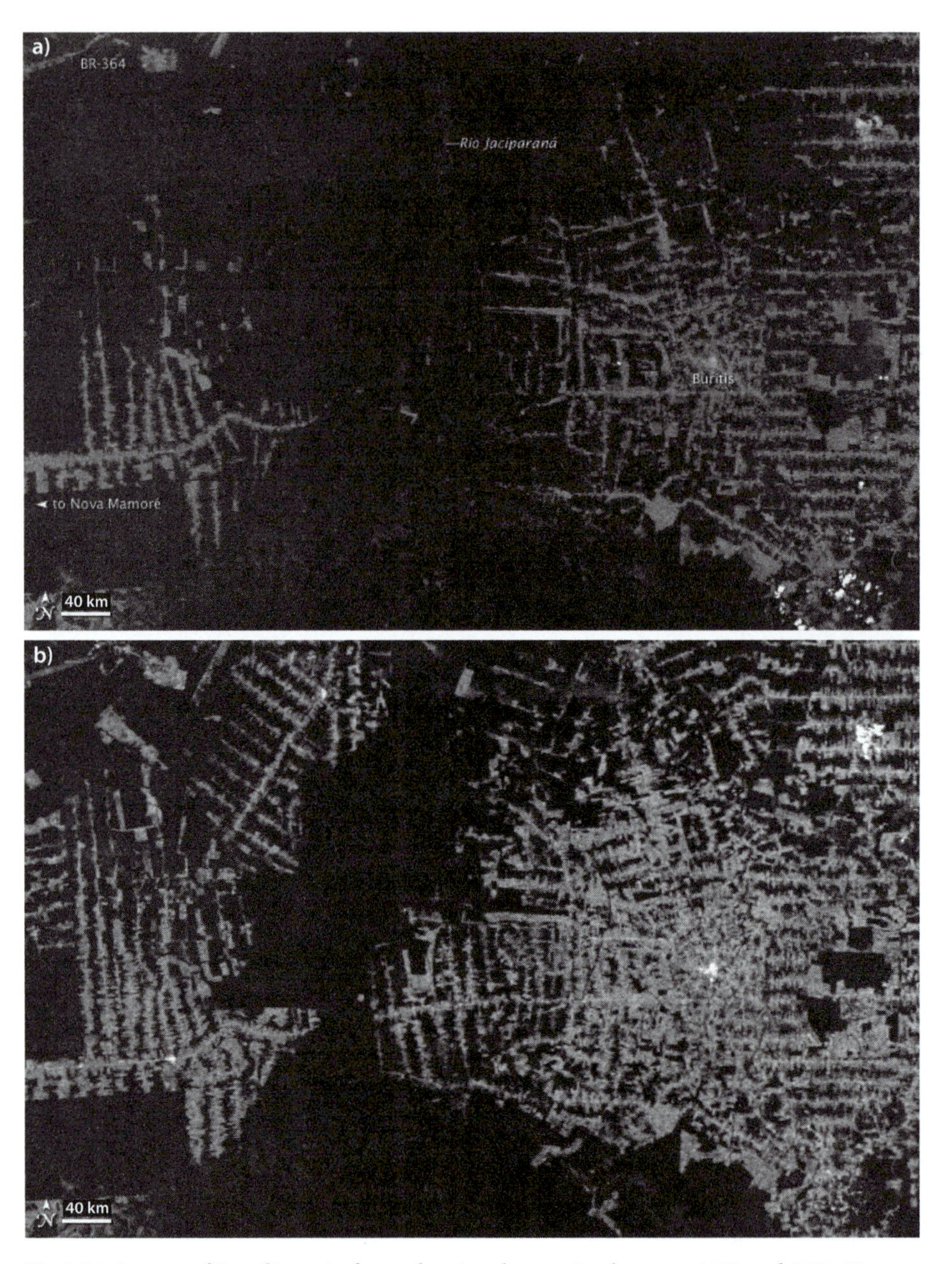

Fig. 20-0: Images of Brazilian rain forest showing destruction between 2000 and 2009. Note the 40 km scale bar in the lower left-hand corner. The area shown is about the size of the state of Wyoming or the country of Poland. (Images from NASA Earth Observatory (earth observatory.nasa.gov/Features/WorldofChange/deforestation.php))

CHAPTER 20

Mankind at the Helm

Human Civilization in a Planetary Context

The rise of human civilization is a transformative event in planetary history. For the first time a single species dominates the entire surface, sits at the top of all terrestrial and oceanic food chains, and has taken over much of the biosphere for its own purposes. We are also influencing the physical environment by changing the composition of the atmosphere and ocean, modifying the water cycle, eliminating soils, and constructing huge communities on a scale never before seen. To an external observer there is also a change in the capabilities of the planet as a whole. Human beings permit access to planet scale sensing from space, land, and sea; we are capable of conscious direction of the planetary systems; and we could communicate with other planetary civilizations, should they exist. We have also vastly increased the rate of and capacity for planetary change through our access to energy, global communication, technology, and our developing ability to direct evolution through modification of DNA. Such changes are monumental in the context of all of Earth's history, and they certainly warrant the name of a new geological epoch—as per the suggestion of Paul Crutzen's "Anthropocene." The vastness of planetary change and capability, however, is more equivalent to the great events of the past that mark boundaries between geological eras or even eons, such as the origin of life, the rise of oxygen, the origin of multicellular life, or the Permo-Triassic extinction. We have entered a potential Anthropozoic era, an era of planetary consciousness and directed evolution where a single species controls the fate of the planet. Eras of the past have lasted hundreds of millions of years—will the Anthropozoic have such longevity? Can we usher in a new era with wisdom and conscience, or will we be a failed and abortive attempt at further planetary evolution?

One practical challenge is to manage planetary resources so that we do not run out of essential materials—and food. Even more important, human activities are now so vast that they impact planetary systems. Energy use has increased greenhouse gases in the atmosphere by 40% for CO_2 *and 100% for* CH_4*, greatly exceeding natural variations, and causing the warming of the atmosphere, the melting of the Arctic ice cap, and an inexorable rise in sea level. Part of the anthropogenic* CO_2 *is absorbed by the oceans, making the oceans more acidic. The effects on the biosphere are even more marked. At sea, the more acid ocean contributes to the decline of coral reefs and impacts the entire oceanic ecosystem. On land, 25% of terrestrial plant and animal productivity now relates to human food production. This has caused massive destruction of habitat for species other than ourselves. Rain forests, the greatest remaining reservoir of biodiversity, are being cut down at the rate of 40,000 km*^{2} *per year, an area larger than the state of Massachusetts (see frontispiece). The diversity of species that can exist depends on contiguous land area; as wild areas diminish, extinctions inevitably follow. Human population has thus far been the driver for loss of biodiversity. Climate change would lead to even greater losses. Loss of biodiversity is a reduction in the genetic potential of the planet and its ability to respond to planetary change.*

We have as yet shown little willingness to take responsibility for our effects on the planet or respond to planetary change. In our current economic model, both energy use and population growth cause negative environmental impact, and few are willing to limit either. The planet is free of charge—environmental costs are rarely meaningfully included in economic models, and including them would require societal readjustments, which are inevitably resisted. Solutions are largely a matter of personal and political choice, because effective steps are possible at a small fraction of what is spent on being able to destroy each other's existence with weapons. While a dependence on fossil fuels is inevitable for coming decades, removal of CO_2 *from the atmo-*

*sphere (*carbon capture and sequestration*) is an emerging capability that could be applied on a global scale. If it were, it could buy enough time for solar, wind, and nuclear power to grow to fill our energy needs. Ultimately fusion power is likely to become economic, providing a source of energy that is inexhaustible from our current perspective. This alone would not avoid planetary crisis, however. A revaluation of the planet that supports us will be necessary to avoid further planetary deterioration. Humanity is the result of billions of years of planetary evolution, and it has flourished from the bounty provided by Earth. Are we entitled, or appreciative? Only a revolution in human attitudes from "planetary user" to "planetary conservator" would permit effective planetary management.*

Introduction

The discussion in the previous chapter presents Earth's resources as if Earth were a gift box containing treasure that needed to be managed. Instead, Earth is a dynamic system whose health is determined by a myriad of biogeochemical cycles involving atmosphere, hydrosphere, biosphere, and the solid Earth. If our use of resources did not influence this system then the problems we face would be a matter of intelligent resource management. But if we are a planetary force on a global scale that changes these cycles, then we need also to consider the overall health of the system and how that is influenced by human activities. The evidence summarized below shows that we have rapidly become a dominant influence on the planetary surface, unlike any other single species in the history of Earth. We have changed climate, modified the chemistry of the oceans, taken control of much of the biosphere, and are extinguishing other species at rates equivalent to a mass extinction.

Because of our influence, we live in a time of profound and rapid planetary change. Human actions have modified climate and the oceans and may end in global catastrophe, not only for other species but also for our own. At the same time, human civilization provides radical new capabilities that could have a beneficent planetary influence, if humans became able to view themselves and act as an integral and responsible part of a planetary system. As human beings we face choices that will forever affect our planet's future. Will we choose planetary decline or planetary husbandry? We cannot avoid these choices, because our impacts are drastic and global in scale. For better or worse, we are now at the helm—the fate of the planet is in our hands.

Human Impacts on the Earth

Human activity influences all the reservoirs of Earth's surface: atmosphere, ocean, soils, and biosphere.

CLIMATE

We learned in Chapter 13 that the stability of Earth's climate has been exquisitely tuned by the greenhouse effect to maintain Earth's surface temperature in a narrow range, despite the increase in solar luminosity over Earth's history. This range has encompassed relatively warm periods, with no glaciers at the poles, which last existed more than 30 million years ago, as well as colder periods when Milankovitch cycles caused oscillations between ice ages and interglacials, as well as more permanent glaciations. While the range in states has been large, changes have been gradual relative to life spans of organisms. The change from a greenhouse state with no ice ages to an icehouse state with periodic ice ages took several million years. Even the "rapid" change from glacial to interglacial is a 10,000-year process, 50 to 100 centuries, encompassing 200 generations even of long-lived species such as humans. Life is forced to adapt to these changes, but the short lifetimes of all species permit gradual movement and modification so that ecosystems have the time to migrate and adjust over thousands to millions of years.

Are human-induced changes significant in this planetary context? To address this question we can examine the detailed record of climate change over the last 700,000 years, preserved in the contents of CO_2 in the air trapped in bubbles in polar ice. CO_2 has oscillated between minimum values near 190 parts per million at glacial maxima to values of about 270 ppm during the brief interglacials, as temperature also varies (Fig. 20-1). CO_2 varies in lockstep with climate change. Prior to the in-

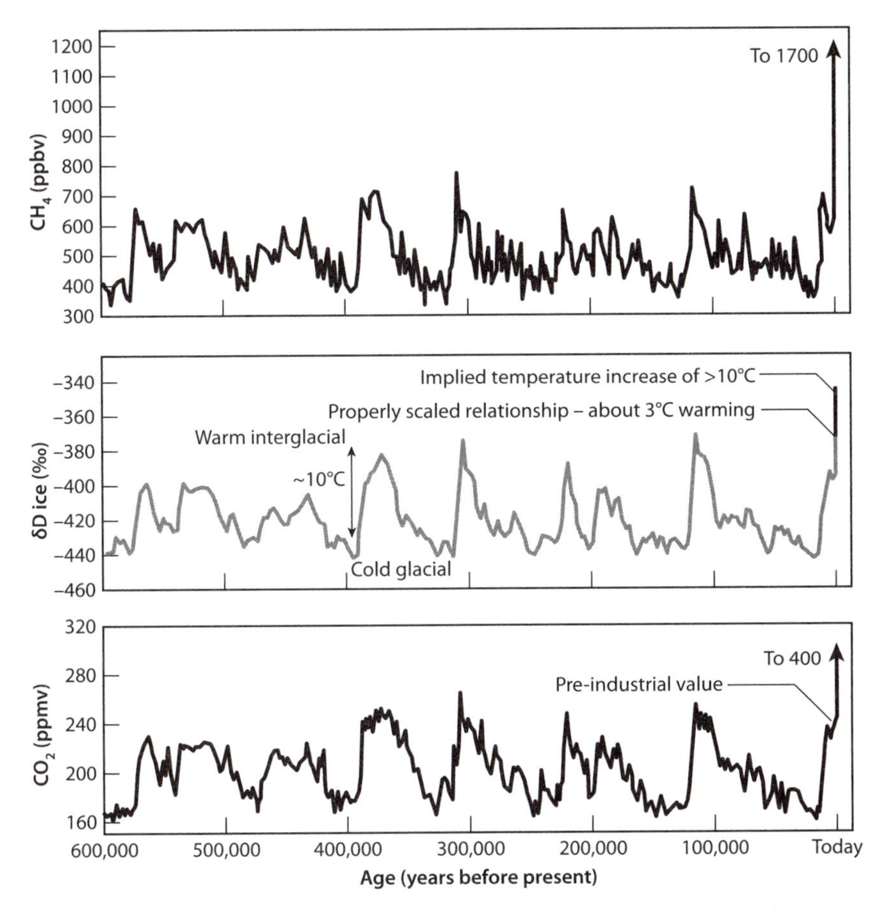

Fig. 20-1: Methane, temperature, and CO_2 from the Dome C ice core showing that CO_2, temperature, and methane have varied in lockstep over the last 700,000 years. During the brief interglacials, all parameters are high, and during glacials all are low. (European Project for Ice Coring in Antarctica (EPICA), project members, 2006)

dustrial revolution, the atmosphere's CO_2 content was consistent with interglacial values at a steady 280 ppm. As industrial emissions began, CO_2 began to increase. Over the last 50 years for which a very accurate record is available, the atmosphere's CO_2 content has risen from 315 to more than 390 ppm, an average rate of 1.2 ppm per year. More recent CO_2 emissions are so large that the rate has increased to 2 ppm per year. The current atmospheric CO_2 content greatly exceeds any values over the last 700,000-year ice-core record and likely exceeds any values for the last several million years. These are large changes on a global scale. But wait—atmospheric CO_2 in the Cretaceous period was likely close to 2,000 ppm, higher even than the most dire predictions for the next century. Of course, at this time Earth was in a "hothouse" state, with no ice sheets and shallow seas that covered much of the continents. But it is natural change, so the total range of CO_2 variations over tens of millions of years is large. How do we know that the recent CO_2 rise is due to human impact, and not simply natural variations?

There are two independent approaches that confirm the human cause of the atmospheric change. The first is simply to add up the amount of CO_2 we have pumped into the atmosphere. The number is so vast that CO_2 in the atmosphere has actually increased less than would be expected from the amount of carbon we have burned. Instead, about 45% of the emitted CO_2 must have been absorbed by the ocean and biosphere, and only the remaining 55% has accumulated in the atmosphere. A second approach turns to the sources of natural variations in atmospheric CO_2. For the CO_2 budget, the elephants in the room are the ocean and solid Earth. The ocean contains fifty times as much CO_2 as the atmosphere, so small variations in the ocean budget could be important. The solid Earth contains even vaster quantities of CO_2, so couldn't solid Earth emissions lead to the rise?

The latter question is addressed by measurements of volcanic emissions of CO_2. Earth's volcanic CO_2 emissions are 0.2 gigatons (Gt) per year. The passage from glacial to interglacial is associated with a temporary increase to as much as 0.5 Gt per year. In comparison, human emissions in 2008 were 30.0 Gt, 150 times the natural background. Solid Earth emissions are trivial in comparison to recent human emissions and produce change only on long timescales.

The ocean as a source of increased CO_2 is also ruled out. First, the ocean must have been a net absorber of CO_2 to account for the fate of all the human emissions. This conclusion can also be tested by measurements of atmospheric O_2 concentrations and the change in carbon isotopes.

Carbon released from the oceans is already in molecules of CO_2, and release of CO_2 from the oceans has no effect on atmospheric oxygen. When we burn carbon, however, the carbon combines with atmospheric oxygen, reducing atmospheric concentrations. If fossil fuels are the cause of increasing CO_2 in the atmosphere, the atmospheric O_2 concentrations should *decrease* in lockstep with CO_2. One difficulty is to make sufficiently precise measurements of O_2, which would vary by only a few parts per million (2 ppm of annual CO_2 rise would only decrease O_2 from 22.9% to 22.8998%). Charles David Keeling mastered such measurements to provide a record of variations in O_2, seen in Figure 20-2. The steady decline of atmospheric O_2 corresponds with carbon burning.

Carbon isotopes are another independent test. Organic carbon in fossil fuels is isotopically light. Burning adds this light carbon to the atmosphere. Indeed, the carbon isotope composition of atmospheric CO_2 has been steadily decreasing, showing the massive input of burned organic carbon (Fig. 20-2). It is a simple fact that atmospheric CO_2 is increasing because of human emissions. It merits a 10 on our theory scale.

One of the intriguing aspects of the glacial record is that coming out of glacial periods, temperatures begin to rise before CO_2 does. Doesn't this prove that CO_2 does not cause warming? In the nonscientific community, this is often interpreted to mean that "atmospheric temperature causes CO_2 to change, and therefore CO_2 is a consequence and not a cause of warming." This misunderstanding results because CO_2 is a positive feedback on warming during deglaciations, whereas it is driving warming today. The increased solar luminosity in the Northern Hemisphere driven by Earth's orbital variations causes a slight temperature rise. This rise causes some CO_2 to be released from the oceans, which causes further warming, which causes CO_2 to rise more—a positive feedback. Melting of the ice sheets then causes increased volcanism that releases further CO_2, augmenting warming. Deglacial warming is the result of multiple feedbacks, with CO_2 playing a pivotal role. The warming effect of greenhouse gases is a fundamental fact of physics, not a matter of "belief" or a political issue.

Another important greenhouse gas is methane. Like CO_2, methane has oscillated regularly with ice age cycles, from 350–400 parts per billion during glacial maxima, spiking to brief peaks near 650 ppb during interglacials, followed by a rapid decline to intermediate values in the subsequent few thousand years (Fig. 20-3). In the last 150 years methane concentrations in the atmosphere have more than doubled, to 1,750 ppb. Human sources of methane to the atmosphere are dominantly from livestock, landfills, and natural gas exploitation, and these exceed natural sources, leading to the atmospheric rise. While the ~1 ppm rise in methane appears small, the effect on atmospheric warming is substantial

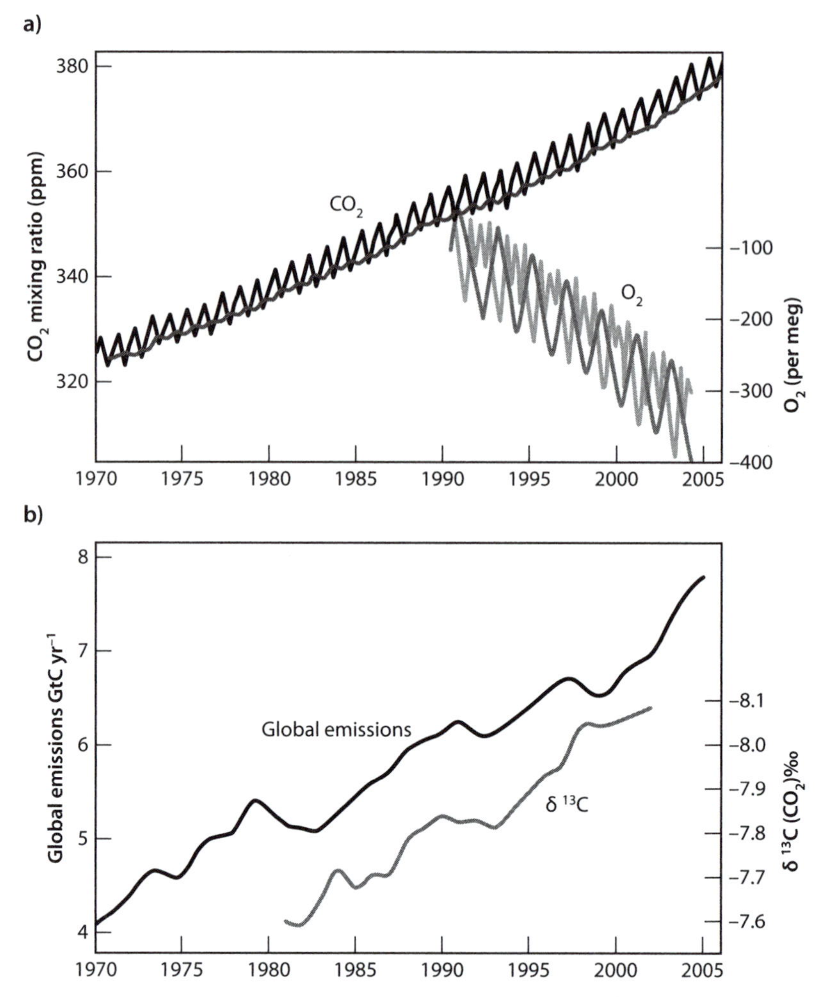

Fig. 20-2: Proof that the increase in CO_2 in the atmosphere is due to human emissions from burning organic carbon. (a) Burning carbon uses up oxygen to produce CO_2, so carbon burning should lead to a decline in O_2. The decline in O_2 is actually greater than would be expected from the atmospheric increase in CO_2, showing that much of the CO_2 from burning is taken up by the oceans and biosphere. This observation is also consistent with simple mass balance constraints—almost twice as much CO_2 has been produced by fossil fuel burning as has accumulated in the atmosphere. The two O_2 curves are data from two different locations. (b) Organic carbon in fossil fuels is preferentially depleted in the heavy carbon isotope ^{12}C leading to low values of $\delta^{13}C$. Adding this isotopically light carbon to the atmosphere decreases the δ^{13} of atmospheric CO_2, as is observed. Note that the y-axis scale is inverted. (©2007 IPCC AR-4 WG1, chap. 2, fig. 2.3)

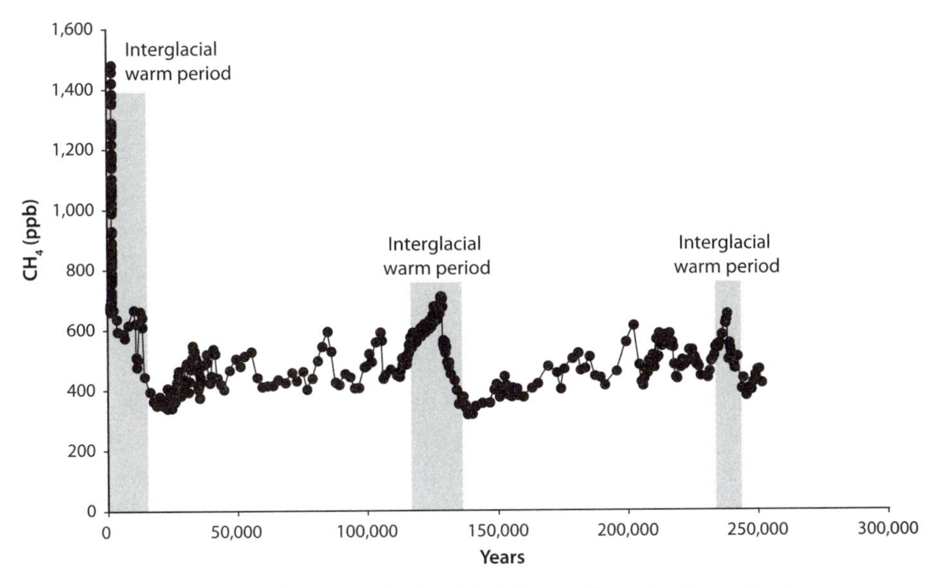

Fig. 20-3: Variations in methane over the last 250,000 years from the Antarctica ice cores, and the near tripling of CH_4 during the last century. (Data from Loulergue et al., *Nature* 453 (2008):383–86, and Etheridge et al., *J. Geophys. Res.* 103 (1998):15,979–93)

because methane is twenty times more powerful as a greenhouse gas than CO_2.

Other greenhouse gases are also of emerging importance. NO_2 has increased markedly and has a significant warming influence. Chlorofluorocarbons (CFCs), the chemical used for refrigerants and other industrial applications in the twentieth century, are also greenhouse gases. CFCs also lead to destruction of Earth's protective ozone shield. The effects of CFCs on ozone were established scientifically in the 1970s, but international action took place only after the discovery of the "ozone hole" over Antarctica in 1985, leading to a 1987 international agreement (the Montreal Protocol) that called for an elimination of CFC emissions. The worldwide agreement, the clear implications for human health, and the expiration of patents on CFCs that reduced their profitability, led to support from the chemical industry, and CFC emissions have been largely eliminated. The cessation has led to a decline in ozone-depleting gases of 10% from their peak in 1995. The ozone hole has not yet recovered, because of the residence time of CFCs in the atmosphere, and is not expected to recover until late in the twenty-first century. The replacement gases for CFCs (notably hydrofluorcarbons, or HFCs) do not destroy the ozone but are nonetheless potent greenhouse gases. Their use is increasing rapidly, particularly in developing countries, and projections suggest they may have an atmospheric influence equivalent to 100 ppm CO_2 by midcentury.

To put these human-induced changes in context, we can recall that an atmospheric greenhouse has been essential for Earth's climate stability over billions of years and that absent a greenhouse, Earth would freeze. Stability has been maintained over the last several million years by 180–280 ppm CO_2 and <650 ppb methane in the atmosphere. On the thousands-of-years timescale, CO_2 and methane have been important parts of the climate oscillations that have taken us in and out of ice ages. We also know that when Earth has been in a "hothouse" state when CO_2 concentrations in the atmosphere were elevated, and sea level was much higher because there were no ice sheets. Earth's history shows us that greenhouse gases matter. They permit and maintain planetary habitability and have participated in changes from ice age to interglacial and icehouse to hothouse. From this perspective, are human impacts significant? We are now in the process of certainly doubling, and potentially quadrupling, the CO_2 content of the atmosphere. We have already more than doubled the methane contents and are adding other potent greenhouse gases for which the atmospheric concentration has been zero since Earth's formation. Human impacts are of larger magnitude than the changes between ice ages and interglacials, and they are approaching the magnitude of hothouse vs. icehouse.

It is not only the magnitude of the atmospheric changes, it is also the rate of change, which is perhaps more important for ecosystem effects. Earth's reservoirs are large and adjust slowly to change. Higher CO_2 in the atmosphere will be absorbed by the oceans over a thousand years or so. Over tens of thousands of years higher CO_2 leads to enhanced weathering, which reduces CO_2. Organisms adapt over thousands of years and evolve over millions of years to changing environments. Earth systems are adapted to slow change. Change over decades is exceedingly rapid, and does not permit these gradual accommodations. Recent changes in CO_2 are 1.8 ppm per year. For comparison, during the glacial terminations CO_2 changes by 100.0 ppm in 5,000 years, or 0.02 ppm per year. Recent changes are ninety times faster! In both magnitude of change

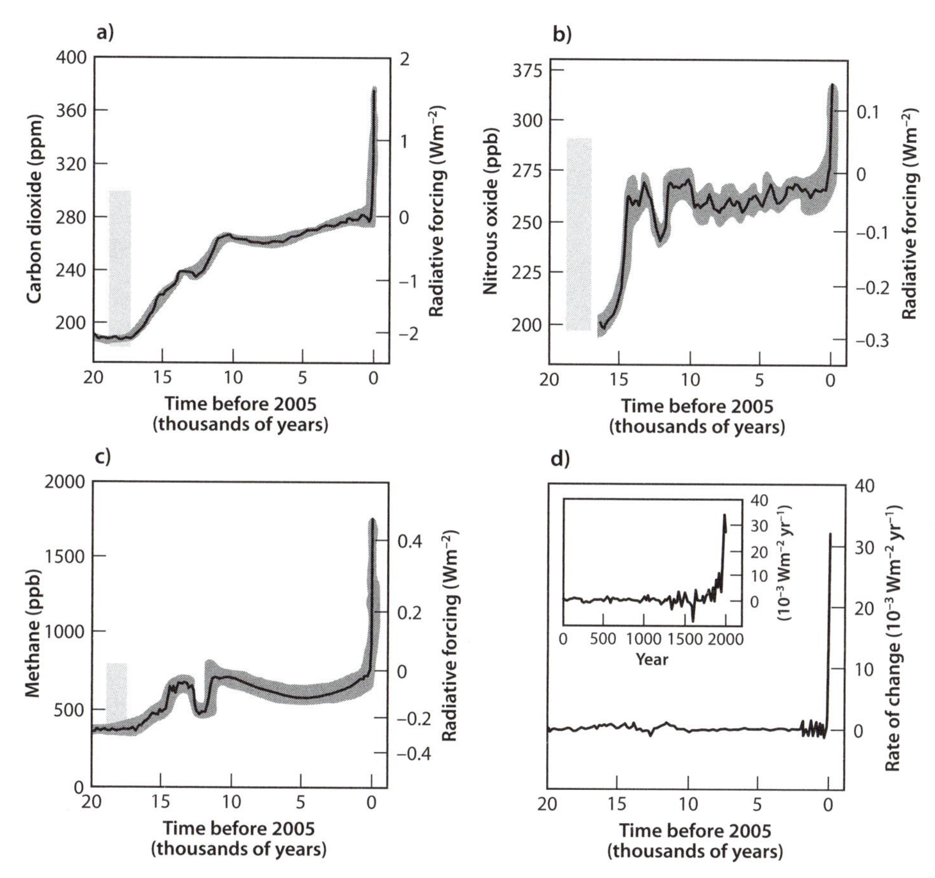

Fig. 20-4: The radiative forcing of the three most important greenhouse gases compared to the changes from the last glacial maximum 20,000 years ago through the present. Light gray bars show the range of natural variations over the last 800,000 years. Significant but gradual changes occurred during the deglaciation, followed by stable values during most of the Holocene. Abrupt changes occur in the Anthropocene. Human-induced forcing of climate of more than 2 watts/m^2 is of the same order as the change from last glacial maximum to the interglacial, but it is occurring over 100 years rather than 10,000 years. Further increase of CO_2 to 500–1,000 ppm will lead to much greater forcing. (©2007 IPCC AR-4 WG1, chap. 6, fig. 6.4)

and rate of change, humans are profoundly influencing the planetary atmosphere and Earth's thermostat, the climate system.

How large will these changes be? From fundamental physics the warming effect on the atmosphere can be rigorously calculated, as shown in Figure 20-4. Human beings are causing the atmosphere to exert a

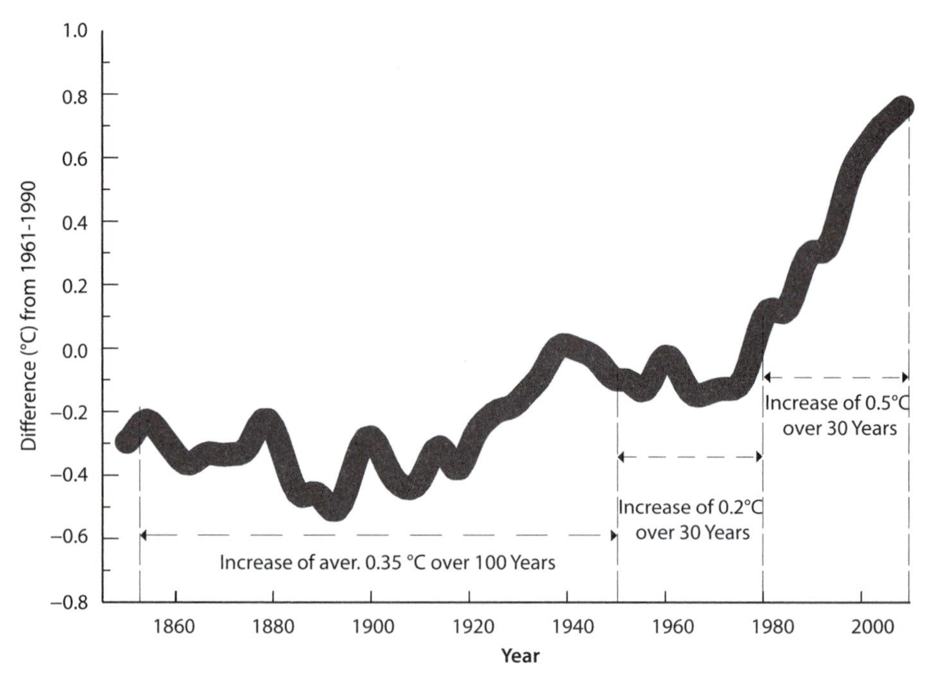

Fig. 20-5: Annual anomalies of global land-surface air temperature (°C), 1850–2005, relative to the 1961–90 mean. (Modified after ©2007 IPCC AR-4 WG1, chap. 3, fig. 3.1)

major warming effect on Earth. Note also that the rate of change of these effects in Figure 20-4d.

The atmospheric changes have in recent decades created measurable increases in the mean temperature of the atmosphere. While global temperature increased from 1900 to 1940, there was little change between 1940 and 1980. The longer record (Fig. 20-5) that is now available shows a marked temperature rise over the last century and also that the rate of rise is increasing, from about 0.035°C per decade between 1850 and 1950, to 0.067°C per decade between 1950 and 1980, to 0.177°C per decade for the last thirty years. The last decade of temperatures is the warmest on record. Temperature has increased about 0.8°C (1.6°F) since 1880; 2009 was the second warmest year on record.

Of course, the increase in global temperature does not preclude an unusually cold day or year in local regions. Temperature increase is also

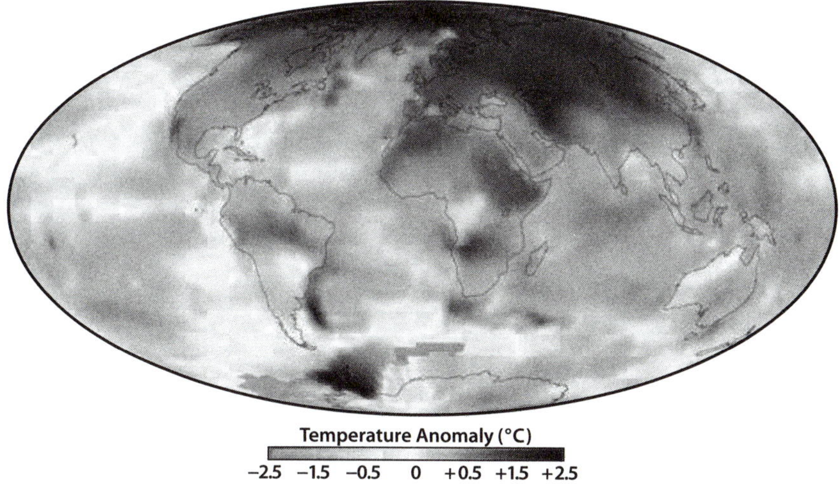

Fig. 20-6: This map illustrates how much warmer temperatures were during the decade 2000–2009, compared to average temperatures recorded between 1951 and 1980 (a common reference period for climate studies). Note that the most extreme warming, shown in dark gray, was in the Arctic. White areas over Africa and parts of the Southern Ocean are places where temperatures were not recorded. A small region near Antarctica has shown slight cooling. See color plate 27. (NASA images by Robert Simmon, based on data from the Goddard Institute for Space Studies)

not evenly distributed around the world. Figure 20-6 shows the map of how warming was distributed in the 2000–2009 decade. Warming has been much more extensive over continents than oceans and is particularly marked at high northern latitudes. In part, this reflects the greater heat capacity of the ocean and its longer equilibration time.

The increase in measured temperature is consistent with a wealth of other studies, including the first date of spring blooming or planting, the length of the growing season, the northern extent of various wildlife, etc. Another direct measure of temperature trends comes from measurements of ice and snow cover. Figure 20-7 summarizes the evidence that, contemporaneously with the temperature increase since 1980, the extent of sea ice, frozen ground, glaciers, and snow cover have all decreased substantially in the Northern Hemisphere. This effect is particularly noteworthy in the Arctic (Fig. 20-8), owing to the concentration of warming at high northern latitudes.

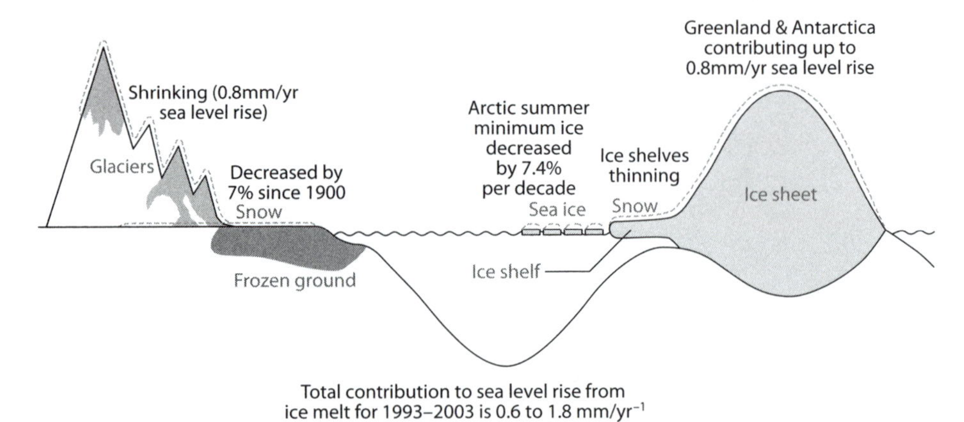

Fig. 20-7: Summary of observed variations in ice, snow, and frozen ground for the period 1993–2003 when global temperatures have been increasing. (Modified from ©2007 IPCC AR-4 WG1, chap. 4, fig. 4.23)

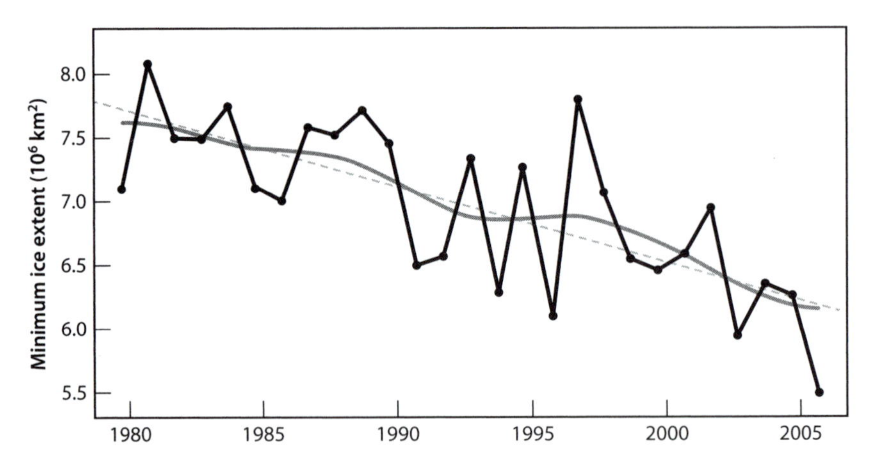

Fig. 20-8: Change in the minimum amount of ice in the Arctic over the last twenty-five years. As global temperature increases, the effect is largest at high northern latitudes. This has caused much greater melting of the Arctic ice through the summer months. The figure shows that the minimum ice extent has decreased by 30% over this time interval. (©2007 IPCC AR-4 WG1, chap. 4, fig. 4.8)

OCEAN ACIDIFICATION

When it is dissolved in water, CO_2 forms carbonic acid, increasing the acidity (lowering the pH) of the water. This experiment, which can be performed simply on a bench top, is being performed on a grand scale by fossil fuel burning. The higher CO_2 in the atmosphere is being partially absorbed by the ocean, lowering its pH. The pH of the surface ocean has been lowered by about 0.1 pH unit, corresponding to a 30% increase in hydrogen ion concentration (Fig. 20-9). This effect is not a consequence of temperature change but is a simple and direct result of increased CO_2 in the atmosphere. As CO_2 continues to increase, the surface oceans where life resides will become increasingly acidic.

The pH of the oceans has a major effect on shell-producing organisms, because the $CaCO_3$ that makes up the shells is less stable in more acidic water. (When geologists want to determine whether a rock is a carbonate, they put a drop of acid on the rock and see if it "fizzes" as the $CaCO_3$ breaks down and CO_2 gas is released.) Such effects will become important if the ocean's acidity continues to increase. Even the small changes that have occurred thus far cause a decrease in the carbonate ion concentration in the ocean that inhibits growth, already contributing to the global decline in coral reefs (Fig. 20-10). The effects for some other calcifying organisms are less clear. The response to increased CO_2 is likely to differ from one organism to another and may not be detrimental for all organisms. For some organisms, higher CO_2 could potentially lead to increased shell thickness, for example, provided adequate other nutrients were available Therefore the situation for the ocean is akin to that of the atmosphere. Increased CO_2 in the atmosphere from fundamental physics leads to warming, but there are other feedbacks that need to be considered for precise calculation of the effects. Similarly, adding CO_2 to the atmosphere from burning of fossil fuels leads to a more acid ocean. What the system-wide effects of such changes are, and what other feedbacks may kick in, will require extensive study. And as with the atmosphere, it is the rate of change as well as the magnitude of change that is the striking human influence.

Note that the effects of ocean acidification occur independently of "global warming" and do not relate to change in atmospheric tempera-

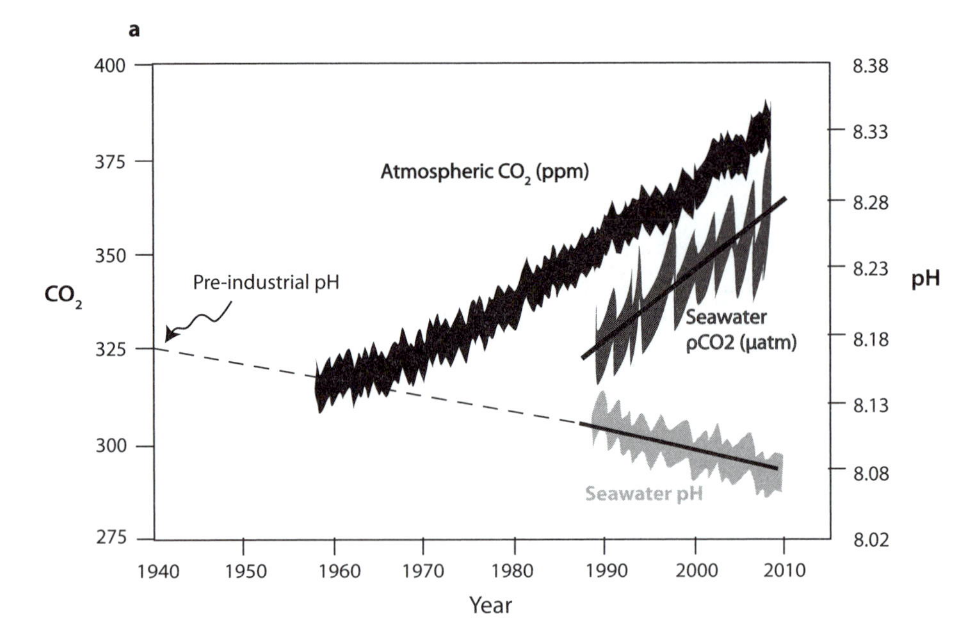

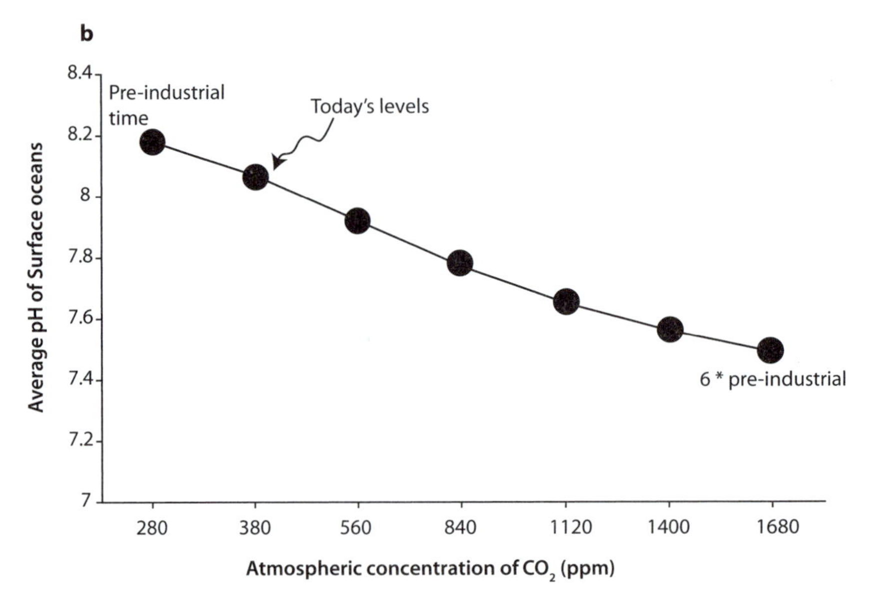

Fig. 20-9: Time series of (a) atmospheric CO_2 at Mauna Loa (in parts per million volume, ppmv), surface ocean pH and pCO_2 in the subtropical North Pacific Ocean. Left-hand scale is for the atmospheric CO_2 data, right-hand scale for the ocean pH. (b) Further change in pH of the surface ocean that will occur for different CO_2 levels in the atmosphere. Note that pH is the logarithm of hydrogen ion, so one pH unit is a factor of ten more acidic. (Doney et al., *Annu. Rev. Mar. Sci.* 1 (2009):169–92. Reprinted, with permission, from the *Annual Review of Marine Science*, Volume 1 © 2009 by Annual Reviews)

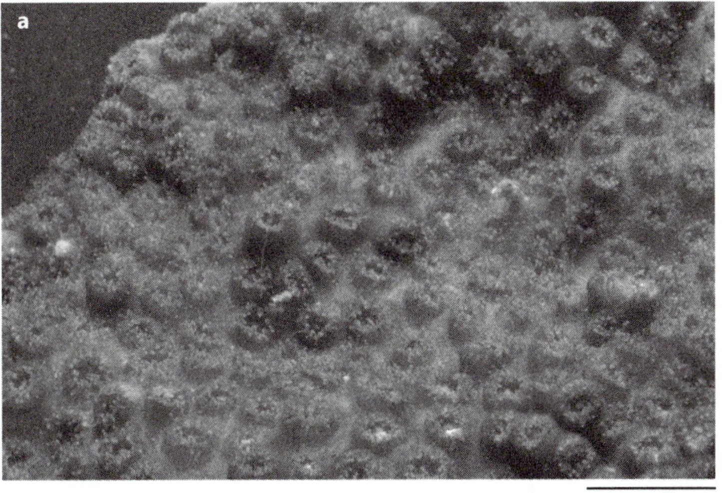

Fig. 20-10: Photos of coral *Oculina patagonica* after being maintained for twelve months in (a) normal seawater (pH = 8.2) and (b) acidified seawater (pH = 7.4). The higher acidity dissolves the protective carbonate around the living coral. (From Fine and Tchernov, *Science* 315 (2007):1811. Reprinted with permission from AAAS. Source: Doney et al. (2009; see Fig. 20-09))

ture. Lower oceanic pH is an inevitable consequence of increased atmospheric CO_2. Warming of the atmosphere will also lead to warming of the surface ocean, with unknown effects on the oceanic biosphere. Both temperature and acidity are significant parameters for ocean ecosystems. Here the rate of emissions is very important relative to natural cycles, because little compensation through dissolution of $CaCO_3$ has time to occur.

BIODIVERSITY

Every animal is driven by a search for food, and *Homo sapiens* is no exception. The trail of human migrations is marked by extinctions, likely associated with hunting. As the ancient Polynesians colonized the Pacific islands, as many as 75% of the native bird species went extinct, likely because they had never seen predators and were an easy source of food. About 45,000 years ago, glacially lowered sea level allowed Aborigines to reach Australia. Their arrival corresponds to, and likely caused, a great reduction in native vegetation (as recorded in the carbon isotope ratios in radiocarbon-dated fossil egg shells) and to the extinction of several species, such as the Tasmanian tiger and many large marsupials. When humans first arrived in North America across the Bering Strait, they found wildlife more diverse even than in Africa—saber-toothed tigers (Fig. 20-11), mammoths, long-horned bison, camels, giant wolves, antelopes, tapirs, and a multitude of large birds. In North and South America, 70–80% of the large mammals are now extinct. Similar extinctions occurred with the first arrival of humans in Madagascar and New Zealand, as well as Australia (Fig. 20-12).

Fig. 20-11: Some of the great mammals that were common before *Homo sapiens* entered their environment.

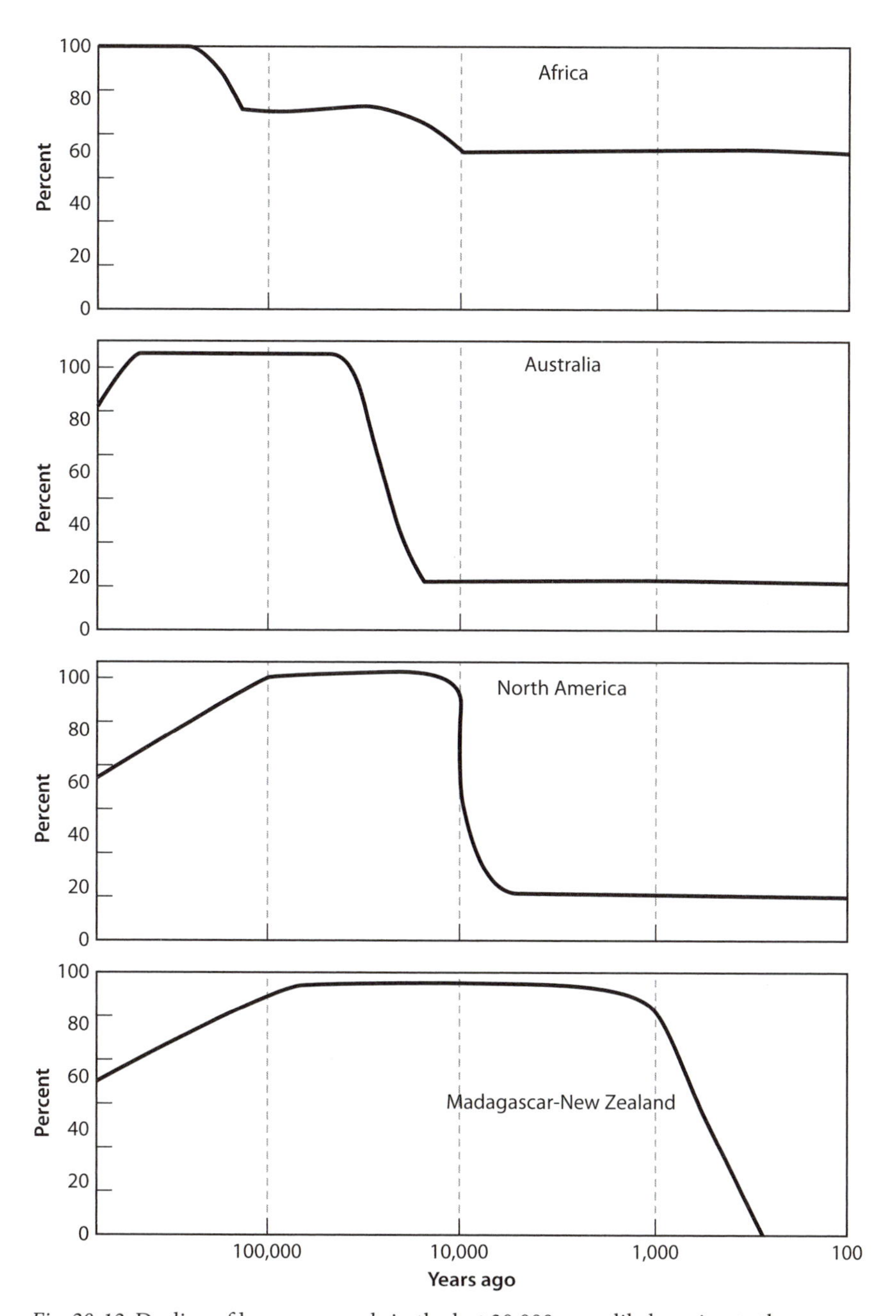

Fig. 20-12: Decline of large mammals in the last 20,000 years likely owing to the presence of *Homo sapiens* on new continents. Rapid extinctions occurred immediately following our arrival. North America used to have a diversity of large mammals even greater than modern Africa. (Modified from E. O. Wilson, ed. *Biodiversity* (Washington, D.C.: National Academy of Sciences, 1988))

The advent of agriculture and domestication of animals led to even greater environmental impact. What grew in our soils and roamed in our grasslands was to an ever-greater extent determined by us. Genetic engineering started at a very early stage by selection of crops and animals for human needs. To keep livestock in and predators out, fences were constructed; to compensate for fluctuations in nature's supply of water, dams and irrigation systems were put in place, all in the name of increased access to food. Estimates are that human activities are now responsible for one-quarter of terrestrial biomass production.

Today the reach of human appetites is not restricted to their local environment but has become global. Scientific data published in 2003 determined that large fish—such as tuna, swordfish, marlin, and the large bottom-dwelling fish such as cod, halibut, skate, and flounder—have declined *throughout the oceans* by 90% compared to 1950 levels. The 1950 levels were likely already diminished relative to preindustrial levels. The decreasing fish supply occurred owing to massive growth in the ocean harvest from factory fishing ships. Stocks are now so low that fish harvest is in decline.

Extinctions have spread to the less visible animals through loss of habitat. Humans have taken land for agriculture, converting an ecosystem with an assemblage of diverse species to large expanses of a single grain, where any other plant is a "weed." Deforestation has been a human footprint. Vast expanses of current landscapes, such as the Scottish highlands and much of Europe, were once rich forests. Land is also converted for animal husbandry, clearing existing plant species and eliminating habitat for wildlife. Global mapping now shows that some 50% of global grasslands, tropical dry forests, and broadleaf forests have been converted to human uses.

The greatest repositories of global biodiversity are the tropical rain forests, where many of the current species are not even known. Modern technology permits very efficient deforestation. Rain forests are being cleared at the rate of 40,000 km^2 per year, ten times the area of the state of Rhode Island—annually (see frontspiece and Fig. 20-13). Some 70% of the Indonesian rain forest is now gone. Rain forest soils are so poor that remedial forestation is problematic.

The destruction of habitat and species is not restricted to land. In some regions coral reefs, which support enormous biodiversity in their

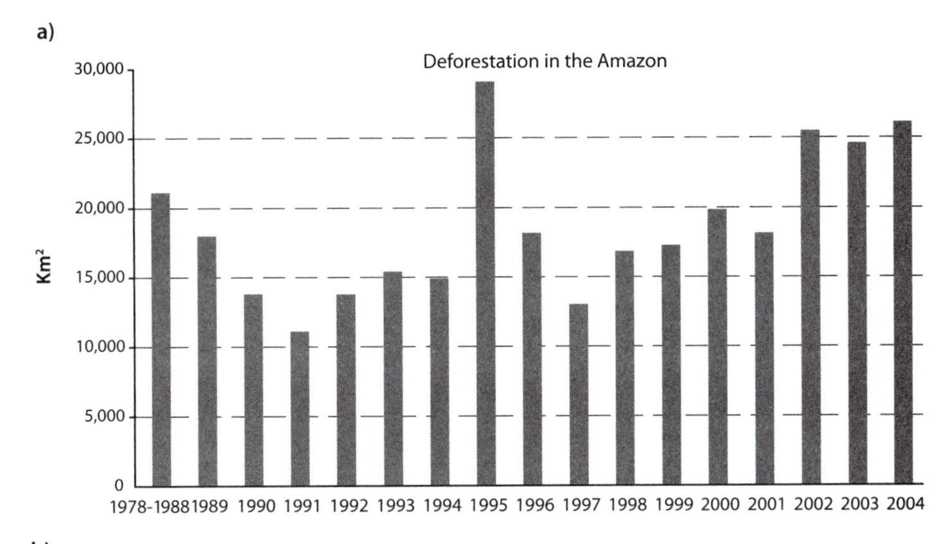

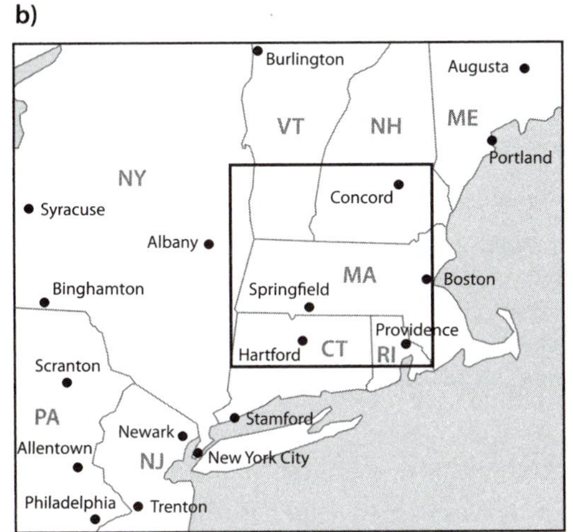

Fig. 20-13: Modern rates of destruction of rain forests. (a) Annual rates of destruction of the Amazon rain forest, which accounts for more than half of global loss. Total global loss per year is 40,000 km^2, an area 200 km on a side. For visualization, (b) shows an area of the northeastern United States equivalent to what is being deforested annually in rain forests. (All figures derived from official National Institute for Amazonian Research (INPA) figures)

environments, have decreased by 90% from levels since the 1970s. Phosphate pollution in lakes and rivers has also created massive "dead zones," where algae thrive but animals cannot.

While extinction of large species is apparent to us, smaller species extinctions are relatively invisible and are more difficult to quantify. General effects can be inferred from laws of the effects of habitat destruction on the abundance of species have been determined by careful studies of specific regions. E. O. Wilson was able to use islands to investigate

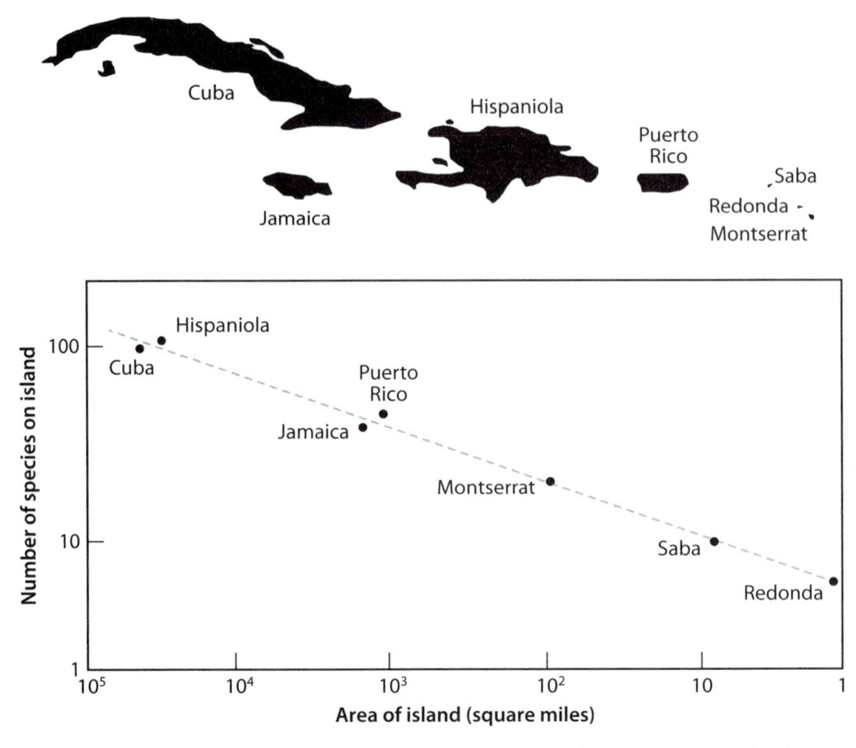

Fig. 20-14: Study of ocean islands shows that the number of species that an island can support decreases progressively with diminishing island size. The same principles apply to "biodiversity islands" on continents. To preserve species, large contiguous areas are necessary, not just isolated pockets. Also, as climate change forces life to change latitudes, the lack of contiguous space leads to extinctions. (Adapted from *Biogeography* by E. O. Wilson)

and understand the importance of habitat size on species diversity. Figure 20-14 shows how the number of species of reptiles and amphibians depends on the contiguous land area available. A reduction in area of 90% leads to a 50% species reduction. As humans take over habitat, even without direct killing, they diminish habitat, causing extinctions and a reduction in global biodiversity. The diminishment is not only at the species extinction level, but also at the important level of variation within a species. Genetic diversity within a species leads to greater survivability and a richer gene pool.

In most ecosystems, species compete and evolve into a diverse mix of organisms that are in an ever-moving balance. Introduction of new species without predators, or against which the indigenous species have no defense, can lead to exponential population growth of the introduced species and extinctions and loss of diversity of local species. Humans have transported nonnative species (often called *invasive species* or *exotic species*) to new continents. Early humans were themselves invasive and also introduced species such as goats, rats, and cats that became a major factor in local ecosystems. In the modern era, the advent of international transport and communication provides easy transport of species from one continent to another. This is apparent with the zebra mussel in North America that eliminates native mussels and now dominates many lake and river ecosystems in the United States and Canada; the African bee in the Americas; the tree snake that has wiped out the bird population on the Pacific island of Guam; the rabbit, sheep, and goat in Australia; bark beetles that ravage forests; and hundreds of plant species.

The combination of factors attributed to humans—overharvesting, habitat takeover, habitat and biological destruction by pollution, introduction of exotic species—all have greatly affected the total biodiversity on Earth. These effects to date have not been caused by climate change but by the growth of human population and our needs for land and food. Climate change is a new factor that will be added on top of these existing effects. It may have even more widespread repercussions for the planet's organisms, because the rate of climate change induced by human activities is so much greater than the natural rate to which ecosystems ordinarily respond.

In a sense, biodiversity is the end consequence of planetary change. As we induce climate change, modify the oceans, destroy habitat, lose soils, and overharvest species for food and sport, we eliminate much of the genetic diversity accumulated by Earth over billions of years. Like the accumulation of oil and coal over hundreds of millions of years, biodiversity is also an accumulating resource, with more species and greater genetic variation through geological time. Genetic diversity is the tool chest of life, and the greater the diversity, the more life is able to adapt to changing conditions, and the greater the opportunity for genetic resiliency. Destruction of biodiversity is the diminishment of the gene pool. By destroying biodiversity, we reduce the evolutionary potential of the planetary biosphere.

Future Prospects

All human beings want to prosper, and that prosperity is generally measured by economic output, of which a common measure is gross domestic product (GDP). The GDP of nations correlates closely with their energy use (Fig. 20-15a), and since the overwhelming proportion of energy comes from fossil fuels, energy use also correlates closely with CO_2 emissions (Fig. 20-15b). Nations and individuals wish to be prosperous; prosperity requires energy; energy production releases CO_2. This fundamental conflict between increasing prosperity and reducing CO_2 emissions makes a very difficult problem.

Compounding the difficulties are the differences among nations. North America and Europe are responsible for the CO_2 problem, having released 70% of integrated CO_2 emissions between 1800 and 2000. These nations also continue to dominate emissions on a per capita basis (Fig. 20-16a). CO_2 emitted per person in the United States is five times higher than the global average, ten times higher than Africa and Asia (minus China) and more than twice as high as Europe. There are also differences in economic efficiency from the point of view of CO_2 emitted per dollar of GDP (Fig. 20-16b). To obtain their GDP, the former Soviet Union and Middle East, both resource-rich regions, emit the most CO_2 per dollar of GDP, followed by the United States, China, Canada, and Australia. The rest of the world is far more CO_2 efficient. The economies of Europe and Japan are 1.6 times more energy efficient than the United States. Despite having about 5% of the world's population, the United States has emitted the most CO_2 over the last two centuries, has the largest present-day per capita emissions of populous nations, and is the least energy-efficient large economy of the developed world.

Against this present situation and history, the *growth* in global CO_2 emissions now comes from the developing world. Asia (less Japan) accounts for more than 70% of the growth in CO_2 emissions from 2001 to 2010. China overtook the United States in terms of total CO_2 emissions in 2006, and China constructs one new coal-fired utility plant per week, with a useable life of more than thirty years. Automobile sales in China are growing at a rapid rate. Because of its large population, poor energy efficiency, reliance on coal, and rapidly growing GDP, growth of CO_2

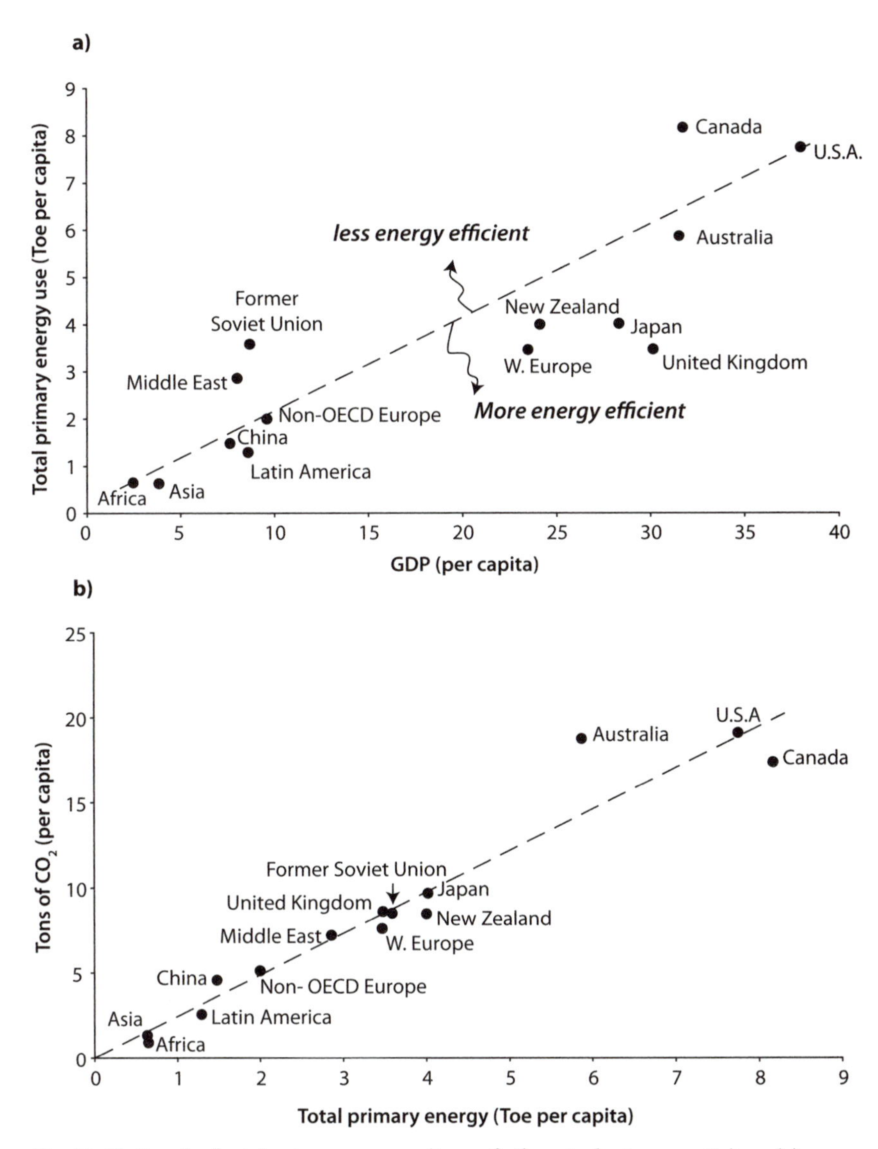

Fig. 20-15: Graph of total primary energy (tons of oil equivalent per capita) vs. (a) gross national product (GDP) per capita and (b) tons of CO_2 emissions per capita. (Data from International Energy Agency, *Key World Energy Statistics*, 2009)

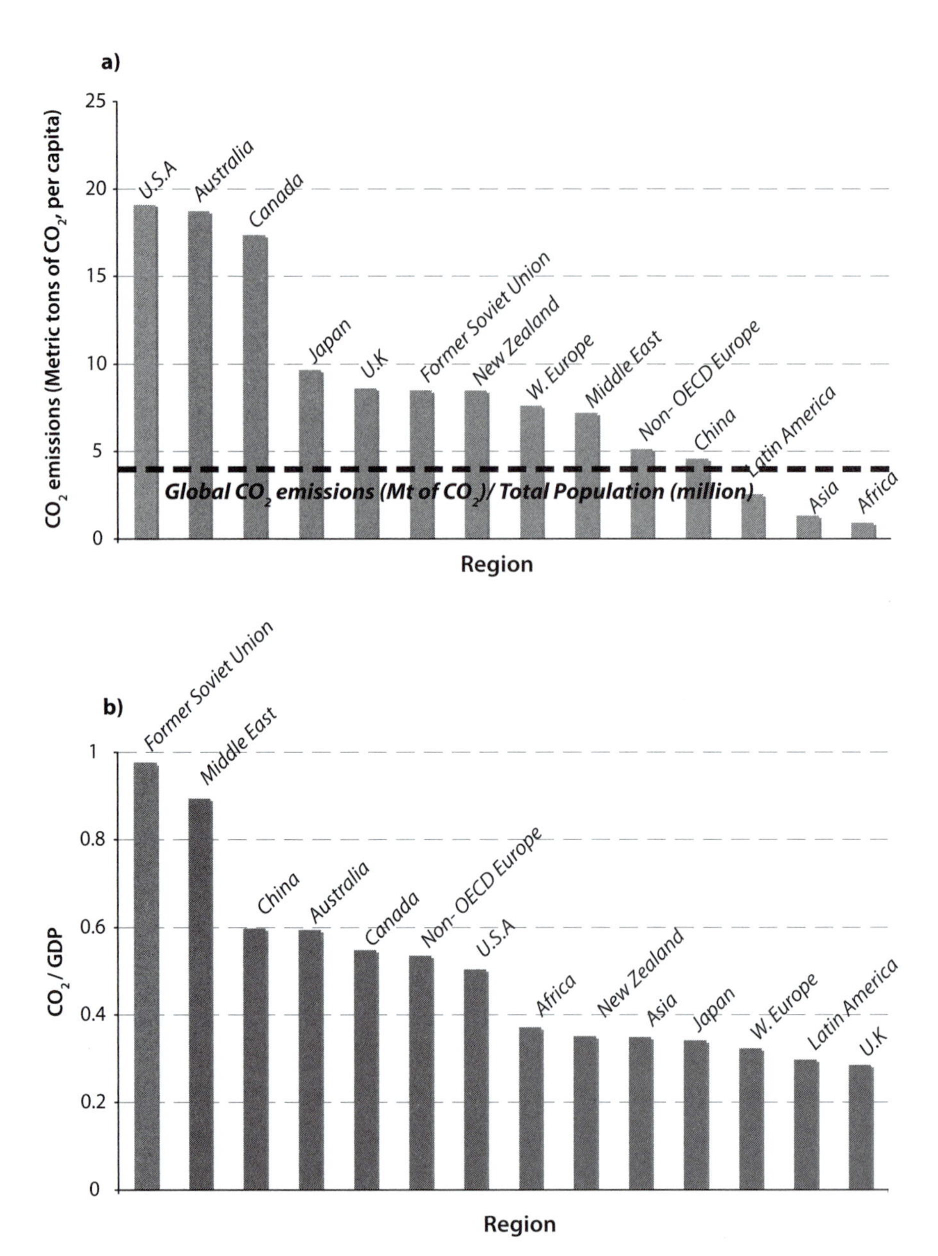

Fig. 20-16: (a) CO_2 emissions vs. region and (b) CO_2/GDP (PPP) versus region. Units are kg CO_2 per dollar. (Data from International Energy Agency, *Key World Energy Statistics*, 2009)

emissions from China is inevitable and will increasingly dominate global CO_2 budgets. India also aspires to such rapid growth.

These facts create a contentious political problem. The world is competitive, and the countries that use the most energy are the most powerful. No country is willing to reduce its economic prosperity or its global power—no politician could be elected on such a platform. The West created the mess, and one could argue that those responsible for a mess should clean it up. The West also still uses a disproportionate share of energy on a per capita basis. But the growth in global CO_2 comes from the developing world. They are not responsible for the current problem, and their citizens do not get their "fair share" of the economic prosperity provided by energy. Why should they have to cut emissions?

This backdrop was encountered in the Kyoto conference on global warming, where there were attempts at a global plan to reduce CO_2 emissions. Europe agreed to reduce emissions, the developing world was exempted, and the United States refused to sign. An unanticipated consequence of the protocol was to export CO_2 emissions from the developed world to the developing world. For example, the United Kingdom reduced its own emissions by 5% but its consumption-based emissions increased by 17%! The net result is that global CO_2 emissions continued to rise by 29% between 2000 and 2008. Growth of CO_2 emissions by China dwarfs any economizing measures taken by the rest of the world. A subsequent conference in Copenhagen in 2009 led to no new agreements, despite a more willing U.S. administration. The politics have made the problem simply intractable, and CO_2 emissions will continue to increase. What then is likely to happen?

As we saw in the previous chapter, the limited quantities of oil make it likely, if not inevitable, that petroleum use will decline over coming decades. There is enough coal, however, to fuel the entire world for another century or two. Further, unlike the reserves of oil that are concentrated in the volatile Middle East, those of coal reside in the United States, Russia, and China. Coal is currently five times cheaper than oil as an energy source for electricity, and as long as economics does not include environmental costs, its use will continue to increase.

There is, of course, much talk about the surge of renewable energy—principally solar and wind. Solar and wind energy today make up less than 1% of energy production in the United States. Both of them are capital-intensive industries, with substantial installation costs. The electrical grid also would need to be redone in order to make good use of their energy. Fossil fuels have the great advantage that they can be easily transported and burned on demand anytime, anywhere. In contrast, the most economical regions for sun and wind power are often far from where electricity is needed. They are also currently not as cheap as fossil fuels. Even with huge growth in renewable energy, it is unlikely to make up more than 10% of energy needs for a very long time.

For all these reasons, our atmosphere is in for further large increases in CO_2 content. Despite optimistic hypothetical scenarios of cutting CO_2 emissions, there is currently no prospect that emissions will decrease on a global basis. This is apparent by comparing actual experience with the various "scenarios" constructed by the International Panel on Climate Change (IPCC). The group made predictions on the basis of the "middle-of-the-road scenario," which would include substantial global efforts to reduce the growth in emissions. What has happened in practice is that global emissions have exceeded the most pessimistic "business as usual" scenario. There is no evidence that our political systems are capable of meaningful reduction in global CO_2 emissions.

The timescale of CO_2 rise may be quite short. In 2008 CO_2 rose by 2.4 parts per million to 390 ppm, the largest annual increase on record. If emissions grow at 2.5% per year as they have for the last decade, 560 ppm would be reached by 2050, and CO_2 rise would continue after that unless emissions suddenly went to zero! To keep the atmosphere below 560 ppm, CO_2 emissions would need to be drastically reduced to almost zero by midcentury.

As CO_2 rises, the acidity of the ocean will continue to increase, with unknown effects on oceanic ecosystems. As population expands and economies grow, biodiversity will continue to decrease. Coral reefs and rain forests may largely disappear.

As concentrations of greenhouse gases rise, so also will Earth's temperature (Fig. 20-17). Estimates of the magnitude of this change require models, and the models are imperfect. The climate system is a complex beast! The models are becoming increasingly complex, can reproduce current global warming, and account for short-term cooling effects

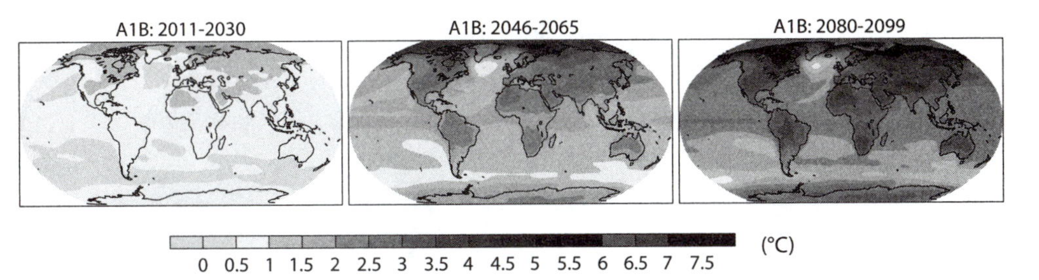

Fig. 20-17: Temperature projections, global map. Multimodel of annual mean surface warming (surface air temperature change, °C) for the A1B model where significant steps are taken to slow the growth in CO_2 emissions and have them decline after 2050. Projections are shown for three time periods, 2011–2030 (*left*), 2046–65 (*middle*), and 2080–99 (*right*). Anomalies are relative to the average of the period 1980–99. See also color plate 28. (©2007 IPCC AR-4 WG1, chap. 10, fig. 10.8)

caused by large volcanic eruptions (Fig. 20-20). The models suggest that at twice the preindustrial content (560 ppm CO_2) mean global warming would be about 3°–5°C (i.e., 6°F). This would give New York City the climate of Atlanta, Georgia. Keeping CO_2 to these levels would require major efforts over the next decades to cut CO_2 emissions. Absent such efforts, a second doubling (i.e., from 560 to 1,020 ppm) would lead to an additional 3.5°C warming. New York would have the climate of Florida.

Not only will our planet become warmer, but the distribution of precipitation will change. The same computer simulations that provide estimates of the magnitude of the coming warming also suggest that as Earth warms, rainfall will be ever more strongly focused on the tropics. If the models are correct, Earth's dry lands will become even more parched (Fig. 20-18).

The paleoclimatic record is consistent with these predictions. The last glacial period provides us with an excellent cold analog. On this colder Earth, the dry lands were much less dry and the tropics less wet than now. The evidence comes from several sources, but the most convincing and easiest to understand is the size of lakes that have no outlet to the sea. The water they receive from the precipitation in their catchments basins is lost entirely by evaporation from the lake's surface. Such closed basin lakes exist only in the dry regions of our planet. This makes them excellent recorders of aridity. If rainfall decreases, they shrink in size. If it increases, they grow in size.

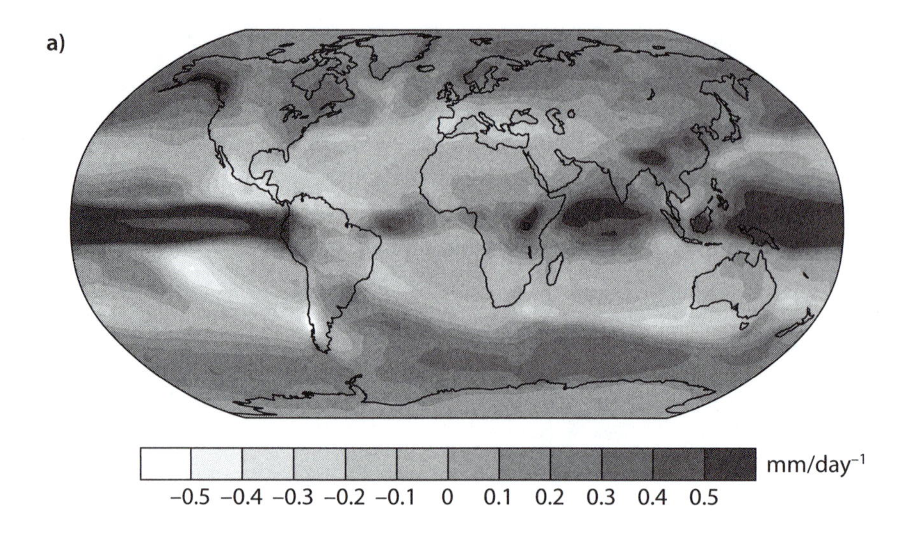

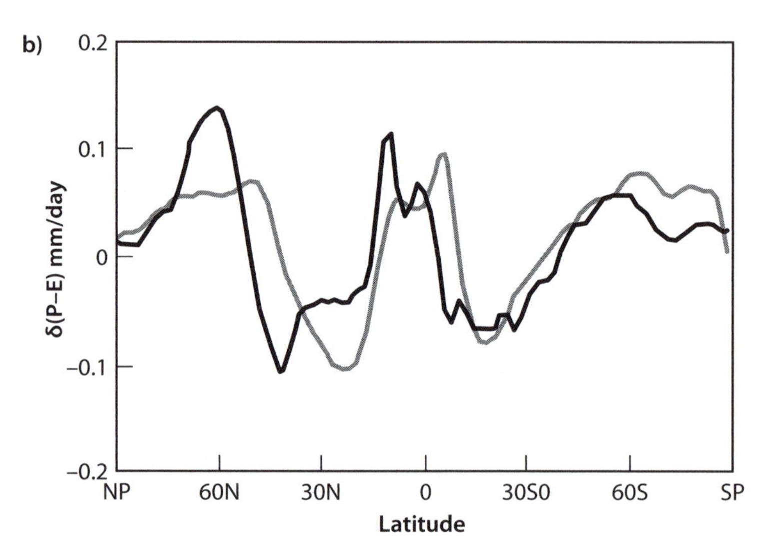

Fig. 20-18: Predicted changes in precipitation. (a) Multimodel mean changes in precipitation. Changes are predicted annual means for the period 2080–99 relative to 1980–99 (©2007 IPCC AR-4 WG1, chap. 10, fig. 10.12). (b) The mean predictions for latitude zones, expressed as the change in precipitation minus evaporation (P-E). Note the predicted drying of midlatitudes (Held and Soden, *J. Climate* 19 (2006): 5686).

Two of these lakes are well known. One is our own Great Salt Lake in the state of Utah. The other is the Dead Sea, which occupies the rift zone separating Israel and Jordan. Both of these lakes were far larger during the last glacial period, as were lakes in New Mexico and Nevada, in northwestern China, and in the Patagonian drylands of Argentina. We know this because the ages of paleoshore lines, which stand out like bathtub rings, have been established using the radiocarbon method.

At the same time that the lakes in Earth's mid-latitude dry zones were several times larger than now, equatorial lakes, such as Africa's Lake Victoria, were dry. We know this because cores penetrating the lake sediments terminate in a soil horizon. Further, the radiocarbon ages of organic matter from the lake sediments deposited on this soil reveal that the lake came back into existence at the onset of the Bolling Allerod warm period, which marked the end of the last glacial period. The record from cold times lends support both to the predictions based on computer simulations that increased CO_2 will warm the planet and also that it will focus its precipitation more strongly in the tropics.

Not only will the warming and the shift in precipitation patterns impact our lives, but these changes will lead to mayhem for our planet's wildlife. As the CO_2-induced warming intensifies, one species after another will be squeezed out of its current habitat and opportunists will move in as replacements. There will no longer be what might be called a steady-state grouping of plants, animals, and insects. Rather, all these residents will be in a state of flux. And of course species adapted to polar conditions (e.g., polar bears) may survive only in artificially cooled "zoos."

As is already happening, Earth's glaciers will continue to shrink. In Peru, where these masses of mountaintop ice serve as reservoirs providing meltwater during dry seasons and droughts, this disappearance may lead to severe water shortages. While the melting of the world's mountain glaciers will lead to a modest rise in sea level, that of the Greenland and Antarctic ice caps surely will lead to very serious changes. If melted, Greenland ice would raise sea level by about 6 m, causing most of southern Florida to disappear beneath the sea. The West Antarctic ice sheet, considered vulnerable to warming, would create another 6 m of rise. Most projections place the timescale required for Greenland's meltdown at several centuries. But the recent pronounced speedup of its main outlet glaciers and the observation of ever more widespread summer melt pools call for a reevaluation of this timescale. Photos of water accumulated in melt pools cascading into two abysses thought to extend to the base of the ice dramatically demonstrate that these outlet glaciers have a self-lubricating mechanism that might greatly increase their rate of flow into the sea. There is simply inadequate experience and understanding of the melting of ice sheets to fully evaluate the appropriate timescale.

For all these consequences, the present is the key to the future. Given the long timescale involved in changing our economy and infrastructure, there is little time to be wasted.

HISTORICAL PERSPECTIVES ON THE FUTURE

Our future prospects also need to take into account the overall condition of Earth during the last several thousand years, which has been a golden age of stability. Based on historical data, we cannot assume such stability will continue.

Figure 20-19 compares global temperatures for the Holocene to comparable periods during interglacials of the last 430,000 years. The length and stability of the current warm period have been the longest and most stable, although the interglacial 400,000 years ago comes close. There have also not been regional droughts of many decades but these are common in the historical record.

This stability is in part a consequence of the fact that the last significant volcanic eruption occurred in the early 1800s, and no truly large volcanic eruption has occurred during the Holocene. This last statement may seem perplexing, since most of us are familiar with the spectacular volcanic eruptions of Mount St. Helens in 1980 and Mount Pinatubo in 1991. Mount St. Helens erupted 2–3 km^3 of material; Pinatubo was significantly larger at 10 km^3 and also ejected 20 million tons of SO_2 into the atmosphere. SO_2 in the stratosphere leads to cooling, and Pinatubo caused about 0.5°C of global cooling between 1991 and 1993 (Fig. 20-20). This cooling also provided a useful calibration of climate models, which were able to predict the cooling effects very precisely. A much earlier basalt eruption from Laki volcano in Iceland (1783–1785) ejected close to 15 km^3 of basaltic lava and likely large amounts of SO_2. At this time the eastern United States recorded winter temperatures almost 5.0°C below average. Iceland lost most of its livestock and one quarter of its

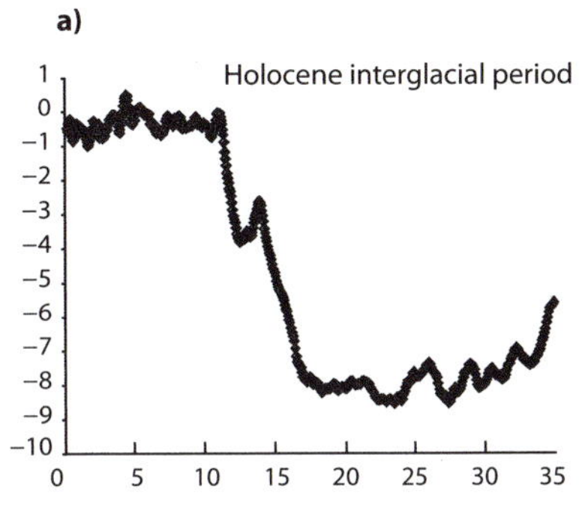

Fig. 20-19: Comparison of global temperatures for the Holocene compared to comparable periods during interglacials of the last 430,000 years. Note the long and stable temperatures during the Holocene that have supported the rise of human civilization and that are very unusual in recent Earth history. Each of the horizontal timescales encompasses 40,000 years. (Data from the Vostok ice core)

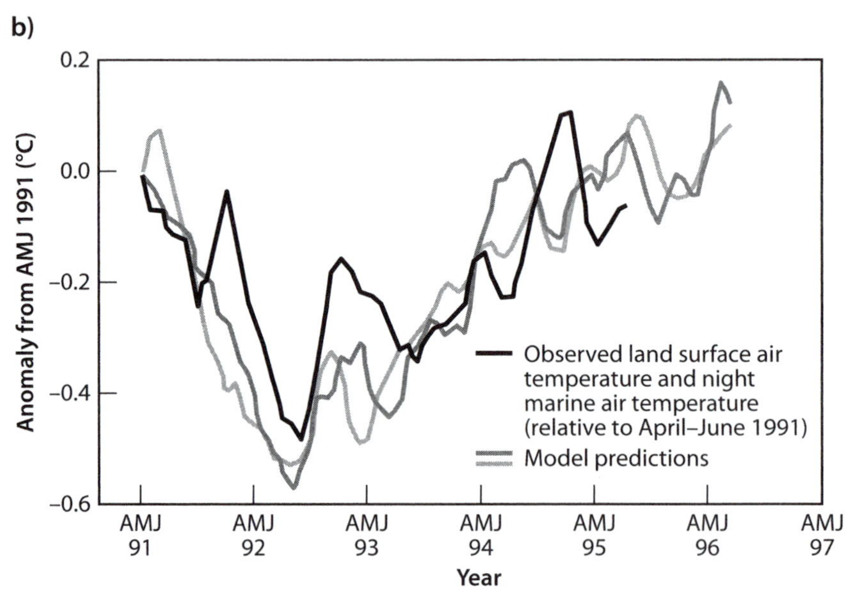

Fig. 20-20: (a) Photograph of the eruption of Mt. Pinatubo in 1991. The eruption injected 20 million tons of SO_2 into the upper atmosphere, causing a temporary cooling of global climate. Much larger eruptions in the past (and future) would have had an even more dramatic effect. (Courtesy of U.S. Geological Survey.) (b) Solid line shows the observed changes in land surface air temperature caused by the Pinatubo eruption. This provided a test for climate models. Light gray shows the model results, which compare well with the observations. (Figure modified from Hansen et al. (1996), A Pinatubo climate modeling investigation, in *The Mount Pinatubo Eruption,* NATO AbI Series, vol. 1, 42, pp. 233–272, Springer Verlag)

population to famine. In 2010 a much smaller Icelandic eruption led to major disruption of European air traffic. It is difficult to fathom what would be the consequences of a Laki eruption were it to occur today.

In the context of the historical record of volcanism, however, even Pinatubo and Laki are babies. The Tambora eruption in Indonesia in 1815 was 160 km^3 of ejecta and led to the "year without a summer" in North America and Europe. Heavy snow and frost gripped New England in June, and most crops were lost in North America, where animals were slaughtered for food. Europe experienced the worst famine of the nineteenth century, and riots, arson, and looting took place in many European cities.

Even Tambora is a modest eruption relative to the massive eruptions of "supervolcanoes." Some 74,000 years ago, Toba in Sumatra erupted 2,800 km^3 of ejecta, almost twenty times the volume of Tambora, leaving a crater (now a lake) that is 100 km long and 30 km wide. Although Tambora is in Indonesia, so much material was erupted that ash layers up to 6 m thick were deposited in India. The record of this eruption is preserved globally, including a marked signal in the Greenland ice core. The amount of SO_2 erupted was likely >100 times larger than Mount Pinatubo. Model simulations of the Toba eruption suggest global cooling of 12°C (20°F) for six years, and destruction of much of the vegetation on the globe, including all the broadleaf trees. The Toba eruption has been called upon to explain a "bottleneck" in human population near that time, when human population is believed to have declined to as few as 10,000 individuals. While volcanic eruptions such as Tambora and Toba occur infrequently, eruptions of that magnitude will happen again periodically. They are the normal volcanic functioning of the solid Earth system.

These examples illustrate the fact that short-term climate change and natural disasters on a global scale are an inevitable aspect of Earth's surface. The smooth and optimal climate conditions of the last two hundred years are exceptional. Our perspective of long-term global stability is an accident of the benign conditions that have prevailed since 1850.

Most of us are also not aware of the delicate balance between food production and supply that characterizes our global stability. Most arable land is cultivated, and there is a percent or two difference between food production and consumption. One billion people are malnourished. There are a couple of months of food reserves globally. We accept that modern civilization is able to feed itself, thanks to the takeover of all easily arable land, extensive use of fertilizer, and the green revolution. What we do not realize is that this has all been possible thanks to a remarkably benign period of planetary history. Between about 1550 and 1850 Earth experienced the "little ice age," where the Thames froze in Britain, skating occurred on Amsterdam canals, and George Washington's troops experienced the deprivations of Valley Forge. The exit from the "little ice age" led to exceptionally recent benign climate conditions for agriculture. So in the growth of human population between 1850 and 2010 there have been no major volcanic eruptions, and climate has been remarkably stable. None of the major short-term oscillations that are so apparent in the ice-core record have occurred. Assuming continued stability and no planetary crisis, all should be well, but we live on the edge of famine. Ultimately, planetary change will occur. Instability is a natural planetary condition. Given this reality, is it prudent to live at the edge of our means and cause perturbations to the planetary systems for which we do not understand the consequences?

Possible Solutions

In the absence of a change in human behavior, many of the negative planetary consequences discussed above will increase greatly in the twenty-first century, because the root causes continue to increase. Our influence scales either directly with population (e.g., food, land use), or exponentially with population owing to the combination of population and economic growth (e.g., energy consumption and associated CO_2 emissions). Controlling population growth is not generally on the table as an acceptable environmental solution. Economic growth is based on free resources and increased energy production, for which the only viable option for decades has been fossil fuels, and this remains the global economic model. As long as biodiversity, soils, and fossil fuels are not recognized as rare and precious planetary resources, they will continue to decline until their scarcity causes crisis. At that point they can never recover on timescales pertinent to us. And if greenhouse gases accumulate sufficiently and climate change ceases to be gradual, nothing can be done because of the long residence time of CO_2 in the atmosphere.

The future need not be so bleak. We have the capacity to change our behavior and manage the planet for the mutual benefit of all species. We simply eliminate population growth by having one to two children per family globally, become largely vegetarian in order to make more efficient use of land and food, make a concerted global effort toward sustainable energy sources, travel in small vehicles and trains, live in only the space we need, wear sweaters in winter and sweat a little in summer, adopt sustainable farming practices, reduce our waste of resources, remove and sequester CO_2 from the atmosphere, and start to promote forests and grasslands instead of destroying them. We can carefully monitor the other life that shares and supports the planet and preserve its capacity for life as well as our own. All of these are within our power of technology and choice. Would these changes drastically reduce our quality of life? Actually, many of them would improve it. Some of these changes would save money. Others could take place only by adding environmental costs to use of resources.

A principal factor that prevents action is that our economic model has no simple way to account for environmental costs. Agriculture does not take into account the costs of soil depletion. Fossil fuel burning does not take into account modification of the atmosphere. People do not pay for the CO_2 they emit. Habitat destruction does not take into account the destruction of species. Fishermen do not pay for the fish they take from the sea; lumber, oil, coal, and mining companies do not pay for their resources, aside from the cost to buy the land. Economic costs are only those of extraction and delivery—Earth is free. It is as if there were a bank in town, filled with gold that has accumulated over vast periods of time, and without asking where it came from or even any appreciation, we were able to walk into the bank, perhaps pay a small entrance fee, and take all the gold we want and use it. Our unquestioned attitude is that Earth's resources are free—the only major costs should be those to extract and ship them. The driving force is "what would make me more money now," irrespective of the long-term consequences for myself or others. Our economic model does not take into account environmental costs and consequences or planetary history.

This economic model of industrial civilization is justifiable when environmental impacts are small enough that they do not have planetary significance and when resources—the Earth bank—are believed to be infinite. If environmental impacts are large enough that they need to be taken into account, we have no free market model for doing so. If resources are finite and environmental impacts detrimental, cost should reflect those facts as well as the cost of extraction.

An additional problem is that we do not know well the cost that would be involved. E. O. Wilson has estimated that $50 billion would be sufficient to save the remaining biodiversity hot spots that contain the majority of threatened species, but this estimate does not include the costs of mitigating climate change that would modify the habitat. Costs to conserve energy can be profitable over time frames of decades; $100 billion dollars per year would make a vast difference in CO_2 emissions. Sustainable farming practices can be economically favorable over the long term. To put these costs in context, these expenses are small relative to the amount we spend globally on the military. Fifteen percent of global expenses on war and its instruments would protect the planet. If the United States were to fund its defense budget with a gas tax, gasoline would cost $8 per gallon, yet even ten cents per gallon is viewed as unacceptable for costs of damage to the planet. So the money is there, the technology can be developed. We have the practical capacity for planetary management. It is a matter of choice. Do we choose to protect the planet and conserve its treasure and thereby also prevent wars over resources, or do we continue on our present course and spend more on military capability and war in order to access the resources?

There is a positive solution where strategic, energy, and climate interests converge. Reducing energy consumption and converting to wind and solar power would reduce the necessity to access oil reserves, lessen the need for costly military interventions, and stem a huge export of national currencies to other nations. Climate, the economy, and national security all benefit.

The lack of action is one of political and personal choice, not one of technological capability. The willingness to act could occur if we were to realize at the level of actions rather than concepts that we are an integral part of the planetary system, not just a user of it. If this choice of planetary conservation were made, what could be done?

SOLVING GREENHOUSE GAS ACCUMULATION

The area of human influence that has received the most attention is what to do about energy and CO_2 emissions. To reduce CO_2 emissions we can:

- increase our productivity per unit of energy (e.g., energy conservation);
- reduce our dependence on fossil fuels for energy and adopt energy sources that do not release CO_2;
- remove CO_2 from the atmosphere by carbon capture and sequestration.

All of these in concert could limit the CO_2 buildup in the atmosphere with its consequences for global warming, sea-level rise, ocean acidification, and biosphere impacts.

One of the easiest means to reduce energy consumption would be to be more efficient. The United States and China in particular are 160% less energy efficient than Europe and Japan per dollar of GDP. Vast changes in this respect would be possible, and would occur quite naturally if the cost of energy were to rise. Such actions would only limit the rise in CO_2 emissions, however, and the ultimate goal must be to have them drop to near zero. This is a much longer-term task, requiring conversion of our entire energy infrastructure. Eighty-five percent of our energy is currently derived from the burning of coal, petroleum, and natural gas. Most electricity globally is produced by coal, which is about one-fifth the cost of oil per unit of energy. To move away from fossil fuels would involve a switch to some combination of other energy sources (nuclear, wind, hydro, solar, geothermal, vegetation . . .). Many of these have their own associated problems. A spread in nuclear power globally could lead to nuclear accidents, nuclear proliferation, and an enhanced terrorist threat, and energy from vegetation competes with crops, raising food prices and limiting food availability. Energy conversion is a multifaceted and complex problem.

ENERGY FROM THE SUN, WIND, AND ATOM

Two major routes exist by which energy from the sun might be harnessed. One involves photovoltaic cells and the other biofuels. The photovoltaic route is presently stalled due to high costs. Further, such an energy source would have to be coupled with an energy storage system that would provide power during nights and periods of heavy cloud cover. While an ideal solution because the energy from the sun is so vast and is everywhere, large-scale use of photovoltaic energy awaits substantial technological advances or a large carbon tax that makes it competitive.

The other form of solar energy is biofuels, which is already being actively pursued. Biofuels are a closed loop for CO_2. The plants remove CO_2 from the atmosphere converting it to organic carbon, which is then burned to return the CO_2 to the atmosphere with no net CO_2 addition. The rub is that production of biofuels takes energy to fertilize the soil, grow and harvest the plants, and produce and transport the fuel. In the United States, biofuel production likely uses more energy than is produced, or at best uses 90% of the energy. At the same time, corn for biofuels competes with food, raising food prices. It is a great boon for agribusiness, because almost all the increase in grain production in the United States has been for ethanol. U.S. biofuels do not contribute meaningfully to CO_2 reduction. In Brazil gasohol made from sugar cane is widely used, and because of the tropical climate and long growing season, such production is energy efficient and the ethanol is cheaper. (Because Brazilian ethanol is cheaper, the United States put major restrictions on its import.) While in principle more land could be devoted to biofuels, that choice would also lead to greatly increased habitat destruction, accompanied by inevitable loss of biodiversity and competition between food and energy that would lead to higher prices for both. The sensible future of biofuels may be to provide liquid energy for transportation once petroleum reserves begin to seriously diminish.

Wind is currently economically competitive, and installation of wind turbines is growing rapidly. Wind has a fundamental limitation in that were it to take over a major segment of the energy supply, it would sap 10–20% of the energy carried by surface winds. The consequent climate change might rival that from a large CO_2 buildup. Wind, like solar, is also a highly distributed energy source, and is not easily stored and cannot be called up on demand. On cloudy and windless days there would be no energy available! Using these sources for more than 10% or so of the energy supply would require a reworking of the national electricity grid. Wind and solar are part of the long-term solution, but the time frame is long and the challenges great.

As things now stand, no combination of these alternative energy sources can make a major contribution. Solar electricity remains far too expensive. Hydro- and geothermal powers are already near their capacity and are too limited. Both wind and solar in large amounts are inconsistent with our current electrical grid.

Nuclear power has supplied significant amounts of electricity in Japan, Switzerland, and particularly France. After a history of safe operation for decades, the 2011 nuclear accident in Japan showed the potential dangers of nuclear power and the widespread consequences that such an accident can have. Even with this accident, however, the long-term environmental costs, amount of radiation released, and industrial safety record are all better for nuclear power than for coal! Construction of new nuclear power plants, however, is such a long and slow process that it could make only a modest increment to our energy needs over the next few decades. The need for enhanced regulation and safety after the accident stemming from the Japan earthquake can only make the process slower. Safety issues may even rule out an important role for nuclear power. Germany, for example, plans to abandon all nuclear power by 2022.

If costs of energy reflected atmospheric change, all these options would become cheaper than fossil fuels, leading to rapid growth in many of these industries. The timescale, however, is long. The pot at the end of the rainbow would be power from nuclear fusion, which has vast potential and is environmentally clean. Advocates claim that fusion power by midcentury may begin to provide portions of our energy needs. If fossil fuel energy became more expensive, therefore, there are long-term prospects of low CO_2 emissions. With conservation, renewable energy, and nuclear energy and fusion, our energy needs could be fulfilled by nonfossil fuels by midcentury. What would be required to make this happen, however, would be coordinated and concerted action now and a revised economic model for energy costs. Even with that optimistic scenario, there is clearly a need for a stopgap over the next decades to moderate CO_2 buildup.

CARBON CAPTURE AND SEQUESTRATION

Fortunately, there is a route by which an unacceptable CO_2 rise can be avoided. It involves the capture of CO_2 coupled with long-term storage. The obvious target would be to scrub CO_2 from the stack effluent of coal-fired electrical power plants. However, as this would likely be prohibitively expensive, a more economical alternative has been identified. Coal gasification runs steam over coal to produce CO and H_2 (i.e., coal + $H_2O \rightarrow H_2 + CO$) instead of burning the coal in atmospheric oxygen. The CO is then oxidized to CO_2, and the H_2 is fed to an electricity-generating fuel cell (i.e., a flow-through battery). Such plants can be far more cheaply modified for CO_2 capture. This route also yields to more efficient use of the chemical energy contained in coal. The top few thousand sites of CO_2 emissions, mostly utility plants, account for more than 30% of total emission. Concentrated CO_2 capture at these locations would make a major contribution.

No matter how it's accomplished, capture of the CO_2 produced in electrical power plants is only part of the answer. CO_2 also has to be removed from small sources. Today roughly two-thirds of the fossil fuels are burned in small units (automobiles, homes, etc.). For each tankful of gasoline (~50 kg) burned in an ordinary automobile, about 150 kg of CO_2 are produced (that's about 1 pound per mile!). Two routes might be taken to abate CO_2 emissions by the transportation fleet, which today accounts for one-third of fossil fuel consumption. One would be to power vehicles with rechargeable batteries or with hydrogen fuel cells. In either case, the energy would ultimately come from electrical power plants. Nice idea, but no cigar. No battery has yet been developed that can power a vehicle for more than short runs, and electrical vehicles still use substantial gasoline. Nor has any feasible means been found to store enough H_2 aboard a vehicle to allow it to operate for many days. So, barring a major breakthrough, this is not the answer.

Klaus Lackner, a scientist at Columbia University, has suggested that removal of CO_2 from the atmosphere may be doable at a modest cost. His case is based on an analogy to wind-generated electricity. In order to supply the energy required to support the average American, a rotor intercepting 100 square meters of brisk wind is required. By contrast, Lackner shows that in order to remove from the atmosphere the CO_2 produced, if this energy were instead obtained by burning fossil fuels, one would have to capture the CO_2 from only 1 square meter of the same wind stream. Capture would either be into a CO_2 absorbing liquid such as $Ca(OH)_2$ or on a plastic that was loaded with chemical receptors able to capture and release CO_2. The advantages of this form of carbon cap-

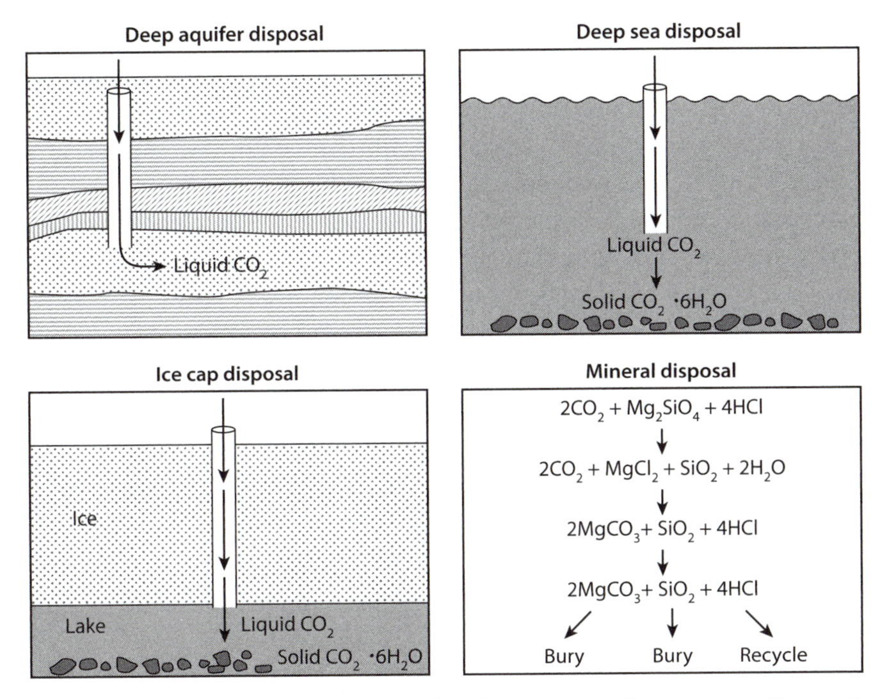

Fig. 20-21: Illustration of the different modes of CO_2 storage after capture, as discussed in the text.

ture are that it can be done anywhere, and it provides a means to capture CO_2 emitted by small sources such as cars and homes. Once captured, the CO_2 would have to be recovered and the absorber recycled. An advantage of air extraction is that it could be done anywhere on the planet rather than close to the place where the energy is produced, as is the case with electrical power plants.

All these mechanisms lead to the capturing of CO_2. The problem then is what to do with it. Several options exist for storing the CO_2 once it's captured (Fig. 20-21).

Deep-Sea Storage

Currently only about one-sixth of the ocean's capacity for CO_2 uptake is being utilized. The reason is that subsurface waters are replaced by waters that have been in contact with the atmosphere only very slowly. The deeper the water, the slower its replacement. As the deeper parts of the ocean will not take up their share for hundreds of years, the idea is to short-circuit the delivery by pumping liquid CO_2 directly into the deep sea. Although liquid CO_2 is less dense than surface ocean water, it is more compressible. At a depth of 3,500 m, the densities of seawater and liquid CO_2 become equal. Below this depth liquid CO_2 is denser than seawater. Hence, if injected below a depth of 3,500 meters, liquid CO_2 would sink to the seafloor. Further, it would not remain a liquid, for under the cold and high-pressure conditions that prevail in the deep sea, the CO_2 would combine with H_2O to form a solid whose formula is $6H_2O \times CO_2$. Chemists refer to this solid as a *clathrate*. As the clathrate is denser than either liquid CO_2 or seawater, it would pile up on the bottom. Of course, over time the clathrate would dissolve and the CO_2 would be dispersed throughout the deep sea ,where it would react with the resident carbonate ions to form bicarbonate ions. In this way, delivery of CO_2 to the deep sea could be greatly speeded.

Storage in Polar Ice Caps

Antarctica's ice cap is underlaid by hundreds of lakes. They form because Earth's internal heat, diffusing up from beneath, warms and in some places melts the basal ice. The idea would be to pipe liquid CO_2 down through the ice into these lakes. Upon arrival, the CO_2 would react with the lake water and form a clathrate, which would sink to the lake bottom. As it would be prohibitively expensive to pipe liquid CO_2 to Antarctica, were this option to be implemented, it would have to be coupled with CO_2 extraction from the air over the ice cap. As the atmosphere mixes extremely rapidly, CO_2 removal could be carried out anywhere on the planet. Just as the air over regions like the New York metropolitan area does not experience a significant buildup in CO_2, neither would the air over Antarctica experience a significant depletion.

Storage in Deep Sediments

The pores in the deep strata of sedimentary basins are invariably filled with very salty waters known as *brines*. As the brines have been trapped in these reservoirs for millions of years, another option is to pump liquid CO_2 into these salty waters. Unlike the deep ocean and the lakes beneath Antarctica, these brines are too warm for CO_2 clathrates to be stable. Hence, the CO_2 would remain in liquid form. This is fortunate, because were clathrates to form, they would clog the sediment pores and prevent the liquid CO_2 from spreading out into the aquifer. Statoil, a Norwegian energy company, is already doing this. They recover methane from a reservoir beneath the North Sea. The 15% CO_2 this gas contains must be separated before the methane can be burned. Normally, this separated CO_2 would be released to the atmosphere. But as Norway has an emission tax of $50 per ton of CO_2, Statoil decided it would be cheaper to liquefy the separated CO_2 and pump it back down into a water-filled stratum. This is now being done routinely. A tiny beginning!

As pressure increases with depth, CO_2 turns from a gas to a liquid and its density becomes greater than seawater. Kurt House and Dan Schrag have proposed that this CO_2 could be injected into sediments below the sea floor and remain there stably with no impact on the deep ocean biosphere and no tendency to escape. This solution would be particularly appropriate for power plants adjacent to a continental shelf, which has sediments at the appropriate depth. Storage in deep-sea sediments is also promising for coastal locations.

Conversion to $MgCO_3$

With somewhat more effort, it is possible to permanently immobilize CO_2. This option involves reacting CO_2 with MgO to form a tough and resistant magnesium carbonate mineral. One option would be to obtain MgO by grinding up and dissolving ultrabasic rock whose dominant mineral is olivine with a chemical formula, Mg_2SiO_4. Hence, the reaction would be:

$$Mg_2SiO_4 + 2CO_2 \rightarrow 2MgCO_3 + SiO_2$$

Although nearly all of Earth's ultrabasic rock resides in its mantle and hence is unavailable to us, surface outcrops do exist in many places. So, large electrical power plants and air extraction facilities would be constructed at the sites of these ultrabasic rock outcrops.

Nature also makes large quantities of carbonate-rich veins in mantle rocks that are exposed at the surface. An intriguing new possibility being proposed by Peter Kelemen is that this reaction might be able to be self-sustaining if the rocks were cracked and injected with CO_2-rich fluids.

None of these storage options is without engineering challenges or environmental impacts. Concern has already been raised about the possible impacts of deep-sea storage on organisms inhabiting the ocean deeps. Greenpeace has already taken a strong stand against this option. In order to implement Antarctic disposal, it would be necessary to modify an existing treaty that bans mining on the continent of Antarctica. Before permitting large quantities of liquid CO_2 to be injected into saline aquifers beneath their homes, people would want to be assured that this activity would not trigger damaging earthquakes or lead to catastrophic releases of CO_2. Finally, even the conversion of CO_2 to $MgCO_3$ is not free of environmental problems. Large mines or injection operations with a substantial infrastructure would need to be constructed.

An additional alternative to CO_2 sequestration is to adopt an approach of "geo-engineering" that would attempt to modify the atmosphere to reduce solar radiation reaching the surface. Models tell us that doubling CO_2 is equivalent to dialing up the sun by about 2%. If so, then to counter a doubling, we would have to reflect away 2% of the sunlight reaching the top of the atmosphere. While several ways to do this have been suggested, one stands out as the cheapest and least environmentally intrusive. It involves injecting SO_2 gas into the stratosphere. As revealed by nature's experiments (i.e., the eruption of El Chichón in 1984 and of Mount Pinatubo in 1991, the SO_2 is rapidly converted to tiny sulfuric acid (H_2SO_4) aerosols. These aerosols backscatter about 10% of the sunlight that encounters them. Thus, to reflect away 2% of the sun's rays, they would have to intercept 20% of the incoming rays. This would require a standing inventory of 32 million tons of SO_2. As the residence time of the aerosols in the stratosphere is only one to two years, they would have to be replaced regularly. This, of course, has the advantage that if the side effects proved unacceptable, the injections could be terminated and the aerosols would soon be gone. It, of course, has the disadvantage that if for some reason the injections were stopped due to war or political upheaval, the warmth would return. Furthermore, such engineering does nothing to halt the acidification of the oceans or the destruction of biodiversity, and there might be unforeseen consequences as well.

Clearly, any solution to the CO_2 problem will have its own set of environmental concerns. As this is unavoidable, the goal would be that the environmental damage stemming from the solution be far smaller than that stemming from the CO_2 itself.

Regardless of the route taken, the solution to the CO_2 problem will be a huge enterprise. Its magnitude is most easily appreciated by considering the amount of liquid CO_2 that would have to be disposed of, if fossil fuels were to continue to dominate the energy market. At today's rate of usage, about 24 km^3 of liquid CO_2 would be produced each year. If in 2060, there are 10 billion people, and if by then poverty has been largely eliminated, the amount would rise to perhaps 64 km^3 each year of liquid CO_2 and some 2,500 km^3 of CO_2 would have been stored, more than the combined volumes of Lakes Erie and Ontario!

Clearly, the capture and storage of CO_2 would raise the cost of fossil fuel energy, but the increases are modest. It is estimated that the cost increase will be on the order of $25 \pm 10\%$. In total, estimates are that controlling CO_2 would reduce global GDP by growth by only 0.1–.15% annually to 2050. This is a very small number, particularly when compared to other governmental costs such as defense or health.

Despite the gravity of the problem and the reasonableness of the cost to take effective action, there is remarkably stiff resistance to any significant steps in the service of planetary conservation. If we are to prevent an objectionable CO_2 rise, we face an uphill battle. The technology required to do the job must be created pretty much from scratch, a plan for payment must be developed, 180 nations must be brought on board, a skeptical public must be convinced.... These tasks will take decades, and implementing the technology and infrastructure decades more. At the moment, there is no indication that the flow of CO_2 into the atmosphere will be brought to a halt.

The Broader Problem

The steps outlined above are possible solutions to the modification of the atmosphere and the acidification of the oceans. They do not address the living environment of soils and biosphere. Atmospheric change is apparent from measurements of gases and in recent decades from increasing temperature and sea level rise. Both of these can be felt and imagined as directly influencing human well-being. Destruction of soils and the biosphere, however, has no immediate impacts and is essentially invisible to the majority of human population that has little contact with land or nature. Soils could be considered as important as the atmosphere to our long-term well-being, but the coverage of soil deterioration in the media is minuscule to absent, as compared to climate change, which appears in most news media on a daily basis. Biosphere destruction is the concern of a small group of conservationists but is peripheral to most of us. For neither soils nor biosphere are there equivalent organizations to the IPCC to bring global attention and scientific expertise to the issues. Both of these problems are as important as climate change in terms of their long-term impact. Both have structural parallels to climate change—the modern economic model does not take into account the long-term costs, which leads to inevitable deterioration. Solving the energy and atmospheric problems could give the impression of a clean slate for further population and economic growth based on human consumption without consideration of consequences. So the problem we face is not simply the technical challenge of climate change; it is the economic model and human attitude toward the planet on which we live.

Another important aspect is the prevalence of food shortages and poverty in the developing world. For the billion people who are the most destitute poor, the main consideration is to have food, shelter, and sufficient fuel to survive. There is no choice of reducing carbon emissions or not cutting down forests for food or fuel. Faced with poverty or famine, none of us is capable of broader questions of planetary health. The poor are also in areas where the population is increasing most rapidly and where biodiversity is most threatened. If we are to solve the planetary crisis, we must also solve the human crisis. The challenge we face is not only our attitude toward our planet, but also our attitude toward our fellow human beings.

An Anthropozoic Era?

The scope and rate of change induced by human civilization is planet altering. This has led Paul Crutzen to suggest that we now live in a new geological epoch, the Anthropocene, and that the Holocene epoch ended as human civilization began. Epoch boundaries are generally minor planetary events, and the change wrought by humans is not minor. Human civilization has led to the first global community of a single species, destruction of billions of years of accumulation of resources, a change in atmospheric composition, a fourth planetary energy revolution, and a mass extinction. Furthermore, our technology now permits directed evolution, a sensing awareness of planetary systems from land and space, and potential communication with other planets in the galaxy. The creation of intelligent life and civilization has produced a fundamentally different planetary system than existed ten thousand years ago. One could argue that the potential for planetary change is almost as great as that caused by origin of life or the rise of oxygen.

The only other boundaries in Earth's history that have encompassed such vast changes are era boundaries. The second rise of O_2 and development of multicellular life were one such boundary. The Permo-Triassic extinction was arguably not as significant, because while it involved a massive change in life, it did not involve an energy revoluton, a change in the character of evolution, or fundamental change in life's capacity for planetary change. It appears, therefore, that we have not simply changed epochs from Holocene to Anthropocene; one could argue that we have changed eras and eons, from Cenozoic and Phanerozoic to Anthropozoic. There is, of course, one aspect of eons that remains a great unknown. Eon and era boundaries of the past have separated spans of time of hundreds of millions of years. We are a few hundred years into the human era. Will we survive? Will we have the wisdom and conscience to realize our potential place in planetary history? There is the potential in human civilization for Earth to pass from "habitable planet" to "inhabited planet," i.e., one that carries intelligence and consciousness on a global scale, for the benefit and further development of the planet and all its life. Or the Anthropozoic era could be an abortive and failed mutation, as the intelligent species destroys itself and its environment. Should we fail and another form of intelligent life comes along in a few tens of millions of years, they would find a planet devoid of most of its treasure chest. A second effort at planetary civilization would be correspondingly more difficult.

Summary

Even in small numbers, human beings exerted a dramatic influence on the planet and its biosphere. Evidence for early interventions comes from extinctions of large animals on the heels of our migration into hitherto "virgin" lands. With the advent of favorable climate conditions and our access to energy, we became able to modify Earth for our needs and to have incomparable advantages compared to other species. Our growth in population and use of resources has enabled all the wonders of modern civilization and also created the potential for environmental catastrophe. Human beings are influencing the atmosphere, ocean, soils, and biosphere on planetary scales. Our success has been possible because of the long evolution of our planet toward increasing habitability and a relatively long period of planetary stability in terms of climate and volcanism. Our recent actions have made the planet more habitable for ourselves in the short term and far less habitable for all other species that we do not protect and harvest for our food. In the long term, our modifications of the physical environment and destruction of other species and entire ecosystems may lead to planetary crises that affect our own preservation. This situation creates a unique problem in the history of life. We have won the battle of biological evolution. We have won it to such an extent that we are able to destroy other life, use up irreplaceable planetary resources, and unconsciously modify the planetary environment. Because of our impacts, planetary evolution now depends on our behavior. The health and future direction of Earth will depend on our actions.

Supplementary Readings

E. O. Wilson. *The Diversity of Life*. 1999. New York: W. W. Norton & Co.

Fourth Assessment Report (AR4) of the Intergovernmental Panel on Climate Change (IPCC). http://www.ipcc.ch/ipccreports/ar4-wg1.htm.

Comprehensive Assessment of Water Management in Agriculture. 2007. *Water for Food, Water for Life: A Comprehensive Assessment of Water Management in*

Agriculture. London: Earthscan, and Colombo: International Water Management Institute. http://www.iwmi.cgiar.org/assessment.
Peter Gleick et al. 2006. *The World's Water 2006–2007*. The Biennial Report on Freshwater Resources. Washington, DC: Island Press.
ISRIC—World Soil Information. http://www.isric.org.

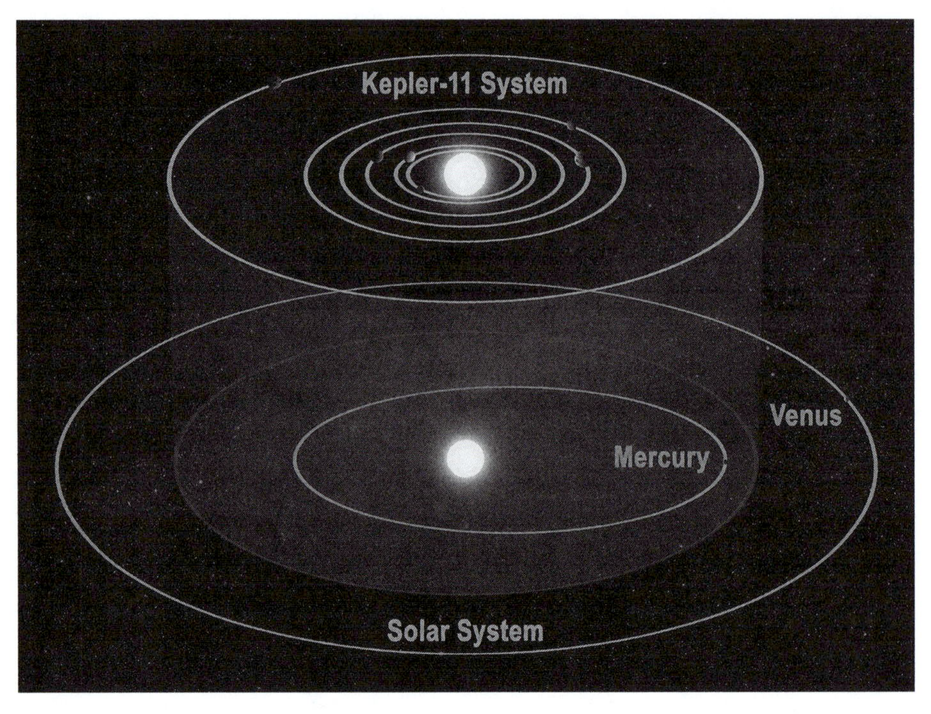

Fig. 21-0: The first discovered solar system other than our own with more than three planets, discovered using the Kepler satellite. All the discovered planets are larger than Earth. (Courtesy of NASA, Tim Pyle)

CHAPTER 21

Are We Alone?

The Question of Habitability in the Universe

One of the most profound questions about the universe is whether life is an emergent property, of which Earth is one example, or whether life is an exception and Earth may be unique. From one point of view, the operation of physical law does not generally lead to uniqueness. It strains common sense to think of our planet as the single inhabited planet given the hundreds of billions of stars in our galaxy alone, particularly if emergence of life may be one of the natural consequences of planetary evolution. An alternative view, however, is that there are so many hurdles that Earth overcame to reach its present state of hosting technological civilization that we may be unique. In the absence of data, we simply do not yet know, but we can question intelligently.

For any solar system there is a "habitable zone" around the star that permits extended planetary evolution with liquid water. Venus may have been within such a habitable zone in its early history, but if life began there it was destroyed by a runaway greenhouse. Venus is too close to the sun and too hot. Mars may also have had early life, since it has water and it conceivably has a small, deep biosphere today. If life did begin there, planetary evolution was halted at the bacterial stage. Mars also lacked sufficient planetary size to maintain a large enough atmosphere and sufficient greenhouse warming. Mars is too small and too cold. Jupiter's moon Europa has a liquid ocean that conceivably may harbor life, but the lack of solar energy necessarily limits its planetary potential for advanced life. Alternative forms of life with a solvent other than water might make colder planets habitable—but this is pure speculation.

New facts about other solar systems promise to bring much greater clarity. Clever astronomical methods are leading to exponential growth in exoplanet discovery. Because of the specifics of how planets are found, most newly discovered planets tend to be large and close to their stars, showing that there is a great diversity of styles of solar systems in the galaxy. With improved data from satellite missions, discovery of Earthlike planets in the habitable zones around other stars is beginning. Investigation of the spectra of their atmospheres will eventually lead to estimates for other life. If we find life in one other place, the statistics are such that life is overwhelmingly abundant in the universe.

The existence of other technological civilizations, however, is another matter. A critical unknown is the fraction of a planetary lifetime that a technological civilization exists. Does such a civilization self-destruct in a few hundred years or last for millions of years? For such a civilization to last, the species driving the technology must sustain and foster planetary habitability rather than ravage planetary resources. In the absence of that understanding, a planet will regress to an earlier stage of planetary development. On the more hopeful side, intelligent life could foster planetary evolution leading to further stages of development that we cannot imagine.

Only if technological civilizations persist for millions of years can the galaxy have a significant number of solar systems with technologically capable life. In this case, since we have had this capability for only a century, all such civilizations would be immeasurably advanced relative to our own. Their survival also would have required a comprehension of and attention to planetary sustainability that is presently lacking on Earth. Only to the extent these challenges have been met successfully elsewhere is our universe inhabited by intelligent life.

Introduction

We have been using the term *a habitable planet* as if the discussion were generic, but of course most of our knowledge is specific to Earth and its planetary experience. In Chapter 17 the possibility was raised that important aspects of the Earth experience are likely to be general phenomena in the Universe. It is then natural to address the popular questions of whether there are other habitable planets in the universe and whether interplanetary communication with other capable civilizations might be possible. Of course, absent such communication in a fully verified and public way, we are inevitably in the realm of speculation. Our views must be so influenced by our own planetary experience that we are incapable of considering the question broadly enough. Nonetheless, there are facts emerging about characteristics of planets around other stars that are highly pertinent, and it is possible by considering the question statistically to at least show where some of the most important questions and sources of uncertainty remain.

Earth's evolution (Table 21-1) involved formation of a rocky planet during accretion from the solar nebula, differentiation into discrete layers, formation of an atmosphere and ocean, development of a stable climate through feedback mechanisms involving subaerial crust and ocean, the origin of life, evolution of photosynthesis, development of an oxygenated atmosphere and eukaryotic cells, a rise of oxygen to ~20% of the atmosphere, the development of multicellular organisms, and recently the evolution of human beings followed by development of civilization and technology. Each of these events is a quantum step in planetary evolution, where the planet has a different capability of function and very different qualities. The entire development has also taken a long time—more than 4 billion years of a stable planetary environment. We can imagine that a planet could be arrested at any of these stages of planetary evolution. Lack of heavy elements from insufficient nucleosynthesis would make rocky planets impossible in some solar systems. Catastrophic events in the solar nebula or interactions with a neighboring star could also inhibit planet creation, or early destruction could be caused by a nearby supernova. An early atmosphere could be lost or the planet could have an insufficient volatile budget to create one. Climate feedbacks could fail. Life could fail to initiate or to undergo the closely coupled coevolution with the planetary environment that permits the development of multicellular life, with its access to the great energy potential of a planetary fuel cell. And it seems evident that a technological civilization could either fail to begin or self-destruct by war, climate change,

Table 21-1
Stages in planetary evolution

Stage 1	Sufficient metallicity to form rocky planet, good galactic environment
Stage 2	Appropriate volatile budget, tectonic circulation permitting exchange between surface and interior
Stage 3	Ocean and stable climate feedback
Stage 4	Origin of life
Stage 5	Photosynthesis and oxygen production
Stage 6	Survival of oxygen catastrophe and development of multicellular life
Stage 7	Intelligent life
Stage 8	Technological civilization with access to external energy

disease, or using up in a short period of time the resources and energy necessary for technology. Given this trail of hurdles, what is the likelihood of other habitable planets in the universe?

COMPARATIVE PLANETOLOGY—LESSONS FROM VENUS AND MARS

In our solar system Venus and Mars are our closest neighbors, and the fact that no life has been discovered on either of them, and that there is no evidence of life on any other object in our solar system, shows that habitability over billions of years leading to technological civilizations requires particular circumstances. Venus and Mars provide our two closest examples of failure of biological planetary evolution, and we can learn much from them as we consider the possibility of life elsewhere in our galaxy. As we learned in Chapter 8, Venus is similar in size to Earth, and with its extensive atmosphere, clearly had a sufficient volatile budget and sufficient internal heat to generate global volcanism. The lack of impacts on the planet's surface shows that some of this volcanism is recent. Since Venus is 20% closer to the sun, through its history it has experienced almost twice the influx of solar energy. It appears that in the odds of planetary evolution, Venus was too hot.

In its planetary characteristics, Mars differs more substantially than Venus from Earth. Its greater distance leads to about half as much solar illumination. In the early solar system with the fainter sun, Mars would have received correspondingly less solar energy. Mars is also much smaller than Earth and Venus. At about half the radius it is only about one-eighth of the mass. The smaller mass makes retention of early volatiles more difficult and atmospheric escape much easier. Evidence from the surface geomorphology of Mars provides strong evidence for periodic periods of abundant water on the surface, and the mineralogy of the Martian soils as determined by the Mars rover expeditions has shown minerals whose formation requires water. There may still be substantial water buried beneath the surface. Mars also has CO_2, and with its cold temperature it has solid CO_2 ice caps that vary with the seasons. The present atmosphere on Mars, however, has a pressure $<1\%$ that of Earth, and it does not have the disequilibrium composition that must have been present on Earth over much of its history. No evidence of life on Mars has been found, though it is conceivable that a small biosphere occurs in the subsurface. Even if Mars has life, its planetary evolution must have been thwarted at an early stage. Mars has the combination of being too cold and too small. (A planet of several Earth masses, for example, might have had a robust enough atmosphere to overcome the greater distance from the sun with a larger greenhouse effect.)

While Mars and Venus are the two most obvious candidates for life in our solar system, and evidence for ancient life on Mars may at some point appear, these are not the only environments where life may be present. The discovery of life at deep-sea hydrothermal vents revealed that life could be supported by planetary heat in the absence of sunlight in a liquid environment. One of the Moons of Jupiter, Europa, has such a planetary environment, and is the only other body of the solar system where the data suggest that both rock and liquid water may be present in abundance. Because of tidal heating from Jupiter, the moon is not cold enough for the ocean to completely freeze, and therefore we know that the temperature environment is also one where life is able to exist on Earth. Since life requires an external energy source, however, the most likely environment for life on Europa would be at deep-sea vents, rather than life distributed throughout the ocean. This makes detection of such life extremely challenging. Furthermore, the available energy is very limited in comparison to the sun-fed photosynthesis that has permitted Earth evolution, and therefore the likelihood of advanced life on Europa is very slim. It would also have been restricted to the very early stages of planetary evolution. If our solar system is an exemplary model, one planet per star had the opportunity for long-term coupled biological-planetary evolution.

These considerations lead to questions of how common planets are around other stars, and how other solar systems may compare with our own in their physical characteristics.

Planet Finding

Until very recently there was no hope of directly observing distant planets around other stars because planets are so small and dim. New techniques, however, have led to spectacular growth in planet finding, with the numbers of discovered planets still increasing exponentially (Fig. 21-1). For such a fast-moving field, this book will be outdated as of its publication date. Fortunately, a website exists that gives up-to-date reports of planet-finding—http://exoplanet.eu—and the reader is encouraged to visit that and other sites to follow the rapid progress of this exciting new area of science.

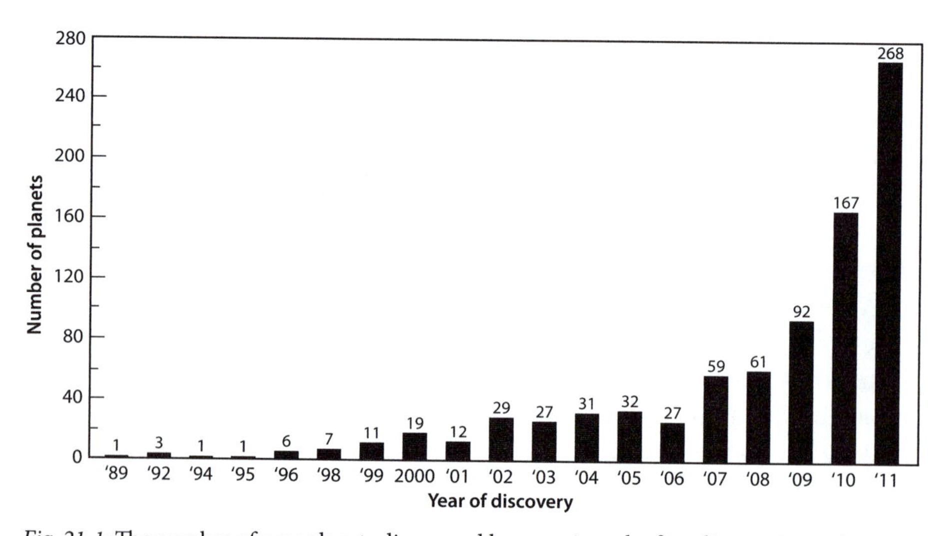

Fig. 21-1: The number of new planets discovered by year since the first discoveries in the mid-1990s. More than 1,000 new candidates have recently been discovered by the Kepler satellite. (Data from http://exoplanet.eu)

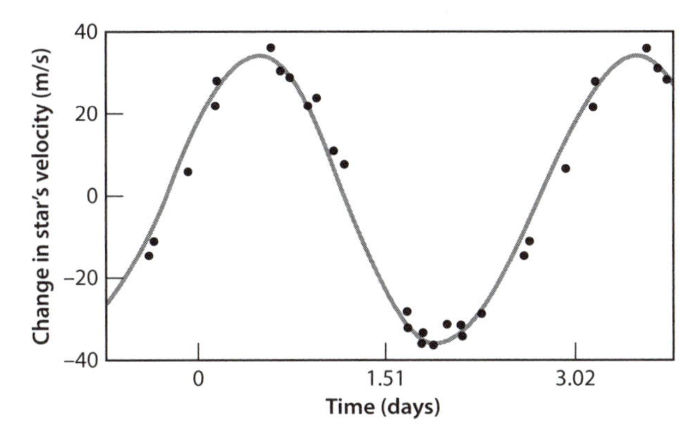

Fig. 21-2: Illustration of the Doppler method of planet detection. The planet exerts a gravitational plug on the star so that its relative velocity changes slightly if the planet is on the near side or the far side. Notice that very small changes in the star's velocity can be measured, making this a very sensitive technique. For example, 10m/sec is only 36 km per hour. (Modified from Marcy, Butler and Vogt, *The Astrophysical Journal* 536 (2000): L43–L46)

The first method to detect planets made use of the Doppler shift of stellar spectra, which varies slightly owing to the gravitational influence of the planet on the star. When the planet is farther away than the star, it exerts a slight gravitational pull away from the Earth, and when it is closer than the star it exhibits a slight pull toward the Earth. These changes lead to slight variations in relative velocity of the star and small changes in the exact position of the lines in the star's visible spectrum (Fig. 21-2). Since the influence of gravity varies as the square of the distance and linearly with mass, this method is most effective for large planets close to their stars; the first detected planets were *hot Jupiters*, massive planets orbiting their stars in tighter orbits even than Mercury in our own solar system.

The second method, called the *transit method*, requires a specific geometrical relationship where the planet crosses in front of its star as viewed from Earth. When the planet "transits" the star, it shadows a small proportion of the star's light, leading to a slight reduction in the luminosity of the star. By observing the changes in luminosity of individual stars, planets can be "seen," as illustrated in Figure 21-3. Over an extended period of observation, the repeat time of the reduction in luminosity gives the period of the planet's rotation around its star. The star's mass and size can be estimated from its luminosity (big stars burn

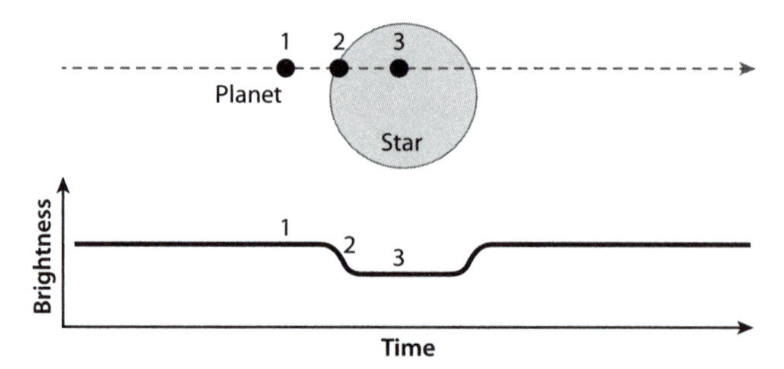

Fig. 21-3: Illustration of the transit method of detection of planets around other stars. As the planet passes in front of the star, a portion of the star is shaded leading to a slight decrease in brightness. The duration of the transit and its periodicity give information about the distance of the planet from the star. The reduction in luminosity then gives information on the planet's size.

brighter). Then the period gives the radius of the orbit, from which the reduction in luminosity gives the size of the planet relative to the star, from which the radius of the planet can be estimated.

The transit method requires that the planet's orbit and the star intersect along the line of sight, so only solar systems that are viewed from the side rather than the top or bottom are susceptible to this method. Simple geometry also shows that the larger the planet the greater will be the change in luminosity, and the more rapid the orbit the more likely the signal will be seen. To detect Jupiter going around our sun, for example, would require capturing the short time interval once every thirty-four years when Jupiter occulted the sun. Closer planets have the additional advantage of a broader angle of sight where transits occur. For example, imagine the (absurd) case of a planet rotating around the surface of its star. All orbits but one, whose plane was perfectly vertical to the line of sight, would transit the star. For this reason, the viewing angle for transit of Mercury is about ninety times wider than the necessary angle for a transit of Neptune. In general the probability of a transit scales with the radius of the star divided by the distance of the planet form the star. To see Earth transiting the Sun from another star, the angle would need to be within 0.05° of horizontal to the ecliptic of the solar system. For Neptune it would have to be within 0.02°! For all these reasons, large planets in tight orbits are more likely to be detected, and the first discovered planets were both large and close to their star, a condition that does not exist in our solar system.

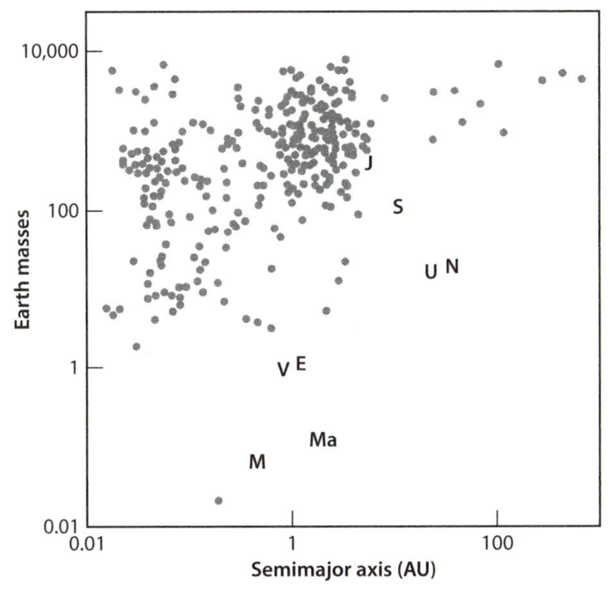

Fig. 21-4: Discoveries of planets around other stars (exoplanets) as of early 2010. Note the log scale, showing the vast differences in size and orbital radius around their stars. Letters give the mass and distance from the sun for planets of our own solar system. Prior to Kepler (see text) the transit method was applicable primarily for large planets very close to their stars. Note that many of the discovered planets are a hundred times or more larger than Earth, and about twenty times closer to their star. (Modified after S. Seager and D. Deming, *Annu. Rev. Astr. Astrophys.* 48 (2010):631–72, with data from http://exoplanet.eu)

If both transit and Doppler shift measurements can be obtained, then the planet's density is constrained, because the change in luminosity reveals the planet's size, and the gravitational pull reveals the planet's mass. Through these clever methods astronomers have been able to extract an amazing amount of information from a small amount of data.

Because of the difficulties of planet finding, discovered planets have a relationship between mass and distance from their star that differs from planets in our own solar system. Figure 21-4 shows the first-discovered planets in this context. While many planets the size of our outer planets have been found, almost all of them are far closer to their stars than the outer planets of our solar system. Small planets at greater distances become progressively more difficult to find.

These discoveries have revealed much about solar systems, showing that the characteristics of our solar system are not ubiquitous. Bode's Law does not apply generally to planetary systems, and large, low-density planets can occur very close to their star. The contrast between inner and outer planets that characterizes our solar system is only one class of solar system, not the general rule. Much more work is required to know the distribution of types of solar system.

Astronomers have also developed a clever method to gain some information about planetary atmospheres. When the planet is in front of the star, the spectrum of light has contributions from both the star and the planet. When the planet is not occulting the star, the spectrum is that of the star alone. By subtracting one from the other, it is possible to infer the planet's influence on the spectrum and to say something about the planetary atmosphere.

Atmospheric composition is one of the critical differences between planets with and without life. Figure 21-5 compares the atmospheric compositions of Mars and Venus with Earth. On Mars and Venus, CO_2 is the dominant gas. Water and O_2 (and therefore O_3) are absent. In our

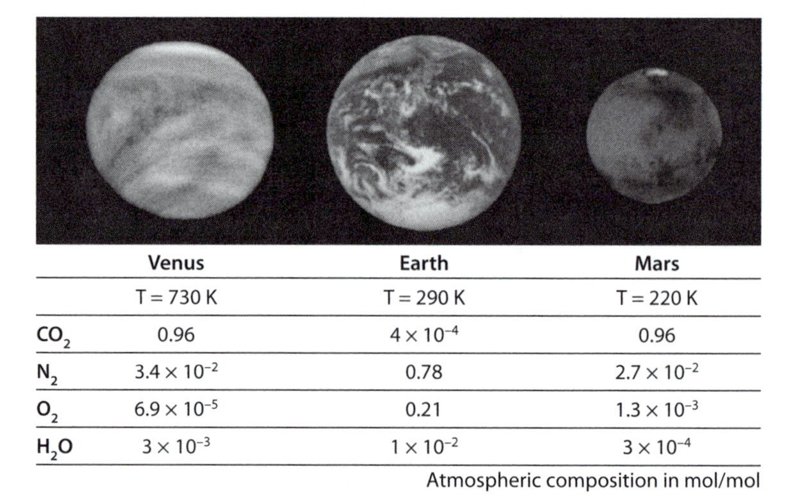

	Venus	Earth	Mars
	T = 730 K	T = 290 K	T = 220 K
CO_2	0.96	4×10^{-4}	0.96
N_2	3.4×10^{-2}	0.78	2.7×10^{-2}
O_2	6.9×10^{-5}	0.21	1.3×10^{-3}
H_2O	3×10^{-3}	1×10^{-2}	3×10^{-4}

Atmospheric composition in mol/mol

Fig. 21-5: Illustration of the relative compositions of the atmospheres of Earth, Mars, and Venus. Notice that Venus and Mars are very poor in oxygen and H_2O, and have similar ratios of CO_2/N_2. Life and the carbon cycle impart a distinct set of gases to Earth's atmosphere, which are detectable in the absorption spectrum of the atmosphere. Spectra of planetary atmospheres are the most likely detection method of life on planets around other stars.

solar system, the planetary composition is responsive to abundant life on a planetary surface. As we saw in Figure 9-7, O_3 has a distinctive absorption in the infrared spectrum. O_3 also is not present in an atmosphere unless there is substantial O_2. The spectrum of a planetary atmosphere thus contains indications of planetary life as we know it. With greater refinement, this is the method that would be most likely to detect evidence of life on other planets. High enough O_2 to produce a significant ozone layer and low CO_2 have been characteristic of Earth's atmosphere for many hundreds of millions of years. Detection of a similar atmosphere for a distant planet would be taken as strong evidence for life elsewhere.

NEW RESULTS FROM KEPLER

The planet-finding satellite named Kepler was launched to discover the frequency of Earthlike planets that would be in the habitable zone of stars similar to the sun. The habitable zone of distant stars depends on the blackbody temperature for planets in orbit around the star. The blackbody radiation of the star gives the temperatures of the stellar surface. Then calculation of how temperature varies with distance from the star is straighforward, as was discussed in Chapter 9 for our solar system. Stars smaller than the sun would have habitable zones much closer to the star, and those larger and hotter than the sun would have habitable zones more distant from the star. Smaller, cooler stars will have habitable planets in tighter radii, and these planets will be the first to be observed because of the observational advantages for planets that are large relative to their star and in tight orbits.

Kepler makes use of the transit method, using an extremely sensitive light detection system to be able to detect minute changes in stellar luminosity. Measurements are made continuously by monitoring more than 150,000 stars in one small part of space, only 1/400 of the visible sky. To find planets as distant from their star as Earth is from the sun will require many years of observations in order to determine the period of the planet's orbit. As of this writing, less than one year of data from Kepler are available and results are still restricted to planets close to their stars. These results are nonetheless remarkable, and show that the new Kepler data will over time greatly expand our understanding of

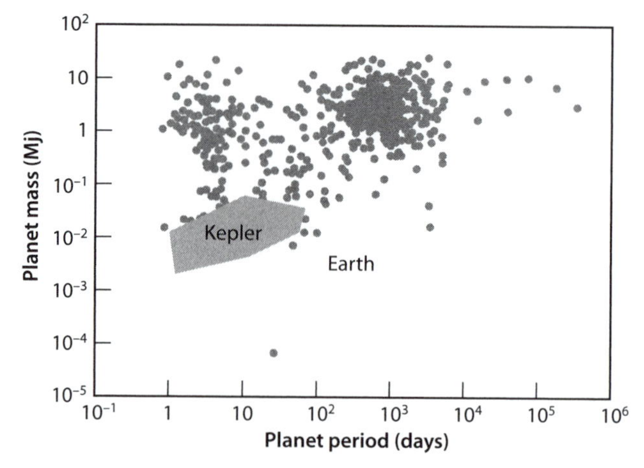

Fig. 21-6: Distribution of planets discovered prior to Kepler. The vertical axis shows mass relative to the mass of Jupiter. While Kepler has also discovered many new large planets, the Kepler data also extend to lower masses, as indicated by the shaded box, and will likely lead to the discovery of planets with characteristics similar to Earth in the near future. (Modified from http://exoplanet.eu)

solar systems and the number of planets around other stars that might be suitable for life.

Kepler has found more than a thousand planet candidates, greatly expanding the number and range of new planets and solar systems. The sensitivity also permits discovery of small planets. Initial results extend to lower limits the detected planetary masses, though the planet periods are as yet limited by the time of observation. Over time, the gray box in Figure 21-6 will gradually extend to longer periods, likely permitting Earthlike planets to be discovered by 2015.

Kepler also made the discovery of a multiplanet solar system where six planets were discovered rotating about a single star. The planets are much closer to their star than is the case for our solar system (see frontispiece), and all of them are larger than Earth. Kepler has also discovered candidate planets whose predicted blackbody temperatures would put them within the habitable zone of their respective stars. Major new discoveries are in store over the next decade, showing that our planet and solar system are one small part of a vast and diverse community of planets and solar systems in our galaxy.

The Number of Other Inhabited Planets in the Galaxy: A Probabilistic Approach

With this prelude of data, we can now turn to more speculative considerations of how many other planets in our galaxy may contain life, and how many might have intelligent life that has developed a technological civilization capable of communicating with our own. A convenient way to explore these questions is to use a statistical method of multiplicative probabilities that is known as the *Drake equation*. The equation can be written as:

$$N = N_s \times f_s \times nL \times f_L \times f_L \times f_{Tech} \times T_{Tech}/T_p \qquad (21\text{-}1)$$

N is the number of planets in our galaxy with technological civilization and therefore the possibility of communicating with us. The number can be estimated by combining a series of numbers and probabilities. N_s is the total number of stars in the galaxy. This is the only number in the equation that is currently knowable; there are about 400 billion stars (4×10^{11}) in the Milky Way. F_s is the fraction of such stars that are suitable for life. Terrestrial experience suggests that life takes a long time to evolve. Large stars with lifetimes of less than a few hundred million years will not have inhabited planets around them. As an extreme example, the largest stars that become supernovae in a million years would have no time for planets and life to develop, and the explosion of the star would destroy any planets in the surroundings.

There are two others aspects to the number of suitable stellar host stars. One is that the nebula that gives rise to the star must have enough C, O, Si, Mg, and Fe to be able to make rocky planets with the necessary ingredients for life. This requires being close enough to the galactic center that there has been a high enough frequency of supernovae to generate sufficient masses of the heavy elements. On the other hand, the center of the galaxy has such a density of stars that the field of galactic radiation and frequency of supernovae would be too intense for life as we know it. So in the overall geography of the galaxy, stars need to be not only the right size, they have to exist in a galactic "habitable zone" (Fig. 21-7), at intermediate locations within the overall galactic structure.

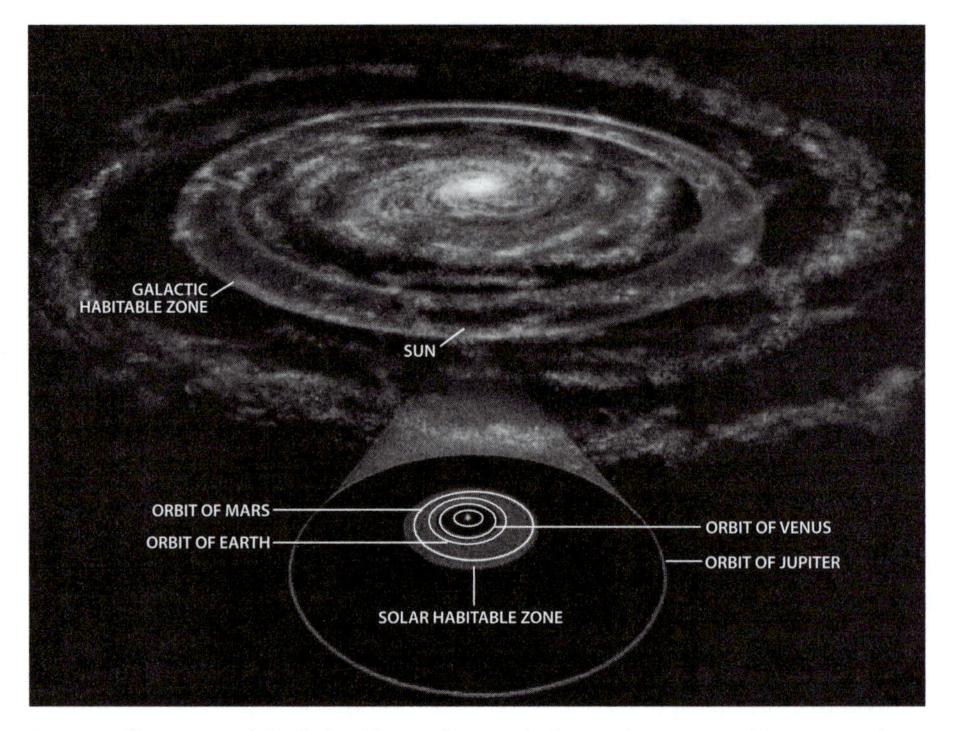

Fig. 21-7: Illustration of the "habitable zone" principle that applies to our solar system and may apply on the galactic scale as well. The interior of the galaxy has too intense a radiation field for life as we know it. The outer realms of the galaxy have not had sufficient supernovae to generate the necessary heavy-element budget for rocky planets and life. (Modified after Lineweaver et al., *Science* 303 (2004), no. 5654: 59–62)

Of course, it is conceivable that life might evolve to take advantage of the high energy of the inner galaxy, but life in that case would be rather different from what we know, and to be conservative in the estimate of inhabited planets, a "habitable zone" restriction seems sensible. These various considerations lead to values of F_s of 0.01 to 0.1.

The rest of the terms in the Drake equation are N_p—the number of planets around such stars that have suitable energy balance for life; f_L—the fraction of those planets on which life arises; f_I—the fraction of those planets on which intelligent life evolves; f_{Tech}—the fraction of those which develop a technological civilization; and T_{tech}/T_p, the fraction of a planet's lifetime during which such technological civilization exists. The game is then to pick numbers for all these terms and come up with the number of communicable civilizations currently present in our galaxy.

Dealing with probabilities is a dicey business, particularly in the complete absence of constraints! For example, if one takes a probabilistic approach to the origin of life, and sets it up like the Drake equation, one could rapidly arrive at a near zero probability. For instance, what is the chance that you make organic molecular precursors in the right locations and proportions, what are the chances of having those combine into polymers, what is the chance of those molecules making up the polymers becoming chiral, what is the chance these would be incorporated into a cellular container, what is the chance of developing replication leading to evolution by natural selection, and so on? We do not know any of these probabilities, and cogent arguments could be made for very small probabilities for many of them, leading to Earth being the unique planet with life.

In the same way that we could argue that life on any particular planet is highly improbable, we can make the case for small enough values of the other terms in the Drake equation that $N << 1$. We can call this the *pessimistic scenario*. For example, one could argue one star in a hundred has an appropriate size and galactic environment, and one of each hundred of these stars has a solar system with a planet in a stable orbit in the habitable zone with an appropriate volatile budget. The odds of life beginning are one in ten thousand, and the odds of planetary evolution leading to intelligent life are one in a thousand (e.g., most planets would be like Mars, Venus, or Europa). Even if we give a probability of 1.0 to the development of a technological civilization, and give the duration of the civilization the 10-million-year average lifetime of a species ($T_{Tech}/T_p = 10^7/4.5 \times 10^9$), the result is that $N = 10^{-2} \times 10^{-2} \times 0^{-4} \times 10^{-3} \times {\sim}2 \times 10^{-3} = 2 \times 10^{-14}$. Multiplying that number by the 4×10^{11} stars in the galaxy leads to $N = 0.08$. There should not be any technological civilizations in the galaxy. Earth is improbable and unique.

The perspective of planetary evolution could provide another set of probabilities that could be additional multiplicative terms to the Drake equation. For example, what is the probability of a planet having just the right volatile budget? What is the probability of the planet having a large moon that provides tides and orbital stability? What is the probability plate tectonics will develop and climate stability will emerge? What is the probability that photosynthesis will evolve? What is the probability that mechanisms will be found to cope with the O_2 poison and convert it to more advanced energy production? What is the probability of endosymbiosis leading to larger and more complex cells? What is the chance these cells will develop partnerships to evolve multicellular life? What is the chance that life will survive the inevitable catastrophes from planet and solar system? And so on. When asked to deal with any of these probabilities, they could always be argued to be far less than 1, and the greater the number of terms included in the chain of probabilities, the smaller the total probability.

What this approach misses is that highly improbable events can often have a 100% certain result given enough time and opportunity. Death is one obvious example. On a more positive note, if the odds of a favorable event are one in a million, but there are millions of chances, then the event will always occur eventually. Then, if that improbable event either has finality or the possibility of replication and amplification, very improbable events ultimately become dominant phenomena. The key ingredient is time, and planets have lots of time. This line of reasoning might apply to the various steps in the origin of life, and also to the various stages of planetary evolution. Yes, O_2 might be a toxic poison for hundreds of millions of years. But one mutation of protection provides evolutionary advantage, and use of O_2 as energy source provides more advantage. O_2 respiring organisms gradually and inevitably come to dominate ecosystems. If planetary evolution ultimately conforms to external forcing functions of energy dissipation and increasing relationship, such phenomena may be universal characteristics, however diverse the manifestations may be in detail.

If we adopt this alternative attitude, then life may be a common result of the dissipation of energy in a solar system on planets with suitable environmental conditions, and planetary evolution toward technological life may be energetically preferred. This could provide an *optimistic scenario*. Perhaps one star in ten could provide the conditions for life, one solar system in ten has the right planet, and one-tenth of planets succeed. Once life begins, intelligent life and technological civilization ultimately develop one time in ten, and persists once it arises for the remaining one half of the planet's life. Then $N = 2 \times 10^7$, 20 million technological civilizations exist in the galaxy, and we have just joined them. And we really do not know which of these scenarios is more likely, because in the absence of data from multiple other solar systems, our reasoning is uninformed.

So we may be alone in our galaxy. Or there may be millions of planets with intelligent life. Of course, these estimates apply only to the Milky Way galaxy. With more than 100 billion galaxies estimated to exist in the universe, the total number of civilizations might be proportionally more. However, given that the nearest neighbor galaxy (Andromeda) is 2 million light years away from us, and the speed of light is our current understanding on the maximum speed of interstellar communication of all kinds, civilizations in neighboring galaxies are not accessible to one another in any way that we can fathom.

We will ultimately be rescued from our guesses and biases by facts. The key fact will be evidence for life on one other planet (or moon) in the galaxy. Because we can only investigate a trivial number of other planets relative to the trillion or so planets that may exist in our galaxy alone, if we find life anywhere else, it means life is overwhelmingly abundant. Even if we investigate 10,000 planets and find only one with life, the implication will be that there are millions of such planets. The most likely evidence will come from the planetary atmospheres of planets in the habitable zones of a nearby star. On the day of that discovery, should it occur, we will know that we are one of a vast community of planetary systems for which biological life is a central aspect.

Human Civilization in the Context of Planetary Evolution and Life in the Universe

Earth has gone through a series of stages in its planetary evolution, with long periods of relatively constant conditions, punctuated by periods of abrupt change that lead to a new condition. The latest such change is the appearance of intelligent life and modern civilization. This change is phenomenally sudden from a planetary perspective. The earliest known civilizations were less than 10,000 years ago, and technological civilization capable of long-distance communication has been present for about

100 years. To put that in a planetary perspective, that is 0.000002 percent of Earth's history. Most mass extinctions that dramatically punctuate the geological record have occupied hundreds of thousands to millions of years, thousands of times as long. From a planetary perspective, human beings are a sudden and recent phenomenon. Relative to the life of a human being, it would be as if such an epochal and transformative event in your personal history happened just as you read this sentence.

So, if someone else were looking for life on planets like Earth, they would find billions of years of unicellular life, a few hundred million years of multicellular life, and a few hundred years of intelligent civilization capable of interplanetary communication. We do not know how long such intelligent civilizations last. We can easily imagine the extinction of our own technological civilization in a short time period from war, famine, disease, exhaustion of resources, or destruction of biodiversity, and there would be great difficulty resuscitating it because we would have used all the readily available energy sources generated by different stages of planetary evolution and over 400 million years of photosynthesis and fossil fuel production.

This gives us an important perspective about searching for life on other planets. Using Earth's history as a guide, the odds of finding a technological civilization on Earth in any particularly year is currently 100 years divided by 4.5 billion (about two-millionths of a percent. The odds of finding a high oxygen atmosphere would be 600:4,500 (13.3%), and the possibility of finding some form of life no matter how primitive would be 3,500:4,500 (85%). These are big differences. Therefore, perhaps the most important term in the Drake equation is the length of time that an intelligent civilization would last. If it lasts only a thousand years, then even the optimistic assessment of 20 million communicable civilizations, based on a 4.5-billion-year duration of a civilized planet, drops to 4 such civilizations in the galaxy. Since the diameter of the galaxy is 100,000 light years, no communication is ever possible, and once again we are alone.

One consequence of this reasoning is the far greater likelihood of finding a planet with microbial life than one with intelligent life. The number for the pessimistic scenario increases to 4,000, and for the optimistic scenario to 400 million. Slightly smaller numbers might apply to the likelihood of finding oxygenated atmospheres elsewhere.

A second and more intriguing consequence is that if intelligent life exists elsewhere in the galaxy, then civilizations must last for a long time, such as the average species length of 10 million years, let alone the likely billions of years left to Earth's history. If we consider the technological advances on Earth that have occurred in a hundred years, then beings on planets with millions of years of civilization would be unimaginably more advanced than we are. We cannot possibly imagine an advanced civilization with thousands of years, let alone millions of years, of exponential growth in knowledge and technological development.

This factor leads also to a perspective on what is known as the *Fermi paradox*—if there is intelligent life throughout the galaxy, then where is everybody? Why don't we at least see evidence of their radio transmissions? The timescale of planetary evolution throws light on this question. It is not clear there would be any point to a million year civilization contacting us in the obvious ways we consider. Presumably they would understand the mysteries of dark energy and dark matter and have a direct knowledge of the universe that makes our understanding more primitive than that of a universal cave man. A perspective very different from our usual one might be necessary. Communication among advanced planetary civilizations may be a rather different phenomenon than anything we can yet imagine.

Summary

Our speculations about life elsewhere in the universe are inevitably biased by our experience of life, planet and solar system, which until very recently were our sole data points. We are also limited by our understanding of the nature of matter and the universe. Based on these biases, leading to phrases such as "life as we know it," life would be restricted to a solar system in a star's habitable zone, where liquid water could persist throughout a planet's history, and to stars within a galactic habitable zone in a similar middle region of galactic geography. The likelihood of life elsewhere in the universe is at the moment partly a philosophical question, since it requires assigning probabilities to many parameters for which we have no reliable data. The latest discoveries of planets in other solar systems hold the promise of providing the necessary data for evaluation of life in the galaxy. Discovery of life in any single other location than Earth would imply that life is pervasive.

The question of other technological civilizations is a more refined question that puts our own planetary behavior in context. Earth as a natural system has gone through billions of years of evolution developing to the present situation of a technological civilization that has appeared in the blink of a planetary eye. For technological civilizations to persist they would need to correspond with the planet as a natural system, following the principles outlined in Chapter 1. To be long lived, a natural system must be sustainable, making use of feedbacks and cycles to preserve resources and living within the limits of the gifts of energy provided by the sun and planet. Natural systems are also in relationship to larger and smaller scales. Human civilization would need to understand its relationship to the planetary scale, to the ecosystems of which we are an important part, and to the smaller scales of other living organisms. This is the challenge of human civilization, to become a part of a natural system to permit and perhaps even to participate in further planetary evolution. Only to the extent that other planetary civilizations have met this challenge are they abundant in the universe. Only if we are able to meet this challenge ourselves will we become a member of this galactic community.

Supplementary Readings

James Kasting. 2010. *How to Find a Habitable Planet*. Princeton, NJ: Princeton University Press.

http://www.exoplanet.eu.

Sara Seager and Drake Deming. 2010. Exoplanet atmospheres. *Annu. Rev. Astron. Astrophys*. 48:631–72.

William J. Borucki, et al. 2010. Kepler planet detection mission: introduction and first results. *Science* 327:977–80.

Jack J. Lissauer, et al. 2011. A closely packed system of low-mass, low-density planets transiting Kepler-11. *Nature* 470:53–57.

GLOSSARY

achondrites Stony meteorites devoid of chondrules, showing they have been modified since the beginning of the solar system.

AGB star An asymptotic giant branch (AGB) star is a low to intermediate mass star in a late stage of its stellar evolution and appears as a red giant.

albedo The proportion of light reflected from a planet. High albedo means high reflectivity.

albite One endmember of the feldspar group of minerals, $NaAlSi_3O_8$. Albite is a primary constituent of granites and the continental crust.

alpha decay One of the three principal forms of radioactive decay. It occurs only in heavy elements such as U and Th. Alpha decay is the emission of the nucleus of a helium atom (i.e., a package of two protons and two neutrons), causing the number of nuclides of the remaining isotope to decrease by four. For example, the decay of ^{238}U to ^{206}Pb includes eight alpha decays.

alpha particle A positively charged particle with an atomic mass of 4 and a charge of +2 (a helium nucleus) produced by alpha decay.

amino acid A complex organic molecule consisting of a central carbon whose four bonds are connected to an "amino" group (NH_2), a carboxyl group (COOH), a hydrogen atom, and a side chain that is called an *R group*; the building blocks of all proteins are the twenty-two amino acids of life.

amphiboles A silicate group of minerals (inosilicates) characterized by double chains of silica (SiO_4) tetrahedra. All amphiboles contain a water molecule and are a primary means by which water is transported in plates.

amphibolite A metamorphic rock with amphibole as a main component.

anaerobic Related to the absence of oxygen. Anaerobic bacteria thrive in the absence of oxygen.

anorthite The Ca-rich endmember of the plagioclase minerals; $CaAl_2Si_2O_8$ forms complete solid solution with albite.

asteroids Small rocky objects that orbit our sun, mostly in the asteroid belt between Mars and Jupiter.

atmophile "Gas-loving" elements and molecules that are volatile and tend to occur as gas or liquid molecules under the conditions on Earth; noble gases (such as helium, neon, and argon), water, carbon dioxide, and nitrogen are common atmophiles; they are overwhelmingly concentrated in the ocean and atmosphere.

atom The basic chemical unit of matter on planets; consists of a dense core of neutrons and protons orbited by a cloud of electrons.

autotrophs Organisms that produce their own organic carbon making use of an external energy source.

banded iron formations Fe-rich rocks, with the acronym BIFs, that occur largely between 2,500 Ma and 1,800 Ma and indicate an important transition in the oxygenation of Earth's surface.

basalt An igneous volcanic rock, derived by melting of the mantle that makes up much of Earth's oceanic crust and an important part of the deep continental crust. It consists mainly of the minerals olivine, pyroxene, and plagioclase. The plutonic equivalent, which has a similar chemical composition, is named *gabbro*.

Benioff zone Dipping plane of earthquakes located along a subduction zone; Benioff zones mark the boundary between the descending ocean crust and the overlying mantle.

bentic foraminifera Tiny-shelled organisms that inhabit the lowest levels of the seafloor.

beta decay One of the three major forms of radioactive decay, where a neutron emits an electron and is converted to a proton.

Big Bang The name given to the event that is the origin of the universe.

bilayer A double layer of molecules that makes up cell membranes and is characterized by hydrophilic (water-loving) ends on the outside and hydrophobic (water-hating) ends on the inside.

binomial nomenclature A Latin nomenclature used to name different species, where the first word is the genus and the second the species. E.g., "*Homo* (genus) *sapiens* (species).

biomarkers Complex organic molecules that do not break down easily and would be the result only of life.

blackbody A body or surface that reflects no light. It has an albedo of zero.

blackbody radiation While a blackbody reflects no light, it has a temperature, and that temperature emits characteristic wavelengths of radiation, which are called *blackbody radiation*. The fact that the universe has a uniform blackbody radiation is one of the lines of evidence for the Big Bang.

black hole An ultradense region where gravity is so intense that even light is unable to escape. Black holes are believed to occupy the central regions of many galaxies.

black smokers High-temperature chimneys in the ocean that can reach the heights of ten-story buildings and belch fluids at temperatures as hot as 400°C. They are sustained by volcanic heat at ocean ridges.

calcite (or calcium carbonate) A mineral consisting of the elements calcium, carbon, and oxygen; a primary constituent of limestones and marbles, and the major inorganic carbon reservoir. Most of Earth's surface CO_2 is locked up in carbonate minerals.

cap carbonates Thick sequences of calcium carbonate (limestones) overlying the glacial deposits that are the evidence for "snowball Earth" episodes.

capture hypothesis An hypothesis proposing that the moon accreted in an orbit similar to Earth, from which it was passively captured to enter orbit around Earth.

carbohydrates Organic molecules that have chemical formulae in the same proportions as carbon matched with one water molecule (CH_2O). For example, glucose. $C_6H_{12}O_6$, could also be written as 6 CH_2O. Carbohydrates are a prime energy source for all cells.

carbonaceous chondrites A type of chondrite containing many mineral phases, including organic molecules, that would be destroyed by even moderate heating.

carbon capture and sequestration (CCS) Phrase descriptive of one solution to rising CO_2 in the atmosphere, i.e., to remove the carbon and sequester it.

carbon cycle The geochemical cycle that links the carbon in organic molecules, the atmosphere, the oceans, and limestone ($CaCO_3$). The carbon cycle involves Earth's interior, the atmosphere, the oceans, and life in a complex process that sustains Earth's climate and permits life.

carbon dioxide A trace gas present in the atmosphere and dissolved in the ocean; because of its strong greenhouse effect, it plays an important role in climate; chemical formula CO_2.

chalcophile Chemical elements with affinity to bond with sulfur ("sulfur-loving"); silver, bismuth, mercury, and zinc are common chalcophile elements.

chemoautotrophs Microorganisms that obtain their energy by chemical reactions rather than from the sun.

chondrites Stony meteorites that contain chondrules and are believed to be preserved objects of earliest solar system history.

chondrules Round grains found in meteorites. Meteorites that contain chondrules are called *chondrites*. The chondrules reflect the time of nebular condensation and subsequent processes of planetary differentiation as they have been preserved in the almost perfect vacuum of space.

codon The three distinct bases defined by the genetic code DNA that are used to specify which amino acid should be placed in a protein.

comets Small icy objects that orbit the sun. Most comets reside in the outer solar system.

convection cells The simplest kind of convection, where hot material ascends at one location, flows laterally, cools off, and descends; the zones of ascent and descent define the cell.

convergent margins Locations where two plates converge, one plate descending beneath the other and returning to the mantle, in a process called *subduction*. A roughly equivalent term is *subduction zone*.

core The central portion of Earth's interior consisting of molten iron and nickel and some unknown light element. It underlies Earth's mantle.

cosmogenic radionuclides Radioactive isotopes that do not have long half-lives and are produced continually by cosmic radiation in Earth's atmosphere. Many of them, such as ^{14}C, are used to provide timescales for young events on Earth.

crust The layer above the Mohorivic discontinuity and below the ocean and atmosphere. Oceanic and continental crust has distinct chemical compositions and densities.

cyanobacteria Bacteria that carry out oxygenic photosynthesis at shallow water levels; they played a key role in the progressive oxygenation of Earth's surface.

daughter isotope The isotope that is the product of decay of its radioactive parent isotope. For example, when ^{87}Rb decays to ^{87}Sr, ^{87}Rb is the parent and ^{87}Sr the daughter.

density The ratio of an object's mass to its volume, e.g., kg/m^3.

depth of compensation Term referring to the depth at which the pressures beneath two columns of material are the same.

D/H ratio The ratio of $^2H/^1H$, or deuterium to hydrogen; these two isotopes have 100% mass difference and can be easily fractionated on the basis of their mass difference at low temperatures, even though they are the same element. The per mil variations in this ratio are referred to as δD.

differential volatility Term expressing the fact that at the same temperature different molecules may be solid, liquid, or gas, with the most volatile molecule being gaseous and the least volatile solid.

diopside A pyroxene mineral with the mineral formula $CaMgSi_2O_6$; a primary constituent of igneous rocks and also a reaction product, when at high temperatures $CaCO_3$ reacts with silicates and breaks down, releasing CO_2.

dolomite A mineral and also the name of a sedimentary rock consisting primarily of a carbonate mineral (akin to calcite) in which half of the metal atoms are magnesium and half are calcium, $CaMg(CO_3)_2$.

eclogite A metamorphic, garnet-bearing rock with a density of 3.35 gm/cm^3; basalt converts to eclogite at high pressures. Eclogite in the down-going slab causes the slab to be heavy and sink into the mantle.

electron One of the basic units of the atom; characterized by a very small mass (compared to the proton and neutron) and by a negative electrical charge.

electron capture One of the three forms of radioactive decay, when a neutron captures an electron and becomes a proton.

element The chemist's subdivision of atoms; based on number of protons in the atom's nucleus. The same element can have many isotopes if the number of neutrons varies.

endosymbiosis The biological process by which partnership between simple cells leads to codependency and ultimately larger and more complex cells.

eukaryotes Refers both to the single-celled organisms that have a nucleus and organelles and also to multicellular organisms where many different cells have become differentiated and specialized. Human beings are eukaryotes.

eutectic The minimum melting point and composition in equilibrium with a fixed number of solid phases, where the liquid composition remains fixed as long as all the solid phases remain present.

eutectic phase diagram A phase diagram that illustrates melting of solid phases that do not form a solid solution, and a liquid phase where all the molecules are fully miscible. The horizontal axis represents the proportion of each mineral in the mixture. Each pure mineral plots at one end of the horizontal axis, and melts at a single temperature. All other compositions along the axis are mixtures of the two minerals. Temperature plots along the vertical axis, increasing upward.

extinct radionuclides Radionuclides with short half-lives so that the parent isotope no longer exists; measurement of the daughter products of extinct radionuclides constrains events in the early history of the solar system.

fayalite Endmember of the olivine solid-solution series with a chemical composition of Fe_2SiO_4; forsterite is the other endmember.

feldspar A group of silicate minerals made up of a three-dimensional framework of both Si and Al tetrahedra, combined with some combination of K, Ca, and Na. Endmembers are orthoclase ($KAlSi_3O_8$), albite ($NaAlSi_3O_8$), and anorthite ($CaAl_2Si_2O_8$). Feldspars are the most abundant minerals in Earth's crust.

felsic Rocks rich in feldspar and quartz (SiO_2, silica) that form most of the upper continental crust. Granites are examples of felsic rocks.

fission The breakup of a nucleus into two major fragments; in some cases the breakup is spontaneous, in others it is induced by the impact of a neutron. Fission occurs only for very large mass isotopes, when repulsive forces become strong enough for the nucleus to break apart.

fission hypothesis The hypothesis proposing that the moon formed by fission from Earth after Earth's core had formed, because of a very fast spin of the early Earth.

forsterite Endmember of the olivine solid-solution series with a chemical composition of Mg_2SiO_4; fayalite is the other endmember. Forsteritic olivine is the mineral that makes up more than half of the upper mantle.

freezing point depression The decrease in the melting temperature of a substance induced by the addition of another compound, as long as the two compounds are miscible in the liquid state.

fusion The merger of two nuclides into a single nuclide. Fusion is the process that produces stellar energy.

gabbro A mafic igneous rock; the plutonic equivalent of basalt, consisting largely of the minerals plagioclase, pyroxene, and olivine.

giant impact hypothesis The hypothesis proposing that another planet about the size of Mars had a grazing impact with Earth, ejecting large amounts of material into space around the joined planets, which condensed to form the moon.

granite An igneous rock that makes up much of the Earth's continental crust; granite consists mainly of the minerals quartz and feldspar.

greenschist A mafic volcanic rock such as basalt or gabbro that has been metamorphosed in the presence of water and contains water-bearing (hydrous) minerals. Greenschists are very common in the ocean crust after it has interacted with seawater at ocean-ridge hydrothermal systems.

Gutenberg discontinuity The discontinuity in density and seismic velocity that occurs at the core/mantle boundary.

half-life The time for half a population of radioactive isotopes to decay. E.g., after ten half-lives, 99.9% of the radioactive isotopes have decayed.

halides A nonsilicate mineral group consisted of a halide ion bearing a negative charge (F^-, Cl^-, Br^-, I^-) and an element such as Na, K, Ag, etc., to form a binary compound (e.g., NaCl).

heterogeneous accretion Hypothesis for the formation of Earth's layers as a sequence of accretion during Earth's formation. First, metals would have gathered to form the core, then silicates were added on top of the core, and finally the gas and water formed the ocean and atmosphere on the surface. The hypothesis is discredited for the solid Earth, but may have merit for the volatile budget.

heterotrophs Organisms that eat organic carbon for their energy and carbon compounds, in contrast to autotrophs.

homogeneous accretion Hypothesis that Earth accretes through undifferentiated objects that are more or less homogeneous in composition. Then separation of Earth into layers would occur subsequently, the core and mantle separating by immiscibility, the crust forming by partial melting, and ocean and atmosphere by degassing. The alternative (and largely accepted) model to heterogeneous accretion.

horizontal gene transfer (HGT) The change in DNA characterized by gene sharing among organisms, leading to lateral changes between different branches of the Tree of Life, rather than linear changes through mutations passed through generations along a single branch of the tree. HGT appears to be common particularly in the bacterial realm.

hot spot A region of excess volcanism and mantle melting that can occur in the middle of a plate as well as occasionally at plate margins. Predominant hypothesis for hot spots is that they are caused by upwelling mantle plumes. Prominent examples include Hawaii, Iceland, and Yellowstone.

human energy revolution The latest energy revolution on Earth, permitting a hundred times more energy production for use by a single species than can be obtained by biological metabolism.

hydrocarbon An organic molecule with a carbon atom connected to hydrogen atoms; methane (CH_4) is the simplest hydrocarbon, and oil and gas are made up of a complex assemblage of hydrocarbons.

hydrogen The first element of the periodic table, but a term also used for the dominant chemical compound in the early solar nebula; now nearly absent from the atmosphere; chemical formula H_2.

hydrothermal vents Locations along spreading axes where seawater circulating through newly created hot basalts spews back into the sea with focused, high-temperature flow. Underwater hot springs.

igneous rock A rock that forms from a silicate liquid (i.e., magma). Magmas form by partial melting, change composition by differentiation, and once cooled and solidified are igneous rocks. Cooling at depth, they produce large crystals and form plutonic rocks. Erupting as lava flows, they are usually fine grained and form volcanic rocks.

immiscibility Property of two phases that do not mix together to form a single phase. The term is usually used in reference to two liquids that do not mix but separate by density to form discrete layers. Examples are oil and water, and metal liquids and silicate liquids.

inorganic molecules A general term for molecules that do not have carbon-hydrogen bonds. They constitute the first building blocks for solid planets and in the solid state are called minerals.

intra-plate volcanism Volcanism happening in the middle of plates and generally accepted to be related to a deep hot mantle source. Hot spots are voluminous examples of intra-plate volcanism.

invasive species Nonnative species that have been generally transported by humans from one geographical location to another; also called *exotic species*.

ion An atom in which the negative charge carried by the orbiting electrons does not match the positive charge of the protons in its nucleus. Ions can be either negatively charged *anions*, or positively charged *cations*.

ionic radius The radius of an atom's ion, determining its size and the geometrical constraints that control how it can combine with other ions to make molecules.

iron meteorites Meteorites consisting of alloys of nickel and iron metal. Fragments of cores of early planetesimals in the solar system.

isobar A term used for the chart of the nuclides, referring to isotopes with constant numbers of nuclear particles. Isotopes along the same isobar are always different elements, because, e.g., an increase in number of neutrons must be accompanied by a corresponding decrease in the number of protons. Electron capture and beta decay produce daughter isotopes along the same isobar as the parent isotope.

isochron A straight line on an isotope ratio diagram indicating that all the materials along the line formed at the same time from a homogeneous reservoir. The isochron then provides the time which is the age of the event. This is a primary method of radioactive dating.

isotone Vertical line on the chart of the nuclides for nuclides with the same neutron number but different proton numbers.

isotope A term that can be used to refer to specific radionuclides, but is generally used to refer to nuclides of the same element that have different masses. Different isotopes of the same element have the same proton number but different neutron numbers.

Jack Hills formation A sedimentary formation located in Australia; the oldest zircon found on Earth (4.4 billion years) comes from this formation.

kimberlite An explosively generated volcanic rock that rises so rapidly from deep in the interior that it contains rock fragments captured at various depths below the surface. Kimberlites bring up fragments of the mantle and are also the source of all natural diamonds.

komatite A very Mg-rich rock that occurs almost exclusively during the Archean. It indicates high temperatures of magma formation in the early Earth.

Kuiper Belt A solar system region located beyond the planets that includes more than 70,000 objects larger than 100 km in size; Pluto is one of the largest of these objects.

late heavy bombardment (LHB) The increased cratering intensity for a short period of time, between 3.9 and 3.8 billion years; also called *terminal cataclysm*. Evidence comes from the lunar cratering record.

Lehmann discontinuity A change in the Earth's interior from a liquid (outer core) to a solid stage (inner core), defined by a change in density and in seismic wave velocities.

limestone A sedimentary rock consisting primarily of the mineral calcite. In young rocks most often this calcite was originally manufactured by marine organisms.

lipids Complex organic molecules with low oxygen content able to store high energy content (e.g., fats); they are also essential building blocks of living cells.

liposome A little spherule formed by bilayers that spontaneously make a sphere; the inner and outer surfaces consist of the hydrophilic ends of the molecules. The bilayers's structure isolates the hydrophobic ends from water.

liquidus The boundary in temperature, pressure and compositional space that separates an all liquid state from the first appearance of a solid

lithophile Rock-loving elements with strong affinity to be in Earth's silicate reservoirs; silicon, magnesium, oxygen, calcium, aluminum, and titanium are common lithophiles. Lithophile trace elements include the rare earth elements, Rb, Sr, and Hf, for example. Lithophile elements are overwhelmingly in Earth's mantle and crust.

mafic Mg-Fe rich materials such as basalt and gabbro. Mantle peridotite is ultramafic.

magmaphile Magma-loving elements and molecules that greatly prefer the liquid to the solid during melting. Magmaphiles include lithophiles such as Rb, K, Cs, Sr, rare earth elements, siderophiles such as W, and volatiles such as H_20 and CO_2.

mantle The major unit of Earth's interior consisting of solid silicate minerals; it overlies the Earth's core.

mantle plume Widely accepted hypothesis for the cause of hot spots. Mantle plumes would form at a deep hot boundary layer and rise as a relatively stationary column that stays fixed in location beneath the moving plates.

marble A common metamorphic rock formed when limestones are recrystallized.

mass dependent fractionation The process by which two stable isotopes of the same element are slightly separated during low temperature processes. The variations are so small that they are generally reported as "per mil" (parts per thousand). Common isotope fractionation occurs during the water and carbon cycles and by biological processes.

mass extinction Catastrophe characterized by the disappearance of a large number of species in Earth's history over very short intervals of time. Mass extinctions are generally used to define the differently named time periods of Earth's history.

mass independent fractionation Small variations in stable isotopes of a single element that are independent of mass. They can be produced by photochemical reactions, for example. Often abbreviated to MIF, mass independent fractionation of sulfur isotopes has been used to constrain the time of the initial rise of oxygen in the atmosphere.

melting regime The region of Earth's mantle that is melted to produce the magmas and igneous rocks that appear at the surface. Most commonly used in reference to ocean ridges.

metamorphic rock A rock that is formed by the recrystallization under heat and pressure of a precursor rock. At the surface, metamorphism often involves a gain in volatiles. At depth where pressures and temperatures are high, the metamorphic reactions cause release of volatiles, which can also transport some other elements that prefer the volatile phase.

metamorphism The high-temperature, high-pressure process that causes minerals to break down and crystallize as other mineral phases.

meteorites Fragments of solar system materials that resided in outer space and have impacted Earth and survived to be recovered.

methane A trace constituent of Earth's atmosphere with a significant greenhouse capacity; chemical formula CH_4.

methanogens Microorganisms that derive energy to make organic matter from chemical reactions that produce methane. For example, $CO_2 + 4H_2 = CH_4 + 2H_2O$.

Milankovitch cycles Orbital cycles that cause the changes in the seasonal and latitudinal distribution of the sunlight reaching Earth.

mineral An inorganic, naturally occurring chemical compound with a chemical composition that can be written as a chemical formula, a crystalline form, and fixed physical properties. The solid Earth is made up of minerals.

Mohorovicic discontinuity An abrupt increase in density from about 2.7–3.3 gm/cm^3, causing a change in seismic velocity at the base of the crust defining the crust/mantle boundary; also called the "Moho."

molecule The chemical combination of two or more atoms into a single species.

monomer An organic chemical compound that can combine with like species to make very extended molecules called polymers.

moon A naturally occurring object that orbits a planet.

moraines A distinctive heap of debris transported into place by the advancing ice.

natural gas Methane produced by natural processes that occur in the earth. Most mined natural gas is biogenic, but some is also produced by rock reactions.

Near-Earth objects Asteroids with Earth-crossing orbits; also called NEO.

neutron One of the basic units of the atom, characterized by the absence of an electrical charge. In isolation neutrons decay to a proton plus an electron with a half-life of 10.3 minutes.

neutron capture The incorporation by a nucleus of a passing neutron.

nitrogen A major constituent of the Earth's atmosphere also present in dissolved form in the ocean; chemical formula N_2.

nuclear fission A highly exothermic nuclear reaction in which the heavy nucleus of an atom splits into smaller parts, two lighter nuclei and neutrons and photons in the form of gamma rays.

nucleic acids A complex organic molecule able to make complementary chains and carry out information, communication and memory functions within living cells; DNA and RNA are nucleic acids that together with proteins make up the most important macromolecules.

nucleobase Essential constituents of nucleotides, DNA and RNA. The bases contain the "letters" of the DNA alphabet. For DNA the four nucleobases are cytosine, guanine, adenine, and thymine.

nucleotide An organic molecule that is an essential constituent of RNA and DNA; it requires the joining of the nucleobases with the sugar backbone and phosphate groups.

nuclide The basic physical unit of matter; consists of a dense package of neutrons and protons.

olivine A mineral with the formula $(Mg,Fe)_2SiO_4$ that is a solid solution between the magnesium endmember forsterite and the iron endmember fayalite. An important mineral in basalts and gabbros, and the predominant mineral in Earth's upper mantle.

Oort Cloud The farthest reaches of the solar system, beyond the Kuiper Belt; billions of potential comets reside in vastly distant orbits extending much of the distance to the nearest star. When their orbits are disturbed by passing stars, they can become the comets that intersect the inner solar system, leading to impacts with the planets.

ophiolite A geological section of oceanic crust and underlying upper mantle, uplifted and exposed at the surface.

ore deposits Concentrations of individual metals sufficiently rich to merit commercial recovery.

organic molecules Molecules that contains carbon and hydrogen. They constitute the first building blocks of life.

orthoclase The K-rich endmember of the feldspar group, $KAlSi_3O_8$. Orthoclase is a primary constituent of granites.

oxidation state A measure of the valence of atoms. The fewer the number of electrons, the greater the positive charge and the higher the oxidation state.

oxides A nonsilicate mineral group where metals combine with oxygen but not silicon; minerals such as magnetite (Fe_3O_4) and hematite (Fe_2O_3) are important oxides.

oxygen A major constituent of Earth's atmosphere also present in dissolved form in the ocean; necessary to animal life; chemical formula O_2.

oxygenic photosynthesis The process of converting CO_2 and H_2O into organic compounds (commonly abbreviated as CH_2O) and producing O_2 as the oxidized by-product. Oxygenic photosynthesis has led to the rise of O_2 over Earth's history and is the base of most food chains to the present day.

parent isotope The radioactive isotope that decays, to produce the daughter isotope.

partial melting The major process that forms the Earth's crust. Materials that have multiple solid phases, such as rocks, melt over a range of temperatures, and therefore can melt partially.

passive upwelling The mantle flow generated at most spreading centers as the mantle rises to fill the gap created by the spreading plates; to be contrasted to dynamically driven active upwelling, which is determined by internal mantle properties.

peak oil The predicted maximum global output of oil. Peak oil is well established for individual fields and well-explored nations. Global peak oil is likely to occur in the coming decade.

peptide bond The bond that combines amino acids and permits the construction of proteins.

peridotite A mantle rock consisting of the mafic minerals olivine and pyroxenes and with a density of about 3.33 gm/cm^3. It is the rock that makes up the Earth's mantle.

phase diagram A two dimensional representation of parameters affecting the identity and proportions of mineral and liquid phases; temperature-composition phase diagrams are commonly used to study the properties of melting mixtures, such as rocks.

photoautotrophs Microorganism able to harvest light energy from the sun to live and produce organic matter

plagioclase A solid solution of the feldspar group with albite and anorthite as endmembers. Plagioclase is the most abundant mineral in basalts, and is also common in granites.

planet A large object that orbits a central star; distinguished from a star in that it is not sufficiently massive to ignite a nuclear fire.

planetesimals The smaller precursors to planets that formed during the process of accretion in the solar nebula. Planetesimals then impacted with each other to form planets.

plasma One of the four states of matter that consists of an electrically conductive ionized gas and free electrical charges.

plate tectonics Accepted theory that Earth's surface is made up of fixed plates that are continually in motion, created at ridges and destroyed at subduction zones.

plates Large sections of the Earth's crust that slide across the underlying mantle as distinct units and are bounded by plate margins.

pluton An igneous body formed by a magma slowly cooled at depth.

polymer A large molecule formed by repetition of smaller and linked units, the monomers.

postglacial rebound Isostatic uplift of continents caused by the removal of mass by melting of ice sheets from the last glacial period.

p-process Nucleosynthetic process leading to the buildup of heavy nuclides by addition of a proton.

pressure release melting The process by which a rock crosses its solidus and begins to melt by mantle upwelling so that the pressure on the rock decreases. It occurs because the melting temperature (solidus) is temperature sensitive, and it decreases markedly as pressure declines.

prokaryotes The simplest single-celled organisms, lacking cell nucleus and organelles. Bacteria and Archea are both prokaryotes. Most of the genetic diversity of the planet resides in prokaryotes. To be contrasted with the larger and more complex eukaryotes.

protein A complex organic molecule made up of combinations of amino acids; proteins are essential building blocks of living cells.

proton One of the basic units of the atom; characterized by a positive electrical charge.

protoplanets Inner planets of the the solar system at its early stage. Intermediate in size between planetesimals and planets.

purines Used to refer to the nucleobases adenine and guanine.

pyrimidines Used to refer to the nucleobases cytosine, thymine, and uracil.

pyroxene A group of minerals consisting mainly of the elements iron, magnesium, silicon, and oxygen. Slightly richer in Si than olivine, pyroxenes are primary constituents of mafic igneous rocks, such as basalt, gabbro, and peridotite.

quartz SiO_2, a primary constituent of granites and sandstones.

radioactive decay The spontaneous transformation of a radionuclide from one element to another through one of its three forms, or by fission.

radionuclide A radiogenic isotope that will decay.

rare earth elements Usually abbreviated as REE, also known as the *lanthanide contraction series*; a group of fourteen chemical elements in the Periodic Table that have the very useful geochemical characteristics that all of them have a common outer electron shell structure and similar chemical behavior that varies slightly but regularly from one element to another.

Rayleigh number A dimensionless quantity that can be used to estimate the likelihood and character of convection.

red giant A very large star that burns fast and hot until its fuel is exhausted, at which time it undergoes an explosive death.

reductant The element that donates an electron in a reduction-oxidation reaction.

r-process The buildup of heavy nuclides during an intense neutron bombardment that occurs during supernova explosions.

salt domes Plugs of solid salt that push up through overlying sediments.

sandstone A coarse-grained rock formed when beach, dune, or river sands become lithified; consists largely of the mineral quartz.

schist A common metamorphic rock formed when shales are recrystallized.

sedimentary rock A rock formed when material deposited from water, ice, or air becomes cemented into a coherent unit.

shadow zone For a given earthquake, the region where there is no arrival of shear waves or compressional waves due to the liquid inner core (for shear waves) or the abrupt change in velocity (for p-waves).

shale A fine-grained sedimentary rock that originates as mud at the bottom of a standing body of water; consists largely of clay minerals from soils.

sheet silicate A silicate group of minerals (phylosilicates) characterized by parallel sheets of silicate tetrahedra. Mica, chlorite, and clay minerals are common sheet silicates.

siderophile Metal-loving elements that prefer to exist in the metallic phase rather than silicate phases; nickel, gold, tungsten, silver, copper, and platinum are common siderophiles. Iron is either lithophile or siderophile, depending on how much oxygen is available.

snowball catastrophes Episodes, named by Joseph Kirschvink, during which Earth became totally frozen; they may have occurred during the Neoproterozoic period from 580 to 750 million years ago.

solar nebula The initial cloud of gas and dust around the sun from which the solar system formed.

solidus The line in pressure-temperature-composition space that delineates the boundary between no melt present and the initiation of melting. Contrast with *liquidus*.

spreading axis The plate margin at an ocean ridge where two plates move directly apart from one another and new ocean crust is formed.

spreading rate A measure of the spreading velocity of the world's plates.

s-process The construction of heavier nuclides through a slow process of neutron addition. Occurs predominately in stellar interiors.

stable isotope fractionation See *mass dependent fractionation*.

star The basic unit of matter within galaxies; distinguished from other objects (i.e., planets, moons, etc.) by the nuclear fires that burn in their interiors.

stishovite The high pressure form of quartz considered definitive evidence of an impact origin for numerous circular, nonvolcanic craters on Earth.

stromatolites Carbonate sediment structures created by photosynthetic microbiological communities that live in shallow seas; they are among the most ancient structures created by living organisms but are also generated in some modern locations.

subduction zones Convergent margins where one plate moves beneath another and is returned to the mantle.

sulfides A nonsilicate mineral group with sulfur as the major anion; pyrite (FeS_2) is the most common and important sulfide.

sulfur mass independent fractionation (SMIF) The most secure line of evidence of a change in atmospheric oxygen about 2.4 Ga.

supernova The explosive demise of very large stars; an event where heavy elements are formed.

tectonic thermostat The hypothesis that a feedback relates subduction and volcanic outgassing of CO_2 to changes in weathering, thereby keeping Earth's climate in a relatively stable state where liquid water is present.

terminal cataclysm The marked increase in cratering intensity for a short period of time, between 3.9 and 3.8 billion years; also called *late heavy bombardment* (LHB).

thermal convection The process by which Earth's mantle and core flow, due to differences in density caused by temperature variations.

T-Tauri Massive ejection of gas by violent winds that occur during a small star's earliest history.

ultramafic nodules Rock fragments from the mantle that are carried to the surface in mantle-derived rocks, such as kimberlites, that are brought to the surface by explosive volcanic eruptions from great depth.

universal common ancestor The forefather from which all life evolved.

volatility A measure of the tendency of a substance to vaporize. Volatile compounds transform to liquid and gas at lower temperatures than refractory compounds.

volcanoes Points on Earth's surface where liquid silicate (i.e., magma) created in the Earth's interior is erupted at the surface.

white dwarf A former star of moderate size that has exhausted its nuclear fuel supply.

INDEX

Note: *f* following a page number denotes a figure, *t* denotes a table, and page numbers in italics indicate the term is found in the glossary.